Lecture Notes in Computer Science 8503

Commenced Publication in 1973
Founding and Former Series Editors:
Gerhard Goos, Juris Hartmanis, and Jan van Leeuwen

R. Ravi   Inge Li Gørtz (Eds.)

Algorithm Theory – SWAT 2014

14th Scandinavian Symposium and Workshops
Copenhagen, Denmark, July 2-4, 2014
Proceedings

 Springer

Volume Editors

R. Ravi
Carnegie Mellon University
Tepper School of Business
Pittsburgh, PA, USA
E-mail: ravi@cmu.edu

Inge Li Gørtz
DTU Compute
Kongens Lyngby, Denmark
E-mail: inge@dtu.dk

ISSN 0302-9743   e-ISSN 1611-3349
ISBN 978-3-319-08403-9   e-ISBN 978-3-319-08404-6
DOI 10.1007/978-3-319-08404-6
Springer Cham Heidelberg New York Dordrecht London

Library of Congress Control Number: 2014941604

LNCS Sublibrary: SL 1 – Theoretical Computer Science and General Issues

Typesetting: Camera-ready by author, data conversion by Scientific Publishing Services, Chennai, India

Printed on acid-free paper

Springer is part of Springer Science+Business Media (www.springer.com)

Preface

This volume contains the papers presented at the 14th Scandinavian Symposium and Workshops on Algorithm Theory (SWAT 2014), held during July 2–4, 2014, in Copenhagen, Denmark. A total of 134 papers were submitted, out of which the Program Committee selected 33 for presentation at the symposium. Each submission was reviewed by at least three members of the Program Committee. In addition, invited lectures were given by Carsten Thomassen from the Technical University of Denmark, Nikhil Bansal from Eindhoven University of Technology and Mikkel Thorup from University of Copenhagen. The Program Committee decided to grant the best student paper award to Keigo Oka and Yoichi Iwata both from the University of Tokyo for the paper titled "Fast Dynamic Algorithms for Parameterized Problems".

SWAT is held biennially in the Nordic countries; it alternates with the Algorithms and Data Structures Symposium (WADS) and is a forum for researchers in the area of design and analysis of algorithms and data structures. The call for papers invited submissions in all areas of algorithms and data structures, including but not limited to approximation algorithms, parameterized algorithms, computational biology, computational geometry and topology, distributed algorithms, external-memory algorithms, exponential algorithms, graph algorithms, online algorithms, optimization algorithms, randomized algorithms, streaming algorithms, string algorithms, sublinear algorithms, and algorithmic game theory. Starting from the first meeting in 1988, previous SWAT meetings have been held in Halmstad, Bergen, Helsinki, Aarhus, Reykjavík, Stockholm, Bergen, Turku, Humlebæk, Riga, Gothenburg, Bergen and Helsinki. Proceedings of all the meetings have been published in the LNCS series, as volumes 318, 447, 621, 824, 1097, 1432, 1851, 2368, 3111, 4059, 5124, 6139 and 7357.

We would like to thank all the people who contributed to making SWAT 2014 a success. We thank the Steering Committee for selecting Copenhagen as the venue for SWAT 2014, and for their help and guidance in different issues. The meeting would not have been possible without the considerable efforts of the local organization teams of SWAT 2014. We thank Otto Mønsteds Fond and the Technical University of Denmark for their financial and organizational support. The EasyChair conference system provided invaluable assistance in coordinating the submission and review process. Finally, we thank the members of the Program Committee and all of our many colleagues whose timely and meticulous efforts helped the committee to evaluate the large number of submissions and select the papers for presentation at the symposium.

May 2014

R. Ravi
Inge Li Gørtz

Organization

Organizing Committee

Philip Bille (chair)	Technical University of Denmark, Denmark
Patrick Hagge Cording	Technical University of Denmark, Denmark
Inge Li Gørtz (chair)	Technical University of Denmark, Denmark
Benjamin Sach	University of Bristol, UK
Hjalte Wedel Vildhøj	Technical University of Denmark, Denmark
Søren Vind	Technical University of Denmark, Denmark

Steering Committee

Lars Arge	Aarhus University, Denmark
Magnús M. Halldórsson	Reykjavík University, Iceland
Andrzej Lingas	Lund University, Sweden
Jan Arne Telle	University of Bergen, Norway
Esko Ukkonen	University of Helsinki, Finland

Program Committee

Boris Aronov	Polytechnic Institute of NYU, USA
Per Austrin	KTH Stockholm, Sweden
Siu-Wing Cheng	HKUST Hong Kong, China
Tamal Dey	Ohio State University, USA
David Eppstein	University of California at Irvine, USA
Sándor Fekete	TU Braunschweig, Germany
Petr Golovach	University of Bergen, Norway
Magnús Halldórsson	Reykjavík University, Iceland
Sariel Har-Peled	University of Illinois at Urbana-Champagne, USA
Satoru Iwata	Tokyo University, Japan
Tibor Jordan	Eötvös Loránd University, Hungary
Inge Li Goertz	Technical University of Denmark, Denmark
Aleksander Madry	EPFL, Switzerland
Daniel Marx	Hungarian Academy of Sciences, Hungary
Marcin Mucha	University of Warsaw, Poland
Viswanath Nagarajan	IBM Research, USA
Alantha Newman	CNRS, Grenoble, France
Andrzej Proskurowski	University of Oregon, USA
Harald Raecke	Technische Universität Munichen, Germany
Venkatesh Raman	IMSc Chennai, India

R. Ravi   Tepper School of Business, Carnegie Mellon
                                   University, USA
Mohammad Salavatipour   University of Alberta, Canada
Emo Welzl   ETH Zurich, Switzerland

Additional Reviewers

Agarwal, Pankaj   Fiala, Jiri
Aichholzer, Oswin   Fiala, Jirka
Alam, Muhammad Jawaherul   Fleiner, Tamas
Ambainis, Andris   Fleszar, Krzysztof
Amit, Mika   Fredriksson, Kimmo
An, Hyung-Chan   Freydenberger, Dominik D.
Antoniadis, Antonios   Friedrichs, Stephan
Aurenhammer, Franz   Fuchs, Moritz
Azar, Yossi   Fukunaga, Takuro
Bansal, Nikhil   Gaertner, Bernd
Bar-Noy, Amotz   Gagie, Travis
Behsaz, Babak   Ganian, Robert
Bienkowski, Marcin   Gaspers, Serge
Bodlaender, Hans L.   Gawrychowski, Pawel
Bonsma, Paul   Ghosh, Subir
Brimkov, Valentin   Grossi, Roberto
Brodal, Gerth   Grünbaum, Branko
Byrka, Jaroslaw   Gyorgyi, Peter
Békési, József   Halldorsson, Magnus
Bérczi, Kristóf   Hemmer, Michael
Cabello, Sergio   Hertli, Timon
Carr, Robert   Hoffmann, Michael
Chalermsook, Parinya   Huang, Sangxia
Chan, Timothy M.   Hüffner, Falk
Chechik, Shiri   I, Tomohiro
Chiu, Man Kwun   Iacono, John
Chrobak, Marek   Ito, Takehiro
Connamacher, Harold   Jacob, Riko
Cording, Patrick Hagge   Jaggi, Martin
Crochemore, Maxime   Jankó, Zsuzsanna
Daruki, Samira   Jeffery, Stacey
Doerr, Carola   Jeż, Artur
Driemel, Anne   Jørgensen, Allan Grønlund
Durocher, Stephane   Karp, Jeremy
Dürr, Christoph   Kavitha, Telikepalli
Erickson, Jeff   Keil, Mark
Ezra, Esther   Kiraly, Tamas
Fagerberg, Rolf   Király, Csaba

Király, Zoltán
Kis-Benedek, Ágnes
Kolay, Sudeshna
Kolliopoulos, Stavros
Kolpakov, Roman
Komusiewicz, Christian
Konrad, Christian
Kopelowitz, Tsvi
Korolova, Aleksandra
Kothari, Robin
Kowalik, Lukasz
Kowaluk, Miroslaw
Kral, Daniel
Kucherov, Gregory
Kumar, Nirman
Kusters, Vincent
Kuszner, Lukasz
Laekhanukit, Bundit
Lancia, Giuseppe
Lau, Man-Kit
Li, Minming
Lokshtanov, Daniel
Lubiw, Anna
Manlove, David
Mathieson, Luke
Meijer, Henk
Meister, Daniel
Melsted, Pall
Mestre, Julian
Meyer, Ulrich
Michalewski, Henryk
Misra, Neeldhara
Miyazaki, Shuichi
Moldenhauer, Carsten
Molinaro, Marco
Morin, Pat
Mouawad, Amer
Muller, Haiko
Mömke, Tobias
Nagano, Kiyohito
Nair, Chandra
Nandy, Subhas
Ng, Ken
Nguyen, Huy
Nicholson, Patrick K.

Nikolov, Aleksandar
Nikzad, Afshin
Nilsson, Stefan
Okamoto, Yoshio
Onak, Krzysztof
Otachi, Yota
Ottaviano, Giuseppe
Paluch, Katarzyna
Panigrahi, Debmalya
Panolan, Fahad
Parekh, Ojas
Paulusma, Daniel
Pemmaraju, Sriram
Pettie, Seth
Pinkau, Chris
Pirwani, Imran
Poláček, Lukáš
Popa, Alexandru
Pruhs, Kirk
Radoszewski, Jakub
Raichel, Benjamin
Rajaraman, Rajmohan
Raman, Rajeev
Razenshteyn, Ilya
Reidl, Felix
Sach, Benjamin
Sadakane, Kunihiko
Salson, Mikaël
Sanders, Peter
Satti, Srinivasa Rao
Saurabh, Nitin
Saurabh, Saket
Schalekamp, Frans
Scheder, Dominik
Schmid, Markus L.
Schmidt, Christiane
Shachnai, Hadas
Shah, Chintan
Simons, Joseph A.
Singh, Mohit
Smid, Michiel
Spalek, Robert
Speckmann, Bettina
Stein, Cliff
Straszak, Damian

Svitkina, Zoya
Szeider, Stefan
Tarjan, Robert
Telikepalli, Kavitha
Thankachan, Sharma
Thilikos, Dimitrios
Thomas, Antonis
Todinca, Ioan
Toman, Stefan
Täubig, Hanjo
Uno, Takeaki
Van 'T Hof, Pim
van Leeuwen, Erik Jan
van Stee, Rob
van Zuylen, Anke
Verbitsky, Oleg
Vildhøj, Hjalte Wedel

Villamil, Fernando Sanchez
Vind, Søren
Vinyals, Marc
Walen, Tomasz
Wang, Yusu
Ward, Justin
Weihmann, Jeremias
Weimann, Oren
Wettstein, Manuel
Wiese, Andreas
Wild, Sebastian
Xiao, Mingyu
Xu, Jinhui
Yan, Lie
Yi, Ke
Ásgeirsson, Eyjólfur Ingi
Łącki, Jakub

Keynote Papers

The Power of Iterated Rounding

Nikhil Bansal

Department of Mathematics and Computer Science
Eindhoven University of Technology

Abstract. In recent years iterated rounding has emerged as a simple, yet extremely powerful technique in algorithm design. In this talk, we will look at various applications of this technique. In particular, we will see how it gives simple new proofs of various classical results, and then consider more recent applications and refinements of the technique.

Orientations and Decompositions of Graphs

Carsten Thomassen

Department of Applied Mathematics and Computer Science
Technical University of Denmark

Abstract. Latin squares, Steiner triple systems and block designs are structures that can be expressed as graph decompositions. A result of Dehn on rigidity of convex polyhedra motivated an early result on claw decompositions of graphs.

In this lecture we focus on the interplay between graph decomposition and graph flow, for example Tutte's flow conjectures. Special emphasis will be on the recent solution of the so-called weak 3-flow conjecture formulated by Jaeger in 1988.

Fast and Powerful Hashing using Tabulation

Mikkel Thorup*

University of Copenhagen

Abstract. Randomized algorithms are often enjoyed for their simplicity, but the hash functions employed to yield the desired probabilistic guarantees are often too complicated to be practical. Here we discuss how simple hashing schemes based on tabulation provide unexpectedly strong guarantees.

Simple tabulation hashing dates back to Zobrist [1970]. Keys are viewed as consisting of q characters and we have precomputed character tables $h_1, ..., h_q$ mapping characters to random hash values. A key $x = (x_1, ..., x_q)$ is hashed to $h_1[x_1] \oplus h_2[x_2] ..... \oplus h_q[x_q]$. This schemes is very fast with character tables in cache.

While simple tabluation is not even 4-independent, we show that it provides many of the guarantees that are normally obtained via higher independence, e.g., linear probing and Cuckoo hashing.

Next we consider *twisted tabulation* where one character is "twisted" with some simple operations. The resulting hash function has powerful distributional properties: Chernoff-Hoeffding type tail bounds and a very small bias for min-wise hashing.

Finally, we consider *double tabulation* where we compose two simple tabulation functions, applying one to the output of the other, and show that this yields very high independence in the classic framework of Carter and Wegman [1977].

While these tabulation schemes are all easy to implement and use, their analysis is not.

The talk surveys result from

- Mihai Pătrașcu and Mikkel Thorup: The power of simple tabulation hashing. J. ACM 59(3): 14 (2012). First announced at STOC 2011: 1-10
- Mihai Pătrașcu and Mikkel Thorup: Twisted Tabulation Hashing. SODA 2013: 209-228
- Mikkel Thorup: Simple Tabulation, Fast Expanders, Double Tabulation, and High Independence. FOCS 2013: 90-99
- Søren Dahlgaard and Mikkel Thorup: Approximately Minwise Independence with Twisted Tabulation. SWAT 2014.

* Research partly supported by an Advanced Grant from the Danish Council for Independent Research under the Sapere Aude research carrier programme.

Table of Contents

I/O-Efficient Range Minima Queries

Peyman Afshani[1,*] and Nodari Sitchinava[2]

[1] MADALGO, Department of Computer Science, University of Aarhus, Denmark
peyman@madalgo.au.dk
[2] Department of Information and Computer Sciences, Univ. of Hawaii – Manoa, USA
nodari.sitchinava@hawaii.edu

Abstract. In this paper we study the *offline (batched) range minima query (RMQ)* problem in the external memory (EM) and cache-oblivious (CO) models. In the *static* RMQ problem, given an array A, a query $\text{RMQ}_A(i,j)$ returns the smallest element in the range $A[i,j]$.

If B is the size of the block and m is the number of blocks that fit in the internal memory in the EM and CO models, we show that Q range minima queries on an array of size N can be answered in $O\left(\frac{N}{B} + \frac{Q}{B}\log_m \frac{Q}{B}\right) = O(\text{scan}(N) + \text{sort}(Q))$ I/Os in the CO model and slightly better $O(\text{scan}(N) + \frac{Q}{B}\log_m \min\{\frac{Q}{B}, \frac{N}{B}\})$ I/Os in the EM model and linear space in both models. Our cache-oblivious result is new and our external memory result is an improvement of the previously known bound. We also show that the EM bound is tight by proving a matching lower bound. Our lower bound holds even if the queries are presorted in any predefined order.

In the batched *dynamic* RMQ problem, the queries must be answered in the presence of the updates (insertions/deletions) to the array. We show that in the EM model we can solve this problem in $O\left(\text{sort}(N) + \text{sort}(Q)\log_m \frac{N}{B}\right)$ I/Os, again improving the best previously known bound.

1 Introduction

Given an array A on N entries, the *range minimum query (RMQ)* $\text{RMQ}(i,j)$, such that $1 \leq i \leq N$, asks for the item in the range $A[i..j]$ with the smallest value.[1] Range minima queries have many practical applications such as data compression, text indexing and graph algorithms and they have been studied extensively. In internal memory, there are many papers that deal with answering range minima queries in constant time and the main basic idea is to use Cartesian trees [12] and to find least common ancestors [10] (see also [7,8,3] for a subset of other results on reducing space and other improvements).

---

* Work supported in part by the Danish National Research Foundation grant DNRF84 through Center for Massive Data Algorithmics (MADALGO).

[1] The query might ask for the index of the item instead, but this variation is an easy adaptation of the known solutions – including the ones in this paper.

R Ravi and I.L. Gørtz (Eds.): SWAT 2014, LNCS 8503, pp. 1–12, 2014.

Table 1. Previous and new results on static and dynamic RMQs in the external memory model (EM) and the cache-oblivious model (CO)

Problem	I/Os	Space	Notes
Static RMQ, EM	$O((n+q)\log_m(n+q))$	$O(Q+N\log_m N)$	[5]
Static RMQ, EM	$O((n+q)\log_m(n+q))$	$O(N+Q)$	[2]
Static RMQ, EM	$O(n+q\log_m \min\{q,n\})$	$O(N+Q)$	new
Static RMQ, EM, CO	$\Omega(n+q\log_m \min\{q,n\})$	-	new
Static RMQ, CO	$O(n+q\log_m q)$	$O(N+Q)$	new
Dynamic RMQ, EM	$O((n+q)\log_m^2(n+q))$	$O(N+Q)$	[2]
Dynamic RMQ, EM	$O((n+q\log_m q)\log_m n)$	$O(N+Q)$	new

In this paper we are interested in the RMQ problem in the *external memory* model. The external memory model (also known as the *I/O model* or *disk access model (DAM)*) was introduced by Aggarwal and Vitter [1] and addresses situations where the data is so big that it can only be stored in slow external storage. The external storage is divided into blocks of size B and all the computations must be done in the internal memory of size M. Each data transfer, an *input/output (I/O)* operation, between the external and internal memory can transfer a single block. The complexity metric of the model, *I/O complexity*, measures the number of such transfers. In this paper we use the common notations $n = N/B$, $m = M/B$, $q = Q/B$, and $\text{sort}(N) = O(n\log_m n)$ – the I/O complexity to sort an array of N elements.

In the external memory model, the online RMQ problem where we require that the answer to each query must be provided immediately, one must spend at least one I/O operation to report the output and, therefore, constant time solutions in the RAM model translate to the optimal solutions in the EM model as well. Instead, Chiang et al. [5] considered the offline version of the problem. In the *offline (batched)* range minima problem we are given a sequence of Q range minima queries $\text{RMQ}(i,j)$ and we are asked to answer each query eventually and in arbitrary order by presenting the output as pairs of the input queries and the corresponding answers.

Previous results in the EM model. Chiang et al. [5] presented an algorithm that answers a batch of Q queries using $O(\text{sort}(N+Q)) = O((n+q)\log_m(n+q))$ I/Os and $O(Q+N\log_m N)$ space. Very recently, Arge et al. [2] improved the space to $O(N+Q)$ while keeping the same I/O complexity. They also showed a solution for the *dynamic* version of the problem where the sequence of queries is intermixed with insertions and deletions of entries to and from the array. Their solution requires $O((n+q)\log_m^2(n+q))$ I/Os. They left a few open questions and in fact they explicitly conjectured that even the static range minima queries should require $\Omega((n+q)\log_m(n+q))$ I/Os in the worse case. The conjecture is non-trivial and interesting because in internal memory, the constant time per query trivially implies $O(N+Q)$ time to answer Q queries in an array of size N.

Our Results. We offer a number of improvements to both static and dynamic batched RMQs. In Section 2, we prove a lower bound of $\Omega(n + q \log_m \min\{q, n\})$ I/Os for the static batched RMQ problem, partially confirming the suspicion of Arge et al. [2] that it is impossible to achieve linear $O(n + q)$ I/O complexity in the EM model. Our lower bound assumes the standard indivisibility of individual records and holds even if the queries are presorted. In the process of proving the lower bound we present an algebraic notation which simplifies the presentation of permutation lower bound proofs and might be of independent interest.

In Section 3 we present a matching upper bound in the EM model, thus proving that our lower bound is asymptotically optimal. Our upper bound immediately implies an improvement to the dynamic version of the RMQ problem by Arge et al. [2], which can be solved in $O(\text{sort}(N) + \text{sort}(Q) \cdot \log_m n)$ I/Os (Section 5).

In Section 4 we present the first solution for the static RMQs in the cache-oblivious model[2]. The *cache-oblivious (CO)* model [9] is similar to the EM model, except the algorithms are not allowed to make use of the parameters M and B. Instead, the data transfer between the external and internal memory is performed automatically by a separate paging algorithm implemented by the system with a reasonable cache replacement strategy, e.g., least recently used (LRU) strategy. Our cache-oblivious solution requires $O(n + q \log_m q)$ beating all the previous results in the EM model.

Table 1 lists our results in comparison with the previous results in the external memory and the cache-oblivious models.

Finally, in Section 5 we discuss some additional simple improvements if some blocks of the input array are not covered by any queries.

2   Lower Bound In Both Models

In this section, we prove a lower bound showing that under a standard assumption of indivisibility of individual items it is impossible to answer Q RMQ queries on a static array of size N in fewer than $\Omega(n + q \log_m \min\{q, n\})$ I/Os.

Atomic elements. We assume each query is accompanied by a label that is a string obtained by concatenating the representation of its left and right boundaries. So, a query $q_i = [\ell_i, r_i]$ is represented by $(s_{\ell_i r_i}, \ell_i, r_i)$, where $s_{\ell_i r_i}$ is its label. Query labels and values in the array A are considered *atomic elements*.

The Model. We conceptually view the external memory as a (horizontal) tape of infinite size consisting of cells arranged from left to right that are also organized into blocks of B cells. Each cell can store one atomic element. Any other information can be stored and accessed for free by the algorithm (i.e., we assume unlimited computational power and full information). The only restriction placed on the algorithm is that it cannot create new atomic elements, but can only make copies of the existing ones. Thus, to manipulate labels or values, the

---

[2] Previously, only *online* results were known. E.g., see [6,11].

algorithm can load one block (containing some atomic elements) from the tape into the internal memory or it can select B atomic elements from the internal memory and write copies of them somewhere on the tape as one block. The algorithm starts with a tape that contains the input values of A in n blocks and the Q queries in the q following blocks and it must end with a tape configuration where each query label is followed by its answer (i.e., a pair $(s_{\ell_i, r_i}, A[j])$ where $A[j]$ is the answer to the query $[\ell_i, r_i]$ labeled s_{ℓ_i, r_j}).

Sequences. In this model, a subset of K cells naturally defines a sequence of K atomic elements, by considering the atomic elements stored in the cells in the left-to-right order. In the rest of this section, we slightly extend the definition of a sequence: a *sequence representation* (*seq-rep* for short) is a sequence of K atomic elements that is stored in $O(K/B)$ blocks[3] on the tape from left to right. Note that we allow some inefficiency in the storage as there could be blocks that store only a few atomic elements. Observe that one sequence can have two different seq-reps S_1 and S_2 and the atomic elements of each block could occupy different addresses within that block. Nonetheless, one can convert one representation into another in $O(K/B)$ I/Os. This implies that for a given sequence, all the seq-reps are essentially equivalent up to an additive term of $O(K/B)$ I/Os.

The Main Idea and Intuition. We prove our lower bound using known hardness results for the problem of permuting array entries. Intuitively, the hard input instance to the RMQ algorithm is a set of queries where the left end points and the right end points correspond to two very "different" permutations; our lower bound follows from the fact that the permutation corresponding to the left end points needs $\Omega(\min\{Q, q\log_m n\})$ I/Os to be transformed into the permutation corresponding to the right end points.

Although our lower bound approach does not introduce fundamentally new techniques, it does require rather complicated logical steps. To follow the argument with greater ease, we introduce a new algebraic notation, which could be considered an interesting way of presenting permutation lower bounds.

An Algebraic Notation. Let $\mathbf{X} := X_1, \ldots, X_N$ be a sequence of N atomic elements. For a given permutation $\pi : \{1, \cdots, N\} \to \{1, \cdots, N\}$, $\pi(\mathbf{X})$ is defined as the sequence $X_{\pi(1)}, \ldots, X_{\pi(N)}$ and we denote $X_{\pi(i)}$ with $\pi^{(i)}(\mathbf{X})$. Furthermore, if we can permute one seq-rep of $\pi(\mathbf{X})$ into another seq-rep of $\varkappa(\mathbf{X})$ using t I/Os, then we can permute *any* seq-rep of $\pi(\mathbf{X})$ into *any* seq-rep of $\varkappa(\mathbf{X})$ using $t + O(n)$ I/Os ($n = N/B$). Note that we can also permute *any* seq-rep of $\varkappa(\mathbf{X})$ into *any* seq-rep of $\pi(\mathbf{X})$ using the same $t + O(n)$ I/Os. We denote such transformation with $\pi(\mathbf{X}) \overset{t+O(n)}{\longleftrightarrow} \varkappa(\mathbf{X})$. Easy to see but important consequences of the indivisibility assumption are summarized below.

---

[3] The O-notation here hides a universal constant that does not depend on any machine or input parameter. We need this constant since during some steps of our proof, we will be working with the sequences that do not necessarily pack B elements in each block.

Observation 1. *Consider two sequences of symbols* $\mathbf{X} := X_1, \ldots, X_N$ *and* $\mathbf{Y} :=$ $Y_1, \ldots, Y_N$. *Let* $\varkappa, \pi$ *and* φ *be three permutations. The following properties hold in the indivisibility model regarding the seq-reps of these sequences.*

(a) *If* $\pi(\mathbf{X}) \overset{t+O(n)}{\longleftrightarrow} \varkappa(\mathbf{X})$ *then* $\pi(\mathbf{Y}) \overset{t+O(n)}{\longleftrightarrow} \varkappa(\mathbf{Y})$

(b) *If* $\pi(\mathbf{X}) \overset{t+O(n)}{\longleftrightarrow} \varkappa(\mathbf{X})$ *then* $\pi(\varphi(\mathbf{X})) \overset{t+O(n)}{\longleftrightarrow} \varkappa(\varphi(\mathbf{X}))$ *and* $\varphi(\pi(\mathbf{X})) \overset{t+O(n)}{\longleftrightarrow} \varphi(\varkappa(\mathbf{X}))$

(c) *If* $\pi(\mathbf{X}) \overset{t+O(n)}{\longleftrightarrow} \varkappa(\mathbf{X})$ *and* $\varkappa(\mathbf{X}) \overset{r+O(n)}{\longleftrightarrow} \varphi(\mathbf{X})$ *then* $\pi(\mathbf{X}) \overset{t+r+O(n)}{\longleftrightarrow} \varphi(\mathbf{X})$.

Remark. The constants hidden in the O-notations above can grow. This is because we are working with *any* seq-rep of permutations rather than specific ones.

The Query Order. We actually prove a stronger lower bound claim. We show that the problem stays hard even if the queries are given in the order of the left end points, or the right end points, or any other ordering that does *not* depend on the input array A. We model this claim precisely. Let Q be the list of queries. Before showing the algorithm the input set A, we allow the algorithm to pick whatever order that it desires for the queries, i.e., the algorithm can permute the queries for free. Once that order is picked (for example, the algorithm can sort the list of queries by the left end points), the algorithm is given an array A. We show that even in this relaxed formulation, the algorithm cannot achieve $O(n + q)$ bound on the number of I/Os.

Observe in the case $Q < N$ we simply need to prove a lower bound for Q queries and an input array of size Q since the upper bound in the previous section has linear dependency on n. Thus, the non-trivial case of the problem is when $Q = \Omega(N)$. Due to this, w.l.o.g, we assume the following in the rest of this section: the range of the queries run from 1 to N and $Q = \alpha N$ where $\alpha \geq 1$ is an integer. We also need the following lemma, which is an easy generalization of the permutation lower bound [1].

Lemma 1. *Let N and α be two integral parameters, and let S_1 be the non-decreasing sequence of length αN composed of α repetitions of i, $i = 1, \cdots, N$. Assuming $2 < B < cM < N$ for a constant c, there exists a permutation S_2 of S_1, s.t., permuting S_1 into S_2 requires $\Omega\left(\min\{\alpha N, \frac{\alpha N}{B} \log_m \frac{N}{B}\}\right)$ I/Os.*

Proof. The proof is almost identical to the one presented by Aggarwal and Vitter [1] for the general permutations. The only difference is that we need to calculate the number of permutations of S_1. Using straightforward combinatorial arguments and Stirling's formula, the number of permutations of S_1 is $\prod_{i=0}^{N-1} \left(\frac{\alpha(N-i)}{\alpha}\right) \geq \left(\frac{1}{\sqrt{\alpha}}\right)^N (N!)^\alpha$. Using this bound instead of $N!$ at the right hand side of the inequality in Section 4 of Aggarwal and Vitter's paper [1] gives the claimed bound. $\square$

Now we are ready to prove our lower bound result.

Theorem 1. *A set of Q range minima queries on a static array of size N requires $\Omega(\min\{Q, q \log_m n\})$ I/Os in the worst case, assuming indivisibility.*

Proof. Consider the sequences S_1 and S_2 defined in Lemma 1; observe both have α repetitions of every value i between 1 and N. We create the sequence of queries $\mathcal{Q}$ based on S_2 in the following way: if the i-th element of S_2 is j, we create the query interval $[\lfloor i/\alpha \rfloor, j]$ with its appropriate label. We present $\mathcal{Q}$ to the algorithm and let $\varkappa(\mathcal{Q})$ be the ordering of the queries picked by the algorithm (remember this is done for free). Note that $\mathcal{Q}$ is sorted by the left end point.

We now define two different input arrays, $A_1[1, \cdots, N]$ and $A_2[1, \cdots, N]$: A_1 is strictly increasing and A_2 is strictly decreasing. This means, the left end points of the queries give the indices of the answers for the queries on A_1, while the right end points do the same on A_2. However, remember that the final answer should contain the labels of the queries. We claim that one of these two inputs should be difficult to solve regardless of choice of $\varkappa$.

Let r_1 be the number of I/Os used by an algorithm to solve the problem when presented with A_1 and σ_1 be the permutation that describes the order of the atomic elements (the query label, value pair) in the output. For simplicity, we assume $\mathcal{Q}$ is the sequence of the query labels. Observe that $\sigma_1(\varkappa(S_1))$ describes the sequence of indices of the answers to the queries in the first input: a query interval $[i, j]$ in the output is followed by $A_1[i]$ and since queries were originally ordered by the left end point, $\sigma_1(\varkappa(S_1))$ gives the ordering of the indices of the answer.

Now consider the input A_2. Let r_2 be the number of I/Os used by an algorithm to solve the problem when presented with A_2 and σ_2 be the permutation that describes the order of the atomic elements (the query label, value pair) in the output. A query interval $[i, j]$ in the output if followed by $A_2[j]$. This means the sequence of indices of the answers to the queries in the second input is described by $\sigma_2(\varkappa(\varphi(S_1)))$ where φ is a permutation such that $\varphi(S_1) = S_2$.

Thus, we have the following (explanations below):

$$\varkappa(\mathcal{Q}) \xleftrightarrow{r_1} \sigma_1(\varkappa(\mathcal{Q})) \tag{1}$$

$$\varkappa(\mathcal{Q}) \xleftrightarrow{r_2} \sigma_2(\varkappa(\mathcal{Q})) \tag{2}$$

$$S_1 \xleftrightarrow{r_1} \sigma_1(\varkappa(S_1)) \tag{3}$$

$$S_1 \xleftrightarrow{r_2} \sigma_2(\varkappa(\varphi(S_1))) \tag{4}$$

The above equations describe how the order of the atomic elements in the output correspond to the order of the atomic elements given to the algorithm, with the difference that (for simplicity) instead of dealing with the values in the arrays A_1 and A_2, we are dealing with their indices; S_1 in the left hand side of the equations correspond to the indices of the values in arrays A_1 and A_2.

Applying Observation 1(c) to (1) through (4), we get

$$\sigma_1(\varkappa(\mathcal{Q})) \xleftrightarrow{r_1+r_2+O(q)} \sigma_2(\varkappa(\mathcal{Q})) \tag{5}$$

$$\sigma_1(\varkappa(S_1)) \xleftrightarrow{r_1+r_2+O(q)} \sigma_2(\varkappa(\varphi(S_1))) \tag{6}$$

Applying Observation 1(a) to (5) we get $\sigma_1(\varkappa(S_1)) \overset{r_1+r_2+O(q)}{\longleftrightarrow} \sigma_2(\varkappa(S_1))$. Finally, with (6) and Observation 1(c) we obtain $\sigma_1(\varkappa(S_1)) \overset{O(r_1+r_2+q)}{\longleftrightarrow} \sigma_1(\varkappa(\varphi(S_1)))$. Set φ as an inverse of σ_1 in Observation 1(b) and this gives $\varkappa(S_1) \overset{O(r_1+r_2+q)}{\longleftrightarrow} \varkappa(\varphi(S_1))$ and similarly $S_1 \overset{O(r_1+r_2+q)}{\longleftrightarrow} \varphi(S_1) = S_2$. Thus, by Lemma 1, we must have $r_1 + r_2 = \Omega(\min\{Q, q \log_m n\})$, so the problem is hard on either A_1 or A_2. $\square$

3 Solution in the External Memory Model

In this section we prove a matching upper bound for the static RMQ problem in the EM model.

Theorem 2. *A set of Q range minima queries on a static array of N elements can be answered in* $O(n + q \cdot \min\{\log_m n, \log_m q\})$ *I/Os and* $O(N + Q)$ *space.*

Note that when $Q = \Theta(N)$ the I/O complexity in the above theorem matches the I/O complexity $O(\text{sort}(N + Q)) = O((n + q) \log_m(n + q))$ of Arge et al. [2]. Thus, we concentrate on two cases: (i) when $N = \omega(Q)$ and (ii) when $Q = \omega(N)$.

Without loss of generality we assume that each query $\text{RMQ}(i, j)$ has a unique identifier – it can be, for example, the initial index in the list of the input queries.

Lemma 2. *The problem of answering a set of Q range minima queries on a static array A of $N = \omega(Q)$ elements can be reduced to the problem of answering Q range minima queries on a static array A' of size $O(Q)$ in $O(n + q \log_m q)$ I/Os and $O(N + Q) = O(N)$ space.*

Proof. Consider any two adjacent array entries $A[i]$ and $A[i+1]$. Observe that if no query starts or ends with an index i and $i+1$, then the larger of the two entries $A[i]$ and $A[i + 1]$ will not be the answer to any of the queries. More generally, for any contiguous region of the array $A[i..j]$, $i < j$, if there are no queries with endpoint indices in the range $[i, j]$, then we can compact the subarray $A[i..j]$ to a single element that is the minimum in the range $A[i..j]$ without affecting the answers to the queries. Since there are $2Q$ query endpoints, the size of the compacted array is $O(Q)$. Obviously, if we compact the input array to a smaller array, we have to adjust the query endpoints appropriately, which we show how to do next.

For each query $\text{RMQ}(i, j)$ we create two items e_i and e_j associated with the two endpoints of the query. Each endpoint e_i (resp. e_j) contains full information about the query $\text{RMQ}(i, j)$, i.e., the unique identifier of the query and the index j (resp. i) of the other endpoint.

We sort the set of endpoints e_x by their indices x. By simultaneously scanning the input array and the sorted set of endpoints we can identify the ranges of array indices that contain no query endpoints. During the scan we can also identify the minimum within each range and copy them into a new array A'. Let $s[i]$ be the number of items among $A[1..i]$ that were not copied to A'. We can compute the values $s[i]$ for all $1 \leq i \leq N$ during the scan. To adjust the queries, we need to update the index of each query endpoint e_i from i to $i - s[i]$. This can be accomplished with a simultaneous scan of the sorted endpoints and the values $s[i]$. Finally, a sort of the endpoints by the query identifiers will place the two endpoints of each query in adjacent memory locations and with a final scan of this sorted sequence we can create the updated queries $\text{RMQ}(i - s[i], j - s[j])$ for each original query $\text{RMQ}(i, j)$. The I/O complexity of the whole process is $O(n + \text{sort}(Q)) = O(n + q \log_m q)$ I/Os because it is just $O(1)$ scans of arrays of size $O(N)$ and $O(1)$ sorts of sets of size $O(Q)$.   □

Lemma 3. *A set of Q range minima queries on a static array A of $N = o(Q)$ elements can be answered in $O(q \log_m n)$ I/Os and $O(N + Q) = O(Q)$ space.*

Proof. In the algorithm of Arge et al. [2], it is difficult to avoid the $O(\text{sort}(N+Q))$ cost; to summarize, they do the following: first they build a full k-ary tree T for $k \in \Theta(m)$ on the array A, with each of $\Theta(N/M)$ leaves associated with a contiguous range of $\Theta(M)$ entries. The algorithm processes the queries down this tree level by level, by computing a running answer for each endpoint of a query and distributing the endpoints to the appropriate children of a node. At the leaves of the tree, the answer to each query $\text{RMQ}(i, j)$ is the minimum of the running answers at the two endpoints e_i and e_j. The two endpoints might be in two different leaves of the tree, i.e., in arbitrary locations in external memory. To compute the minimum of each pair I/O-efficiently, the algorithm sorts the endpoints by the query identifier, which results in the two endpoints being in adjacent memory locations and the minimum can be computed with a simple scan. The I/O complexity of this solution consists of $O(\text{sort}(N))$ I/Os to build the tree, $O(q \log_m(N/M))$ to propagate all queries down to the leaves of the tree (the distribution involves scanning Q queries at each of $O(\log_m(N/M))$ levels of the tree), and $O(\text{sort}(Q))$ I/Os to compute the minima of pairs of endpoints at the leaves of the tree. Note, when $N = o(Q)$ the I/O complexity of this solution reduces to $O(\text{sort}(Q)) = O(q \log_m q)$ I/Os.

To improve the I/O complexity to $O(q \log_m n)$ we show how to compute the minima of the pairs of endpoints at the leaves of the tree more efficiently. In particular, we observe that the distribution of the query endpoints to the children nodes of the tree is performed stably – that is, the relative order of the queries distributed to each child node is the same as in the (parent) node itself. Thus, we maintain the invariant that at each node the query endpoints are sorted by the initial order of the input queries. Initially, at the root of the node, the invariant is trivially true and the stability of the distribution ensures that the invariant is maintained at each consequent level.

Once the query endpoints reach the leaf level, we do the following. We load $O(M)$ array entries associated with a leaf into internal memory and scan the endpoints within that leaf, finding and reporting the answers to queries that contain both endpoints within that leaf. Once a query answer is determined unambiguously, we stop considering it any further. At this point, instead of sorting the remaining endpoints, we propagate them up the tree, merging them by comparing the original indices of the query in the input set. Since the query endpoints at each node are sorted in this order, we can perform this merge I/O efficiently and if the two endpoints e_i and e_j of a query RMQ(i,j) are present in the subtrees rooted at two children w_k and $w_{k'}$ of some tree node v, the merging process at node v will place them next to each other and we can compute the minima among both endpoints, report it as the answer to query RMQ(i,j) and stop considering the two endpoints any further.

The I/O complexity of the merging process is $O(n+q)$ to process the leaves and $O(q \log_m(N/M))$ I/Os to perform the merge up the tree. Thus the total I/O complexity of the whole algorithm adds up to $O(q \log_m n)$ I/Os. $\qquad\square$

The proof of the Theorem 2 follows from Lemma 2 and Lemma 3.

4 Solution in the Cache-Oblivious Model

In this section we will prove the following result:

Theorem 3. *In the cache-oblivious model a set of Q range minima queries on an array of size N can be answered in $O(n + q \log_m q)$ I/Os, assuming $M = \Omega(B^{1+\epsilon})$.*

First, note that Lemma 2 holds in the cache-oblivious model because the reduction consists of a constant number of scans and sorts, which can be accomplished cache-obliviously [9]. Thus, it only remains to show how to answer Q range minima queries on an array of size $N = O(Q)$ in $O(q \log_m q)$ I/Os.

The static solution of Arge et al. [2] can be viewed as using the top-down distribution sweeping approach, where at each node of the recursive tree the queries are considered in some predetermined order (a sweep of queries) and distributed to the $\Theta(M/B)$ children of the node. Brodal and Fagerberg [4] presented a framework to implement distribution sweeping paradigm cache-obliviously by a bottom-up recursive process, where at each recursive level the objects of the children nodes are merged. We will show how to answer the range minima queries by merging the queries bottom up instead, thus allowing us to use the cache-oblivious distribution sweeping framework of Brodal and Fagerberg.

Again, without loss of generality, we assume that each query RMQ(i,j) has a unique identifier.

We proceed as follows. For each query RMQ(i,j) we create two items e_i and e_j associated with the two endpoints of the query. Each endpoint e_i (resp. e_j) contains full information about the query RMQ(i,j), i.e., the unique identifier of the query and the index j (resp. i) of the other endpoint. Each endpoint e_x we will maintain a running answer rmq_{e_x}. At the end of the computation, both rmq_{e_i} and rmq_{e_j} will hold the answer to the query RMQ(i,j).

Initially, we sort the endpoints e_x by its index x and initialize $rmq_{e_i} = A[i]$ and $rmq_{e_j} = A[j]$. Next we perform the following merging algorithm. Conceptually, we can visualize a merge tree built on top of the sorted list of endpoints with a single endpoint at each leaf of the merge tree. A node v of the tree represents a contiguous range $\mathcal{R}(v)$ of the indices in the array, such that $\mathcal{R}(v) = \mathcal{R}(w_L) \cup$

$\mathcal{R}(w_R)$, where w_L and w_R are the two children of v. Each node v of the tree maintains $minS(v)$ – the smallest array entry among the indices in its range $\mathcal{R}(v)$. This value is defined as rmq_e at the leaf node e and can be computed at each internal node v as $minS(v) = \min\{minS(w_L), minS(w_R)\}$ and is updated as the first step before the merging at that node begins.

During the merge up the tree the endpoints are compared by the unique identifiers of the queries associated with that endpoint. For the merge step at each internal node v with the two children w_L and w_R we do the following. If the next two smallest endpoints e_i and e_j are for the same query $\text{RMQ}(i,j)$, we set $rmq_i = rmq_j = \min\{rmq_i, rmq_j\}$ (the final answer to query $\text{RMQ}(i,j)$) and the two endpoints are discarded and never considered again in the merging process. If the next two smallest endpoints are not of the same query, assume the smallest endpoint e is the left endpoint of a query (the right endpoints are treated symmetrically). Then if e is coming from the right child w_R, we propagate e to the output of node v without altering it. If e is coming from the left child w_L, we set $rmq_e = \min\{rmq_e, minS(w_R)\}$ and then propagate it to the output of v.

Lemma 4. *At the end of the merging process, all pairs of items e_i and e_j associated with each query $\text{RMQ}(i,j)$ store the answer to the query in rmq_{e_i} and rmq_{e_j}.*

Proof. The proof is by induction on the level of recursion. First observe that at node v, if $e_i \in \mathcal{R}(w_L)$ and $e_j \in \mathcal{R}(w_R)$ and they represent the same query $\text{RMQ}(i,j)$, they will be considered together at some point during merging at node v, because the comparisons are performed by the query identifiers, which are unique and equal for e_i and e_j. Thus an item $e \in R(v)$ is propagated to the output of a node v iff e's other endpoint is not in $\mathcal{R}(v)$'s subtree. Let A_v represent the subarray of A which is defined by the indices in the range $\mathcal{R}(v)$. Then the correctness of the algorithm follows from the fact that for $i < j$, if $e_i \in w_L$ and $e_j \notin w_R$, $\text{RMQ}_{A_{w_R} \cup A_{w_L}}(i, +\infty) = \min\{\text{RMQ}_{A_{w_L}}(i, +\infty), \text{RMQ}_{A_{w_R}}(-\infty, +\infty)\}$, and if $e_i \in w_R$ and $e_j \notin w_R$, $\text{RMQ}_{A_{w_R} \cup A_{w_L}}(i, +\infty) = \text{RMQ}_{A_{w_R}}(i, +\infty)$. The case of e_j is symmetrical. $\square$

Now we are ready to prove Theorem 3 stated at the beginning of this section.

Proof (of Theorem 3). Creation of the items can be performed with a single scan of the queries, which is trivially cache-oblivious. The initial sorting of the items is implemented using one of the cache-oblivious sorting algorithms [9]. The initialization of rmq_e is implemented using a simultaneous scan of the input array and the sorted items. Finally, the merging is implemented using the lazy funnels [4]. Note, that we compute the value $minS(v)$ only once – the first time a merger at node v is invoked. Both cache-oblivious sorting and lazy funnels require the tall cache assumption $M = \Omega(B^{1+\epsilon})$. The I/O complexity follows from [4]. $\square$

Note, we can extend the above merging algorithm to solve the problem in the external memory model using merging, rather than distribution, which might be of independent interest. This is accomplished by performing $\Theta(M/B)$-way merging at each node and maintaining at each node $\Theta(M/B)$ minima of all its children.

5   Additional Improvements

The techniques described in the previous sections are quite simple and can be applied to other contexts. We briefly discuss some of these in this section.

Towards an adaptive analysis: In special cases when large portions of the input array do not overlap with the query ranges, one can achieve better I/O complexity than of the algorithms presented here. Let $n' \leq n$ denote the number of blocks of the input which overlap with the union of ranges defined by the queries. Then both our upper bounds and lower bounds can easily be extended to show that the support of batched RMQ queries is within $\Theta(n' + q \log_m \min\{q, n'\})$ accesses and linear space, in both the external memory model and the cache oblivious model.

Dynamic batched RMQ problem: In the dynamic RMQ problem we are given a sequence that contains Q queries and N update operations (insertions and deletions). There are many different ways to model the behavior of the insertions and deletions with respect to the array indices. As discussed by Arge et al. [2], one can consider an *array version* in which the updates shift the indices (an insertion at position i increases all the succeeding indices by one; a deletion reduces them by one), or a *geometric version* in which such shifting does not occur and the indices are in fact x-coordinates, or a *linked list* version where indices are in fact pointers. All these formulations are equivalent up to an additive $O(\text{sort}(N+Q))$ term. Previously, Arge et al. had shown how to solve such dynamic problems in $O(\text{sort}(N+Q) \log_m(n+q))$ I/Os. Using our static $O(n + q \log_m \min\{q, n\})$ solution as the base case in their solution, we can easily improve the I/O complexity of the dynamic batched RMQ solution to $O(\text{sort}(N) + \text{sort}(Q) \log_m n)$ I/Os.

6   Conclusions

In this paper, we investigate batched range minimum query (RMQ) problem in the external memory (EM) and the cache-oblivious (CO) models. Improving on the previous papers, we obtain matching upper and lower bounds for the static version of the problem in the EM model. Interestingly, our lower bound shows that the problem cannot be solved in linear I/O complexity (in the number of queries) even if we allow the algorithm to reorder the queries in any arbitrary order for free before it is presented with the input array. We also present the first cache-oblivious solution to the problem and although we do not know if it is optimal, it is faster than the previous external memory solutions.

Open problems. Although our work closes the case of the static version of the problem in the EM model, there are still several interesting open problems remaining. There is no better lower bound known for the dynamic version of the problem than the lower bound that we presented here for the static version. And although we improved the upper bound of the dynamic version of the problem, there is still a gap of $O(\log_m n)$ I/Os remaining between the upper and lower bounds. Closing this gap remains an open problem.

In the cache-oblivious model, the merge-based solution presented here seems to require sorting all the queries. Our EM model solutions on the other hand show that when $Q \gg N$ we can avoid the complexity of sorting the queries. It would be interesting to see if similar bound can be shown in the cache-oblivious model or the sorting of the queries is inherently required in the cache-oblivious model.

Acknowledgements. The authors would like to thank Jérémy Barbay for many useful discussions that inspired and motivated us in this work.

References

1. Aggarwal, A., Vitter, J.S.: The input/output complexity of sorting and related problems. Communications of the ACM 31, 1116–1127 (1988)
2. Arge, L., Fischer, J., Sanders, P., Sitchinava, N.: On (dynamic) range minimum queries in external memory. In: Dehne, F., Solis-Oba, R., Sack, J.-R. (eds.) WADS 2013. LNCS, vol. 8037, pp. 37–48. Springer, Heidelberg (2013)
3. Bender, M.A., Farach-Colton, M.: The LCA problem revisited. In: Proc. 4th Latin American Theoretical Informatics Symposium, pp. 88–94 (2000)
4. Brodal, G.S., Fagerberg, R.: Cache oblivious distribution sweeping. In: Proc. 29th International Colloquium on Automata, Languages, and Programming, pp. 426–438 (2002)
5. Chiang, Y.J., Goodrich, M.T., Grove, E.F., Tamassia, R., Vengroff, D.E., Vitter, J.S.: External-memory graph algorithms. In: Proc. 6th ACM/SIAM Symposium on Discrete Algorithms, pp. 139–149 (1995)
6. Demaine, E.D., Landau, G.M., Weimann, O.: On cartesian trees and range minimum queries. Algorithmica 68(3), 610–625 (2014)
7. Fischer, J., Heun, V.: Space-efficient preprocessing schemes for range minimum queries on static arrays. SIAM Journal on Computing 40(2), 465–492 (2011)
8. Fischer, J.: Optimal succinctness for range minimum queries. In: Proc. 9th Latin American Theoretical Informatics Symposium, pp. 158–169 (2010)
9. Frigo, M., Leiserson, C.E., Prokop, H., Ramachandran, S.: Cache-oblivious algorithms. In: Proc. 40th IEEE Symposium on Foundations of Computer Science, pp. 285–297 (1999)
10. Gabow, H.N., Bentley, J.L., Tarjan, R.E.: Scaling and related techniques for geometry problems. In: Proc. 16th ACM Symposium on Theory of Computation, pp. 135–143 (1984)
11. Hasan, M., Moosa, T.M., Rahman, M.S.: Cache oblivious algorithms for the RMQ and the RMSQ problems. Mathematics in Computer Science 3(4), 433–442 (2010)
12. Vuillemin, J.: A unifying look at data structures. Comm. ACM 23(4), 229–239 (1980)

Online Makespan Minimization with Parallel Schedules*

Susanne Albers[1] and Matthias Hellwig[2]

[1] Technische Universität München
albers@in.tum.de
[2] Humboldt-Universität zu Berlin
hub1@matthias-hellwig.de

Abstract. Online makespan minimization is a classical problem in which a sequence of jobs $\sigma = J_1, \ldots, J_n$ has to be scheduled on m identical parallel machines so as to minimize the maximum completion time of any job. In this paper we investigate the problem in a model where extra power/resources are granted to an algorithm. More specifically, an online algorithm is allowed to build several schedules in parallel while processing σ. At the end of the scheduling process the best schedule is selected. This model can be viewed as providing an online algorithm with extra space, which is invested to maintain multiple solutions.

As a main result we develop a $(4/3 + \varepsilon)$-competitive algorithm, for any $0 < \varepsilon \leq 1$, that uses a constant number of schedules. The constant is equal to $1/\varepsilon^{O(\log(1/\varepsilon))}$. We also give a $(1 + \varepsilon)$-competitive algorithm, for any $0 < \varepsilon \leq 1$, that builds a polynomial number of $(m/\varepsilon)^{O(\log(1/\varepsilon)/\varepsilon)}$ schedules. This value depends on m but is independent of the input σ. The performance guarantees are nearly best possible. We show that any algorithm that achieves a competitiveness smaller than $4/3$ must construct $\Omega(m)$ schedules. On the technical level, our algorithms make use of novel guessing schemes that (1) predict the optimum makespan of σ to within a factor of $1 + \varepsilon$ and (2) guess the job processing times and their frequencies in σ. In (2) we have to sparsify the universe of all guesses so as to reduce the number of schedules to a constant.

1 Introduction

Makespan minimization is a fundamental and extensively studied problem in scheduling theory. Consider a sequence of jobs $\sigma = J_1, \ldots, J_n$ that has to be scheduled on m identical parallel machines. Each job J_t is specified by a processing time $p_t > 0$, $1 \leq t \leq n$. Preemption of jobs is not allowed. The goal is to minimize the makespan, i.e. the maximum completion time of any job in the constructed schedule. We focus on the online version of the problem, initially introduced by Graham [18]. Here the jobs of σ arrive one by one as elements of a list. Each incoming job J_t has to be assigned immediately to one of the machines without knowledge of any future jobs $J_{t'}$, $t' > t$. Once all jobs have arrived, the execution of the constructed schedule starts.

---

* Work supported by the German Research Foundation, grant AL 464/7-1.

R Ravi and I.L. Gørtz (Eds.): SWAT 2014, LNCS 8503, pp. 13–25, 2014.

Online algorithms for makespan minimization have been studied since the 1960s. In his early paper Graham [18] showed that the famous *List* scheduling algorithm is $(2 - 1/m)$-competitive. The best online strategy currently known achieves a competitiveness of about 1.92 [16]. Makespan minimization has also been studied with various types of *resource augmentation*, giving an online algorithm additional information or power while processing σ. The following scenarios were considered. (1) A online algorithm may use more machines than an offline algorithm. (2) An online algorithm knows the optimum makespan or the sum of the processing times of σ. (3) An online strategy has a buffer that can be used to reorder σ. Whenever a job arrives, it is inserted into the buffer; then one job of the buffer is removed and placed in the current schedule. (4) An online algorithm may migrate a certain number or volume of jobs.

In this paper we investigate makespan minimization assuming that an online algorithm is allowed to build several schedules in parallel while processing a job sequence σ. Each incoming job is sequenced in each of the schedules. At the end of the scheduling process the best schedule is selected. We believe that this is a sensible form of resource augmentation: In the classical online makespan minimization problem, studied in the literature so far, an algorithm constructs a schedule while jobs arrive one by one. Only when all jobs have arrived, the schedule is executed. Hence there is a priori no reason why an algorithm should not be able to construct several solutions, the best of which is finally chosen.

The investigated setting can be viewed as providing an online algorithm with extra space, which is used to maintain several solutions. Very little is known about the value of extra space in the design of online algorithms. Makespan minimization with parallel schedules is of particular interest in parallel processing environments where each processor can take care of a single or a small set of schedules. We develop algorithms that require hardly any coordination or communication among the schedules. Moreover, the proposed setting is interesting w.r.t. the foundations of scheduling theory, giving insight into the value of multiple candidate solutions. Our study complements work along another line of research, investigating online algorithms with advice, see e.g. [11,13,24]. In that scenario an online algorithm, at any time, can query some information about future input and thereby achieve an improved solution.

Makespan minimization with parallel schedules was also addressed by Kellerer et al. [22]. However, the paper focused on the restricted setting with $m = 2$ machines. In this paper we explore the problem for a general number m of machines. As a main result we show that a constant number of schedules suffices to achieve a significantly improved competitiveness, compared to the standard setting without resource augmentation. The competitive ratios obtained are at least as good and in most cases better than those attained in the other models of resource augmentation mentioned above.

Problem Definition: We investigate the problem *Makespan Minimization with Parallel Schedules (MPS)*. As always, the jobs of a sequence $\sigma = J_1, \ldots, J_n$ arrive one by one and must be scheduled non-preemptively on m identical parallel machines. Each job J_t has a processing time $p_t > 0$. In MPS, an online algorithm $\mathcal{A}$ may maintain a set $\mathcal{S} = \{S_1, \ldots, S_l\}$ of schedules during the scheduling process while jobs of σ arrive. Each job J_t is sequenced in each schedule S_k, $1 \leq k \leq l$. At the end of σ, algorithm $\mathcal{A}$ selects a schedule $S_k \in \mathcal{S}$ having the smallest makespan and outputs this solution. The other schedules of $\mathcal{S}$ are deleted.

As we shall show MPS can be reduced to the problem variant where the optimum makespan of the job sequence to the processed is known in advance. Hence let MPS_{opt} denote the variant of MPS where, prior to the arrival of the first job, an algorithm $\mathcal{A}$ is given the value of the optimum makespan $\text{OPT}(\sigma)$ for the incoming job sequence σ. An algorithm $\mathcal{A}$ for MPS or MPS_{opt} is ρ-competitive if, for every job sequence σ, it outputs a schedule whose makespan is at most ρ times $\text{OPT}(\sigma)$.

Our Contribution: We present a comprehensive study of MPS. We develop a $(4/3 + \varepsilon)$-competitive algorithm, for any $0 < \varepsilon \leq 1$, that uses a constant number of $1/\varepsilon^{O(\log(1/\varepsilon))}$ schedules. Furthermore, we give a $(1 + \varepsilon)$-competitive algorithm, for any $0 < \varepsilon \leq 1$, that uses a polynomial number of schedules. The number is $(m/\varepsilon)^{O(\log(1/\varepsilon)/\varepsilon)}$, which depends on m but is independent of the job sequence σ. These performance guarantees are nearly best possible. The algorithms are obtained via some intermediate results that may be of independent interest.

First, in Section 2 we show that the original problem MPS can be reduced to the variant MPS_{opt} in which the optimum makespan is known. More specifically, given any ρ-competitive algorithm $\mathcal{A}$ for MPS_{opt} we construct a $(\rho + \varepsilon)$-competitive algorithm $\mathcal{A}^*(\varepsilon)$, for any $0 < \varepsilon \leq 1$. If $\mathcal{A}$ uses l schedules, then $\mathcal{A}^*(\varepsilon)$ uses $l \cdot \lceil \log(1 + \frac{6\rho}{\varepsilon}) / \log(1 + \frac{\varepsilon}{3\rho}) \rceil$ schedules. The construction works for any algorithm $\mathcal{A}$ for MPS_{opt}. In particular we could use a 1.6-competitive algorithm by Chen et al. [12] that assumes that $\text{OPT}(\sigma)$ is known and builds a single schedule.

We proceed to develop algorithms for MPS_{opt}. In Section 3 we give a $(1 + \varepsilon)$-competitive algorithm, for any $0 < \varepsilon \leq 1$, using $(\lfloor 2m/\varepsilon \rfloor + 1)^{\lceil \log(2/\varepsilon)/\log(1+\varepsilon/2) \rceil}$ schedules. In Section 4 we devise a $(4/3 + \varepsilon)$-competitive algorithm, for any $0 < \varepsilon \leq 1$, that uses $1/\varepsilon^{O(\log(1/\varepsilon))}$ schedules. Combining these algorithms with $\mathcal{A}^*(\varepsilon)$, we derive the two algorithms for MPS mentioned in the above paragraph; see also Section 5. The number of schedules used by our strategies depends on $1/\varepsilon$ and exponentially on $\log(1/\varepsilon)$ or $1/\varepsilon$. Such a dependence seems inherent if we wish to explore the full power of parallel schedules. The trade-offs resemble those exhibited by PTASs in offline approximation. Recall that the PTAS by Hochbaum and Shmoys [20] for makespan minimization achieves a $(1 + \varepsilon)$-approximation with a running time of $O((n/\varepsilon)^{1/\varepsilon^2})$.

In Section 6 we present lower bounds. We show that any online algorithm for MPS that achieves a competitive ratio smaller than $4/3$ must construct more than $\lfloor m/3 \rfloor$ schedules. Hence the competitive ratio of $4/3$ is best possible using a constant number of schedules. We show a second lower bound that implies that the number of schedules of our $(1 + \varepsilon)$-competitive algorithm is nearly optimal, up to a polynomial factor.

Our algorithms make use of novel guessing schemes. $\mathcal{A}^*(\varepsilon)$ works with guesses on the optimum makespan. Guessing and *doubling* the value of the optimal solution is a technique that has been applied in other load balancing problems, see e.g. [6]. However here we design a refined scheme that carefully sets and readjusts guesses so that the resulting competitive ratio increases by a factor of $1 + \varepsilon$ only, for any $\varepsilon > 0$. Moreover, the readjustment and job assignment rules have to ensure that scheduling errors, made when guesses were too small, are not critical. Our $(4/3 + \varepsilon)$-competitive algorithm works with guesses on the job processing times and their frequencies in σ. In order to achieve a constant number of schedules, we have to sparsify the set of all possible guesses in an appropriate/novel way.

All our algorithms have the property that the parallel schedules are constructed basically independently. The algorithms for $\mathrm{MPS_{opt}}$ require no coordination at all among the schedules. In $\mathcal{A}^*(\varepsilon)$ a schedule only has to report when it fails, i.e. when a guess on the optimum makespan is too small.

The competitive ratios achieved with parallel schedules are considerably smaller than the best ratios of about 1.92 known for the scenario without resource augmentation. Our ratio of $(4/3+\varepsilon)$, for small ε, is lower than the competitiveness of about 1.46 obtained in the settings where a reordering buffer of size $O(m)$ is available or $O(m)$ jobs may be reassigned [2,14]. Sanders et al. [27] gave an online algorithm that is $(1 + \varepsilon)$-competitive if, before the assignment of any job J_t, jobs of processing volume $2^{O((1/\varepsilon) \log^2 (1/\varepsilon))} p_t$ may be migrated. Hence the total amount of extra resources used while scheduling σ depends on the input sequence. As for online computation with advice, Renault et al. [24] devised an algorithm that is $(1 + \varepsilon)$-competitive and, per incoming job, queries $O(\frac{1}{\varepsilon} \log \frac{1}{\varepsilon})$ bits about future input, for any $0 < \varepsilon < 1/2$.

Remark: Due to space limitations the proofs of the theorems and corollaries developed in this paper are given in the full version of this article.

Related Work: Makespan minimization with parallel schedules was first studied by Kellerer et al. [22]. They assume that $m = 2$ machines are available and two schedules may be constructed. They show that in this case the optimal competitive ratio is $4/3$.

We summarize results known for online makespan minimization without resource augmentation. As mentioned before, *List* is $(2 - 1/m)$-competitive. Deterministic online algorithms with a smaller competitive ratio were presented in [1,10,16,17,21]. The best algorithm currently known is 1.9201-competitive [16]. Lower bounds on the performance of deterministic strategies were given in [1,9,15,19,25,26]. The best bound currently known is 1.88, see [25].

We next review the results for the various models of resource augmentation. Azar et al. [8] devise an online algorithm attaining a competitiveness of $1 + (1/2)^{m'/m(1-o(1))}$ assuming that the algorithm may use $m' \geq m$ machines. Articles [3,4,5,7,12,22] study makespan minimization assuming that an online algorithm knows the optimum makespan $\text{OPT}(\sigma)$ or the sum of the processing times of σ. Chen et al. [12] developed a 1.6-competitive algorithm. Azar and Regev [7] showed that no online algorithm can attain a competitive ratio smaller than $4/3$ if $\text{OPT}(\sigma)$ is known. The setting in which an online algorithm is given a reordering buffer was explored in [14,22]. Englert et al. [14] presented an algorithm that, using a buffer of size $O(m)$, achieves a competitive ratio of $W_{-1}(-1/e^2)/(1+W_{-1}(-1/e^2)) \approx 1.46$, where W_{-1} is the Lambert W function. No algorithm using a buffer of size $o(n)$ can beat this ratio.

Makespan minimization with job migration was addressed in [2,27]. An algorithm that achieves again a competitiveness of $W_{-1}(-1/e^2)/(1+W_{-1}(-1/e^2)) \approx 1.46$ and uses $O(m)$ job reassignments was devised in [2]. No algorithm using $o(n)$ reassignments can obtain a smaller competitiveness. We refer again to Sanders et al. [27] for a study of scenario in which before the assignment of each job J_t, jobs up to a total processing volume of βp_t may be migrated, for some constant β. Specifically, for $\beta = 4/3$, they also present a 1.5-competitive algorithm.

As for memory in online algorithms, Sleator and Tarjan [28] studied the paging problem assuming that an online algorithm has a larger fast memory than an offline strategy. Raghavan and Snir [23] traded memory for randomness in online caching.

Notation: Throughout this paper it will be convenient to associate schedules with algorithms, i.e. a schedule S_k is maintained by an algorithm A_k that specifies how to assign jobs to machines in S_k. Thus an algorithm $\mathcal{A}$ for MPS or MPS_{opt} can be viewed as a family $\{A_k\}_{k \in \mathcal{K}}$ of algorithms that maintain the various schedules. We will write $\mathcal{A} = \{A_k\}_{k \in \mathcal{K}}$. If $\mathcal{A}$ is an algorithm for MPS_{opt}, then the value $\text{OPT}(\sigma)$ is of course given to all algorithms of $\{A_k\}_{k \in \mathcal{K}}$. Furthermore, the *load* of a machine always denotes the sum of the processing times of the jobs already assigned to that machine.

2   Reducing MPS to MPS$_{\text{opt}}$

In this section we will show that any ρ-competitive algorithm $\mathcal{A}$ for MPS_{opt} can be used to construct a $(\rho + \varepsilon)$-competitive algorithm $\mathcal{A}^*(\varepsilon)$ for MPS, for any $0 < \varepsilon \leq 1$. The main idea is to repeatedly execute $\mathcal{A}$ for a set of guesses on the optimum makespan. The initial guesses are small and are increased whenever a guess turns out to be smaller than $\text{OPT}(\sigma)$. The increments are done in small steps so that, among the final guesses, there exists one that is upper bounded by approximately $(1 + \varepsilon)\text{OPT}(\sigma)$. In the analysis of this scheme, we will have to bound machine loads caused by scheduling "errors" made when guesses were too small. Unfortunately the execution of $\mathcal{A}$, given a guess $\gamma \neq \text{OPT}(\sigma)$, can lead to undefined algorithmic behavior. As we shall show, guesses $\gamma \geq \text{OPT}(\sigma)$ are not critical. However, guesses $\gamma < \text{OPT}(\sigma)$ have to be handled carefully.

So let $\mathcal{A} = \{A_k\}_{k \in \mathcal{K}}$ be a ρ-competitive algorithm for MPS_{opt} that, given guess γ, is executed on a job sequence σ. Upon the arrival of a job J_t, an algorithm $A_k \in \mathcal{A}$ may *fail* because the scheduling rules of A_k do not specify a machine where to place J_t in the current schedule S_k. We define two further conditions when an algorithm A_k fails. The first one identifies situations where a makespan of $\rho\gamma$ is not preserved and hence ρ-competitiveness may not be guaranteed. More precisely, A_k would assign J_t to a machine M_j such that $\ell(j) + p_t > \rho\gamma$, where $\ell(j)$ denotes M_j's machine load before the assignment. The second condition identifies situations where γ is not consistent with lower bounds on the optimum makespan, i.e. γ is smaller than the average machine load or the processing time of J_t. Formally, an algorithm A_k *fails* if a job J_t, $1 \leq t \leq n$, has to be scheduled and one of the following conditions holds: (i) A_k does not specify a machine where to place J_t in the current schedule S_k. (ii) There holds $\ell(j) + p_t > \rho\gamma$, for machine M_j to which A_k would assign J_t in S_k. (iii) There holds $\gamma < \sum_{t' \leq t} p_{t'}/m$ or $\gamma < p_t$.

Algorithm for MPS: We describe our algorithm $\mathcal{A}^*(\varepsilon, h)$ for MPS, where $0 < \varepsilon \leq 1$ and $h \in \mathbb{N}$ may be chosen arbitrarily. The construction takes as input any algorithm $\mathcal{A} = \{A_k\}_{k \in \mathcal{K}}$ for MPS$_{\text{opt}}$. For a proper choice of h, $\mathcal{A}^*(\varepsilon, h)$ will be $(\rho + \varepsilon)$-competitive, provided that $\mathcal{A}$ is ρ-competitive.

At any time $\mathcal{A}^*(\varepsilon, h)$ works with h guesses $\gamma_1 < \ldots < \gamma_h$ on the optimum makespan for the incoming job sequence σ. These guesses may be adjusted during the processing of σ; the update procedure will be described in detail below. For each guess γ_i, $1 \leq i \leq h$, $\mathcal{A}^*(\varepsilon, h)$ executes $\mathcal{A}$. Hence $\mathcal{A}^*(\varepsilon, h)$ maintains a total of $h|\mathcal{K}|$ schedules, which can be partitioned into subsets $\mathcal{S}_1, \ldots, \mathcal{S}_h$. Subset $\mathcal{S}_i$ contains those schedules generated by $\mathcal{A}$ using γ_i, $1 \leq i \leq h$. Let $S_{ik} \in \mathcal{S}_i$ denote the schedule generated by A_k using γ_i.

A job sequence σ is processed as follows. Initially, upon the arrival of the first job J_1, the guesses are initialized as $\gamma_1 = p_1$ and $\gamma_i = (1+\varepsilon)\gamma_{i-1}$, for $i = 2, \ldots, h$. Each job J_t, $1 \leq t \leq n$, is handled in the following way. Of course each such job is sequenced in every schedule S_{ik}, $1 \leq i \leq h$ and $1 \leq k \leq |\mathcal{K}|$. Algorithm $\mathcal{A}^*(\varepsilon, h)$ checks if A_k using γ_i fails when having to sequence J_t in S_{ik}. This check can be performed easily by just verifying if one of the conditions (i–iii) holds. If A_k using γ_i does not fail and has not failed since the last adjustment of γ_i, then in S_{ik} job J_t is assigned to the machine specified by A_k using γ_i. The initialization of a guess is also regarded as an adjustment. If A_k using γ_i does fail, then J_t and all future jobs are always assigned to a least loaded machine in S_{ik} until γ_i is adjusted the next time.

Suppose that after the sequencing of J_t all algorithms of $\mathcal{A} = \{A_k\}_{k \in \mathcal{K}}$ using a particular guess γ_i have failed since the last adjustment of this guess. Let i^* be the largest index i with this property. Then the guesses $\gamma_1, \ldots, \gamma_{i^*}$ are adjusted. Set $\gamma_1 = (1 + \varepsilon) \max\{\gamma_h, p_t, \sum_{1 \leq t' \leq t} p_{t'}/m\}$ and $\gamma_i = (1+\varepsilon)\gamma_{i-1}$, for $i = 2, \ldots, i^*$. For any readjusted guess γ_i, $1 \leq i \leq i^*$, algorithm $\mathcal{A}$ using γ_i ignores all jobs $J_{t'}$ with $t' < t$ when processing future jobs of σ. Specifically, when making scheduling decisions and determining machine loads, algorithm A_k using γ_i ignores all job $J_{t'}$ with $t' < t$ in its schedule S_{ik}. These jobs are also ignored when $\mathcal{A}^*(\varepsilon, h)$ checks if A_k using guess γ_i fails on the arrival of a job. Furthermore, after the assignment of J_t, machines in S_{ik} machines are renumbered so that J_t is located on a machine it would occupy if it were the first job of an input sequence.

When guesses have been adjusted, they are renumbered, together with the corresponding schedule sets $\mathcal{S}_i$, such that again $\gamma_1 < \ldots < \gamma_h$. Hence at any time $\gamma_1 = \min_{1 \leq i \leq h} \gamma_i$ and $\gamma_i \geq (1 + \varepsilon)\gamma_{i-1}$, for $i = 2, \ldots, h$. We also observe that whenever a guess is adjusted, its value increases by a factor of at least $(1 + \varepsilon)^h$.

Theorem 1. *Let $\mathcal{A} = \{A_k\}_{k \in \mathcal{K}}$ be a ρ-competitive algorithm for MPS$_{opt}$. Then for any $0 < \varepsilon \leq 1$ and $h = \lceil \log(1 + \frac{6\rho}{\varepsilon}) / \log(1 + \frac{\varepsilon}{3\rho}) \rceil$, algorithm $\mathcal{A}^*(\varepsilon) = \mathcal{A}^*(\varepsilon/(3\rho), h)$ for MPS is $(\rho + \varepsilon)$-competitive and uses $h|\mathcal{K}|$ schedules.*

3   A $(1 + \varepsilon)$-Competitive Algorithm for MPS$_{opt}$

We present an algorithm $\mathcal{A}_1(\varepsilon)$ for MPS$_{opt}$ that attains a competitive ratio of $1 + \varepsilon$, for any $\varepsilon > 0$. The algorithms will yield a $(1 + \varepsilon)$-competitive strategy for MPS and will be useful in the next section where we develop a $(4/3 + \varepsilon)$-competitive algorithm for MPS$_{opt}$. There $\mathcal{A}_1(\varepsilon)$ will be used as subroutine for a small, constant number of m.

Description of $\mathcal{A}_1(\varepsilon)$: Let $\varepsilon > 0$ be arbitrary. Assume without loss of generality that OPT$(\sigma) = 1$. Then all job processing times are in $(0, 1]$. Set $\varepsilon' = \varepsilon/2$. First we partition the range of possible job processing times into intervals $I_0, \ldots, I_l$ such that, within each interval I_i with $i \geq 1$, the values differ by a factor of at most $1 + \varepsilon'$. Such a partitioning is standard and has been used e.g. in the PTAS for offline makespan minimization [20]. Let $l = \lceil \log(1/\varepsilon') / \log(1 + \varepsilon') \rceil$. Set $I_0 = (0, \varepsilon']$ and $I_i = ((1 + \varepsilon')^{i-1}\varepsilon', (1 + \varepsilon')^i \varepsilon']$, for $i = 1, \ldots, l$. Obviously $I_0 \cup \ldots \cup I_l = (0, (1 + \varepsilon')^l \varepsilon']$ and $(0, 1] \subseteq (0, (1 + \varepsilon')^l \varepsilon']$. A job is *small* if its processing time is at most ε' and hence contained in I_0; otherwise the job is *large*.

Each σ with OPT$(\sigma) = 1$ contains at most $\lfloor m/\varepsilon' \rfloor$ large jobs. For each possible distribution of large jobs over the processing time intervals $I_1, \ldots, I_l$, algorithm $\mathcal{A}_1(\varepsilon)$ prepares one algorithm/schedule. Let $V = \{(v_1, \ldots, v_l) \in \mathbb{N}_0^l \mid v_i \leq \lfloor m/\varepsilon' \rfloor\}$. There holds $|V| = (\lfloor m/\varepsilon' \rfloor + 1)^l$. Let $\mathcal{A}_1(\varepsilon) = \{A_v\}_{v \in V}$. For any vector $v = (v_1, \ldots, v_n) \in V$, algorithm A_v works as follows. It assumes that the incoming job sequence σ contains exactly v_i jobs with a processing time in I_i, for $i = 1, \ldots, l$. Moreover, it pessimistically assumes that each processing time in I_i takes the largest possible value $(1 + \varepsilon')^i \varepsilon'$. Hence, initially A_v computes an optimal schedule S_v^* for a job sequence consisting of v_i jobs with a processing time of $(1 + \varepsilon')^i \varepsilon'$, for $i = 1, \ldots, l$. Small jobs are ignored. Let $n_i^*(j)$ denote the number of jobs with a processing time of $(1 + \varepsilon')^i \varepsilon' \in I_i$ assigned to machine M_j in S_v^*, where $1 \leq i \leq l$ and $1 \leq j \leq m$. Moreover, let $\ell^*(j) = \sum_{i=1}^l n_i^*(j)(1 + \varepsilon')^i \varepsilon'$ be the load on machine M_j in S_v^*, $1 \leq j \leq m$.

When processing the actual job sequence σ and constructing a real schedule S_v, A_v uses S_v^* as a guideline to make scheduling decisions. At any time during the scheduling process, let $n_i(j)$ be the number of jobs with a processing time in I_i that have already been assigned to machine M_j in S_v, where again $1 \le i \le l$ and $1 \le j \le m$. Each incoming job J_t, $1 \le t \le n$, is handled as follows. If J_t is large, then let I_i with $1 \le i \le l$ be the interval such that $p_t \in I_i$. Algorithm A_v checks if there is a machine M_j such that $n_i^*(j) - n_i(j) > 0$, i.e. there is a machine that can still accept a job with a processing time in I_i as suggested by the optimal schedule S_v^*. If such a machine M_j exists, then J_t is assigned to it; otherwise J_t is scheduled on an arbitrary machine. If J_t is small, then J_t is assigned to a machine M_j with the smallest current value $\ell^*(j) + \ell_s(j)$. Here $\ell_s(j)$ denotes the current load on machine M_j caused by small jobs in S_v.

Theorem 2. *For any* $\varepsilon > 0$, $\mathcal{A}_1(\varepsilon)$ *is* $(1 + \varepsilon)$-*competitive and uses at most* $(\lfloor 2m/\varepsilon \rfloor + 1)^{\lceil \log(2/\varepsilon)/\log(1+\varepsilon/2) \rceil}$ *schedules.*

4   A $(4/3 + \varepsilon)$-Competitive Algorithm for $\mathrm{MPS_{opt}}$

We develop an algorithm $\mathcal{A}_2(\varepsilon)$ for $\mathrm{MPS_{opt}}$ that is $(4/3 + \varepsilon)$-competitive, for any $0 < \varepsilon \le 1$, if the number m of machines is not too small. We then combine $\mathcal{A}_2(\varepsilon)$ with $\mathcal{A}_1(\varepsilon)$, presented in the last section, and derive a strategy $\mathcal{A}_3(\varepsilon)$ that is $(4/3 + \varepsilon)$-competitive, for arbitrary m. The number of required schedules is $1/\varepsilon^{O(\log(1/\varepsilon))}$, which is a constant independent of n and m.

Before describing $\mathcal{A}_2(\varepsilon)$ in detail, we explain the main ideas of the algorithm. One concept is identical to that used by $\mathcal{A}_1(\varepsilon)$: Partition the range of possible job processing times into intervals or *job classes* and consider distributions of jobs over these classes. However, in order to achieve a constant number of schedules, we have to refine this scheme and incorporate new ideas. First, the job classes have to be chosen properly so as to allow a compact packing of jobs on the machines. An important, new aspect in the construction of $\mathcal{A}_2(\varepsilon)$ is that we will not consider the entire set V of tuples specifying how large jobs of an input sequence σ are distributed over the job classes. Instead we will define a suitable sparsification V' of V. Each $v \in V'$ represents an *estimate* or *guess* on the number of large jobs arising in σ. More specifically, if $v = (v_1, \dots, v_l)$, then it is assumed that σ contains at least v_i jobs with a processing time of job class i.

The job sequence σ may contain more than v_i jobs of class i, $1 \le i \le l$, the exact number of which is unknown. Furthermore, it is unknown which portion of the total processing time of σ will arrive as small jobs. In order to cope with these uncertainties $\mathcal{A}_2(\varepsilon)$ has to construct robust schedules. To this end the number of machines is partitioned into two sets $\mathcal{M}_c$ and $\mathcal{M}_r$. For the machines of $\mathcal{M}_c$, the algorithm initially determines a good assignment or *configuration* assuming that v_i jobs of job class i will arrive. The machines of $\mathcal{M}_r$ are reserve machines and will be assigned additional large jobs as they arise in σ. Small jobs will always be placed on machines in $\mathcal{M}_c$. The initial configuration determined for these machines has the property that, no matter how many small jobs arrive, a machine load never exceeds $4/3 + \varepsilon$ times the optimum makespan.

We next describe $\mathcal{A}_2(\varepsilon)$ in detail. Let $0 < \varepsilon \leq 1$ and set $\varepsilon' = \varepsilon/8$. Again we assume without loss of generality that, for an incoming job sequence, there holds $\text{OPT}(\sigma) = 1$. Hence the processing time of any job is upper bounded by 1.

Job Classes: A job J_t, $1 \leq t \leq n$, is *small* if $p_t \leq 1/3 + 2\varepsilon'$; otherwise J_t is *large*. We divide the range of possible job processing times into job classes. Let $I_s = (0, 1/3 + 2\varepsilon']$ be the interval containing the processing times of small jobs. Let $\lambda = \lceil \log(\frac{3}{8} + \frac{1}{48\varepsilon'}) \rceil$ and $l = \lambda + 2$, where the logarithm is taken to base 2. For $i = 1, \ldots, l$, let

$$a_i = \max\{\tfrac{1}{3} - 2\varepsilon' + (\tfrac{1}{12} + \tfrac{3}{2}\varepsilon')\tfrac{1}{2^{\lambda+1-i}}, \tfrac{1}{3} + 2\varepsilon'\} \quad \text{and} \quad b_i = \tfrac{1}{3} - 2\varepsilon' + (\tfrac{1}{12} + \tfrac{3}{2}\varepsilon')\tfrac{1}{2^{\lambda-i}}.$$

It is easy to verify that $a_1 = 1/3 + 2\varepsilon'$ and $a_i < b_i$, for $i = 1, \ldots, l$. Furthermore $b_{l-1} = 1/2 + \varepsilon'$ and $b_l = 2/3 + 4\varepsilon'$. For $i = 1, \ldots, l$ define $I_i = (a_i, b_i]$. There holds $\bigcup_{1 \leq i \leq l} I_i = (1/3 + 2\varepsilon', 2/3 + 4\varepsilon']$. Moreover, for $i = 1, \ldots, l-1$, let $I_{l+i} = (2a_i, 2b_i]$. Intuitively, I_{l+i} contains the processing times that are twice as large as those in I_i, $1 \leq i \leq l-1$. There holds $\bigcup_{1 \leq i \leq l-1} I_{l+i} = (2/3 + 4\varepsilon', 1 + 2\varepsilon']$. Hence $I_s \cup I_1 \cup \ldots \cup I_{2l-1} = (0, 1 + 2\varepsilon']$. In the following I_i represents *job class i*, for $i = 1, \ldots, 2l-1$. We say that J_t is a *class-i job* if $p_t \in I_i$, where $1 \leq i \leq 2l-1$.

Definition of Target Configurations: As mentioned above, for any incoming job sequence σ, $\mathcal{A}_2(\varepsilon)$ works with estimates on the number of class-i jobs arising in σ, $1 \leq i \leq 2l - 1$. For each estimate, the algorithm initially determines a virtual schedule or *target configuration* on a subset of the machines, assuming that the estimated set of large jobs will indeed arrive. Hence we partition the m machines into two sets $\mathcal{M}_c$ and $\mathcal{M}_r$. Let $\mu = \lceil \frac{1+\varepsilon'}{1+2\varepsilon'} m \rceil$. Moreover, let $\mathcal{M}_c = \{M_1, \ldots, M_\mu\}$ and $\mathcal{M}_r = \{M_{\mu+1}, \ldots, M_m\}$. Set $\mathcal{M}_c$ contains the machines for which a target configuration will be computed; $\mathcal{M}_r$ contains the reserve machines. The proportion of $|\mathcal{M}_r|$ to $|\mathcal{M}_c|$ is roughly $1 : 1 + 1/\varepsilon'$.

A target configuration has the important property that any machine $M_j \in \mathcal{M}_c$ contains large jobs of only one job class i, $1 \leq i \leq 2l - 1$. Therefore, a target configuration is properly defined by a vector $c = (c_1, \ldots, c_\mu) \in \{0, \ldots, 2l-1\}^\mu$. If $c_j = 0$, then M_j does not contain any large jobs in the target configuration, $1 \leq j \leq \mu$. If $c_j = i$, where $i \in \{1, \ldots, 2l-1\}$, then M_j contains class-i jobs, $1 \leq j \leq \mu$. The vector c implicitly also specifies how many large jobs reside on a machine. If $c_j = i$ with $1 \leq i \leq l$, then M_j contains two class-i jobs. Note that there exist $i \in \{1, \ldots, l\}$ such that a third job cannot be placed on the machine without exceeding a load bound of $4/3 + \varepsilon$. If $c_j = i$ with $l + 1 \leq i \leq 2l - 1$, then M_j contains one class-i job. Again, the assignment of a second job is not feasible in general. Given a configuration c, M_j is referred to as a *class-i machine* if $c_j = i$, where $1 \leq j \leq \mu$ and $1 \leq i \leq 2l - 1$.

With the above interpretation of target configurations, each vector $c = (c_1, \ldots, c_\mu)$ encodes inputs containing $2|\{c_j \in \{c_1, \ldots c_\mu\} : c_j = i\}|$ class-i jobs, for $i = 1, \ldots, l$, as well as $|\{c_j \in \{c_1, \ldots c_\mu\} : c_j = i\}|$ class-i jobs, for $i = l+1, \ldots, 2l - 1$. Hence, for an incoming job sequence, instead of considering estimates on the number of class-i jobs, for any $1 \leq i \leq 2l - 1$, we can equivalently consider target configurations. Unfortunately, it will not be possible to work with all target configurations $c \in \{0, \dots, 2l - 1\}^\mu$ since the resulting number of schedules to be constructed would be $(2l)^\mu = (\log(1/\varepsilon))^{\Omega(m)}$. Therefore, we will work with a suitable sparsification of the set of all configurations.

Sparsification of the Set of Target Configurations: Let $\kappa = \lceil 2(2 + 1/\varepsilon')(2l - 1) \rceil$ and $U = \{0, \dots, \kappa\}^{2l-1}$. We can show that $\kappa \lfloor (m - \mu)/(2l - 1) \rfloor \geq m$ if m is not too small. This property in turn will ensure that any job sequence σ can be mapped to a $u \in U$. For any vector $u = (u_1, \dots, u_{2l-1}) \in U$, we define a target configuration $c(u)$ that contains $u_i \lfloor (m - \mu)/(2l - 1) \rfloor$ class-i machines, for $i = 1, \dots, 2l - 1$, provided that $\sum_{i=1}^{2l-1} u_i \lfloor (m - \mu)/(2l - 1) \rfloor$ does not exceed μ. More specifically, for any $u = (u_1, \dots, u_{2l-1}) \in U$, let $\pi_0 = 0$ and $\pi_i = \sum_{j=1}^{i} u_j \lfloor (m - \mu)/(2l - 1) \rfloor$, be the partial sums of the first i entries of u, multiplied by $\lfloor (m - \mu)/(2l - 1) \rfloor$, for $i = 1, \dots, 2l - 1$. Let $\mu' = \pi_{2l-1}$. First construct a vector $c'(u) = (c'_1, \dots, c'_{\mu'})$ of length μ' that contains exactly $u_i \lfloor (m - \mu)/(2l - 1) \rfloor$ class-i machines. That is, for $i = 1, \dots, 2l - 1$, let $c'_j = i$ for $j = \pi_{i-1} + 1, \dots, \pi_i$. We now truncate or extend $c'(u)$ to obtain a vector of length μ. If $\mu' \geq \mu$, then $c(u)$ is the vector consisting of the first μ entries of $c'(u)$. If $\mu' < \mu$, then $c(u) = (c'_1, \dots, c'_{\mu'}, 0, \dots, 0)$, i.e. the last $\mu - \mu'$ entries are set to 0. Let $C = \{c(u) \mid u \in U\}$ be the set of all target configurations constructed from vectors $u \in U$.

The Algorithm Family: Let $\mathcal{A}_2(\varepsilon) = \{A_c\}_{c \in C}$. For any $c \in C$, algorithm A_c works as follows. Initially, prior to the arrival of any job of σ, A_c determines the target configuration specified by $c = (c_1, \dots, c_\mu)$ and uses this virtual schedule for the machines of $\mathcal{M}_c$ to make scheduling decisions. Consider a machine $M_j \in \mathcal{M}_c$ and suppose $c_j > 0$, i.e. M_j is a class-i machine for some $i \geq 1$. Let $\ell^-(j)$ and $\ell^+(j)$ be the targeted minimal and maximal loads caused by large jobs on M_j, according to the target configuration. More precisely, if $i \in \{1, \dots, l\}$, then $\ell^-(j) = 2a_i$ and $\ell^+(j) = 2b_i$. Recall that in a target configuration a class-i machine contains two class-i jobs if $1 \leq i \leq l$. If $i \in \{l+1, \dots, 2l - 1\}$ and hence $i = l + i'$ for some $i' \in \{1, \dots, l - 1\}$, then $\ell^-(j) = 2a_{i'}$ and $\ell^+(j) = 2b_{i'}$. If $M_j \in \mathcal{M}_c$ is a machine with $c_j = 0$, then $\ell^-(j) = \ell^+(j) = 0$. While the job sequence σ is processed, a machine $M_j \in \mathcal{M}_c$ may or may not be *admissible*. Again assume that M_j is a class-i machine with $i \geq 1$. If $i \in \{1, \dots, l\}$, then at any time during the scheduling process M_j is admissible if it has received less than two class-i jobs so far. Analogously, if $i \in \{l + 1, \dots, 2l - 1\}$, then M_j is admissible if it has received no class-i job so far. Finally, at any time during the scheduling process, let $\ell(j)$ be the current load of machine M_j and let $\ell_s(j)$ be the load due too small jobs, $1 \leq j \leq m$.

Algorithm A_c schedules each incoming job J_t, $1 \leq t \leq n$, in the following way. First assume that J_t is a large job and, in particular, a class-i job, $1 \leq i \leq 2l - 1$. The algorithm checks if there is a class-i machine in $\mathcal{M}_c$ that is admissible. If so, J_t is assigned to such a machine. If there is no admissible class-i machine available, then J_t is placed on a machine in $\mathcal{M}_r$. There jobs are scheduled according to the *Best-Fit* policy. More specifically, A_c checks if there exists a machine $M_j \in \mathcal{M}_r$ such that $\ell(j) + p_t \leq 4/3 + \varepsilon$. If this is the case, then J_t is assigned to such a machine with the largest current load $\ell(j)$. If no such machine exists, J_t is assigned to an arbitrary machine in $\mathcal{M}_r$. Next assume that J_t is small. The job is a assigned to a machine in $\mathcal{M}_c$, where preference is given to machines that have already received small jobs. Algorithm $\mathcal{A}_c$ checks if there is an $M_j \in \mathcal{M}_c$ with $\ell_s(j) > 0$ such that $\ell^+(j) + \ell_s(j) + p_t \leq 4/3 + \varepsilon$. If this is the case, then J_t is assigned to any such machine. Otherwise $\mathcal{A}_c$ considers the machines of $\mathcal{M}_c$ which have not yet received any small jobs. If there exists an $M_j \in \mathcal{M}_c$ with $\ell_s(j) = 0$ such that $\ell^+(j) + p_t \leq 4/3 + \varepsilon$, then among these machines J_t is assigned to one having the smallest targeted load $\ell^-(j)$. If again no such machine exists, J_t is assigned to an arbitrary machine in $\mathcal{M}_c$.

Theorem 3. $\mathcal{A}_2(\varepsilon)$ *is* $(4/3+\varepsilon)$*-competitive, for any* $0 < \varepsilon \leq 1$ *and* $m \geq 2l/(\varepsilon')^2$. *It uses* $1/\varepsilon^{O(\log(1/\varepsilon))}$ *schedules. Again* $l = \lambda + 2$ *and* $\lambda = \lceil \log(\frac{3}{8} + \frac{1}{48\varepsilon'}) \rceil$.

$\mathcal{A}_2(\varepsilon)$ is $(4/3 + \varepsilon)$-competitive if, for the chosen ε, the number of machines is at least $2l/(\varepsilon')^2$. If the number of machines is smaller, we can simply apply algorithm $\mathcal{A}_1(\varepsilon)$ with an accuracy of $\varepsilon_0 = 1/3$. Let $\mathcal{A}_3(\varepsilon)$ be the following combined algorithm. If for the chosen ε, $m < 2l/(\varepsilon')^2$, execute $\mathcal{A}_1(1/3)$. Otherwise execute $\mathcal{A}_2(\varepsilon)$.

Corollary 1. $\mathcal{A}_3(\varepsilon)$ *is* $(4/3 + \varepsilon)$*-competitive, for any* $0 < \varepsilon \leq 1$, *and uses* $1/\varepsilon^{O(\log(1/\varepsilon))}$ *schedules.*

5   Algorithms for MPS

We derive our algorithms for MPS. The strategies are obtained by simply combining $\mathcal{A}^*(\varepsilon)$, presented in Section 2, with $\mathcal{A}_1(\varepsilon)$ and $\mathcal{A}_3(\varepsilon)$. In order to achieve a precision of ε in the competitive ratio, the strategies are combined with a precision of $\varepsilon/2$ in its parameters. For any $0 < \varepsilon \leq 1$, let $\mathcal{A}_3^*(\varepsilon)$ be the algorithm obtained by executing $\mathcal{A}_3(\varepsilon/2)$ in $\mathcal{A}^*(\varepsilon/2)$. For any $0 < \varepsilon \leq 1$, let $\mathcal{A}_1^*(\varepsilon)$ be the algorithm obtained by executing $\mathcal{A}_1(\varepsilon/2)$ in $\mathcal{A}^*(\varepsilon/2)$.

Corollary 2. $\mathcal{A}_3^*(\varepsilon)$ *is a* $(4/3 + \varepsilon)$*-competitive algorithm for MPS and uses* $1/\varepsilon^{O(\log(1/\varepsilon))}$ *schedules, for any* $0 < \varepsilon \leq 1$.

Corollary 3. $\mathcal{A}_1^*(\varepsilon)$ *is a* $(1 + \varepsilon)$*-competitive algorithm for MPS and uses* $(m/\varepsilon)^{O(\log(1/\varepsilon)/\varepsilon)}$ *schedules, for any* $0 < \varepsilon \leq 1$.

6   Lower Bounds

We present lower bounds that apply to both MPS and MPS_{opt}.

Theorem 4. *Let $\mathcal{A}$ be a deterministic online algorithm for MPS or MPS_{opt}. If $\mathcal{A}$ achieves a competitive ratio smaller than $4/3$, then it must maintain at least $\lfloor m/3 \rfloor + 1$ schedules.*

Theorem 5 gives a lower bound on the number of schedules required by a $(1+\varepsilon)$-competitive algorithm, where $0 < \varepsilon < 1/4$. It implies that, for any fixed ε, the number asymptotically depends on $m^{\Omega(1/\varepsilon)}$, as m increases. For instance, any algorithm with a competitive ratio smaller than $1 + 1/12$ requires $\Omega(m^2)$ schedules. Any algorithm with a competitiveness smaller than $1 + 1/16$ needs $\Omega(m^3)$ schedules.

Theorem 5. *Let $\mathcal{A}$ be a deterministic online algorithm for MPS or MPS_{opt}. If $\mathcal{A}$ achieves a competitive ratio smaller than $1 + \varepsilon$, where $0 < \varepsilon \leq 1/4$, then it must maintain at least $\binom{m'+h-1}{h-1}$ schedules, where $m' = \lfloor m/2 \rfloor$ and $h = \lfloor 1/(4\varepsilon) \rfloor$. The binomial coefficient increases as ε decreases and is at least $\Omega((\varepsilon m)^{\lfloor 1/(4\varepsilon)\rfloor - 1/2}/\sqrt{m})$.*

References

1. Albers, S.: Better bounds for online scheduling. SIAM J. Comput. 29, 459–473 (1999)
2. Albers, S., Hellwig, M.: On the value of job migration in online makespan minimization. In: Epstein, L., Ferragina, P. (eds.) ESA 2012. LNCS, vol. 7501, pp. 84–95. Springer, Heidelberg (2012)
3. Angelelli, E., Nagy, A.B., Speranza, M.G., Tuza, Z.: The on-line multiprocessor scheduling problem with known sum of the tasks. J. Scheduling 7, 421–428 (2004)
4. Angelelli, E., Speranza, M.G., Tuza, Z.: Semi-on-line scheduling on two parallel processors with an upper bound on the items. Algorithmica 37, 243–262 (2003)
5. Angelelli, E., Speranza, M.G., Tuza, Z.: New bounds and algorithms for on-line scheduling: two identical processors, known sum and upper bound on the tasks. Discrete Mathematics & Theoretical Computer Science 8, 1–16 (2006)
6. Azar, Y.: On-line load balancing. In: Fiat, A., Woeginger, G.J. (eds.) Online Algorithms 1996. LNCS, vol. 1442, pp. 178–195. Springer, Heidelberg (1998)
7. Azar, Y., Regev, O.: On-line bin-stretching. Theor. Comput. Sci. 268, 17–41 (2001)
8. Azar, Y., Epstein, L., van Stee, R.: Resource augmentation in load balancing. In: Halldórsson, M.M. (ed.) SWAT 2000. LNCS, vol. 1851, pp. 189–199. Springer, Heidelberg (2000)
9. Bartal, Y., Karloff, H., Rabani, Y.: A better lower bound for on-line scheduling. Infomation Processing Letters 50, 113–116 (1994)
10. Bartal, Y., Fiat, A., Karloff, H., Vohra, R.: New algorithms for an ancient scheduling problem. Journal of Computer and System Sciences 51, 359–366 (1995)
11. Böckenhauer, H.-J., Komm, D., Královič, R., Královič, R.: On the advice complexity of the k-server problem. In: Aceto, L., Henzinger, M., Sgall, J. (eds.) ICALP 2011, Part I. LNCS, vol. 6755, pp. 207–218. Springer, Heidelberg (2011)
12. Cheng, T.C.E., Kellerer, H., Kotov, V.: Semi-on-line multiprocessor scheduling with given total processing time. Theor. Comput. Sci. 337, 134–146 (2005)
13. Emek, Y., Fraigniaud, P., Korman, A., Rosén, A.: Online computation with advice. Theor. Comput. Sci. 2412(24), 2642–2656 (2011)
14. Englert, M., Özmen, D., Westermann, M.: The power of reordering for online minimum makespan scheduling. In: Proc. 49th IEEE FOCS, pp. 603–612 (2008)
15. Faigle, U., Kern, W., Turan, G.: On the performance of on-line algorithms for partition problems. Acta Cybernetica 9, 107–119 (1989)

16. Fleischer, R., Wahl, M.: Online scheduling revisited. J. Scheduling 3, 343–353 (2000)
17. Galambos, G., Woeginger, G.: An on-line scheduling heuristic with better worst case ratio than Graham's list scheduling. SIAM J. Comput. 22, 349–355 (1993)
18. Graham, R.L.: Bounds for certain multi-processing anomalies. Bell System Technical Journal 45, 1563–1581 (1966)
19. Gormley, T., Reingold, N., Torng, E., Westbrook, J.: Generating adversaries for request-answer games. In: Proc. 11th ACM-SIAM SODA, pp. 564–565 (2000)
20. Hochbaum, D.S., Shmoys, D.B.: Using dual approximation algorithms for scheduling problems: Theoretical and practical results. J. ACM 34, 144–162 (1987)
21. Karger, D.R., Phillips, S.J., Torng, E.: A better algorithm for an ancient scheduling problem. Journal of Algorithms 20, 400–430 (1996)
22. Kellerer, H., Kotov, V., Speranza, M.G., Tuza, Z.: Semi on-line algorithms for the partition problem. Operations Research Letters 21, 235–242 (1997)
23. Raghavan, P., Snir, M.: Memory versus randomization in on-line algorithms. IBM Journal of Research and Development 38, 683–708 (1994)
24. Renault, M.P., Rosén, A., van Stee, R.: Online Algorithms with advice for bin packing and scheduling problems. CoRR abs/1311.7589 (2013)
25. Rudin III., J.F.: Improved bounds for the on-line scheduling problem. Ph.D. Thesis (2001)
26. Rudin III., J.F., Chandrasekaran, R.: Improved bounds for the online scheduling problem. SIAM J. Comput. 32, 717–735 (2003)
27. Sanders, P., Sivadasan, N., Skutella, M.: Online scheduling with bounded migration. Mathematics of Operations Reseach 34(2), 481–498 (2009)
28. Sleator, D.D., Tarjan, R.E.: Amortized efficiency of list update and paging rules. Communications of the ACM 28, 202–208 (1985)

Expected Linear Time Sorting for Word Size $\Omega(\log^2 n \log \log n)$

Djamal Belazzougui[1,*], Gerth Stølting Brodal[2], and Jesper Sindahl Nielsen[2]

[1] Helsinki Institute for Information Technology (hiit),
Department of Computer Science, University of Helsinki
`dbelaz@liafa.univ-paris-diderot.fr`
[2] MADALGO**, Department of Computer Science, Aarhus University, Denmark
`{gerth,jasn}@cs.au.dk`

Abstract. Sorting n integers in the word-RAM model is a fundamental problem and a long-standing open problem is whether integer sorting is possible in linear time when the word size is $\omega(\log n)$. In this paper we give an algorithm for sorting integers in expected linear time when the word size is $\Omega(\log^2 n \log \log n)$. Previously expected linear time sorting was only possible for word size $\Omega(\log^{2+\varepsilon} n)$. Part of our construction is a new packed sorting algorithm that sorts n integers of w/b-bits packed in $\mathcal{O}(n/b)$ words, where b is the number of integers packed in a word of size w bits. The packed sorting algorithm runs in expected $\mathcal{O}(\frac{n}{b}(\log n + \log^2 b))$ time.

1 Introduction

Sorting is one of the most fundamental problems in computer science and has been studied widely in many different computational models. In the comparison based setting the worst case and average case complexity of sorting n elements is $\Theta(n \log n)$, and running time $\mathcal{O}(n \log n)$ is e.g. achieved by Mergesort and Heapsort [19]. The lower bound is proved using decision trees, see e.g. [4], and is also valid in the average case.

In the word-RAM model with word size $w = \Theta(\log n)$ we can sort n w-bit integers in $\mathcal{O}(n)$ time using radix sort. The exact bound for sorting n integers of w bits each using radix sort is $\Theta(n \frac{w}{\log n})$. A fundamental open problem is if we can still sort in linear time when the word size is $\omega(\log n)$ bits. The RAM dictionary of van Emde Boas [17] allows us to sort in $\mathcal{O}(n \log w)$ time. Unfortunately the space usage by the van Emde Boas structure cannot be bounded better than $\mathcal{O}(2^w)$. The space usage can be reduced to $\mathcal{O}(n)$ by using the Y-fast trie of Willard [18], but the time bound for sorting becomes expected. For polylogarithmic word sizes, i.e. $w = \log^{\mathcal{O}(1)} n$, this gives sorting in time $\mathcal{O}(n \log \log n)$. Kirkpatrick and Reisch gave an algorithm achieving $\mathcal{O}(n \log \frac{w}{\log n})$ [11], which also

---

* Work done while visiting MADALGO.
** Center for Massive Data Algorithmics, a Center of the Danish National Research Foundation (grant DNRF84).

R Ravi and I.L. Gørtz (Eds.): SWAT 2014, LNCS 8503, pp. 26–37, 2014.

gives $\mathcal{O}(n\log\log n)$ for $w = \log^{\mathcal{O}(1)} n$. Andersson et al. [3] showed how to sort in expected $\mathcal{O}(n)$ time for word size $w = \Omega(\log^{2+\varepsilon} n)$ for any $\varepsilon > 0$. The result is achieved by exploiting word parallelism on "signatures" of the input elements packed into words, such that a RAM instruction can perform several element operations in parallel in constant time. Han and Thorup [10] achieved running time $\mathcal{O}(n\sqrt{\log(w/\log n)})$, implying the best known bound of $\mathcal{O}(n\sqrt{\log\log n})$ for sorting integers that is independent of the word size. Thorup established that maintaining RAM priority queues and RAM sorting are equivalent problems by proving that if we can sort in time $\mathcal{O}(n \cdot f(n))$ then there is a priority queue using $\mathcal{O}(f(n))$ time per operation [15].

Our results. We consider for which word sizes we can sort n w-bit integers in the word-RAM model in expected linear time. We improve the previous best word size of $\Omega(\log^{2+\varepsilon} n)$ [3] to $\Omega(\log^2 n \log\log n)$. Word-level parallelism is used extensively and we rely on a new packed sorting algorithm (see Section 5) in intermediate steps. The principal idea for the packed sorting algorithm is an implementation of the randomized Shell-sort of Goodrich [7] using the parallelism in the RAM model. The bottleneck in our construction is $\mathcal{O}(\log\log n)$ levels of packed sorting of $\mathcal{O}(n)$ elements each of $\Theta(\log n)$ bits, where each sorting requires time $\mathcal{O}(n\frac{\log^2 n}{w})$. For $w = \Omega(\log^2 n \log\log n)$, the overall time becomes $\mathcal{O}(n)$.

This paper is structured as follows: Section 2 contains a high level description of the ideas and concepts used by our algorithm. In Section 3 we summarize the RAM operations adopted from [3] that are needed to implement the algorithm outlined in Section 2. In Section 4 we give the details of implementing the algorithm on a RAM and in Section 5 we present the packed sorting algorithm. Finally, in Section 6 we discuss how to adapt our algorithm to work with an arbitrary word size.

2   Algorithm

In this section we give a high level description of the algorithm. The input is n words $x_1, x_2, \ldots, x_n$, each containing a w-bit integer from $U = \{0, 1, \ldots, 2^w - 1\}$. We assume the elements are distinct. Otherwise we can ensure this by hashing the elements into buckets in expected $\mathcal{O}(n)$ time and only sorting a reduced input with one element from each bucket. The algorithm uses a Monte Carlo procedure, which sorts the input with high probability. While the output is not sorted, we repeatedly rerun the Monte Carlo algorithm, turning the main sorting algorithm into a Las Vegas algorithm.

The Monte Carlo algorithm is a recursive procedure using geometrically decreasing time in the recursion, ensuring $\mathcal{O}(n)$ time overall. We view the algorithm as building a Patricia trie over the input words by gradually refining the Patricia trie in the following sense: on the outermost recursion level characters are considered to be w bits long, on the next level $w/2$ bits, then $w/4$ bits and so on. The main idea is to avoid considering all the bits of an element to decide its rank. To avoid looking at every bit of the bit string e at every level of the recursion, we either consider the MSH(e) (Most Significant Half, i.e. the $\frac{|e|}{2}$ most significant bits of e) or LSH(e) (Least Significant Half) when moving one level down in the recursion (similar to the recursion in van Emde Boas trees).

The input to the ith recursion is a list $(id_1, e_1), (id_2, e_2), \ldots, (id_m, e_m)$ of length m, where $n \leq m \leq 2n-1$, id_j is a $\log n$ bit id and e_j is a $w/2^i$ bit element. At most n elements have equal id. The output is a list of ranks $\pi_1, \pi_2, \ldots, \pi_m$, where the j'th output is the rank of e_j among elements with id identical to id_j using $\log n$ bits. There are $m(\log n + \frac{w}{2^i})$ bits of input to the ith level of recursion and $m \log n$ bits are returned from the ith level. On the outermost recursion level we take the input $x_1, x_2, \ldots, x_n$ and produce the list $(1, x_1), (1, x_2), \ldots, (1, x_n)$, solve this problem, and use the ranks $\pi_1, \pi_2, \ldots, \pi_n$ returned to permute the input in sorted order in $\mathcal{O}(n)$ time.

To describe the recursion we need the following definitions.

Definition 1 ([6]). *The* Patricia trie *consists of all the branching nodes and leaves of the corresponding compacted trie as well as their connecting edges. All the edges in the Patricia trie are labeled only by the first character of the corresponding edge in the compacted trie.*

Definition 2. *The Patricia trie of $x_1, x_2, \ldots, x_n$ of detail i, denoted T^i, is the Patricia trie of $x_1, \ldots, x_n$ when considered over the alphabet $\Sigma^i = \{0,1\}^{w/2^i}$.*

The input to the ith recursion satisfies the following invariants, provided the algorithm has not made any errors so far:

  i. The number of bits in an element is $|e| = \frac{w}{2^i}$.
 ii. There is a bijection from id's to non leaf nodes in T^i.
iii. The pair (id, e) is in the input if and only if there is an edge from a node $v \in T^i$ corresponding to id to a child labeled by a string in which $e \in \Sigma^i$ is the first character.

That the maximum number of elements at any level in the recursion is at most $2n-1$ follows because a Patricia trie on n strings has at most $2n-1$ edges.

The recursion. The base case of the recursion is when $|e| = \mathcal{O}(\frac{w}{\log n})$ bits, i.e. we can pack $\Omega(\log n)$ elements into a single word, where we use the packed sorting algorithm from Section 5 to sort (id_j, e_j, j) pairs lexicographically by (id, e) in time $\mathcal{O}(\frac{n}{\log n}(\log n + (\log\log n)^2)) = \mathcal{O}(n)$. Then we generate the ranks π_j and return them in the correct order by packed sorting pairs (j, π_j) by j.

When preparing the input for a recursive call we need to halve the number of bits the elements use. To maintain the second invariant we need to find all the branching nodes of T^{i+1} to create a unique id for each of them. Finally for each edge going out of a branching node v in T^{i+1} we need to make the pair (id, e), where id is v's id and e is the first character (in Σ^{i+1}) on an edge below v. Compared to level i, level $i+1$ may have two kinds of branching nodes: *inherited nodes* and *new nodes*, as detailed below (Figure 1).

In Figure 1 we see T^i and T^{i+1} on 5 bit-strings. In T^i characters are 4 bits and in T^{i+1} they are 2 bits. Observe that node a is not going to be a branching

Detail i Detail $i+1$

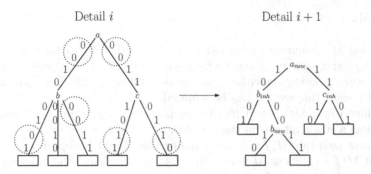

Fig. 1. Example of how nodes are introduced and how they disappear from detail i to $i+1$. The bits that are marked by a dotted circle are omitted in the recursion.

node when characters are 2 bits because "00" are the first bits on both edges below it. Thus the "00" bits below a should not appear in the next recursion – this is captured by Invariant iii. A similar situation happens at the node b, however since there are *two different* 2-bit strings below it, we get the inherited node b_{inh}. At the node c we see that the order among its edges is determined by the first two bits, thus the last two bits can be discarded. Note there are 7 elements in the ith recursion and 8 in the next – the number of elements may increase in each recursion, but the maximum amount is bounded by $2n - 2$.

By invariant ii) every *id* corresponds to a node v in T^i. If we find all elements that share the same *id*, then we have all the outgoing edges of v. We refine an edge labeled e out of v to have the two characters $\text{MSH}(e)\text{LSH}(e)$ both of $w/2^{i+1}$ bits. Some edges might then share their MSH. The node v will appear in level $i+1$ if and only if at least two outgoing edges do not share MSH – these are the *inherited nodes*. Thus we need only count the number of unique MSHs out of v to decide if v is also a node in level $i+1$. The edges out of v at level $i+1$ will be the unique MSH characters (in Σ^{i+1}) on the edges down from v at level i.

If at least two edges out of v share the same first character c (MSH), but not the second, then there is a branching node following c – these are the *new nodes*. We find all new nodes by detecting for each MSH character $c \in \Sigma^{i+1}$ going out of v if there are two or more edges with c as their first character. If so, we have a branching node following c and the labels of the edges are the LSHs. At this point everything for the recursion is prepared.

We receive for each id/node of T^{i+1} the ranks of all elements (labels on the outgoing edges) from the recursion. A *relative* rank for an element at level i is created by concatenating the rank of $\text{MSH}(e)$ from level $i+1$ with the rank of $\text{LSH}(e)$ from level $i+1$. All edges branching out of a new node needs to receive the rank of their MSH (first character). If the MSH was not used for the recursion, it means it did not distinguish any edges, and we can put an arbitrary value as the rank (we use 0). The same is true for the LSHs. Since each relative rank consists of $2 \log n$ bits we can sort them fast using packed sorting (Section 5) and finally the actual ranks can be returned based on that.

3 Tools

This section is a summary of standard word-parallel algorithms used by our sorting algorithm; for an extensive treatment see [12]. In particular the prefix sum and word packing algorithms can be derived from [13]. For those familiar with "bit tricks" this section can be skipped.

We adopt the notation and techniques used in [3]. A w-bit word can be interpreted as a single integer in the range $0, \ldots, 2^w - 1$ or the interpretation can be parameterized by (M, f). A word under the (M, f) interpretation uses the rightmost $M(f+1)$ bits as M fields using $f+1$ bits each and the most significant bit in each field is called the *test bit* and is 0 by default.

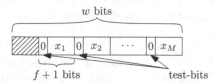

We write $X = (x_1, x_2, \ldots, x_M)$ where x_i uses f bits, meaning the word X has the integer x_1 encoded in its leftmost field, x_2 in the next and so on. If $x_i \in \{0, 1\}$ for all i we may also interpret them as boolean values where 0 is false and 1 is true. This representation allows us to do "bit tricks".

Comparisons. Given a word $X = (x_1, x_2, \ldots, x_M)$ under the (M, f) interpretation, we wish to check $x_i > 0$ for $1 \le i \le M$, i.e. we want a word $Z = [X > 0] = (z_1, z_2, \ldots, z_M)$, in the (M, f) interpretation, such that $z_i = 1$ (true) if $x_i > 0$ and $z_i = 0$ (false) otherwise. Let $k_{M,f}$ be the word where the number k is encoded in each field where $0 \le k < 2^f$. Create the word $0_{M,f}$ and set all test bits to 1. Evaluate $\neg(0_{M,f} - X)$, the ith test bit is 1 if and only if $x_i > 0$. By masking away everything but the test bit and shifting right by f bits we have the desired output. We can also implement more advanced comparisons, such as comparing $[X \le Y]$ by setting all test bits to 1 in Y and 0 in X and subtracting the word X from Y. The test bits now equal the result of comparing $x_i \le y_i$.

Hashing. We will use a family of hash functions that can hash n elements in some range $0, \ldots, m - 1$ with $m > n^c$ to $0, \ldots n^c - 1$. Furthermore a family of hash functions that are injective on a set with high probability when chosen uniformly at random, can be found in [5]. Hashing is roughly just multiplication by a random odd integer and keeping the most significant bits. The integer is at most f bits. If we just multiply this on a word in (M, f) interpretation one field might overflow to the next field, which is undesirable. To implement hashing on a word in (M, f) representation we first mask out all even fields, do the hashing, then do the same for odd fields. The details can be found in [3]. In [5] it is proved that if we choose a function h_a uniformly at random from the family $H_{k,\ell} = \{h_a \mid 0 < a < 2^k, \text{ and } a \text{ is odd}\}$ where $h_a(x) = (ax \bmod 2^k) \operatorname{div} 2^{k-\ell}$ for $0 \le x < 2^k$ then $\Pr[h_a(x) = h_a(y)] \le \frac{1}{2^{\ell-1}}$ for distinct x, y from a set of size n. Thus choosing $\ell = c \log n + 1$ gives collision probability $\le 1/n^c$.

The probability that the function is not injective on n elements: $\Pr[\exists x, y : x \neq y \wedge h_a(x) = h_a(y)] \leq \frac{n^2}{n^c}$ (union bound on all pairs).

Prefix sum. Let $A = (a_1, \ldots, a_M)$ be the input with $M = b$, $f = w/b$ and $a_i \in \{0, 1\}$. In the output $B = (b_1, \ldots, b_M)$, $b_i = 0$ if $a_i = 0$ and $b_i = \sum_{j=1}^{i-1} a_j$ otherwise. We describe an $\mathcal{O}(\log b)$ time algorithm. The invariant is that in the jth iteration a_i has been added to its 2^j immediately right adjacent fields. Compute B_1, which is A shifted right by f bits and added to itself[1]: $B_1 = A + (A \downarrow f)$. Let $B_i = (B_{i-1} \downarrow 2^{i-1}f) + B_{i-1}$. This continues for $\log b$ steps. Then we keep all fields i from $B_{\log b}$ where $a_i = 1$, subtract 1 from all of these fields and return it.

Packing words. We are given a word $X = (x_1, \ldots, x_M)$ in $(M, f) = (b, w/b)$ representation. Some of the fields are zero fields, i.e. a field only containing bits set to 0. We want to produce a "packed word", such that reading from left to right there are no zero fields, followed only by zero fields. The fields that are nonzero in the input must be in the output and in the same order. This problem is solved by Andersson et al. [3, Lemma 6.4]

Expanding. Given a word with fields using b' bits we need to expand each field to using b bits i.e., given $X = (x_1, \ldots, x_k)$ where $|x_i| = b'$ we want $Y = (y_1, \ldots, y_k)$ such that $y_i = x_i$ but $|y_i| = b$. We assume there are enough zero fields in the input word such that the output is only one word. The general idea is to just do packing backwards. The idea is to write under each field the number of bits it needs to be shifted right, this requires at most $\mathcal{O}(\log b)$ bits per field. We now move items based on the binary representation. First we move those who have the highest bit set, then we continue with those that have the second highest bit set and so on. The proof that this works is the same as for the packing algorithm.

Creating index. We have a list of n elements of w/b bits each, packed in an array of words $X_1, X_2, \ldots, X_{n/b}$, where each word is in $(b, w/b)$ representation and $w/b \geq \lceil \log n \rceil$. Furthermore, the rightmost $\lceil \log n \rceil$ bits in every field are 0. The index of an element is the number of elements preceding it and we want to put the index in the rightmost bits of each field. First we will spend $\mathcal{O}(b)$ time to create the word $A = (1, 2, 3, \ldots, b)$ using the rightmost bits of the fields. We also create the word $B = (b, b, \ldots, b)$. Now we run through the input words, update $X_i = X_i + A$, then update $A = A + B$. The time is $\mathcal{O}(n/b + b)$, which in our case always is $\mathcal{O}(n/b)$, since we always have $b = \mathcal{O}(\log n \log \log n)$.

4 Algorithm – RAM Details

In this section we describe how to execute each step of the algorithm outlined in Section 2. We first we describe how to construct T^{i+1} from T^i, i.e. advance one level in the recursion. Then we describe how to use the output of the recursion

---

[1] We use $\uparrow$ and $\downarrow$ as the shift operations where $x \uparrow y$ is $x \cdot 2^y$ and $x \downarrow y$ is $\lfloor x \text{ div } 2^y \rfloor$.

for T^{i+1} to get the ranks of the input elements for level i. Finally the analysis of the algorithm is given.

The input to the ith recursion is a list of pairs: (id, e) using $\log n + \frac{w}{2^i}$ bits each and satisfying the invariants stated in Section 2. The list is packed tightly in words, i.e. if we have m input elements they occupy $\mathcal{O}(\frac{m \cdot (\log n + w/2^i)}{w})$ words. The returned ranks are also packed in words, i.e. they occupy $\mathcal{O}(\frac{m \cdot \log n}{w})$ words. The main challenge of this section is to be able to compute the necessary operations, even when the input elements and output ranks are packed in words. For convenience and simplicity we assume tuples are not split between words.

Finding branching nodes. We need to find the branching nodes (*inherited* and *new*) of T^{i+1} given T^i. For each character e_j in the input list (i.e. T^i) we create the tuple (id_j, H_j, j) where id_j corresponds to the node e_j branches out of, $H_j = h(\mathrm{MSH}(e_j))$ is the hash function applied to the MSH of e_j, and j is the index of e_j in the input list. The list L consists of all these tuples and L is sorted. We assume the hash function is injective on the set of input MSHs, which it is with high probability if $|H_j| \geq 4 \log n$ (see the analysis below). If the hash function is not injective, this step may result in an error which we will realize at the end of the algorithm, which was discussed in Section 2. The following is largely about manipulating the order of the elements in L, such that we can create the recursive sub problem, i.e. T^{i+1}.

To find *Inherited nodes* we find all the edges out of nodes that are in both T^i and T^{i+1} and pair them with unique identifiers for their corresponding nodes in T^{i+1}. Consider a fixed node a which is a branching node in T^i – this corresponds to an id in L. There is a node a_{inh} in T^{i+1} if a and its edges satisfy the following condition: When considering the labels of the edges from a to its children over the alphabet Σ^{i+1} instead of Σ^i, there are at least two edges from a to its children that do not share their first character. When working with the list L the node a and its edges correspond to the tuples where the id is the id that corresponds to a. This means we need to compute for each id in L whether there are at least 2 unique MSHs, and if so we need to extract precisely all the unique MSHs for that id.

The list L is sorted by (id_j, H_j, j), which means all edges out of a particular node are adjacent in L, and all edges that share their MSH are adjacent in L (with high probability), because they have the same hash value which is distinct from the hash value of all other MSHs (with high probability). We select the MSHs corresponding to the first occurrence of each unique hash value with a particular id for the recursion (given that it is needed). To decide if a tuple contains a first unique hash value, we need only consider the previous tuple: did it have a different hash value from the current, or did it have a different id? To decide if $\mathrm{MSH}(e_j)$ should be extracted from the corresponding tuple we also need to compute whether there are at least two unique hash values with id id_j. This tells us we need to compute two things for every tuple (id_j, H_j, j) in L:

1. Is j the first index such that $(id_{j-1} = id_j \wedge H_{j-1} \neq H_j) \vee id_{j-1} \neq id_j$?
2. Is there an i such that $id_i = id_j$ and $H_i \neq H_j$?

To accomplish the first task we do parallel comparison of id_j and H_j with id_{j-1} and H_{j-1} on L and L shifted left by one tuple length (using the word-level parallel comparisons described in Section 3). The second task is tedious but conceptually simple to test: count the number of unique hash values for each id, and test for each id if there are at least two unique hash values.

The details of accomplishing the two tasks are as follows (keep in mind that elements of the lists are bit-strings). Let B be a list of length $|L|$ and consider each element in B as being the same length as a tuple in L. Encode 1 in element j of B if and only if $id_{j-1} \neq id_j$ in L. Next we create a list C with the same element size as B. There will be a 1 in element j of C if and only if $H_j \neq H_{j-1} \wedge id_j = id_{j-1}$ (this is what we needed to compute for task 1). The second task is now to count how many 1s there are in C between two ones in B. Let CC be the prefix sum on C (described in Section 3) and keep only the values where there is a corresponding 1 in B, all other elements become 0 (simple masking). Now we need to compute the difference between each non-zero value and the next non-zero value in CC – but these are varying lengths apart, how do we subtract them? The solution is to pack the list CC (see Section 3) such that the values become adjacent. Now we compute the difference, and by maintaining some information from the packing we can unpack the differences to the same positions that the original values had. Now we can finally test for the first tuple in each id if there are at least two different hash values with that id. That is, we now have a list D with a 1 in position j if j is the first position of an id in L and there are at least two unique MSHs with that id. In addition to completing the two tasks we can also compute the unique identifiers for the inherited nodes in T^{i+1} by performing a prefix sum on D.

Finding the *new nodes* is simpler than finding the *inherited nodes*. The only case where an LSH should be extracted is when two or more characters out of a node share MSH, in which case all the LSHs with that MSH define the outgoing edges of a *new node*. Observe that if two characters share MSH then their LSHs must differ, due to the assumption of distinct elements propagating through the recursion. To find the relevant LSHs we consider the sorted list L. Each *new node* is identified by a pair (id, MSH) where $(id_j, h(\text{MSH}(e_j)), \cdot)$ appears at least twice in L, i.e. two or more tuples with the same id and hash of MSH. For each *new node* we find the leftmost such tuple j in L.

Technically we scan through L and evaluate $(H_{j-1} \neq H_j \vee id_{j-1} \neq id_j) \wedge (H_{j+1} = H_j \wedge id_{j+1} = id_j)$. If this evaluates to true then j is a *new node* in T^{i+1}. Using a prefix sum we create and assign all ids for *new nodes* and their edges. In order to test if LSH_j should be in the recursion we evaluate $(H_{j-1} = H_j \wedge id_{j-1} = id_j) \vee (H_{j+1} = H_j \wedge id_{j+1} = id_j)$. This evaluates to true only if the LSH should be extracted for the recursion because we assume distinct elements.

Using results from the recursion. We created the input to the recursion by first extracting all MSHs, packing them and afterwards extracting all LSHs and then packing them. Finally concatenate the two packed arrays. Now we simply have to reverse this process, first for the MSHs, then the LSHs. Technically after the recursive call the array of tuples $(j, rank_{\text{MSH}_j}, rank_{\text{LSH}_j}, H_j, id_j, rank_{new})$ is filled out. Some of the fields are just additional fields to the array L. The three ranks use $\log n$ bits each and are initialized to 0. First $rank_{\text{MSH}_j}$ is filled out and afterwards $rank_{\text{LSH}_j}$. The same procedure is used for both.

For retrieving $rank_{\text{MSH}_i}$, we know how many MSHs were extracted for the recursion, so we separate the ranks of MSHs and LSHs and now only consider MSHs ranks. We first expand the MSH ranks as described in Section 3 such that each rank uses the same number of bits as an entire tuple. Recall that the MSHs were packed and we now need to unpack them. If we saved information on how we packed elements, we can also unpack them. The information we need to retain is how many elements each word contributed and for each element in a word its initial position in that word. Note that for each unique H_j we only used one MSH for the recursion, thus we need to propagate its rank to all other elements with the same hash and id. Fortunately the hash values are adjacent, and by noting where the hash values change we can do an operation similar to a prefix sum to copy the ranks appropriately.

Returning. As this point the only field not filled out is $rank_{new}$. To fill it out we sort the list by the concatenation of $rank_{\text{MSH}_j}$ and $rank_{\text{LSH}_j}$. In this sorted list we put the current position of the elements in $rank_{new}$ (see Section 3 on creating index). The integer in $rank_{new}$ is currently not the correct rank, but by subtracting the first $rank_{new}$ in an id from the other $rank_{new}$s with that id we get the correct rank. Then we sort by j, mask away everything except $rank_{new}$, pack the array and return. We are guaranteed the ranks from the recursion use $\log n$ bits each, which means the concatenation uses $2\log n$ bits so we can sort the array efficiently.

Analysis. We argue that the algorithm is correct and runs in linear time.

Lemma 1. *Let n be the number of integers we need to sort then the maximum number of elements in any level of the recursion is $2n - 1$.*

Proof. This follows immediately from the invariants.   □

Theorem 1. *The main algorithm runs in $\mathcal{O}(n)$ time.*

Proof. At level i of the recursion $|e| = \frac{w}{2^i}$. After $\log \log n$ levels we switch to the base case where there are $b = 2^{\log \log n} = \log n$ elements per word. The time used in the base case is $\mathcal{O}(\frac{n}{b}(\log^2 b + \log n)) = \mathcal{O}(\frac{n}{\log n}((\log \log n)^2 + \log n)) = \mathcal{O}(n)$.

At level i of the recursion we have $b = 2^i$ elements per word and the time to work with each of the $\mathcal{O}(\frac{n}{b})$ words using the methods of Section 3 is $\mathcal{O}(\log b)$. The packed sorting at each level sorts elements with $\mathcal{O}(\log n)$ bits, i.e. $\mathcal{O}\left(\frac{w}{\log n}\right)$ elements per word in time $\mathcal{O}\left(\frac{n}{w/\log n}\left(\log^2 \frac{w}{\log n} + \log n\right)\right)$. Plugging in our assumption $w = \Omega(\log^2 n \log \log n)$, we get time $\mathcal{O}\left(\frac{n}{\log \log n}\right)$. For all levels the total time becomes $\sum_{i=0}^{\log \log n} \left(\frac{n}{2^i}i + \frac{n}{\log \log n}\right) = \mathcal{O}(n)$.   □

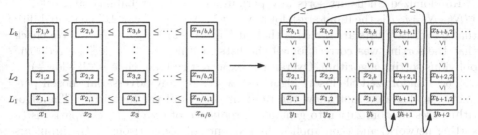

Fig. 2. Transposing and concatenating blocks

The probability of doing more than one iteration of the algorithm is the probability that there is a level in the recursion where the randomly chosen hash function was not injective. The hash family can be designed such that the probability of a hash function not being injective when chosen uniformly at random is less than $1/n^2$ [5]. We need to choose $\log\log n$ such functions. The probability that at least one of the functions is not injective is $\mathcal{O}(\log\log n/n^2) < \mathcal{O}(1/n)$. In conclusion the sorting step works with high probability, thus we expect to repeat it $\mathcal{O}(1)$ times.

5 Packed Sorting

We are given n elements of $\frac{w}{b}$ bits packed into $\frac{n}{b}$ words using $(M, f) = (b, w/b)$ representation that we need to sort. Albers and Hagerup [2] describe how to perform a deterministic packed sorting in time $\mathcal{O}(\frac{n}{b} \log n \cdot \log b)$. We describe a simple randomized word-level parallel sorting algorithm running in time $\mathcal{O}(\frac{n}{b}(\log n + \log^2 b))$. Packed sorting proceeds in four steps described in the following sections. The idea is to implement b sorting networks in parallel using word-level parallelism. In sorting networks one operation is available: compare the elements at positions i and j then swap i and j based on the outcome of the comparison. Denote the ℓth element of word i at any point by $x_{i,\ell}$. First we use the ℓth sorting network to get a sorted list L_ℓ: $x_{1,\ell} \leq x_{2,\ell} \leq \cdots \leq x_{n/b,\ell}$ for $1 \leq \ell \leq b$. Each L_ℓ then occupies field ℓ of every word. Next we reorder the elements such that each of the b sorted lists uses n/b^2 consecutive words, i.e. $x_{i,j} \leq x_{i,j+1}$ and $x_{i,w/b} \leq x_{i+1,1}$, where $n/b^2 \cdot k < i \leq n/b^2 \cdot (k+1)$ and $0 \leq k \leq b-1$ (See Figure 2). From that point we can merge the lists using the RAM implementation of bitonic merging (see below). The idea of using sorting networks or *oblivious* sorting algorithms is not new (see e.g. [9]), but since we need to sort in sublinear time (in the number of elements) we use a slightly different approach.

Data-oblivious sorting. A famous result is the AKS deterministic sorting network which uses $\mathcal{O}(n \log n)$ comparisons [1]. Other deterministic $\mathcal{O}(n \log n)$ sorting networks were presented in [2,8]. However, in our application randomized sorting suffices so we use the simpler randomized Shell-sort by Goodrich [7]. An alternative randomized sorting-network construction was given by Leighton and Plaxton [14].

Randomized Shell-sort sorts any permutation with probability at least $1 - 1/N^c$ ($N = n/b$ is the input size), for any $c \geq 1$. We choose $c = 2$. The probability that b arbitrary lists are sorted is then at least $1 - b/N^c \geq 1 - N^{c-1}$. We check that the sorting was correct for all the lists in time $\mathcal{O}(\frac{n}{b})$. If not, we redo the oblivious sorting algorithm. Overall the expected running time is $\mathcal{O}(\frac{n}{b} \log \frac{n}{b})$.

The Randomized Shell-sort algorithm works on any adversarial chosen permutation that does not know the random choices of the algorithm. The algorithm uses randomization to generate a sequence of $\Theta(n \log n)$ comparisons (a sorting network) and then applies the sequence of comparisons to the input array. We start the algorithm of Goodrich [7] to get the sorting network. We run it with $N = n/b$ as the input size. When the network compares i and j, we compare words i and j field-wise. That is, the first element of the two words are compared, the second element of the words are compared and so on. Using the result we can implement the swap that follows. After this step we have $x_{1,\ell} \leq x_{2,\ell} \leq \cdots \leq x_{n/b,\ell}$ for all $1 \leq \ell \leq b$.

The property of Goodrich' Shellsort that makes it possible to apply it in parallel is its data obliviousness. In fact any sufficiently fast data oblivious sorting algorithm would work.

Verification step. The verification step proceeds in the following way: we have n/b words and we need to verify that the words are sorted field-wise. That is, to check that $x_{i,\ell} \leq x_{i+1,\ell}$ for all i, ℓ. One packed comparison will be applied on each pair of consecutive words to verify this. If the verification fails, then we redo the oblivious sorting algorithm.

Rearranging the sequences. The rearrangement in Figure 2 corresponds to looking at b words as a $b \times b$ matrix (b words with b elements in each) and then transposing this matrix. Thorup [16, Lemma 9] solved this problem in $\mathcal{O}(b \log b)$ time. We transpose every block of b consecutive words. The transposition takes overall time $\mathcal{O}(\frac{n}{b} \log b)$. Finally, we collect in correct order all the words of each run. This takes time $\mathcal{O}(\frac{n}{b})$. Building the ith run for $1 \leq i \leq b$ consists of putting together the ith words of the blocks in the block order. This can be done in a linear scan in $\mathcal{O}(n/b)$ time.

Bitonic merging. The last phase is the bitonic merging. We merge pairs of runs of $\frac{n}{b^2}$ words into runs of $\frac{2n}{b^2}$ words, then runs of $\frac{2n}{b^2}$ words into runs of size $\frac{4n}{b^2}$ and so on, until we get to a single run of n/b words. We need to do $\log b$ rounds, each round taking time $\mathcal{O}(\frac{n}{b} \log b)$ making for a total time of $\mathcal{O}(\frac{n}{b} \log^2 b)$ [2].

6   General Sorting

In this section we tune the algorithm slightly and state the running time of the tuned algorithm in terms of the word size w. We see that for some word sizes we can beat the $\mathcal{O}(n\sqrt{\log \log n})$ bound. We use the splitting technique of [10, Theorem 7] that given n integers can partition them into sets $X_1, X_2, \ldots X_k$ of at most $\mathcal{O}(\sqrt{n})$ elements each, such that all elements in X_i are less than all elements in X_{i+1} in $\mathcal{O}(n)$ time. Using this we can sort in $\mathcal{O}(n \log \frac{\log n}{\sqrt{w/\log w}})$ time.

The algorithm repeatedly splits the set S of inital size n_0 into smaller subsets of size $n_j = \sqrt{n_{j-1}}$ until we get $\log n_j \leq \sqrt{w/\log w}$ where it stops and sorts each subset in linear time using our sorting algorithm. The splitting is performed $\log((\log n)/(\sqrt{w/\log w})) = \frac{1}{2}\log\frac{\log^2 n \log w}{w} = \mathcal{O}(\log\frac{\log^2 n \log\log n}{w})$ times. An interesting example is to sort in time $\mathcal{O}(n\log\log\log n)$ for $w = \frac{\log^2 n}{(\log\log n)^c}$ for any constant c. When $w = \frac{\log^2 n}{2^{\Omega(\sqrt{\log\log n})}}$, the sorting time is $\Omega(n\sqrt{\log\log n})$.

References

1. Ajtai, M., Komlós, J., Szemerédi, E.: An $\mathcal{O}(n\log n)$ sorting network. In: STOC, pp. 1–9 (1983)
2. Albers, S., Hagerup, T.: Improved parallel integer sorting without concurrent writing. Inf. Comput. 136(1), 25–51 (1997)
3. Andersson, A., Hagerup, T., Nilsson, S., Raman, R.: Sorting in linear time? Journal of Computer and System Sciences 57, 74–93 (1998)
4. Cormen, T.H., Leiserson, C.E., Rivest, R.L., Stein, C.: Introduction to Algorithms, 3rd edn. MIT Press and McGraw Hill (2009)
5. Dietzfelbinger, M., Hagerup, T., Katajainen, J., Penttonen, M.: A reliable randomized algorithm for the closest-pair problem. J. Algorithms 25(1), 19–51 (1997)
6. Ferragina, P., Grossi, R.: The string B-tree: A new data structure for string search in external memory and its applications. J. ACM 46(2), 236–280 (1999)
7. Goodrich, M.T.: Randomized shellsort: A simple data-oblivious sorting algorithm. J. ACM 58(6), 27 (2011)
8. Goodrich, M.T.: Zig-zag sort: A simple deterministic data-oblivious sorting algorithm running in $\mathcal{O}(n\log n)$ time. CoRR, abs/1403.2777 (2014)
9. Hagerup, T.: Sorting and searching on the word RAM. In: STACS, pp. 366–398 (1998)
10. Han, Y., Thorup, M.: Integer sorting in $\mathcal{O}(n\sqrt{\log\log n})$ expected time and linear space. In: FOCS, pp. 135–144 (2002)
11. Kirkpatrick, D., Reisch, S.: Upper bounds for sorting integers on random access machines. Theoretical Computer Science 28(3), 263–276 (1983)
12. Knuth, D.E.: The Art of Computer Programming, volume 4A: Combinatorial Algorithms. Addison-Wesley Professional (2011)
13. Leighton, F.T.: Introduction to Parallel Algorithms and Architectures: Arrays, Trees, Hypercubes. In: Packing, Spreading, and Monotone Routing Problems, ch. 3.4.3, Morgan Kaufmann Publishers, Inc. (1991)
14. Leighton, T., Plaxton, C.G.: Hypercubic sorting networks. SIAM Journal on Computing 27(1), 1–47 (1998)
15. Thorup, M.: On RAM priority queues. SIAM J. Comput. 30(1), 86–109 (2000)
16. Thorup, M.: Randomized sorting in $\mathcal{O}(n\log\log n)$ time and linear space using addition, shift, and bit-wise boolean operations. J. Alg. 42(2), 205–230 (2002)
17. van Emde Boas, P.: Preserving order in a forest in less than logarithmic time. In: FOCS, pp. 75–84 (1975)
18. Willard, D.E.: Log-logarithmic worst-case range queries are possible in space $\Theta(n)$. Inf. Process. Lett. 17(2), 81–84 (1983)
19. Williams, J.W.J.: Algorithm 232: Heapsort. CACM 7(6), 347–348 (1964)

Amortized Analysis of Smooth Quadtrees in All Dimensions[*]

Huck Bennett and Chee Yap

Department of Computer Science, Courant Institute, New York University
{hbennett,yap}@cs.nyu.edu

Abstract. Quadtrees are a well-known data structure for representing geometric data in the plane, and naturally generalize to higher dimensions. A basic operation is to expand the tree by splitting a given leaf. A quadtree is *smooth* if adjacent leaf boxes differ by at most one in height.

In this paper, we analyze quadtrees that maintain smoothness with each split operation and also maintain neighbor pointers. Our main result shows that the smooth-split operation has an amortized cost of $O(1)$ time for quadtrees of any fixed dimension D. This bound has exponential dependence on D which we show is unavoidable via a lower bound construction. We additionally give a lower bound construction showing an amortized cost of $\Omega(\log n)$ for splits in a related quadtree model that does not maintain smoothness.

1 Introduction

Quadtrees [dBCvKO08, FB74, Sam90b] are a well-known data structure for representing geometric data in two dimensions. In this case there exists a natural one-to-one correspondence between quadtree nodes v and boxes B in an underlying subdivision of a square which allows us to refer to boxes and nodes interchangeably. Here we consider the extension to a subdivision of a D-dimensional box in which an internal node is a box containing 2^D congruent subboxes. We refer the reader to [dBCvKO08, Chap. 14] whose nomenclature we largely follow.

Two boxes (or nodes in a quadtree) are *adjacent* if the boxes share a $(D-1)$-dimensional facet, but have disjoint interiors. The *neighbors* of a box B are those boxes adjacent to B. We call a quadtree *smooth* if any two adjacent leaf boxes differ by at most one in height. Other sources use the term *balanced* to refer to this condition, which we avoid in order to avoid conflation with the standard meaning of balanced trees in computer science.

We study three operations on quadtrees: split, smooth, and neighbor_query as well as the hybrid operation ssplit which combines a split and a smooth.

A basic operation is a *split* of a leaf box B, written split(B). This divides B into 2^D congruent subboxes which become its children (B is no longer a leaf). A split operation is a useful abstraction of many common operations performed on quadtrees including point insertion and mesh refinement. A smooth operation

---

[*] This work was supported by NSF Grant CCF-0917093.

performs the minimum sequence of splits necessary to restore smoothness. A *smooth split* operation or `ssplit`(B) is a split `split`(B) followed by a `smooth` of the resulting tree.

Let $d \in \{\pm 1, \pm 2, , \ldots, , \pm D\}$ identify one of the $2D$ semi-axis directions. If box B' is a neighbor of B, and the depth of B' is maximal subject to $depth(B') \leq depth(B)$ over all neighbors of B in direction d, then we call B' the *principal d-neighbor* of B. We note that the principal d neighbor of a box B is unique if it exists (it may not if B is on the boundary of the subdivision). A *neighbor query* operation `neighbor_query`(B, d) returns the principal d-neighbor of B. This operation has been used in [dBCvKO08, ABCC06].

In many quadtree applications one is interested in the set of leaf neighbors of a box [WCY13]. The goal is to enumerate these in time $O(1)$ per leaf neighbor. This is achieved by giving each box a set of pointers to its principal neighbors. We can then enumerate all leaf neighbors of a box by going to each principal neighbor, and enumerating all leaf neighbors in the corresponding subtree. Without such pointers, neighbor queries require $\Theta(h)$ time in order to traverse to the nearest common ancestor. We also show that a tree with neighbor pointers must maintain smoothness to ensure amortized $O(1)$ splits.

This neighbor enumeration functionality makes smooth quadtrees useful in motion planning [WCY13]. They are also useful in other domains including good mesh generation [dBCvKO08, BEG94].

1.1   The Smooth Quadtree Model

In this paper we present and analyze a quadtree model that we call the *smooth quadtree* which maintains smoothness as an invariant between splits via the smooth split operation, and maintains principal neighbor pointers. This model has been proposed before such as in Exercise 14.8 in [dBCvKO08], but to the best of our knowledge the complexity of smooth splits has never been studied rigorously. To provide context for our smooth quadtree model, we discuss two options in designing quadtrees:

1. A quadtree can either maintain or not maintain neighbor pointers. The indicator for this option is P (Pointer) or N (No Pointer).
2. A quadtree can either maintain or not maintain smoothness as an invariant. It maintains smoothness by supporting `ssplit` instead of `split`. The indicator for this option is S (Smooth) or U (Unsmooth).

If a quadtree maintains neighbor pointers, we assume the pointers are to the $2D$ principal neighbor. Then the `neighbor_query` operation requires worst case $O(1)$ time. These considerations give rise to four models of quadtrees: PS, PU, NS, NU, where our smooth quadtree corresponds to the PS quadtree model because it maintains both pointers and smoothness. We also refer to the NU quadtree model as the *simple* quadtree model. The simple quadtree model is frequently used as the primary definition of quadtrees [dBCvKO08]. Intermediate between these two extreme models are the PU and NS models. NS quadtrees are similar

Table 1. Comparison of operational costs in three quadtree models where h denotes height and n the number of nodes in the quadtree. Costs are worst-case unless otherwise noted. All four models have $\Theta(n)$ space complexity.

	Smooth (PS) Quadtrees	PU Quadtrees	Simple (NU) Quadtrees
neighbor_query	$\Theta(1)$	$\Theta(1)$	$\Theta(h)$
ssplit/split	Amortized $\Theta(1)$	Amortized $\Omega(\log n)$	$\Theta(1)$
smooth	(Maintained as invariant)	$\Omega(n \log n)$	$O((h+1)n)$

Algorithm 1. Smooth Split (ssplit)

Input: Smooth quadtree T, Leaf $v \in T$ to split
Output: Smooth quadtree T'
split(v)
foreach $v' \in v.principal\_neighbors \setminus v.siblings$ do
    if $v'.depth < v.depth$ then
        | ssplit(v')
    end
end

---

to PS quadtrees, but may lose a factor of h in the cost of neighbor_query and ssplit because traversing to the nearest common ancestor requires $O(h)$ time.

Table 1 compares the cost of our three main operations on these quadtree models. Here, n is the number of nodes in a quadtree and h is its height.

The smooth (PS) quadtree achieves improvements to the neighbor_query and smooth operations at the cost of split operations requiring amortized rather than worst-case $O(1)$ time. The $O(1)$ time bounds for the ssplit and split operations are for the local operations, i.e., when the algorithm already has a pointer to the box it wishes to split such as in the scenario described in [WCY13].

Algorithm 1 shows the simplicity of the smooth split algorithm: recursively check whether any neighbors of a node need to be split to regain smoothness. Nevertheless, the amortized analysis of the algorithm is subtle.

1.2   Our Results

The primary contribution of this paper is a proof that amortized $O(1)$ additional split operations are sufficient for each smooth split operation in quadtrees of any fixed dimension. We prove this result in section 2, and give a self-contained, elementary proof of the 2-dimensional case in the full version of the paper [BY]. More formally we have:

Theorem 1 (Main Theorem). *Starting from an initially trivial subdivision consisting of one box, the total cost of any sequence of smooth splits* $ssplit(B_1)$, $\ldots, ssplit(B_n)$ *is* $O(n)$. *Thus the amortized cost of a smooth split is* $O(1)$.

Additionally, we give lower bounds motivating our data structure and analysis. We first show that without smoothing we cannot achieve an amortized $O(1)$ cost for both splits and neighbor queries.

We also address the dependence on dimension in the $O(1)$ amortized bound on the number of splits per smooth split. In dimension D, the proof of Theorem 1 gives an upper bound of $O(2^D(D+1)!)$, while the proof of Lemma 6 gives a lower bound of $\Omega(D2^D)$.

1.3   Related Work

The following theorem is a well-known result, saying that a simple quadtree can be smoothed using $O(n)$ splits:

FACT 1 (Theorem 14.4 in [dBCvKO08], Theorem 3 in [Moo95]). *Let T be a simple quadtree with n nodes and of height h. Then the smooth version of T has $O(n)$ nodes and can be constructed in $O((h+1)n)$ time.*

Fact 1 gives a bound for *monolithic* tree smoothing, the operation that we call smooth in Table 1. It says that given an arbitrary quadtree we can smooth it all at once in $O(n)$ time. Here we study *dynamic* tree smoothing in which we smooth the tree after each split, instead of performing an arbitrary number of splits before smoothing.

Intuitively a single splitting operation does not unsmooth a quadtree much, so only a few additional splits should be required to resmooth a tree after one split. To show this formally one might try applying the analysis given by Fact 1 to a sequence of smooth splits $\mathtt{ssplit}(B_1), \ldots, \mathtt{ssplit}(B_n)$. However that analysis does not consider any measure of how smooth the starting tree is, and only gives a worst-case linear time bound of $O(i)$ for smoothing after the ith split in a sequence $\mathtt{split}(B_1), \ldots, \mathtt{split}(B_n)$ where B_1 is the root. This analysis shows that a sequence of smooth splits $\mathtt{ssplit}(B_1), \ldots, \mathtt{ssplit}(B_n)$ requires $\sum_{i=1}^{n} O(i) = O(n^2)$ time for an amortized bound of $O(n)$ which is no better than the worst-case bound. Therefore although Theorem 1 implies Fact 1, the converse implication is unclear.

1.4   Other Results

In recent work Löffler et al. [LSS13] recognize that maintaining smoothness "could cause a linear 'cascade' of cells needing to be split." This cascading behavior – what we define formally in terms of *forcing chains* – is the focus of our analysis and main result.

A natural question asks whether there exists a *worst-case* $O(1)$-time algorithm for smooth splitting a box B. The most natural such algorithm would recursively check whether neighbors of a split box must themselves be split, as in Algorithm 1, but would only recurse to some fixed depth. However, a forcing chain may be arbitrarily long in general meaning that this approach does not work in our model.

We may generalize the notion of smoothness as follows: call two neighbors *k-smooth* if the the boxes differ in height by at most k in the quadtree. In two dimensions this is equivalent to having at most 2^k neighbors in a given direction. We have used the term "smoothness" to denote 1-smoothness. A natural question asks whether the relaxed smoothness constraint induced by increasing k would lead to a worst-case $O(1)$ algorithm. In general, this does not help because a forcing chain may still be arbitrarily long.

However, Löffler et al. [LSS13] sketch an $O(1)$ worst-case algorithm for performing smooth splits in a related quadtree model. The most important distinction in their model comes from defining two tiers of quadtree nodes – *true* cells which would be present in any unsmoothed quadtree, and B-cells which are only present to ensure smoothness. Different smoothness invariants hold for these two tiers of cells – true cells are required to be 1-smooth with respect to their neighbors while B-cells are only required to be 2-smooth. The splitting operation is defined on true cells whose children are not true cells. If a true cell A has B-cells as children then $\mathtt{ssplit}(A)$ promotes the children of A to true cells.

The algorithm sketched in the paper and private correspondence [Sim] omits details and a proof of correctness for several key points, such as the promotion of B-cells to true cells, however it appears to be correct. The model differs from ours in that it only allows splits on "true" nodes, maintains a weaker balance invariant, and requires more complicated algorithms. Our result, although requiring involved analysis, shows that smoothing is efficient using a simple algorithm and quadtree model.

Moore [Moo92, Moo95] proves that "monolithic" smoothing of arbitrary quadtrees requires $O(n)$ splits as given in Fact 1. Although this result seems to have been known earlier, Moore reproves this result in [Moo95] for basic quadtrees using a gadget called a "barrier", and then extends the result to generalizations of quadtrees including triangular quadtrees, higher degree quadtrees, and higher dimensional quadtrees. Fact 1 states this result in the standard setting.

In [dBRS12], de Berg et al. study *refinement* of compressed quadtrees. They consider a refinement T_1 of a quadtree T_0 to be an extension of T_0 in which all boxes that were in T_0 have $O(1)$ neighbors in T_1. This is a relaxation of the notion of balancing both in terms of the precise number of neighbors that a box may have (which is simply assumed to be bounded, but not by a particular constant) and in the sense that boxes in T_1 need not be smooth with respect to each other. The authors prove that a refinement of a compressed quadtree may be performed in $O(n)$ time, where n is the size of the quadtree. This result has a similar flavor to the "monolithic" balancing result described in Fact 1.

Amortized analysis of quadtree operations has appeared in previous work. Park and Mount [PM12] introduce the *splay quadtree*, in which they use amortized analysis to analyze the cost of a sequence of data accesses in a quadtree whose balance is dynamically updated using rotations in a similar manner to standard splay trees. Overmars and van Leeuwen [OvL82] analyze dynamic quadtrees, studying the amortized (what they call average-case) cost of insertions into quadtrees.

Recently Sheehy [She] proposed extending results in his previous work on optimal mesh sizes [She12] to prove the efficient smoothing results presented in this paper. A reviewer proposed a similar proof strategy based on Ruppert's work on local feature size [Rup93]. Future work involves studying these continuous techniques, and determining whether the approach is both viable and leads to better bounds than those given by the combinatorial approach.

1.5   Neighbor Pointers without Smoothing

The motivation for studying the quadtree model presented in this paper comes from the ineffectiveness of other natural models to support both efficient `split` and `neighbor_query` operations. Here we analyze what happens if we use our model but without smoothing.

Suppose that we maintain principal neighbor pointers in an unsmoothed subdivision. This is the `UP` quadtree model in Table 1. The following result gives an amortized $\Omega(\log n)$ lower bound on the time complexity of a split in this model, based on the high number of neighbor pointer updates required:

Lemma 1. Let B_1 denote the root box. In the worst case, both a sequence of n splits $split(B_1), \ldots, split(B_n)$ and a $smooth$ in the UP quadtree model require $\Omega(n \log n)$ time.

2   Analysis of Forcing Chains

In order to prove Theorem 1 for quadtrees in arbitrary dimensions, we will need to develop some notation and concepts. The idea behind the proof is to analyze what conditions lead to balancing splits propagating through the data structure, and to show that a suitably defined cost-potential invariant is only violated a bounded number of times per smooth split. Due to space constraints and for clarity some proofs are omitted. All missing proofs, a detailed lower bound construction, and a self-contained proof of the 2-dimensional case are contained in the full version of this paper [BY].

2.1   Basic Terminology

We give a brief summary of the concepts needed. The appendix contains full details. We consider subdivision of the standard cube $[-1, 1]^D$ in $D \geq 1$ dimensions. A (box) *subdivision tree* $\mathbb{T}$ is a finite tree rooted at $[-1, 1]^D$ whose nodes are subboxes of $[-1, 1]^D$, and where each internal node has 2^D congruent children. The set of leaves of $\mathbb{T}$ constitute a *subdivision* of $[-1, 1]^D$. Nodes of $\mathbb{T}$ are also called *aligned boxes*, and every aligned box has a natural depth. Conversely, given any subdivision $\mathbb{S}$ of aligned boxes, there is a unique subdivision tree $\mathbb{T}(\mathbb{S})$. Henceforth, we use the terminology "subdivision tree" synonymously with "quadtree in any dimension".

Let $j \in \{0, \ldots, D\}$. We say that boxes B, B' are j-*adjacent* if $B \cap B'$ is a j-dimensional box. Two special cases are noteworthy: if they are D-adjacent, we say B and B' *overlap* and if they are $(D-1)$-adjacent, we say they are *neighbors*.

FACT 2. *Let B, B' be overlapping aligned boxes. Then either $B \subseteq B'$ or $B' \subseteq B$.*

By an *indicator* we mean an element $d \in \{1, 0, -1\}^D$. If d has exactly one non-zero component, we call it a *direction indicator*; if it has no zero components, we call it a *child indicator* (we do not need child indicators in this paper, but it will be useful in coding these algorithms). Two directions d and d' are *opposite* if $d = -d'$, and *adjacent* if $d \neq \pm d'$. If B is a child of B', then we write $B \prec B'$, and denote the parent by $\mathrm{p}(B) = B'$. For example, $\mathrm{p}^2(B)$ is the grandparent of B.

If B and B' are $(D-1)$-adjacent, there is a unique direction indicator d such that B' is *adjacent to B in direction d*, which we denote by $B \overset{d}{\longrightarrow} B'$. Moreover, $B \overset{d}{\longrightarrow} B'$ if and only if $B' \overset{-d}{\longrightarrow} B$. We may simply write $B \longrightarrow B'$ if there exists some d such that $B \overset{d}{\longrightarrow} B'$. See the appendix for the formal definition of this relation.

Given a box B, we can project and co-project it in one of D directions: let $i \in \{1, \ldots, D\}$.

- (Projection) $\mathrm{Proj}_i(B) := \prod_{j=1, j \neq i}^{D} I_j$ is a $(D-1)$ dimensional box.
- (Co-Projection) $\mathrm{Coproj}_i(B) := I_i$ is the ith interval of $B = \prod_{j=1}^{D} I_j$.

2.2 Forcing Chains

Let S be a subdivision of the standard cube $[-1, 1]^D$. We say S is *smooth* if any two neighboring boxes B, B' in S differ in depth by at most 1. We are interested in maintaining smooth subdivisions. More precisely, if S is smooth, and we split a box in S, there is minimal set of additional boxes in S that must be split in order to maintain smoothness.

If $B \overset{d}{\longrightarrow} B'$, and the $depth(B) > depth(B')$ then we denote this relationship by $B \overset{d}{\Longrightarrow} B'$. We say B *d-forces B'* (or simply, B forces B'). Intuitively it means that if B, B' are boxes in a subdivision and we split B, then we are forced to split B' if we want to make the subdivision smooth. Because we maintain smoothness as an invariant, $B \Longrightarrow B'$ means $depth(B) = depth(B') + 1$.

A sequence of such forcing relations $C : B_0 \overset{d_1}{\Longrightarrow} B_1 \overset{d_2}{\Longrightarrow} \cdots \overset{d_k}{\Longrightarrow} B_k$ is called a *forcing chain* or simply *chain* with *k links*. The set $\{d_1, \ldots, d_k\}$ are the *directions* of C; we say C is *monotone* if its direction set does not contain any pair of opposite directions.

The following lemma follows from the definition of forcing:

Lemma 2 (Forcing). *The forcing relationship $B \overset{d}{\Longrightarrow} B'$ is equivalent to the following two conditions:*

(i) $\mathrm{Proj}_d(B) \prec \mathrm{Proj}_d(B')$
(ii) $\mathrm{Coproj}_d(B) \Longrightarrow \mathrm{Coproj}_d(B')$

Note that conditions (i) and (ii) refer to child and forcing relationships in dimensions $D - 1$ and 1, respectively.

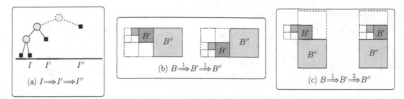

Fig. 1. Analysis of 2-Link Chains

2.3   Analysis of 2-Link Chains

In this part, we consider chains with 2-links: $B\overset{d}{\Longrightarrow}B'\overset{d'}{\Longrightarrow}B''$. There are two cases to understand: when $d = d'$ and when $d \neq d'$. We first have

Theorem 2 (Single Direction). *Suppose* $B\overset{d}{\Longrightarrow}B'\overset{d}{\Longrightarrow}B''$ *holds for boxes in a smooth subdivision. Then* $\mathrm{p}^2(B) = \mathrm{p}(B')$.

It is useful to understand the idiom "$\mathrm{p}^2(B) = \mathrm{p}(B')$" as telling us that $\mathrm{p}(B)$ and B' are siblings. Figure 1(b) illustrates two cases of Theorem 2 when $D = 2$.

Next, consider the chain $B\overset{d}{\Longrightarrow}B'\overset{d'}{\Longrightarrow}B''$ where $d \neq d'$:

Theorem 3 (Two Directions). *Consider boxes in a smooth subdivision of* $[-1, 1]^D$ *($D \geq 2$). Suppose* $B\overset{d}{\Longrightarrow}B'\overset{d'}{\Longrightarrow}B''$ *holds where* $d \neq d'$. *Then* $\mathrm{p}^2(B) \neq \mathrm{p}(B')$.

Two cases in the 2-dimensional case are illustrated by figure Figure 1(c): in both cases, we have that $B\overset{1}{\Longrightarrow}B'\overset{2}{\Longrightarrow}B''$. In the first case, the subdivision is smooth and $\mathrm{p}^2(B) \neq \mathrm{p}(B')$. In the second case, $\mathrm{p}^2(B) = \mathrm{p}(B')$ but the subdivision is not smooth, thus confirming the theorem in the contrapositive.

The next result is a kind of commutative diagram argument. Its proof depends on the Two Directions result (Theorem 3).

Theorem 4 (Commutative Diagram). *Consider boxes in a smooth subdivision* $\mathbb{S}$ *of* $[-1,1]^D$ *for* $D \geq 2$. *Suppose* $B\overset{d}{\Longrightarrow}B'\overset{d'}{\Longrightarrow}B''$ *holds for some* $d \neq d'$. *Then there exists a box* A' *in* $\mathbb{S}$ *such that* $A'\overset{d}{\Longrightarrow}B''$.

This theorem is best understood in terms of a commutative diagram as shown in Figure 2. It says that there exists some A where $\mathrm{p}(A) = \mathrm{p}(B)$ and some A' such that $A\overset{d}{\Longrightarrow}B'\overset{d'}{\Longrightarrow}B''$ and $A\overset{d'}{\Longrightarrow}A'\overset{d}{\Longrightarrow}B''$. Intuitively we can project a higher dimensional subdivision into the plane spanned by directions d, d' and then apply the reasoning shown in Figure 2.

2.4   Monotonicity in Smooth Subdivisions

Theorem 4 motivates the following notions for boxes in a subdivision $\mathbb{S}$: for all $B \in \mathbb{S}$, if there exists $A \in \mathbb{S}$ such that $A\overset{d}{\Longrightarrow}B$ then we say B is *d-forced*, and write $*\overset{d}{\Longrightarrow}B$. Furthermore, let $R(B)$ denote the set of directions d such that B is d-forced, and

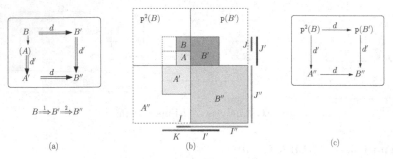

Fig. 2. Commutative Diagram for Forcing

let $r(B) = |R(B)|$ be its cardinality. Note that $0 \leq r(B) \leq 2D$. Similarly, we write $B \overset{d}{\Longrightarrow} *$ if there exists $A \in \mathbb{S}$ such that $B \overset{d}{\Longrightarrow} A$, and let $S(B)$ denote the set of directions d such that $B \overset{d}{\Longrightarrow} *$; let $s(B) = |S(B)|$. Clearly, $0 \leq s(B) \leq D$.

Note that $A \overset{d}{\Longrightarrow} B$ and $B \overset{-d}{\Longrightarrow} B'$ would imply $p^2(A) \subseteq B'$. This is impossible since A, B' are boxes of a subdivision. In other words, $d \in R(B)$ implies $-d \notin S(B)$, and conversely $d \in S(B)$ implies $-d \notin R(B)$. Thus:

$$R(B) \cap -S(B) = \emptyset. \tag{1}$$

The following follows directly from Theorem 4:

Theorem 5. *For boxes in a smooth subdivision, $B \Longrightarrow B'$ implies $R(B) \subseteq R(B')$ and hence $r(B) \leq r(B')$.*

In a general subdivision, we could have non-monotone chains (i.e., a chain whose directions include both d and $-d$ for some d). We show that smoothness implies monotone chains:

Theorem 6. *Chains in a smooth subdivision are monotone.*

Proof. Consider any chain $C : B_0 \overset{d_1}{\Longrightarrow} B_1 \overset{d_2}{\Longrightarrow} \cdots \overset{d_k}{\Longrightarrow} B_k$. It follows from the above corollary that $\{d_1, \ldots, d_i\} \subseteq R(B_i)$ for each i. It suffices to show that $-d_{i+1} \notin R(B_i)$. Note that $d_{i+1} \subseteq S(B_i)$. Therefore (1) implies $-d_{i+1} \notin R(B_i)$.
Q.E.D.

If $A \Longrightarrow B$ and $p^2(A) = p(B)$, then $p(A)$ is called a *split adjacent sibling* of B. The next lemma upper bounds $s(B)$ when B has split adjacent siblings:

Lemma 3.
(i) If B has exactly one split adjacent sibling, then $s(B) \leq 1$.
(ii) If B has at least two split adjacent siblings, then $s(B) = 0$.

The next result is critical. It shows that $r(B)$ must increase whenever B can force in more than one direction:

Lemma 4. *Let $B \Longrightarrow B'$ in a smooth subdivision. If $s(B) > 1$ then $r(B) < r(B')$.*

The next lemma shows that an increase in $r(B)$ implies a decrease in $s(B)$:

Lemma 5. *For any non-root,* $s(B) \leq \begin{cases} 0 & \text{if } r(B) > D, (CASE\ 0) \\ 1 & \text{if } r(B) = D, (CASE\ 1) \\ D - r(B) & \text{if } r(B) < D. (CASE\ 2) \end{cases}$

Let $B \in \mathbb{S}(\mathbb{T})$. The *forcing graph* $F(B)$ of B is the directed acyclic graph rooted at B, whose maximal paths are all the maximal chains beginning at B. Note that the nodes in $F(B)$ belong to $\mathbb{S}(\mathbb{T})$. The smooth split of B amounts to splitting every node in $F(B)$. Each node B' in $F(B)$ has $s(B')$ children; so B' is a leaf (or sink) if and only if $s(B') = 0$. If $s(B') > 1$, we call B' a *branching node*. Note that $F(B)$ would be a tree rooted at B if all the maximal chains are disjoint except at B. However, in general, maximal chains can merge.

Using the preceding two lemmas (Lemma 4 and Lemma 5) we can prove the following about $F(B)$:

Theorem 7. *Let B be a box in a smooth subdivision. There are at most $(D - r(B))!$ maximal paths in the forcing graph $F(B)$ where we define $x! = 1$ for $x \leq 0$.*

3   Amortized Bounds for Smooth Splits

The analysis of forcing chains in the last section will now be used to obtain an upper bound on the amortized complexity of smooth splits. We also provide a lower bound construction.

3.1   Potential of Subdivision Tree

Let $\mathbb{S}$ be a smooth subdivision. Denote by $\mathbb{T} = \mathbb{T}(\mathbb{S})$ the subdivision tree whose leaves constitute $\mathbb{S}$. Define the *potential* $\Phi(\mathbb{T})$ of the subdivision tree $\mathbb{T}$ to be the sum of the potential $\Phi(B)$ of all the nodes $B \in \mathbb{T}$. The potential of node B is

$$\Phi(B) := \begin{cases} 0 & \text{if } B \text{ has no split children,} \\ \# \text{ of unsplit children of } B \text{ otherwise.} \end{cases} \tag{2}$$

Note that $\Phi(B) = 0$ if and only if it has no split children or all its children are split. Otherwise, $1 \leq \Phi(B) \leq 2^D - 1$. Intuitively, each unit of potential pays for the cost of a single split.

For $B \in \mathbb{S}(\mathbb{T})$, let $c(B)$ denote the number of nodes B' in $F(B)$ such that $\Phi(p(B')) = 0$. But $\Phi(p(B')) = 0$ if and only if $p(B')$ has no split children or all of its children is split. Since B' is a leaf in $\mathbb{T}$, $\Phi(p(B')) = 0$ implies that B' has no split siblings. Thus, $c(B)$ is counting the number of nodes in $F(B)$ with no split siblings.

THEOREM 1[MAIN THEOREM]. *Starting from an initially trivial subdivision consisting of one box, the total cost of any sequence of smooth splits $\mathtt{ssplit}(B_1)$, ..., $\mathtt{ssplit}(B_n)$ is $O(n)$. Thus the amortized cost of a smooth split is $O(1)$.*

Proof. We show that starting from the initial box $[-1,1]^D$, a sequence of n smooth splits produces at most $(2^D(D+1)!)n$ splits. Therefore for fixed D each smooth split produces amortized $O(1)$ splits.

The smooth split of B amounts to splitting each node in its forcing graph $F(B)$. Recall that $c(B)$ is the number of nodes $B' \in F(B)$ with $\Phi(\mathtt{p}(B')) = 0$. We show that $c(B) \leq (D+1)!$.

By Theorem 7 we know that there are at most $D!$ maximal paths in $F(B)$. We need to show that each maximal chain has at most $D+1$ indices $i \in [k]$ such that $\Phi(\mathtt{p}(B_i)) = 0$. For such an i, we claim that $r(B_i) < r(B_{i+1})$. To show this, it suffices to prove that $d_{i+1} \notin R(B_i)$ because $d_{i+1} \in R(B_{i+1})$. Among the D adjacent siblings of B_i, there is one, say A, such that $A \overset{d_{i+1}}{\longrightarrow} B_i$. If $d_{i+1} \in R(B_i)$ then $A' \overset{d_{i+1}}{\Longrightarrow} B_i$ for some child A' of A. Since $\Phi(\mathtt{p}(B_i)) = 0$, A has not been split and so A' does not exist. Therefore, if there are $\geq D+1$ such indices, the $(D+1)$-st index i has the property that $r(B_{i+1}) \geq D+1$. Then $s(B_{i+1}) = 0$ by Lemma 5. Hence B_{i+1} must be the last node B_k in the chain. This proves our claim.

The smooth split of B amounts to splitting each box $B' \in F(B)$. There are two cases of B':

- $\Phi(\mathtt{p}(B')) > 0$. Then splitting B' can be charged to the corresponding unit decrease in potential $\Phi(\mathbb{T})$, since $\Phi(\mathtt{p}(B'))$ decreases by one when B' is split.
- $\Phi(\mathtt{p}(B')) = 0$. Then splitting of B' will be charged 2^D, corresponding to one unit for splitting B' and $2^D - 1$ units for increase in $\Phi(\mathtt{p}(B'))$.

It follows that the total charge for the smooth split of B is at most $2^D c(B) \leq 2^D(D+1)!$, as claimed.   **Q.E.D.**

3.2   Lower Bound on Smooth Split Complexity

We also give a lower bound showing that the amortized cost of smooth splits is exponential in the dimension D in dynamically smoothed quadtrees. The construction and a proof are given in the full version of the paper.

Lemma 6. *The cost of a sequence* $\mathtt{ssplit}(B_1), \ldots, \mathtt{ssplit}(B_n)$ *of smooth split operations in the worst case is* $\Omega(nD2^D)$. *This implies that the amortized cost of the smooth split operation is* $\Omega(D2^D)$.

This shows that although our amortized cost upper bound of $O((D+1)!2^D)$ is likely not tight, the exponential dependence on dimension D is unavoidable. We analyze the exact amortized asymptotic cost of a split in more detail in the full version of this paper.

4   Conclusion

We have given a combinatorial proof that for any fixed dimension the amortized cost of performing a smooth split is $O(1)$. We did this by defining a suitable potential function based on the number of split siblings of a node, and by presenting a sequence of lemmas reasoning about how smooth splitting can propagate through the data structure. Our smooth quadtree model is useful in applications, and we have implemented it in the `Core Library` [Cor].

Acknowledgements. We would like to thank Don Sheehy for discussions, Joe Simons for answering questions about [LSS13], and the anonymous reviewers for helpful references and clarifications.

References

[ABCC06]   Aronov, B., Bronnimann, H., Chang, A.Y., Chiang, Y.-J.: Cost predic-
           tion for ray shooting in octrees. Computational Geometry: Theory and
           Applications 34(3), 159–181 (2006)
[BEG94]   Bern, M.W., Eppstein, D., Gilbert, J.R.: Provably good mesh genera-
           tion. J. Comput. Syst. Sci. 48(3), 384–409 (1994)
[BY]   Bennett, H., Yap, C.: Amortized Analysis of Smooth Quadtrees in All
           Dimensions,
           http://www.cs.nyu.edu/exact/doc/smoothSubdiv2014.pdf
[CLRS09]   Cormen, T.H., Leiserson, C.E., Rivest, R.L., Stein, C.: Introduction to
           Algorithms, 3rd edn. MIT Press (2009)
[Cor]   Core Library homepage. Software download, source, documentation and
           links: http://cs.nyu.edu/exact/core/
[dBCvKO08] de Berg, M., Cheong, O., van Kreveld, M., Overmars, M.: Computa-
           tional Geometry: Algorithms and Applic., 3rd edn. Springer (2008)
[dBRS12]   de Berg, M., Roeloffzen, M., Speckmann, B.: Kinetic compressed
           quadtrees in the black-box model with applications to collision detection
           for low-density scenes. In: [EF 2012], pp. 383–394
[EF12]   Epstein, L., Ferragina, P. (eds.): ESA 2012. LNCS, vol. 7501. Springer,
           Heidelberg (2012)
[FB74]   Finkel, R.A., Bentley, J.L.: Quad trees: A data structure for retrieval
           on composite keys. Acta Inf. 4, 1–9 (1974)
[LSS13]   Löffler, M., Simons, J.A., Strash, D.: Dynamic planar point location
           with sub-logarithmic local updates. In: Dehne, F., Solis-Oba, R., Sack,
           J.-R. (eds.) WADS 2013. LNCS, vol. 8037, pp. 499–511. Springer,
           Heidelberg (2013)
[Moo92]   Moore, D.: Simplicial Mesh Generation with Applications. PhD thesis,
           Cornell University (1992)
[Moo95]   Moore, D.: The cost of balancing generalized quadtrees. In: Symposium
           on Solid Modeling and Applications, pp. 305–312 (1995)
[OvL82]   Overmars, M.H., van Leeuwen, J.: Dynamic multi-dimensional data
           structures based on quad- and k-d trees. Acta Inf. 17, 267–285 (1982)
[PM12]   Park, E., Mount, D.M.: A self-adjusting data structure for multidimen-
           sional point sets. In: [EF 2012], pp. 778–789
[Rup93]   Ruppert, J.: A new and simple algorithm for quality 2-dimensional mesh
           generation. In: SODA, pp. 83–92. ACM/SIAM (1993)
[Sam90a]   Samet, H.: Applications of spatial data structures - computer graphics,
           image processing, and GIS. Addison-Wesley (1990)
[Sam90b]   Samet, H.: The Design and Analysis of Spatial Data Structures.
           Addison-Wesley (1990)
[She]   Sheehy, D.R.: Private correspondence
[She12]   Sheehy, D.R.: New Bounds on the Size of Optimal Meshes. Computer
           Graphics Forum 31(5), 1627–1635 (2012)
[Sim]   Simons, J.A.: Private correspondence
[WCY13]   Wang, C., Chiang, Y.-J., Yap, C.: On soft predicates in subdivision
           motion planning. In: 29th SoCG, pp. 349–358. ACM (2013)

New Approximability Results for the Robust k-Median Problem

Sayan Bhattacharya, Parinya Chalermsook,
Kurt Mehlhorn, and Adrian Neumann

Max-Planck Institut für Informatik
{bsayan,parinya,mehlhorn,aneumann}@mpi-inf.mpg.de

Abstract. We consider a variant of the classical k-median problem, introduced by Anthony et al. [1]. In the *Robust k-Median problem*, we are given an n-vertex metric space (V, d) and m client sets $\{S_i \subseteq V\}_{i=1}^m$. We want to open a set $F \subseteq V$ of k facilities such that the worst case connection cost over all client sets is minimized; that is, minimize $\max_i \sum_{v \in S_i} d(F, v)$. Anthony et al. showed an $O(\log m)$ approximation algorithm for any metric and APX-hardness even in the case of uniform metric. In this paper, we show that their algorithm is nearly tight by providing $\Omega(\log m / \log \log m)$ approximation hardness, unless $\mathsf{NP} \subseteq \bigcap_{\delta > 0} \mathsf{DTIME}(2^{n^\delta})$. This result holds even for uniform and line metrics. To our knowledge, this is one of the rare cases in which a problem on a line metric is hard to approximate to within logarithmic factor. We complement the hardness result by an experimental evaluation of different heuristics that shows that very simple heuristics achieve good approximations for realistic classes of instances.

1 Introduction

In the classical k-median problem, we are given a set of clients located on a metric space with distance function $d : V \times V \to \mathbb{R}$. The goal is to open a set of facilities $F \subseteq V$, $|F| = k$, so as to minimize the sum of the connection costs of the clients in V, i.e., their distances from their nearest facilities in F. This is a central problem in approximation algorithms, and has received a large amount of attention in the past two decades [4,6,7,11,12].

At SODA 2008 Anthony et al. [1] introduced a generalization of the k-median problem. In their setting, the set of clients that are to be connected to some facility is not known in advance, and the goal is to perform well in spite of this uncertainty about the future. They formulated the problem as follows.

Definition 1 (Robust k-Median). *An instance of this problem is a triple $(V, \mathcal{S}, d)$. This defines a set of locations V, a collection of m sets of clients $\mathcal{S} = \{S_1, \ldots, S_m\}$, where $S_i \subseteq V$ for all $i \in \{1, \ldots, m\}$, and a metric distance function $d : V \times V \to \mathbb{R}$. We have to open a set of k facilities $F \subseteq V$, $|F| = k$, and the goal is to minimize the cost of the most expensive set of clients, i.e. minimize $\max_{i=1}^m \sum_{v \in S_i} d(v, F)$. Here, $d(v, F)$ denotes the minimum distance of the client v from any location in F, i.e. $d(v, F) = \min_{u \in F} d(u, v)$.*

R Ravi and I.L. Gørtz (Eds.): SWAT 2014, LNCS 8503, pp. 50–61, 2014.

Robust k-Median is a natural generalization of the classical k-median problem (for $m = 1$). Additionally, we can think of it as capturing a notion of *fairness*. To see this, interpret each set S_i as a *group* of clients who pay $\sum_{v \in S_i} d(v, F)$ for connecting to a facility. The objective ensures that no single group pays too much, while minimizing the cost. Anthony et al. [1] gave an $O(\log m)$-approximation algorithm for this problem, and a lower bound of $(2 - \epsilon)$ by a reduction from Vertex Cover. The lower bound was improved to $\log^\alpha n$ for small constant $\alpha > 0$ in [5]. Note that their lower bound does not hold in the line metric.

Our Results. We prove nearly tight hardness of approximation for Robust k-Median. We show that, unless $\mathsf{NP} \subseteq \cap_{\delta > 0} \mathsf{DTIME}(2^{n^\delta})$, it admits no poly-time $o(\log m / \log \log m)$-approximation, *even on uniform and line metrics*.

Our first hardness result is tight up to a constant factor, as a simple rounding scheme gives a matching upper bound on uniform metrics (Sect. 3.1). Our second result shows that Robust k-Median is a rare problem with super-constant hardness of approximation even on line metrics. This surprising result puts Robust k-Median in sharp contrast to most other geometric optimization problems which admit polynomial time approximation schemes, e.g. [2, 10].

Experimentally we show that simple heuristics provide good performance on a realistic class of instances. The details appear in the full paper.

Our Techniques. First, we note that Robust k-Median on uniform metrics is equivalent to the following variant of the set cover problem: Given a set U of ground elements, a collection of sets $\mathcal{X} = \{X \subseteq U\}$, and an integer $t \leq |\mathcal{X}|$, our goal is to select t sets from $\mathcal{X}$ in order to minimize the number of times an element from U is hit (Lemma 2). We call this problem Minimum Congestion Set Packing (MCSP). This characterization allows us to focus on proving the hardness of MCSP, and to employ the tools developed for the set cover problem.

We now revisit the reduction used by Feige [8], building on results of Lund and Yannakakis [13], to prove the hardness of the set cover problem and discuss how our approach differs. Intuitively, they compose the Label Cover instance with a set system that has some desirable properties. Informally speaking, in the Label Cover problem, we are given a graph where each vertex v can be assigned a label from a set L, and each edge e is equipped with a constraint $\Pi_e \subseteq L \times L$ specifying the accepting pairs of labels for e. Our goal is to find a labeling of vertices that maximizes the number of accepting edges. This problem is known to be hard to approximate to within a factor of $2^{\log^{1-\epsilon} |E|}$ [3, 14], where $|E|$ is the number of edges. Thus, if we manage to reduce Label Cover to MCSP, we would hopefully obtain a large hardness of approximation factor for MCSP as well.

From the Label Cover instance, [13] creates an instance of Set Cover by having sets of the form $S(v, \ell)$ for each vertex v and each label $\ell \in L$. Intuitively the set $S(v, \ell)$ means choosing label ℓ for vertex v in the Label Cover instance. Now, if we assume that the solution is well behaved, in the sense that for each vertex v, only one set of the form $S(v, \ell)$ is chosen in the solution, we would be immediately done (because each set indeed corresponds to a label). However, solutions need not have this form, e.g. choosing sets $S(v, \ell)$ and $S(v, \ell')$ translates to having two labels ℓ, ℓ' for the Label Cover instance. To prevent an ill-behaved solution, *partition systems* were introduced and used in both [13] and [8]. Feige considers the hypergraph version of Label Cover to obtain a sharper hardness result of $\ln n - O(\ln \ln n)$ instead of $\frac{1}{4} \ln n$ in [13]; here n denotes the size of the universe.

Now we highlight how our reduction is different. The high level idea stays the same, i.e. we have sets of the form $S(v, \ell)$ that represent assigning label ℓ to vertex v. However, we need a different partition system and a totally different analysis. Moreover, while a reduction from standard Label Cover gives nearly tight $O(\log n)$ hardness for Set Cover, it can (at best) only give a $2 - \epsilon$ hardness for MCSP. For our results, we do need a reduction from Hypergraph Label Cover. This suggests another natural distinction between MCSP and Set Cover.

Finally, to obtain the hardness result for the line metric, we embed the instance created from the MCSP reduction onto the line while preserving values of optimal solutions. This way we get the same hardness gap for line metrics.

2   Preliminaries

We will show that Robust k-Median is $\Omega(\log m / \log \log m)$ hard to approximate, even for the special cases of *uniform metrics* (Sect. 3) and *line metrics* (Sect. 4). Recall that d is a uniform metric iff we have $d(u, v) \in \{0, 1\}$ for all locations $u, v \in V$. Further, d is a line metric iff the locations in V can be embedded into a line in such a way that $d(u, v)$ equals the euclidean distance between u and v, for all $u, v \in V$. Throughout this paper, we will denote any set of the form $\{1, 2, \ldots, i\}$ by $[i]$. Our hardness results will rely on a reduction from the *r-Hypergraph Label Cover* (HGLC) problem, which is defined as follows.

Definition 2 (r-Hypergraph Label Cover (HGLC)). *An instance of this problem is a triple (G, π, r), where $G = (\mathcal{V}, \mathcal{E})$ is a r-partite hypergraph with vertex set $\mathcal{V} = \bigcup_{j=1}^{r} \mathcal{V}_j$ and edge set $\mathcal{E}$. Each edge $h \in \mathcal{E}$ contains one vertex from each part of $\mathcal{V}$, i.e. $|h \cap \mathcal{V}_j| = 1$ for all $j \in [r]$. Every set $\mathcal{V}_j$ has an associated set of labels L_j. Further, for all $h \in \mathcal{E}$ and $j \in [r]$, there is a mapping $\pi_h^j : L_j \to C$ that projects the labels from L_j to a common set of colors C.*

The problem is to assign to every vertex $v \in \mathcal{V}_j$ some label $\sigma(v) \in L_j$. We say that an edge $h = (v_1, \ldots, v_r)$, where $v_j \in \mathcal{V}_j$ for all $j \in [r]$, is strongly satisfied under σ iff the labels of all its vertices are mapped to the same element in C, i.e. $\pi_h^j(\sigma(v_j)) = \pi_h^{j'}(\sigma(v_{j'}))$ for all $j, j' \in [r]$. In contrast, we say that the edge is weakly satisfied iff there exists some pair of vertices in h whose labels are mapped to the same element in C, i.e. $\pi_h^j(\sigma(v_j)) = \pi_h^{j'}(\sigma(v_{j'}))$ for some $j, j' \in [r]$, $j \neq j'$.

For ease of exposition, we will often abuse the notation and denote by $j(v)$ the part of $\mathcal{V}$ to which a vertex v belongs, i.e. if $v \in \mathcal{V}_j$ for some $j \in [r]$, then we set $j(v) \leftarrow j$. The next theorem will be crucial in deriving our hardness result. The proof of this theorem follows from Feige's r-Prover system [8].

Theorem 1. *Let $r \in \mathbb{N}$ be a parameter. There is a polynomial time reduction from n-variable 3-SAT to r-HGLC with the following properties:*

- *(Yes-Instance) If the formula is satisfiable, there is a labeling that strongly satisfies every edge in G.*
- *(No-Instance) If the formula is not satisfiable, every labeling weakly satisfies at most a $2^{-\gamma r}$ fraction of the edges in G, for some universal constant γ.*
- *The number of vertices in the graph is $|\mathcal{V}| = n^{O(r)}$ and the number of edges is $|\mathcal{E}| = n^{O(r)}$. The sizes of the label sets are $|L_j| = 2^{O(r)}$ for all $j \in [r]$, and $|C| = 2^{O(r)}$. Further, we have $|\mathcal{V}_j| = |\mathcal{V}_{j'}|$ for all $j, j' \in [r]$, and each vertex $v \in \mathcal{V}$ has the same degree $r|\mathcal{E}|/|\mathcal{V}|$.*

We use a *partition system* that is motivated by the hardness proof of the Set Cover problem [8] but uses a different construction.

Definition 3 (Partition System). *Let $r \in \mathbb{N}$ and let C be any finite set. An (r, C)-partition system is a pair $(Z, \{p_c\}_{c \in C})$, where Z is an arbitrary (ground) set, such that the following properties hold.*

- *(Partition) For all $c \in C$, $p_c = (A_c^1, \ldots, A_c^r)$ is a partition of Z, that is $\bigcup_{j=1}^r A_c^j = Z$, and $A_c^j \cap A_c^{j'} = \emptyset$ for all $j, j' \in [r], j \neq j'$.*
- *(r-intersecting) For any r distinct indices $c_1, \ldots, c_r \in C$ and not-necessarily distinct indices $j_1, \ldots, j_r \in [r]$, we have that $\bigcap_{i=1}^r A_{c_i}^{j_i} \neq \emptyset$. In particular, $A_c^j \neq \emptyset$ for all c and j.*

In order to achieve a good lower bound on the approximation factor, we need partition systems with *small* ground sets. The most obvious way to build a partition system is to form an r-hypercube: Let $Z = [r]^{|C|}$, and for each $c \in C$ and $j \in [r]$, let A_c^j be the set of all elements in Z whose c-th component is j. It can easily be verified that this is an (r, C)-partition system with $|Z| = r^{|C|}$. With this construction, however, we would only get a hardness of $\Omega(\log \log m)$ for our problem. The following lemma shows that it is possible to construct an (r, C)-partition system probabilistically with $|Z| = r^{O(r)} \log |C|$.

Lemma 1. *There is an (r, C)-partition system with $|Z| = r^{O(r)} \log |C|$ elements. Further, such a partition system can be constructed efficiently with high probability.*

Proof. Let Z be any set of $r^{O(r)} \log |C|$ elements. We build a partition system $(Z, \{p_c\}_{c \in C})$ as described in Algorithm 1. By construction each p_c is a partition of Z, i.e. the first property stated in Def. 3 is satisfied. We bound the probability that the second property is violated.

Fix any choice of r *distinct* indices $c_1, \ldots, c_r \in C$ and *not necessarily distinct* indices $j_1, \ldots, j_r \in [r]$. We say that a *bad event* occurs when the intersection of the corresponding sets is empty, i.e. $\bigcap_{i=1}^r A_{c_i}^{j_i} = \emptyset$. To upper bound the probability of a bad event, we focus on events of the form $E_{e,i}$ – this occurs when an element $e \in Z$ is included in a set $A_{c_i}^{j_i}$. Since the indices $c_1 \ldots c_r$ are distinct, it follows that the events $\{E_{e,i}\}$ are mutually independent. Furthermore, note that we have $\Pr[E_{e,i}] = 1/r$ for all $e \in Z, i \in [r]$. Hence, the probability that an element $e \in Z$ does not belong to the intersection $\bigcap_{i=1}^r A_{c_i}^{j_i}$ is given by

---

Algorithm 1. A randomized construction of an (r, C)-partition system.

---

input : A ground set Z, a parameters $r \in \mathbb{N}$, and a set C.
foreach $c \in C$ **do**

  /* Construct the partition $p_c = (A_c^1, \ldots, A_c^r)$ */
  Initialize A_c^j to the empty set for all $j \in [r]$
  foreach *ground element* $e \in Z$ **do**
    | Pick a $j \in [r]$ independently and uniformly at random and add e to A_c^j

---

$1 - \Pr[\bigcap_{i=1}^r E_{e,i}] = 1 - 1/r^r$. Accordingly, the probability that no element $e \in Z$ belongs to the intersection, which defines the bad event, is equal to $(1 - 1/r^r)^{|Z|}$.

Now, the number of choices for r distinct indices $c_1, \ldots, c_r$ and r not-necessarily distinct indices $j_1, \ldots, j_r$ is equal to $\binom{|C|}{r} \cdot r^r$. Hence, by a union-bound over all bad events, the second property stated in Def. 3 is violated with probability at most $\binom{|C|}{r} \cdot r^r \cdot (1 - r^r)^{|Z|} \le (|C| r)^r \cdot \exp(-|Z|/r^r)$. If we set $|Z| = d \cdot r^{d \cdot r} \log |C|$ with large enough constant d, the property is satisfied with high probability. $\square$

3 Hardness of Robust k-Median on Uniform Metrics

First, we define *Minimum Congestion Set Packing* (MCSP), and then show a reduction from MCSP to Robust k-Median on uniform metrics. In Sect. 3.2, we will then show that MCSP is hard to approximate by reducing HGLC to MCSP.

Definition 4 (Minimum Congestion Set Packing (MCSP)). *An instance of this problem is a triple* $(U, \mathcal{X}, t)$, *where* U *is a universe of* m *elements, i.e.* $|U| = m$, $\mathcal{X}$ *is a collection of sets* $\mathcal{X} = \{X \subseteq U\}$ *such that* $\bigcup_{X \in \mathcal{X}} X = U$, *and* $t \in \mathbb{N}$ *and* $t \le |\mathcal{X}|$. *The objective is to find a collection* $\mathcal{X}' \subseteq \mathcal{X}$ *of size* t *that minimizes* $\mathrm{CONG}(\mathcal{X}') = \max_{e \in U} \mathrm{CONG}(e, \mathcal{X}')$. *Here,* $\mathrm{CONG}(\mathcal{X}')$ *refers to* the congestion of the solution $\mathcal{X}'$, and $\mathrm{CONG}(e, \mathcal{X}') = |\{X \in \mathcal{X}' : e \in X\}|$ *is the congestion of the element* $e \in U$ *under the solution* $\mathcal{X}'$.

Lemma 2. *Given any MCSP instance* $(U, \mathcal{X}, t)$, *we can construct a Robust k-Median instance* $(V, \mathcal{S}, d)$ *with the same objective value in* $\mathrm{poly}(|U|, |\mathcal{X}|)$ *time, such that* $|U| = |\mathcal{S}|$, $|\mathcal{X}| = |V|$, d *is a uniform metric, and* $k = |V| - t$.

Proof. We construct the Robust k-Median instance $(V, \mathcal{S}, d)$ as follows. For every $e \in U$ we create a set of clients $S(e)$, and for each $X \in \mathcal{X}$ we create a location $v(X)$. Thus, we get $V = \{v(X) : X \in \mathcal{X}\}$, and $\mathcal{S} = \{S(e) : e \in U\}$. We place the clients in $S(e)$ at the locations of the sets that contain e, i.e. $S(e) = \{v(X) : X \in \mathcal{X}, e \in X\}$ for all $e \in U$. The distance is defined as $d(u, v) = 1$ for all $u, v \in V, u \ne v$, and $d(v, v) = 0$. Finally, we set $k \leftarrow |V| - t$.

Now, it is easy to verify that the Robust k-Median instance $(V, \mathcal{S}, d)$ has a solution with objective ρ iff the corresponding MCSP instance $(U, \mathcal{X}, t)$ has a solution with objective ρ. The intuition is that a location $v(X) \in V$ is *not* included in the solution F to the Robust k-Median instance iff the corresponding set X is included in the solution $\mathcal{X}'$ to the MCSP instance. Indeed, let F be any subset of $\mathcal{X}$ of size k (= the set of open facilities) and let $\mathcal{X}' = \mathcal{X} - F$. Further, let $[X \in \mathcal{X}']$ be an indicator variable that is set to 1 iff $X \in \mathcal{X}'$. Then

$$\mathrm{CONG}(\mathcal{X}') = \max_{e \in U} \mathrm{CONG}(e, \mathcal{X}') = \max_{e \in U} \sum_{X; e \in X} [X \in \mathcal{X}']$$

$$= \max_{e \in U} \sum_{X; e \in X} \min_{Y \in F} d(X, Y) = \max_{S(e) \in \mathcal{S}} \sum_{v(X) \in S(e)} d(v(X), F).$$

$\square$

We devote the rest of Sect. 3 to MCSP and show that it is $\Omega(\log |U| / \log \log |U|)$ hard to approximate. This, in turn, will imply a $\Omega(\log |\mathcal{S}| / \log \log |\mathcal{S}|)$ hardness of approximation for Robust k-Median on uniform metrics. We will prove the hardness result via a reduction from HGLC.

3.1 Integrality Gap

Before proceeding to the hardness result, we show that a natural LP relaxation for the MCSP problem [1] has an integrality gap of $\Omega(\log m / \log \log m)$, where $m = |U|$ is the size of the universe of elements. In the LP, we have a variable $y(X)$ indicating that the set $X \in \mathcal{X}$ is chosen, and a variable z which represents the maximum congestion among the elements.

$$\min \quad z$$

$$\text{s.t.} \quad \sum_{X \in \mathcal{X}: e \in X} y(X) \le z \text{ for all } e \in U$$

$$\sum_{X \in \mathcal{X}} y(X) = t$$

The Instance: Now, we construct a bad integrality gap instance $(U, \mathcal{X}, t)$. Let d be the intended integrality gap, let $\eta = d^2$, and let $U = \{I : I \subseteq [\eta], |I| = d\}$ be all subsets of $[\eta]$ of size d. The collection $\mathcal{X}$ consists of η sets $X_1, \ldots, X_\eta$, where $X_i = \{I : I \in U \text{ and } i \in I\}$. Note that the universe U consists of $|U| = m = \binom{\eta}{d}$ elements, and each element I is contained in exactly d sets, namely $I \in X_i$ if and only if $i \in I$. Finally, we set $t \leftarrow \eta/d$.

Analysis: The fractional solution simply assigns a value of $1/d$ to each variable $y(X_i)$; this ensures that the total (fractional) number of sets selected is $\eta/d = t$. Furthermore, each element is contained (fractionally) in exactly one set, so the fractional solution has cost one. Since $t = \eta/d = d$, any integral solution must choose d sets, say $X_{i_1}, \ldots, X_{i_d}$. Now consider $I = \{i_1, \ldots, i_d\}$ which belongs to set X_{i_λ} for all $\lambda \in [d]$ and hence the congestion of I is d. Finally, since $|U| = m \le \eta^d \le (d^2)^d$, we have $d = \Omega(\log m / \log \log m)$.

Tightness of the Result: The bound on the hardness and integrality gap is tight for the uniform metric case, as there is a simple $O(\log m / \log \log m)$-approximation algorithm. Pick each set X with probability equal to $\min(1, 2y(X))$.

The expected congestion is $2z$ for each element. By Chernoff's bound [9], an element is covered by no more than $z \cdot O(\log m / \log \log m)$ sets with high probability. A similar algorithm gives the same approximation guarantee for Robust k-Median on uniform metrics.

3.2 Reduction from r-Hypergraph Label Cover to Minimum Congestion Set Packing

The input is an instance (G, π, r) of r-HGLC (Def. 2). From this we construct the following instance $(U, \mathcal{X}, t)$ of MCSP (Def. 4).

- We define the universe U as a union of disjoint sets. For each edge $h \in \mathcal{E}$ in the hypergraph we have a set U_h. All these sets have the same size m^* and are pairwise disjoint, i.e. $U_h \cap U_{h'} = \emptyset$ for all $h, h' \in \mathcal{E}$, $h' \neq h$. The universe U is then the union of these sets $U = \bigcup_{h \in \mathcal{E}} U_h$. Since the U_h are mutually disjoint, we have $m = |U| = |\mathcal{E}| \cdot m^*$. Recall that C is the target set of π. Each set U_h is the ground set of an (r, C)-partition system (Def. 3) as given by Lemma 1. In particular we have $m^* = r^{O(r)} \log |C|$. We denote the r-partitions associated with U_h by $\{p_c(h)\}_{c \in C}$, where $p_c(h) = \left(A_c^1(h), \ldots, A_c^r(h)\right)$.
- We construct the collection of sets $\mathcal{X}$ as follows. For each $j \in [r]$, $v \in \mathcal{V}_j$ and $\ell \in L_j$, $\mathcal{X}$ contains the set $X(v, \ell)$, where $X(v, \ell) = \bigcup_{h:v \in h} A_{\pi_h^j(\ell)}^j(h)$. That is, $X(v, \ell) \cap U_h$ is empty if $v \notin h$ and is equal to $A_{\pi_h^j(\ell)}^j(h)$ if $v \in h$. Intuitively, choosing the set $X(v, \ell)$ corresponds to assigning label ℓ to the vertex v.
- We define $t \leftarrow |\mathcal{V}|$. Intuitively, this means each vertex in $\mathcal{V}$ gets one label.

We assume for the sequel that the r-HGLC instance is chosen according to Thm. 1. We assume that the parameter r satisfies $r^7 2^{-\gamma r} < 1$. In the proof of the main theorem, we will fix r to a specific value.

3.3 Analysis

We show that the reduction from HGLC to MCSP satisfies two properties. In Lemma 3, we show that for Yes-Instances (see Thm. 1) the corresponding MCSP instance admits a solution with congestion one. For No-Instances, Lemma 4 shows that any solution to the corresponding MCSP instance has congestion at least r.

Lemma 3 (Yes-Instance). *If the HGLC instance (G, π, r) admits a labeling that strongly satisfies every edge, then the MCSP instance $(U, \mathcal{X}, t)$ as in Sect. 3.2 admits a solution where the congestion of every element in U is exactly one.*

Proof. Suppose that there is a labeling σ that strongly satisfies every edge $h \in \mathcal{E}$. We will show how to pick $t = |\mathcal{V}|$ sets from $\mathcal{X}$ such that each element in U is contained in exactly one set. This implies that the maximum congestion is one. For each $j \in [r]$ and each vertex $v \in \mathcal{V}_j$, we choose the set $X(v, \sigma(v))$. Thus, the total number of sets chosen is exactly $|\mathcal{V}|$.

To see that the congestion is one, we concentrate on the elements in U_h, where $h = (v_1, \ldots, v_r)$, $v_j \in \mathcal{V}_j$ for all $j \in [r]$, is one of the edges in $\mathcal{E}$. The picked sets that intersect U_h are $X(v_j, \sigma(v_j))$, where $j \in [r]$. Since h is strongly satisfied, π_h maps all vertex labels in h to a common $c \in C$, i.e. $\pi_h^j(\sigma(v_j)) = c$ for all $j \in [r]$. Thus $U_h \cap X(v_j, \sigma(v_j)) = A_c^j(h)$. By definition (Def. 3), the sets $A_c^1(h) \ldots A_c^r(h)$ partition the elements in U_h. This completes the proof. $\qquad \square$

Now, we turn to the proof of Lemma 4. Towards this end, we fix a collection $\mathcal{X}' \subseteq \mathcal{X}$ of size t and show that some element in U has congestion at least r under $\mathcal{X}'$. The intuition being that many edges in $G = (\mathcal{V}, \mathcal{E})$ are not even weakly satisfied, and the elements in U corresponding to those edges incur large congestion. Recall that for a $v \in \mathcal{V}$, we define $j(v) \in \mathbb{N}$ to be such that $v \in \mathcal{V}_{j(v)}$.

Claim 2. *For $v \in \mathcal{V}$, let $\mathcal{L}_v = \{\ell \in L_{j(v)} : X(v, \ell) \in \mathcal{X}'\}$. For $h \in \mathcal{E}$, let $\Lambda_h = \{X(v, \ell) \in \mathcal{X}' : v \in h\}$. If $\mathrm{CONG}(\mathcal{X}') < r$, then $|\mathcal{L}_v| < r^2$ and $|\Lambda_h| < r^3$.*

Proof. Since $\Lambda_h = \bigcup_{v \in h} \mathcal{L}_v$, it suffices to prove $|\mathcal{L}_v| < r^2$ for all v. Assume otherwise, i.e., $|\mathcal{L}_v| \geq r^2$ for some $v \in \mathcal{V}_j$, $j \in [r]$. Let h be any hyper-edge with $v \in h$. Consider the images of the labels in $\mathcal{L}_v$ under π_h^j. Either there are at least r distinct images or at least r elements in L_v are mapped to the same $c \in C$.

In the former case, we have r pairwise distinct labels ℓ_1 to ℓ_r in $\mathcal{L}_v$ and r pairwise distinct labels c_1 to c_r in C such that $\pi_h^j(\ell_i) = c_i$ for $i \in [r]$. The set $X(v, \ell_i)$ contains $A_{c_i}^j(h)$ and $\bigcap_{i \in [r]} A_{c_i}^j(h) \neq \emptyset$ by property (2) of partition systems (Def. 3). Thus some element has congestion at least r.

In the latter case, we have r pairwise distinct labels ℓ_1 to ℓ_r in $\mathcal{L}_v$ and a label c in C such that $\pi_h^j(\ell_i) = c$ for $i \in [r]$. The set $X(v, \ell_i)$ contains $A_c^j(h)$ and hence every element in this non-empty set (property (2) of partition systems) has congestion at least r. $\qquad \square$

Definition 5 (Colliding Edge). *We say that an edge $h \in \mathcal{E}$ is colliding iff there are sets $X(v, \ell), X(v', \ell') \in \mathcal{X}'$ with $v, v' \in h$, $v \neq v'$, and $\pi_h^{j(v)}(\ell) = \pi_h^{j(v')}(\ell')$.*

Claim 3. *Suppose that the solution $\mathcal{X}'$ has congestion less than r, and more than a $r^4 2^{-\gamma r}$ fraction of the edges in $\mathcal{E}$ are colliding. Then there is a labeling σ for G that weakly satisfies at least a $2^{-\gamma r}$ fraction of the edges in $\mathcal{E}$.*

Proof. For each $v \in \mathcal{V}$, we define $\mathcal{L}_v = \{\ell \in L_{j(v)} : X(v, \ell) \in \mathcal{X}'\}$. Then $|\mathcal{L}_v| < r^2$ by Claim 2. We construct a labeling function σ using Algorithm 2.

Now we bound the expected fraction of weakly satisfied edges under σ from below. Take any colliding edge $h \in \mathcal{E}$. Then there are vertices $v \in \mathcal{V}_j$, $v' \in \mathcal{V}_{j'}$ with

---

Algorithm 2. An algorithm for constructing a labeling function.

---

foreach *vertex* $v \in \mathcal{V}$ **do**
    if $\mathcal{L}_v \neq \emptyset$ **then**
        Pick a color $\sigma(v)$ uniformly and independently at random from $\mathcal{L}_v$
    else
        Pick an arbitrary color $\sigma(v)$ from $L_{j(v)}$

$j \neq j'$, and colors $\ell \in \mathcal{L}_v$, $\ell' \in \mathcal{L}_{v'}$ such that $v, v' \in h$ and $\pi_h^j(\ell) = \pi_h^{j'}(\ell')$. By Claim 2, $|\mathcal{L}_v|$ and $|\mathcal{L}_{v'}|$ are both at most r^2. Since the colors $\sigma(v)$ and $\sigma(v')$ are chosen uniformly and independently at random from their respective palettes $\mathcal{L}_v$ and $\mathcal{L}_{v'}$, we have $\Pr[\sigma(v) = \ell \text{ and } \sigma(v') = \ell'] \geq 1/r^4$. In other words, every colliding edge is weakly satisfied with probability at least $1/r^4$. Since more than a $r^4 2^{-\gamma r}$ fraction of the edges in $\mathcal{E}$ are colliding, from linearity of expectation we infer that the expected fraction of edges weakly satisfied by σ is at least $2^{-\gamma r}$.   □

Claim 4. *Let* $\Lambda_h = \{X(v, \ell) \in \mathcal{X}' : v \in h\}$ *and* $\lambda(h) = |\Lambda_h|$. $\sum_{h \in \mathcal{E}} \lambda(h) = r|\mathcal{E}|$.

Proof. This is a simple counting argument. Consider a bipartite graph H with vertex set $A \dot\cup B$, where each vertex in A represents a set $X(v, \ell)$, and each vertex in B represents an edge $h \in \mathcal{E}$. There is an edge between two vertices iff the set $X(v, \ell)$ contains some element in U_h. The quantity $\sum_{h \in \mathcal{E}} \lambda(h)$ counts the number of edges in H where one endpoint is included in the solution $\mathcal{X}'$. Since $\mathcal{X}'$ picks $t = |\mathcal{V}|$ sets and each set has degree $r|\mathcal{E}|/|\mathcal{V}|$ in H (Thm. 1), the total number of edges that are chosen is exactly $|\mathcal{V}| \times (r|\mathcal{E}|/|\mathcal{V}|) = r|\mathcal{E}|$.   □

Let $\mathcal{E}' \subseteq \mathcal{E}$ denote the set of colliding edges, and define $\mathcal{E}'' = \mathcal{E} - \mathcal{E}'$. Suppose that we are dealing with a No-Instance (Thm. 1), i.e. the solution $\mathcal{X}'$ has congestion less than r and every labeling weakly satisfies at most a $2^{-\gamma r}$ fraction of the edges in $\mathcal{E}$. Then $\lambda(h) \leq r^3$ for all $h \in \mathcal{E}$ by Claim 2, and no more than $r^4 2^{-\gamma r}|\mathcal{E}|$ edges are colliding, i.e. $|\mathcal{E}'| \leq r^4 2^{-\gamma r}|\mathcal{E}|$, by Claim 3. Using these facts we conclude that $\sum_{h \in \mathcal{E}'} \lambda(h) \leq r^7 2^{-\gamma r}|\mathcal{E}| < |\mathcal{E}|$, as by assumption $r^7 2^{-\gamma r} < 1$. Now, applying Claim 4, we get $\sum_{h \in \mathcal{E}''} \lambda(h) = r|\mathcal{E}| - \sum_{h \in \mathcal{E}'} \lambda(h) > (r-1)|\mathcal{E}|$. In particular, there is an edge $h \in \mathcal{E}''$ with $\lambda(h) \geq r$.

Recall that $\Lambda_h = \{X(v, \ell) \in \mathcal{X}' : v \in h\}$ are the sets in $\mathcal{X}'$ that intersect U_h and note that $|\Lambda_h| = \lambda(h) \geq r$. Let $\mathcal{X}^* \subseteq \Lambda_h$ be a *maximal* collection of sets with the following property: For every two distinct sets $X(v, \ell), X(v', \ell') \in \mathcal{X}^*$ we have $\pi_h^{j(v)}(\ell) \neq \pi_h^{j(v')}(\ell')$. Hence, from the definition of a partition system (Def. 3), it follows that the intersection of the sets in $\mathcal{X}^*$ and the set U_h is nonempty.

Now, consider any set $X(v, \ell) \in \Lambda_h - \mathcal{X}^*$. Since the collection $\mathcal{X}^*$ is maximal, there must be at least one set $X(v', \ell')$ in $\mathcal{X}^*$ with $\pi_h^{j(v)}(\ell) = \pi_h^{j(v')}(\ell')$. Since h is not colliding, we must have $j(v) = j(v')$. Consequently we get $X(v, \ell) \cap U_h = X(v', \ell') \cap U_h$. In other words, for every set $X \in \Lambda_h - \mathcal{X}^*$, there is some set $X' \in \mathcal{X}^*$ where $X \cap U_h = X' \cap U_h$. Thus, $U_h \cap (\bigcap_{X \in \Lambda_h} X) = U_h \cap (\bigcap_{X \in \mathcal{X}^*} X) \neq \emptyset$. Every element in the intersection of the sets in Λ_h and U_h will have congestion $|\Lambda_h| \geq r$. This leads to the following lemma.

Lemma 4 (No-Instance). *If every labeling weakly satisfies at most a $2^{-\gamma r}$ fraction of the edges in the hypergragph Label Cover instance (G, π, r), for some universal constant γ and $r^7 2^{-\gamma r} < 1$ then the congestion incurred by every solution to the MCSP instance $(U, \mathcal{X}, t)$ constructed in Sect. 3.2 is at least r.*

We are now ready to prove the main theorem of this section.

Theorem 5. *Robust k-Median $(V, \mathcal{S}, d)$ is $\Omega(\log m/\log\log m)$ hard to approximate on uniform metrics, where $m = |\mathcal{S}|$, unless $\mathsf{NP} \subseteq \bigcap_{\delta > 0} \mathsf{DTIME}(2^{n^\delta})$.*

Proof. Assume that there is a polynomial time algorithm for Robust k-Median that guarantees an approximation ratio in $o(\log |\mathcal{S}| / \log \log |\mathcal{S}|)$. Then, by Lemma 2, there is an approximation algorithm for the Minimum Congestion Set Packing problem with approximation guarantee $o(\log |U| / \log \log |U|)$.

Let $\delta > 0$ be arbitrary and set $r = \lfloor n^\delta \rfloor$, where n is the number of variables in the 3-SAT instance (Thm. 1). Then $r^7 2^{-\gamma r} < 1$ for all sufficiently large n. We first bound the size of the MCSP instance $(U, \mathcal{X}, t)$ constructed in Sect. 3.2. By Lemma 1, the size of an (r, C)-partition system is $|Z| = r^{O(r)} \log |C|$. By Thm. 1, we have $|C| = 2^{O(r)}$. So each set U_h has cardinality at most $r^{O(r)} \cdot r = r^{O(r)}$. Also recall that the number of sets in the MCSP instance is $|\mathcal{X}| = \sum_{j \in [r]} |\mathcal{V}_j| \cdot |L_j| = n^{O(r)}$, and that the number of elements is $|U| = m = |\mathcal{E}| \cdot r^{O(r)} \le (nr)^{O(r)} = n^{O(r)} = n^{O(n^\delta)} = 2^{O(r \log r)}$. Thus $r \ge \Omega(\log m / \log \log m)$.

The gap in the optimal congestion between the Yes-Instance and the No-Instance is at least r (Thm. 1 and Lemmas 3, 4). More precisely, for Yes-instances the congestion is at most one and for No-instances it is at least r. Since the approximation ratio of the alleged algorithm is $o(\log m / \log \log m)$, it is better than r for all sufficiently large n and hence it can be used to decide SAT.

The running time is polynomial in the size of the MCSP instance, i.e., is $\text{poly}(n^{O(n^\delta)}) = n^{O(n^\delta)} = 2^{O(n^{2\delta})}$. Since δ is arbitrary, the theorem follows.   □

4   Hardness of Robust k-Median on Line Metrics

We modify the reduction from r-HGLC to Minimum Congestion Set Packing (MCSP) to give a $\Omega(\log m / \log \log m)$ hardness of approximation for Robust k-Median on line metrics as well, where $m = |\mathcal{S}|$ is the number of client-sets. For this section, it is convenient to assume that the label-sets are the initial segments of the natural numbers, i.e., $L_j = \{1, \ldots, |L_j|\}$ and $C = \{1, \ldots, |C|\}$.

Given a HGLC instance (G, π, r), we first construct a MCSP instance $(U, \mathcal{X}, t)$ in accordance with the procedure outlined in Sect. 3.2. Next, from this MCSP instance, we construct a Robust k-Median instance $(V, \mathcal{S}, d)$ as described below.

- We create a location in V for every set $X(v, \ell) \in \mathcal{X}$. To simplify the notation, the symbol $X(v, \ell)$ will represent both a set in the instance $(U, \mathcal{X}, t)$, and a location in the instance $(V, \mathcal{S}, d)$. Thus, we have $V = \{X(v, \ell) \in \mathcal{X}\}$. Furthermore, we create a set of clients $S(e)$ for every element $e \in U$, which consists of all the locations whose corresponding sets in the MCSP instance contain the element e. Thus, we have $\mathcal{S} = \{S(e) : e \in U\}$, where $S(e) = \{X(v, \ell) \in \mathcal{X} : e \in X(v, \ell)\}$ for all $e \in U$. This step is same as in Lemma 2.
- We now describe how to embed the locations in V on a given line. For every vertex $v \in \mathcal{V}_j, j \in [r]$, the locations $X(v, 1), \ldots, X(v, |L_j|)$ are placed next to one another in sequence, in such a way that the distance between any two consecutive locations is exactly one. Formally, this gives $d(X(v, \ell), X(v, \ell')) = |\ell' - \ell|$ for all $\ell, \ell' \in L_j$. Furthermore, we ensure that any two locations corresponding to two different vertices in $\mathcal{V}$ are *not close to each other*. To be more specific, we have the following guarantee: $d(X(v, \ell), X(v', \ell')) \ge 2$ whenever $v \ne v'$. It is easy to verify that d is a line metric.

– We define $k \leftarrow |\mathcal{X}| - t$.

Note that as $k = |\mathcal{X}| - t$, there is a one to one correspondence between the solutions to the MCSP instance and the solutions to the Robust k-Median instance. Specifically, a set in $\mathcal{X}$ is picked by a solution to the MCSP instance iff the corresponding location is *not* picked in the Robust k-Median instance.

Lemma 5 (Yes-Instance). *Suppose that there is a labeling strategy σ that strongly satisfies every edge in the HGLC instance (G, π, r). Then there is a solution to the Robust k-Median instance $(V, \mathcal{S}, d)$ with objective one.*

Proof. Recall the proof of Lemma 3. We construct a solution $\mathcal{X}' \subseteq \mathcal{X}, |\mathcal{X}'| = t$, to the MCSP instance $(U, \mathcal{X}, t)$ as follows. For every $v \in \mathcal{V}_j, j \in [r]$, the solution $\mathcal{X}'$ contains the set $X(v, \sigma(v))$. Now, focus on the corresponding solution $F_{\mathcal{X}'} \subseteq V$ to the Robust k-Median instance, which picks a location X iff $X \notin \mathcal{X}'$. Hence, for every vertex $v \in \mathcal{V}_j, j \in [r]$, all but one of the locations $X(v, 1), \ldots, X(v, |L_j|)$ are included in $F_{\mathcal{X}'}$. Since any two consecutive locations in such a sequence are unit distance away from each other, the cost of connecting any location in V to the set $F_{\mathcal{X}'}$ is either zero or one, i.e., $d(X, F_{\mathcal{X}'}) \in \{0, 1\}$ for all $X \in V = \mathcal{X}$.

For the rest of the proof, fix any set of clients $S(e) \in \mathcal{S}, e \in U$. The proof of Lemma 3 implies that the element e incurs congestion one under $\mathcal{X}'$. Hence, the element belongs to exactly one set in $\mathcal{X}'$, say X^*. Again, comparing the solution $\mathcal{X}'$ with the corresponding solution $F_{\mathcal{X}'}$, we infer that $S(e) - F_{\mathcal{X}'} = \{X^*\}$. In other words, every location in $S(e)$, except X^*, is present in the set $F_{\mathcal{X}'}$. The clients in such locations require zero cost for getting connected to $F_{\mathcal{X}'}$. Thus, the total cost of connecting the clients in $S(e)$ to the set $F_{\mathcal{X}'}$ is at most: $\sum_{X \in S(e)} d(X, F_{\mathcal{X}'}) = d(X^*, F_{\mathcal{X}'}) \leq 1$.

Thus, every set of clients in $\mathcal{S}$ requires unit cost for connecting to $F_{\mathcal{X}'}$. So the solution $F_{\mathcal{X}'}$ to the Robust k-Median instance indeed has objective one.   □

Lemma 6 (No-Instance). *If every labeling weakly satisfies at most a $2^{-\gamma r}$ fraction of the edges in the HGLC instance (G, π, r), for some constant γ then every solution to the Robust k-Median instance $(V, \mathcal{S}, d)$ has objective at least r.*

Proof. Fix any solution $F \subseteq V$ to the Robust k-Median instance $(V, \mathcal{S}, d)$, and let $\mathcal{X}'_F \subseteq \mathcal{X}$ denote the corresponding solution to the MCSP instance $(U, \mathcal{X}, t)$. By Lemma 4 there is some element $e \in U$ with congestion at least r under $\mathcal{X}'_F$. In other words, there are at least r sets $X_1, \ldots, X_r \in \mathcal{X}'_F$ that contain the element e. The locations corresponding to these sets are not picked by the solution F. Furthermore, the way the locations have been embedded on a line ensures that the distance between any location and its nearest neighbor is at least one. Hence, we have $d(X_i, F) \geq 1$ for all $i \in [r]$. Summing over these distances, the total cost of connecting the clients in $S(e)$ to F is at least $\sum_{i \in [r]} d(X_i, F) \geq r$. Thus, the solution F to the Robust k-Median instance has objective at least r.   □

Finally, applying Lemmas 5, 6, and an argument similar to the proof of Thm. 5, we get the following result.

Theorem 6. *The Robust k-Median problem $(V, \mathcal{S}, d)$ is $\Omega(\log m / \log \log m)$ hard to approximate even on line metrics, where $m = |\mathcal{S}|$, unless $\mathsf{NP} \subseteq \cap_{\delta > 0} \mathsf{DTIME}(2^{n^\delta})$.*

5   Conclusion and Future Work

We show a logarithmic lower bound for Robust k-median on the uniform and line metrics. However, the empirical results suggest that real-world instances are much easier, so it is interesting to see if realistic assumptions can be added to the problem in order to obtain constant approximation. For instance, one may assume that the diameter of each set S_i is small compared to the real diameter. This captures the "locality" of communities. Our hardness results do not apply in this case. Also, one may attack the problem from parameterized complexity's angle: Can we obtain an $O(1)$ approximation algorithm in time $g(k)\operatorname{poly}(n)$?

References

1. Anthony, B.M., Goyal, V., Gupta, A., Nagarajan, V.: A plant location guide for the unsure: Approximation algorithms for min-max location problems. Math. Oper. Res. 35(1), 79–101 (2010) (Also in SODA 2008)
2. Arora, S.: Polynomial time approximation schemes for euclidean traveling salesman and other geometric problems. J. ACM 45(5), 753–782 (1998)
3. Arora, S., Lund, C., Motwani, R., Sudan, M., Szegedy, M.: Proof verification and the hardness of approximation problems. J. ACM 45(3), 501–555 (1998)
4. Arya, V., Garg, N., Khandekar, R., Meyerson, A., Munagala, K., Pandit, V.: Local search heuristics for k-median and facility location problems. SIAM J. Comput. 33(3), 544–562 (2004)
5. Bansal, N., Khandekar, R., Könemann, J., Nagarajan, V., Peis, B.: On generalizations of network design problems with degree bounds. Math. Program. 141(1-2), 479–506 (2013)
6. Charikar, M., Guha, S.: Improved combinatorial algorithms for the facility location and k-median problems. In: FOCS, pp. 378–388. IEEE Computer Society (1999)
7. Charikar, M., Guha, S., Tardos, É., Shmoys, D.B.: A constant-factor approximation algorithm for the k-median problem. J. Comput. Syst. Sci. 65(1), 129–149 (2002)
8. Feige, U.: A threshold of ln n for approximating set cover. J. ACM 45(4), 634–652 (1998)
9. Hagerup, T., Rüb, C.: A guided tour of Chernoff bounds. Information Processing Letters 33(6), 305–308 (1990),
   http://www.sciencedirect.com/science/article/pii/002001909090214I
10. Kolliopoulos, S.G., Rao, S.: A nearly linear-time approximation scheme for the euclidean k-median problem. In: Nešetřil, J. (ed.) ESA 1999. LNCS, vol. 1643, pp. 378–389. Springer, Heidelberg (1999)
11. Li, S., Svensson, O.: Approximating k-median via pseudo-approximation. In: Boneh, D., Roughgarden, T., Feigenbaum, J. (eds.) STOC, pp. 901–910. ACM (2013)
12. Lin, J.H., Vitter, J.S.: Approximation algorithms for geometric median problems. Inf. Process. Lett. 44(5), 245–249 (1992)
13. Lund, C., Yannakakis, M.: On the hardness of approximating minimization problems. J. ACM 41(5), 960–981 (1994)
14. Raz, R.: A parallel repetition theorem. SIAM J. Comput. 27(3), 763–803 (1998)

Trees and Co-trees with Bounded Degrees
in Planar 3-connected Graphs*

Therese Biedl

David R. Cheriton School of Computer Science, University of Waterloo,
Waterloo, Ontario N2L 1A2, Canada

Abstract. This paper considers the conjecture by Grünbaum that every
planar 3-connected graph has a spanning tree T such that both T and
its co-tree have maximum degree at most 3. Here, the *co-tree* of T is the
spanning tree of the dual obtained by taking the duals of the non-tree
edges. While Grünbaum's conjecture remains open, we show that every
planar 3-connected graph has a spanning tree T such that both T and
its co-tree have maximum degree at most 5. It can be found in linear
time.

Keywords: Planar graph, canonical ordering, spanning tree, maximum
degree.

1 Introduction

In 1966, Barnette showed that every planar 3-connected graph has a spanning
tree with maximum degree at most 3 [2]. (In the following, a *k-tree* denotes a
tree with maximum degree at most k.) Since the dual of a 3-connected planar
graph is also 3-connected, the dual graph G^* also has a spanning 3-tree. In 1970,
Grünbaum [11] conjectured that there are spanning 3-trees in the graph and
its dual that are simultaneous in the sense of being tree and co-tree. For any
spanning tree T in a planar graph, define the *co-tree* to be the subgraph of the
dual graph formed by taking the dual edges of the edges in $G - T$. Since cuts
in planar graphs correspond to union of cycles in the dual graph, it is easy to
see that the co-tree is a spanning tree of G^*. Grünbaum conjecture is hence the
following:

Conjecture 1. [11] Every planar 3-connected graph has a spanning 3-tree for
which the co-tree is a spanning 3-tree of the dual graph.

This conjecture was still open in 2007 [12], and to our knowledge remains open
today. This paper proves a slightly weaker statement: Every planar 3-connected
graph has a spanning 5-tree for which the co-tree is a spanning 5-tree of the dual
graph.

---

* Supported by NSERC and the Ross and Muriel Cheriton Fellowship. Research initiated while participating at Dagstuhl seminar 13421.

R Ravi and I.L. Gørtz (Eds.): SWAT 2014, LNCS 8503, pp. 62–73, 2014.

Our approach is to read this spanning 5-tree from the *canonical ordering*, a decomposition that exists for all 3-connected planar graphs [14] and that has properties useful for many algorithms for graph drawing (see e.g. [6,14,16]) and other applications (see e.g. [13]). This will be formally defined in Section 2. There are readily available implementations for finding a canonical ordering (see for example [4,7]), and getting our tree from the canonical ordering is nearly trivial, so our trees not only can be found in linear time, but it would be very easy to implement the algorithm.

The canonical ordering is useful for Barnette's theorem as well. Barnette's proof [2] is constructive, but the algorithm that can be derived from the proof likely has quadratic run-time (he did not analyze it). With a slightly more structured proof and suitable data structures, it is possible to find the 3-tree in linear time [18]. But in fact, the 3-tree can be directly read from the canonical ordering. This was mentioned by Chrobak and Kant in their technical report [5], but no details were given as to why the degree-bound holds, and they did not include the result in their journal version [6]. We provide these details in Section 3, somewhat as a warm-up and because the key lemma will be needed later. Then we prove the weakened version of Grünbaum's conjecture in Section 4.

2   Background

Assume that $G = (V, E)$ is a planar graph, i.e., it can be drawn in the plane without crossing. Also assume that G is 3-connected, i.e., for any two vertices $\{u, v\}$ the graph resulting from deleting u and v is still connected. By Whitney's theorem a 3-connected planar graph G has a unique *combinatorial embedding*, i.e., in any planar drawing of G the circular clockwise order of edges around each vertex v is the same, up to reversal of all these orders. Given a planar drawing Γ, a *face* is a maximal connected region of $\mathbb{R}^2 - \Gamma$. The unbounded face is called the *outer-face*, all other faces are *interior faces*.

Define the *dual graph* G^* as follows. For every face f in G, add a vertex f^* to G^*. If e is an edge of G with incident faces f_ℓ and f_r, then add edge $e^* := (f_\ell^*, f_r^*)$ to G^*; e^* is called the *dual edge* of e.

De Fraysseix, Pach and Pollack [9] were the first to introduce a canonical ordering for triangulated planar graphs. Kant [14] generalized the canonical ordering to all 3-connected planar graphs.

Definition 1. *[14] A canonical ordering of a planar graph G with a fixed combinatorial embedding and outer-face is an ordered partition $V = V_1 \cup \cdots \cup V_K$ that satisfies the following:*

- *V_1 consists of two vertices v_1 and v_2 where v_2 is the counter-clockwise neighbour of v_1 on the outer-face.*
- *V_K is a singleton $\{v_n\}$ where v_n is the clockwise neighbour of v_1 on the outer-face.*
- *For each k in $2, \ldots, K$, the graph $G[V_1 \cup \cdots \cup V_k]$ induced by $V_1 \cup \cdots \cup V_k$ is 2-connected and contains edge (v_1, v_2) and all vertices of V_k on the outer-face.*

- For each k in $2, \ldots, K - 1$ one of the two following conditions hold:
  1. V_k contains a single vertex z that has at least two neighbours in $V_1 \cup \cdots \cup V_{k-1}$ and at least one neighbour in $V_{k+1} \cup \cdots \cup V_K$.
  2. V_k contains $\ell \geq 2$ vertices that induce a path $z_1 - z_2 - \cdots - z_\ell$, enumerated in clockwise order around the outer-face of $G[V_1 \cup \cdots \cup V_k]$. Vertices z_1 and z_ℓ have exactly one neighbour each in $V_1 \cup \cdots \cup V_{k-1}$, while $z_2, \ldots, z_{\ell-1}$ have no such neighbours. Each z_i, $1 \leq i \leq \ell$ has at least one neighbour in $V_{k+1} \cup \cdots \cup V_K$.

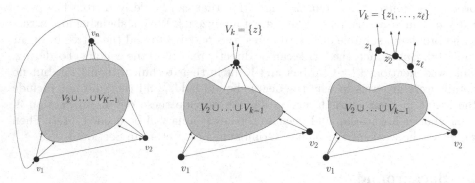

Fig. 1. The canonical ordering with its implied edge directions (defined in Section 2.1)

Figure 1 illustrates this definition. A set V_k, $k = 1, \ldots, K$ is called a *group* of the canonical ordering; a group with one vertex is a *singleton-group*, all other groups are *chain-groups*. Edges with both ends in the same group are called *intra-edges*, all others are *inter-edges*. Notice that when adding group V_k for $k \geq 2$, there exists some faces (one for a chain-group, one or more for a singleton-group) that are interior faces of $G[V_1 \cup \cdots \cup V_k]$ but were not interior faces of $G[V_1 \cup \cdots \cup V_{k-1}]$; these faces are called the *faces completed by group V_k*.

Kant [14] showed that any 3-connected planar graph has such a canonical ordering, even if the outer-face and the 2-path $v_n - v_1 - v_2$ on it to be used for the canonical ordering have been fixed. Furthermore, it can be found in linear time.

2.1   Edge Directions

Given a canonical ordering, one naturally directs inter-edges from the lower-indexed to the higher-indexed group. For proving Barnette's theorem, it will be useful to direct intra-edges as well as follows:

Definition 2. *Given a canonical ordering, enumerate the vertices as $v_1, \ldots, v_n$ as follows. Group V_1 consists of v_1 and v_2. For $2 \leq k \leq K$, let $s = |V_1| + \cdots + |V_{k-1}|$.*

- If V_k is a singleton group $\{z\}$, then set $v_{s+1} := z$.
- If V_k is a chain-group $z_1, \ldots, z_\ell$, then let v_h and v_i be the neighbours of z_1 and z_ℓ in $V_1 \cup \cdots \cup V_{k-1}$, respectively. If $h < i$, then set $v_{s+j} := z_j$ for $j = 1, \ldots, \ell$, else set $v_{s+j} := z_{\ell-j+1}$ for $j = 1, \ldots, \ell$.

Let $\text{idx}(v)$ be the index of vertex v in this enumeration. Consider edges to be directed from the lower-indexed to the higher-indexed vertex, with the exception of edge (v_1, v_n), which we direct $v_n \to v_1$. These edge directions are illustrated in Figure 1, with higher-indexed vertices drawn with larger y-coordinate.

Observation 1 *(1) Every vertex has, in its clockwise order of incident edges, a non-empty interval of incoming edges followed by a non-empty interval of outgoing edges.*
(2) The edges on each of the two faces incident to (v_1, v_n) form a directed cycle.
(3) For every face not incident to (v_1, v_n), the incident edges form two directed paths.

Proof. For purposes of this proof only, consider edge (v_1, v_n) to be directed $v_1 \to v_n$. Then by properties of the canonical ordering, every vertex except v_1 has at least one incoming edge, and every vertex except v_n has at least one outgoing inter-edge. Therefore this orientation is *bi-polar*: it is acyclic with a single source v_1 and a single sink v_n. It is known [19] that property (1) holds for all vertices $\neq v_1, v_n$ in a bi-polar orientation in a planar graph. Orienting edge (v_1, v_n) as $v_n \to v_1$ also makes (1) hold at v_1 and v_n, since they then have exactly one incoming/one outgoing edge.

In the bi-polar orientation, property (3) holds for any face f [19]. Orienting edge (v_1, v_n) as $v_n \to v_1$ will not change the property unless f is incident to (v_1, v_n). If f is incident to (v_1, v_n), then v_1 (as a source) was necessarily the beginning and v_n was necessarily the end of the two directed paths. Orienting edge (v_1, v_n) as $v_n \to v_1$ therefore turns the two directed paths into one directed cycle. So (2) holds.

Define the *first* and *last* outgoing edge to be the first and last edge in the clockwise order around v that is outgoing; this is well-defined by Observation 1(1). Also define the following:

Definition 3. *For any vertex v_i, $i \geq 2$, let the* parent-edge *be the incoming edge $v_h \to v_i$ for which h is maximized.*

If $e = v \to w$ is a directed edge, then w is the *head* of e, v is the *tail* of e, and v is a *predecessor* of w. The *left face* of e is the face to the left when walking from the tail to the head, and the *right face* of e is the other face incident to e. The predecessor at the parent-edge of w is called the *parent* of w. The *predecessors of group V_k* are all vertices that are predecessors of some vertex in V_k.

2.2   Edge Labels

To read trees from the canonical ordering, it helps to assign labels to the edges incident to a vertex. They are very similar to Felsner's triorientation derived from Schnyder labellings [8] (which in turn can easily be derived from the canonical ordering [15]), but differ slightly in the handling of intra-edges and edge (v_1, v_n).

Definition 4. *Given a canonical ordering, label the edge-vertex-incidences as follows:*

- *If V_k is a singleton-group $\{z\}$ with $2 \leq k \leq K$, then the first incoming edge of z (in clockwise order) is labelled SE, the last incoming edge of z (in clockwise order) is labelled SW, and all other incoming edges of z are labelled S.*
- *If V_k is a chain-group $\{z_1, \ldots, z_\ell\}$ with $2 \leq k < K$, then the incoming inter-edge of z_1 is labelled SW at z_1, the incoming inter-edge of z_ℓ is labelled SE at z_ℓ, and any intra-edge (z_i, z_{i+1}) is labelled E at z_i and W at z_{i+1}.*
- *Edge $v_1 \rightarrow v_2$ is labelled E at v_1 and W at v_2.*
- *Edge $v_n \rightarrow v_1$ is labelled S at v_1.*
- *If an inter-edge $v \rightarrow w$ is labelled SE / S / SW at w, then label it NW / N / NE at v.*

Call an edge an $\mathcal{L}$-edge (for $\mathcal{L} \in \{S, SW, W, NW, N, NE, E, SE\}$) if it is labelled $\mathcal{L}$ at one endpoint.

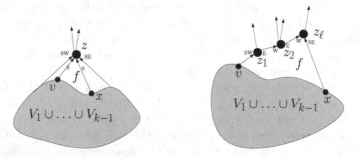

Fig. 2. The canonical ordering with its implied edge labelling. We also illustrate notations for the proof of Lemma 2.

See Figure 2 for an illustration of this labelling. The following properties are easily verified (see also [6] and [8] for similar results):

Lemma 1.   – *At each vertex there are, in clockwise order, some edges labelled S, at most one edge labelled SW, at most one edge labelled W, some edges labelled NW, at most one edge labelled N, some edges labelled NE, at most one edge labelled E, and at most one edge labelled SE.*
- *An edge is an intra-edge if and only if it is labelled E at one endpoint and W at the other.*
- *No vertex has an edge labelled W and an edge labelled SW.*
- *No vertex has an edge labelled E and an edge labelled SE.*

3   Barnette's Theorem via the Canonical Ordering

We now show that Barnette's theorem has a proof where the tree can be read directly from a canonical ordering.

Theorem 1. *Let G be a planar graph with a canonical ordering. Then the parent-edges forms a spanning tree of maximum degree 3.*

Proof. Let T be the set of parent edges. First note that each vertex $v_2, \ldots, v_n$ has exactly one incoming edge in T, and there is no directed cycle since (v_1, v_n) is not a parent-edge and therefore edges are directed according to indices. So T is indeed a spanning tree. To see the bound on the maximum degree, the following lemma suffices:

Lemma 2. *Assume $v \to w$ is a parent-edge of w. Then either $v \to w$ is the first outgoing edge at v and labelled W or NW or N at v, or $v \to w$ is the last outgoing edge at v and labelled E or NE or N at v.*

Proof. $w = v_1$ is impossible since v_1 has no parent. If $w = v_2$, then its parent-edge $v_1 \to v_2$ is the last outgoing edge of v_1 and labelled W, so the claim holds. Now consider $w = v_i$ for some $i \geq 3$, which means that w belongs to some group V_k for $k \geq 2$. There are two cases:

- V_k is a chain-group $z_1 - \cdots - z_\ell$, which implies $k < K$. Assume that the chain is directed $z_1 \to \cdots \to z_\ell$; the other case is symmetric. Refer to Figure 2(right). Note that z_i is the parent of z_{i+1} for $1 \leq i < \ell$, and $z_i \to z_{i+1}$ is the last outgoing edge of z_i and labelled E, so the claim holds for $w \in \{z_2, \ldots, z_\ell\}$.
  Consider $w = z_1$. The parent v of z_1 is the predecessor of V_k adjacent to z_1. Let x be the other predecessor of V_k (it is adjacent to z_ℓ). The direction of the chain implies $\mathrm{idx}(v) > \mathrm{idx}(x)$. Let f be the face completed by V_k and observe that it does not contain (v_1, v_n). By Observation 1(3) the boundary of f consists of two directed paths, which both end at z_ℓ. The vertex where these two paths begin cannot be v, otherwise there would be a directed path from v to x and therefore $\mathrm{idx}(x) > \mathrm{idx}(v)$. So v has at least one incoming edge on face f, and hence $v \to z_1$ is its last outgoing edge. Also, this edge is labelled SW at z_1, hence NE at v, as desired.
- V_k is a singleton-group $\{z\}$ with $z = w$. Refer to Figure 2(left). Let $x \to w$ be an incoming edge of w that comes before or after $v \to w$ in the clockwise order of edges at w. Such an edge must exist since w has at least two incoming edges (this holds for $w = v_n$ by 3-connectivity). Assume that the clockwise order at w contains $x \to w$ followed by $v \to w$; the other case is similar.
  Let f be the face incident to edges $v \to w$ and $x \to w$. By construction f is not incident to (v_1, v_n), and by Observation 1(3) the boundary of f consists of two directed paths, which both end at w. The vertex where these two paths begin cannot be v, otherwise there would be a directed path from v to x, hence $\mathrm{idx}(x) > \mathrm{idx}(v)$ contradicting the definition of parent-edge $v \to w$.

So v has at least one incoming edge on face f. hence $v \to w$ is the last outgoing edge at v. Furthermore, $v \to w$ cannot be labelled SE at w (since $x \to w$ comes clockwise before it), so it is labelled SW or S at w, hence NE or N at v as desired.

So in T, every vertex is incident to at most three edges: the parent-edge, the first outgoing edge, and the last outgoing edge. This finishes the proof of Theorem 1.

In a later paper [3], Barnette strengthened his own theorem to show that in addition one can pick one vertex and require that it has degree 1 in the spanning tree. Using the canonical ordering allows us to strengthen this result even further: All vertices on one face have degree at most 2, and two of them can be required to have degree 1.

Corollary 1. *Let G be a planar graph with vertices u, w on a face f, and assume that the graph that results from adding edge (u, w) to G is 3-connected. Then G has a spanning tree T with maximum degree 3 such that $\deg_T(u) = 1 = \deg_T(w)$, and all other vertex x on face f have $\deg_T(x) \leq 2$.*

Proof. Let $G^+ = G \cup (u, w)$ and find a canonical ordering of G^+ with $u = v_1$ and $w = v_n$. Let T be the spanning 3-tree of G^+ obtained from the parent-edges; this will satisfy all properties.

Observe that (v_1, v_n) is not a parent-edge, so T is a spanning tree of G as well. Let f_ℓ and f_r be the left and right face of $v_n \to v_1$. Both faces are completed by $V_K = \{v_n\}$. It follows that any edge on f_ℓ (except $v_n \to v_1$) is a SW-edge, because only such edges may have a not-yet-completed face on their left. Therefore for any vertex $x \neq v_n$ on f_ℓ the first outgoing edge is labelled NE and by Lemma 2 it does not belong to T. So $\deg_T(x) \leq 2$ for all $x \in f_\ell$. Similarly one shows that $\deg_T(x) \leq 2$ for all $x \in f_r$. Finally, $\deg_T(v_n) = 1$ since v_n has no outgoing parent-edges, and $\deg_T(v_1) = 1$ since all vertices other than v_2 have higher-indexed predecessors.

4   On Grünbaum's Conjecture

One can easily find an example of a graph where the 3-tree from Theorem 1 yields a co-tree with unbounded degree. So unfortunately the proof of Theorem 1 does not help to solve Grünbaum's conjecture. In this section, we show that every planar 3-connected graph G has a spanning tree T such that both T and its co-tree T^* are 5-trees. Tree T will again be read from the canonical ordering, but with a different approach. Assume throughout this section that a canonical order of G has been fixed.

A crucial insight is that a canonical ordering implies a *dual canonical ordering*, i.e., a canonical ordering of the dual graph G^*. This was shown, for example, by Badent et al. [1]. An inspection of the construction shows also that the edge labels of G and G^* relate as follows:

Theorem 2. *For any canonical ordering of a 3-connected planar graph G, there exists a canonical ordering of the dual graph G^* such that the following hold:*

- *The dual of any intra-edge of G is a S-edge in G^*.*
- *The dual of any S-edge of G is an intra-edge in G^*.*
- *The dual of any SW-edge e of G is a SE-edge in G^*, and directed from the left face of e to the right face of e.*
- *The dual of any SE-edge e of G is a SW-edge in G^*, and directed from the right face of e to the left face of e.*

Now define a subgraph of G from the labels of its edges. If a vertex has NW-edges, then let the last one (in clockwise order around v) be the *NNW-edge*. Similarly define the *NNE-edge* as the first NE-edge in clockwise order.

Definition 5. *Presume a canonical ordering of a planar graph G is fixed. An edge e of G is called an H-edge if it satisfies one of the following:*

(H1) e is an intra-edge,
(H2) e is the NNW-edge of its tail,
(H3) e is the NNE-edge of its tail,
(H4) e is the parent-edge of its head and the N-edge of its tail.

The graph formed by the H-edges of G is denoted $H(G)$.

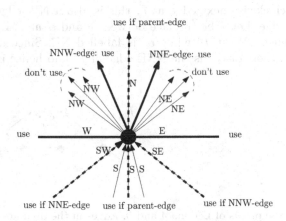

Fig. 3. Illustration of H-edges. Solid edges are H-edges; thick dashed edges may be H-edges depending on the other endpoint.

Lemma 3. *Any vertex v has at most 5 incident H-edges.*

Proof. Observe first that v has at most two incident H-edges that are outgoing inter-edges. For no such edge is added under rule (H1). Rules (H2), (H3) and (H4) add at most one such H-edge each. But if rule (H4) adds edge e, then e is the N-edge of v. By Lemma 2 it also is the first or last outgoing edge of v.

Therefore if rule (H4) applies then v has no NW-edge or no NE-edge, and so one of rules (H2) and (H3) does not apply.

Next consider the group of edges at v consisting of the intra-edges at v, and the SW-edge and SE-edge. Clearly this group has at most four edges, but actually they are only two edges by Lemma 1. So v has at most two incident H-edges in this group.

All edges at v that are neither outgoing inter-edges nor in the above group are incoming edges labelled S. Only one such edge (namely, the parent-edge of v) can be an H-edge. So v has at most 5 incident H-edges.

Let $H(G^*)$ be the graph formed by the H-edges of G^*, using the dual canonical ordering. $H(G^*)$ also has maximum degree 5. Neither $H(G)$ nor $H(G^*)$ is necessarily a tree, and it is not even obvious that they are connected. The plan is now to find a spanning tree of $H(G)$ for which the co-tree belongs to $H(G^*)$. Two lemmas are needed for this.

Lemma 4. *Let e be an edge in $G - H(G)$. Then the dual edge e^* of e belongs to $H(G^*)$.*

Proof. If e is a N-edge, then its dual is an intra-edge and hence belongs to $H(G^*)$. Edge e cannot be a NNW-edge or NNE-edge or intra-edge since it is not in $H(G)$. The remaining case is hence that e is a NW-edge of its tail v, but not the NNW-edge. (The case of a NE-edge that is not the NNE-edge is similar.) Figure 4 (left) illustrates this case.

Let e' be the clockwise next edge at v; this is also a NW-edge of v since e is not the NNW-edge. Let f be the face between e and e' at v. By Theorem 2, edge $(e')^*$ is labelled SW at f^* while e^* is labelled NE. Since e^* and e'^* are consecutive at f^*, therefore e^* is the NNE-edge of f^* and hence in $H(G^*)$.

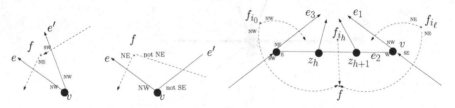

Fig. 4. For the proofs of Lemma 4 and 5. Edges in the dual are dashed.

Lemma 5. *Let C be a cycle of edges in $H(G)$. Then there exists an edge $e \in C$ such that e^* belongs to $H(G^*)$.*

Proof. There are three cases where e can be found easily; the bulk of the proof deals with the more complicated situation where none of them applies.

Case (C1): C contains a N-edge e. Then e^* is an intra-edge and belongs to $H(G^*)$ by rule (H1).

Case (C2): C contains a NW-edge e such that the clockwise next edge e' at the tail v of e is not a SE-edge. This case is illustrated in Figure 4(middle). Let f be the face between e and e'. Since e is a NW-edge, e^* is a NE-edge. Since e' is not a SE-edge, $(e')^*$ is not a NE-edge. So e^* is the NNE-edge of f^* and belongs to $H(G^*)$ by rule (H2).

Case (C3): C contains a NE-edge e such that the counter-clockwise next edge at e's tail is not a SW-edge. With a symmetric argument to (C2) one then shows that e^* is a NNW-edge and belongs to $H(G^*)$ by rule (H3).

Case (C4): None of the above cases applies. Since intra-edges form paths, cycle C must contain some inter-edges. Let e_1 be the inter-edge of C that minimizes the index of its tail v. e_1 is not a N-edge, otherwise (C1) would apply. So e_1 is either a NW-edge or a NE-edge of v. By definition of H-edges, therefore e_1 is the NNW-edge or the NNE-edge of v. Assume the former, the other case is symmetric. We will show that the situation is as in Figure 4(right).

Let e_2 be the other edge in C incident to v. Edge e_2 cannot be a N-edge at v, otherwise (C1) would apply. It also cannot be a NE-edge or E-edge at v, otherwise the clockwise edge after e_1 at v is not a SE-edge and (C2) would apply. Edge e_2 also cannot be a SE-edge or S-edge or SW-edge at v, otherwise it would be an incoming inter-edge and its tail would have a smaller index than v, contradicting the choice of e_1. Also e_2 cannot be a NW-edge at v, because the NNW-edge e_1 is the only NW-edge that is an H-edge at v. Thus edge e_2 must be an intra-edge labelled W at v.

Let $V_k = \{z_1, \ldots, z_\ell\}$ be the chain-group containing edge e_2. Notice that v has no E-edge (otherwise (C2) would apply), so $v = z_\ell$. Let a be the minimal index such that that path $z_a - z_{a+1} - \cdots - z_\ell$ is part of C. Let e_3 be the edge incident to z_a that is on C and different from (z_a, z_{a+1}). Observe that e_3 is an inter-edge, for if it were an intra-edge then its other endpoint would be z_{a-1}, contradicting the definition of a. Also observe that e_3 cannot be incoming at z_a, for otherwise the index of its tail would be smaller than all indices in V_k, and in particular smaller than the index of $v = z_\ell$; this contradicts the choice of e_1.

So e_3 is an outgoing inter-edge at z_a. If e_3 were a N-edge then (C1) would apply. If it were a NW-edge, then (due to E-edge (z_a, z_{a+1})) (C2) would apply. So e_3 is a NE-edge. Since it is an H-edge, it is the NNE-edge of z_a. Since (C3) does not apply, z_a cannot have a W-edge, which shows that $a = 1$.

Let f be the face completed by the chain-group V_k, and let $f^*_{i_0}, \ldots, f^*_{i_\ell}$ be the predecessors of f^* in the dual canonical order. By the correspondence of edge-label of Theorem 2, f_{i_0} shares the SW-edge of z_1 with f, face f_{i_h} (for $1 \leq h < \ell$) shares (z_i, z_{i+1}) with f, and f_{i_ℓ} shares the SE-edge of z_ℓ with f.

Let $f^*_{i_p} \to f^*$ be the parent-edge of f^* in the dual canonical ordering. Observe that $p \neq 0$. For edge $(f^*_{i_0}, f^*)$ is a NW-edge at $f^*_{i_0}$, as is e^*_3. Thus $(f^*_{i_0}, f^*)$ is not the first outgoing edge at $f^*_{i_0}$, and by Lemma 2 hence not a parent-edge. Likewise one shows $p \neq \ell$. So $1 \leq p < \ell$ and the parent-edge of f^* is a N-edge. By rule (H4) the parent-edge of f^* is in $H(G^*)$. Setting $e = (z_p, z_{p+1})$ yields the result.

4.1 Putting It All Together

Theorem 3. *Every planar 3-connected graph G has a spanning tree T such that both T and its co-tree have maximum degree at most 5. T can be found in linear time.*

Proof. First observe that $H(G)$ is connected. For if it were disconnected, then there would exist a non-trivial cut with all cut-edges in $G - H(G)$. By Lemma 4 the duals of the cut-edges belong to $H(G^*)$. Since cuts in a planar graph correspond to unions of cycles in the dual, hence the duals of the cut-edges contain a non-empty cycle C of edges in $H(G^*)$. By Lemma 5 one edge of C has its dual in $H(G)$, contradicting the definition of the cut.

Let H_0 be all those edges in $H(G)$ for which the dual edge does not belong to $H(G^*)$. By Lemma 5 H_0 contains no cycle, so it is a forest. Assign a weight of 0 to all edges in H_0, a weight of 1 to all edges in $H(G) - H_0$, and a weight of ∞ to all edges in $G - H(G)$. Then compute a minimum spanning tree T of G. Since H_0 is a forest, all its edges are in T. Since $H(G)$ is connected, no edge in $G - H(G)$ belongs to T. So T is a subgraph of $H(G)$ and has maximum degree at most 5. All edges in the co-tree T^* of T are duals of edges that are in $G - H_0$, and by definition of H_0 and Lemma 4 these edges belong to $H(G^*)$. So T^* is a subgraph of $H(G^*)$ and has maximum degree at most 5.

It remains to analyze the time complexity. One can compute a canonical ordering in linear time, and from it, obtain the dual canonical ordering and the edge-sets $H(G)$ and $H(G^*)$ in linear time. The bottleneck is hence the computation of the minimum spanning tree. But there are only 3 different weights, and using a bucket-structure, rather than a priority queue, in Prim's algorithm, we can find the next vertex to add to the tree in constant time. Hence the minimum spanning tree can be found in linear time.

5 Conclusion

In this paper, we showed that every planar 3-connected graph has a spanning tree of maximum degree 5 such that the co-tree also has a spanning tree of maximum degree 5. This is a first step towards proving Grünbaum's conjecture.

Barnette's theorem has as easy consequence that every planar 3-connected graph has a *3-walk*: a walk that visits every vertex at most 3 times. But in fact, one can show a stronger statement: Every planar 3-connected graph has a 2-walk [10]. The results in the paper imply similar results: every planar 3-connected graph has a walk that alternates between faces and incident vertices and visits every vertex and every face at least once and at most 5 times. (Here by "visit v" we mean that the walk alternates between v and incident faces, and similarly for "visiting f".) An interesting open problem is, as a first step towards Grünbaum's conjecture, to try to reduce this "5" to a smaller number.

A second open problem concerns generalizations to other surfaces. Barnette's theorem generalizes to 3-connected graphs on the projective plane, torus or the Klein bottle [3]; see also a recent survey [17] for many related results. For what

k can one find a spanning k-tree in, say, a toroidal 3-connected graph such that the duals of the non-tree edges form a graph of maximum degree at most k?

References

1. Badent, M., Baur, M., Brandes, U., Cornelsen, S.: Leftist canonical ordering. In: Eppstein, D., Gansner, E.R. (eds.) GD 2009. LNCS, vol. 5849, pp. 159–170. Springer, Heidelberg (2010)
2. Barnette, D.W.: Trees in polyhedral graphs. Canad. J. Math. 18, 731–736 (1966)
3. Barnette, D.W.: 3-trees in polyhedral maps. Israel Journal of Mathematics 79, 251–256 (1992)
4. boost C + + libraries on planar graphs (2013), http://www.boost.org/ (last accessed December 2, 2013)
5. Chrobak, M., Kant, G.: Convex grid drawings of 3-connected planar graphs. Technical Report RUU-CS-93-45, Rijksuniversiteit Utrecht (1993)
6. Chrobak, M., Kant, G.: Convex grid drawings of 3-connected planar graphs. Internat. J. Comput. Gcom. Appl. 7(3), 211 223 (1997)
7. de Fraysseix, H., Ossona de Mendez, P.: P.I.G.A.L.E., Public Implementation of Graph Algorithm Libeary and Editor (2013), http://pigale.sourceforge.net/ (last accessed December 2, 2013)
8. Felsner, S.: Convex drawings of planar graphs and the order dimension of 3-polytopes. Order 18, 19–37 (2001)
9. de Fraysseix, H., Pach, J., Pollack, R.: How to draw a planar graph on a grid. Combinatorica 10, 41–51 (1990)
10. Gao, Z., Richter, R.B.: 2-walks in circuit graphs. J. Comb. Theory, Ser. B 62(2), 259–267 (1994)
11. Grünbaum, B.: Polytopes, graphs, and complexes. Bull. Amer. Math. Soc. 76, 1131–1201 (1970)
12. Grünbaum, B.: Graphs of polyhedra; polyhedra as graphs. Discrete Mathematics 307(3-5), 445–463 (2007)
13. He, X., Kao, M.-Y., Lu, H.-I.: Linear-time succinct encodings of planar graphs via canonical orderings. SIAM J. Discrete Math. 12(3), 317–325 (1999)
14. Kant, G.: Drawing planar graphs using the canonical ordering. Algorithmica 16, 4–32 (1996)
15. Miura, K., Azuma, M., Nishizeki, T.: Canonical decomposition, realizer, Schnyder labeling and orderly spanning trees of plane graphs. Int. J. Found. Comput. Sci. 16(1), 117–141 (2005)
16. Nishizeki, T., Rahman, M.S.: Planar Graph Drawing. Lecture Notes Series on Computing, vol. 12. World Scientific (2004)
17. Ozeki, K., Yamashita, T.: Spanning trees: A survey. Graphs and Combinatorics 27(1), 1–26 (2011)
18. Strothmann, W.-B.: Bounded-degree spanning trees. PhD thesis, FB Mathematik/Informatik und Heinz-Nixdorf Institute, Universität-Gesamthochschule Paderborn (1997)
19. Tamassia, R., Tollis, I.: A unified approach to visibility representations of planar graphs. Discrete Computational Geometry 1, 321–341 (1986)

Approximating the Revenue Maximization Problem with Sharp Demands*

Vittorio Bilò[1], Michele Flammini[2,3], and Gianpiero Monaco[2]

[1] Department of Mathematics and Physics "Ennio De Giorgi", University of Salento
Provinciale Lecce-Arnesano, P.O. Box 193, 73100 Lecce, Italy
vittorio.bilo@unisalento.it
[2] Department of Information Engineering Computer Science and Mathematics,
University of L'Aquila, Via Vetoio, Coppito, 67100 L'Aquila, Italy
{flammini,gianpiero.monaco}@di.univaq.it
[3] Gran Sasso Science Institute, L'Aquila, Italy

Abstract. We consider the revenue maximization problem with sharp multi-demand, in which m indivisible items have to be sold to n potential buyers. Each buyer i is interested in getting exactly d_i items, and each item j gives a benefit v_{ij} to buyer i. We distinguish between unrelated and related valuations. In the former case, the benefit v_{ij} is completely arbitrary, while, in the latter, each item j has a quality q_j, each buyer i has a value v_i and the benefit v_{ij} is defined as the product $v_i q_j$. The problem asks to determine a price for each item and an allocation of bundles of items to buyers with the aim of maximizing the total revenue, that is, the sum of the prices of all the sold items. The allocation must be envy-free, that is, each buyer must be happy with her assigned bundle and cannot improve her utility. We first prove that, for related valuations, the problem cannot be approximated to a factor $O(m^{1-\epsilon})$, for any $\epsilon > 0$, unless $\mathsf{P} = \mathsf{NP}$ and that such result is asymptotically tight. In fact we provide a simple m-approximation algorithm even for unrelated valuations. We then focus on an interesting subclass of "proper" instances, that do not contain buyers a priori known not being able to receive any item. For such instances, we design an interesting 2-approximation algorithm and show that no $(2 - \epsilon)$-approximation is possible for any $0 < \epsilon \leq 1$, unless $\mathsf{P} = \mathsf{NP}$. We observe that it is possible to efficiently check if an instance is proper, and if discarding useless buyers is allowed, an instance can be made proper in polynomial time, without worsening the value of its optimal solution.

1 Introduction

A major decisional process in many business activities concerns whom to sell products (or services) to and at what price, with the goal of maximizing the

---

* This work was partially supported by the PRIN 2010–2011 research project ARS TechnoMedia: "Algorithmics for Social Technological Networks" funded by the Italian Ministry of University.

R. Ravi and I.L. Gørtz (Eds.): SWAT 2014, LNCS 8503, pp. 74–85, 2014.
© Springer International Publishing Switzerland 2014

total revenue. On the other hand, consumers would like to buy at the best possible prices and experience fair sale criteria.

In this work, we address such a problem from a computational point of view, considering a two-sided market in which the supply side consists of m indivisible items and the demand one is populated by n potential buyers (in the following also called consumers or customers), where each buyer i has a demand d_i (the number of items that i requests) and valuations v_{ij} representing the benefit i gets when owing item j. As several papers on this topic (see for instance [19,9,16]), we assume that, by means of market research or interaction with the consumers, the seller knows each customer's valuation for each item. The seller sets up a price p_j for each item j and assigns (i.e., sells) bundle of items to buyers with the aim of maximizing her revenue, that is the sum of the prices of all the sold items. When a consumer is assigned (i.e., buys) a set of items, her utility is the difference between the total valuation of the items she gets (valuations being additive) and the purchase price. The sets of the sold items, the purchasing customers and their purchase prices are completely determined by the allocation of bundles of items to customers unilaterally decided by the seller. Nevertheless, we require such an allocation to meet two basic fairness constraints: (i) each customer i is allocated at most one bundle not exceeding her demand d_i and providing her a non-negative utility, otherwise she would not buy the bundle; (ii), the allocation must be envy-free [25], i.e., each customer i does not prefer any subset of d_i items different from the bundle she is assigned. Notice that in our scenario a trivial envy-free solution always exists that lets $p_j = \infty$ for each item j and does not assign any item to any buyer.

Many papers (see the *Related Work* section for a detailed reference list) considered the *unit demand case* in which $d_i - 1$ for each consumer i. Arguably, the *multi-demand case*, where $d_i \geq 1$ for each consumer i, is more general and finds much more applicability. To this aim, we can identify two main multi-demand schemes. The first one is the *relaxed multi-demand model*, where each buyer i requests at most $d_i \geq 1$ items, and the second one is the *sharp multi-demand model*, where each buyer i requests exactly $d_i \geq 1$ items and, therefore, a bundle of size less than d_i has no value for buyer i. For relaxed multi-demand models, a standard technique can reduce the problem to the unit demand case in the following way: each buyer i with demand d_i is replaced by d_i copies of buyer i, each requesting a single item. However, such a trick does not apply to the sharp demand model. Moreover, as also pointed out in [9], the sharp multi-demand model exhibits a property that unit demand and relaxed multi-demand ones do not posses. In fact, while in the latter model any envy-free pricing is such that the price p_j is always at most the value of v_{ij}, in the sharp demand model, a buyer i may pay an item j more than her own valuation for that item, i.e., $p_j > v_{ij}$ and compensate her loss with profits from the other items she gets (see section 3.1 of [9]). Such a property, also called *overpricing*, clearly adds an extra challenge to find an optimal revenue.

The sharp demand model is quite natural in several settings. Consider, for instance, a scenario in which a public organization has the need of buying a fixed quantity of items in order to reach a specific purpose (i.e. locations for offices, cars for services, bandwidth, storage, or whatever else), where each item might have a different valuation for the organization because of its size, reliability, position, etc. Yet, suppose a user wants to store on a remote server a file of a given size s and there is a memory storage vendor that sells slots of fixed size c, where each cell might have different features depending on the server location and speed and then yielding different valuations for the user. In this case, a number of items smaller than $\lceil \frac{s}{c} \rceil$ has no value for the user. Similar scenarios also apply to cloud computing. In [9], the authors used the following applications for the sharp multi-demand model. In TV (or radio) advertising [21], advertisers may request different lengths of advertising slots for their ads programs. In banner (or newspaper) advertising, advertisers may request different sizes or areas for their displayed ads, which may be decomposed into a number of base units. Also, consider a scenario in which advertisers choose to display their advertisement using medias (video, audio, animation) [2,22] that would usually need a fixed number of positions, while text ads would need only one position each. An example of formulation sponsored search using sharp multi-demands can be found in [14]. Other results concerning the sharp multi-demand model in the Bayesian setting can be found in [13].

Related Work. Pricing problems have been intensively studied in the literature, see e.g., [23,1,18] just to cite a few, both in the case in which the consumers' preferences are unknown (mechanism design [24]) and in the case of full information that we consider in this paper. In fact, our interest here is in maximizing the seller's profit assuming that consumers' preferences are gathered through market research or conjoint analysis [19,9,16]. From an algorithmic point of view, [19] is the first paper dealing with the problem of computing the envy-free pricing of maximum revenue. The authors considered the limited supply unit demand and the unlimited supply single minded cases for which they gave $O(\log n)$ and $O(\log n + \log m)$ approximation algorithms, respectively. An $\Omega(\log^{\epsilon} n)$ hardness result has been showed in [4] for the unit demand case, and a tight hardness results of $\Omega(\log^{1-\epsilon} n)$ recently appeared in [5,6,7]. For the single minded case an $\Omega(\log n)$ hardness result can be found in [11]. The subcase in which every buyer positively evaluates at most two items has been studied in [8]. The authors proved that the problem is solvable in polynomial time and it becomes NP-hard if some buyer gets interested in at least three items. For the multi-demand model, Chen et. al. [10] gave an $O(\log D)$ approximation algorithm when there is a metric space behind all items, where D is the maximum demand, and Briest [4] showed that the problem is hard to approximate within a ratio of $O(n^{\varepsilon})$ for some $\varepsilon > 0$.

To the best of our knowledge, [9] is the first paper explicitly dealing with the sharp multi-demand model. The authors considered a particular valuation scheme (also used in [15] for keywords advertising scenarios) where each item j has a parameter q_j measuring the quality of the item and each buyer i has a value v_i representing the benefit that i gets when owing an item of unit quality. Thus, the benefit that i obtains from item j is given by $v_i q_j$. For such a problem, the authors proved that computing the envy-free pricing of maximum revenue is

NP-hard. Moreover, they showed that if the demand of each buyer is bounded by a constant, the problem becomes solvable in polynomial time. We remark that this valuation scheme is a special case of the one in which the valuations v_{ij} are completely arbitrary and given as an input of the problem. Throughout the paper, we will refer to the former scheme as to *related valuations* and to the latter as to *unrelated valuations*. Recently [12] considered the sharp multi-demand model with the additional constraint in which items are arranged as a sequence and buyers want items that are consecutive in the sequence.

Finally [16] studied the pricing problem in the case in which buyers have a budget, but no demand constraints. The authors considered a special case of related valuations in which all qualities are equal to 1 (i.e., $q_j = 1$ for each item j). They proved that the problem is still NP-hard and provided a 2-approximation algorithm. Such algorithm assigns the same price to all the sold items. Many of the papers listed above deal with the case of limited supply. Another stream of research considers unlimited supply, that is, the scenario in which each item j exists in e_j copies and it is explicitly allowed that $e_j = \infty$. The limited supply setting seems generally more difficult than the unlimited supply one. In this paper we consider the limited supply setting. Interesting results for unlimited supply can be found in [19,11].

Our Contribution. We consider the revenue maximization problem with sharp multi-demand and limited supply. We first prove that, for related valuations, the problem cannot be approximated to a factor $O(m^{1-\epsilon})$, for any $\epsilon > 0$, unless P = NP and that such result is asymptotically tight. In fact we provide a simple m-approximation algorithm even for unrelated valuations. Our inapproximability proof relies on the presence of some buyers not being able to receive any bundle of items in any envy-free outcome. Thus, it becomes natural to ask oneself what happens for instances of the problem, that we call *proper*, where no such pathological buyers exist. For proper instances, we design an interesting 2-approximation algorithm and show that the problem cannot be approximated to a factor $2 - \epsilon$ for any $0 < \epsilon \le 1$ unless P = NP. Therefore, also in this subcase, our results are tight. We remark that it is possible to efficiently decide whether an instance is proper. Moreover, if discarding useless buyers is allowed, an instance can be made proper in polynomial time, without worsening the value of its optimal solution.

Paper Organization. Next section contains the necessary definitions and some preliminary results, while Section 3 defines a useful pricing scheme of fundamental importance for our analysis. Finally, Sections 4 and 5 contains our results for general and proper instances, respectively. Due to space constraints, some details and almost all proofs have been removed and can be found in the full version of the paper [3].

2   Model and Preliminaries

In the *Revenue Maximization Problem with Sharp Multi-Demands* (RMPSD) investigated in this paper, we are given a *market* made up of a set $M = \{1, 2, \ldots, m\}$

of *items* and a set $N = \{1, 2, \ldots, n\}$ of *buyers*. Each item $j \in M$ has unit supply (i.e., only one available copy). We consider both unrelated and related valuations. In the former each buyers i has valuations v_{ij} representing the benefit i gets when owing item j. In the latter each item is characterized by a *quality* (or desirability) $q_j > 0$, while each buyer $i \in N$ has a *value* $v_i > 0$, measuring the benefit that she gets when receiving a unit of quality, thus, the valuation that buyer i has for item j is $v_{ij} = v_i q_j$. We notice that related is a special case of unrelated valuations. Throughout the paper, when not explicitly indicated, we refer to related valuations. Finally each buyer i has a *demand* $d_i \in \mathbb{Z}^+$, which specifies the exact number of items she wants to get. In the following we assume items and bidders ordered in non-increasing order, that is, $v_i \geq v_{i'}$ for $i < i'$ and $q_j \geq q_{j'}$ for $j < j'$.

An *allocation vector* is an n-tuple $\mathbf{X} = (X_1, \ldots, X_n)$, where $X_i \subseteq M$, with $|X_i| \in \{0, d_i\}$, $\sum_{i \in N} |X_i| \leq m$ and $X_i \cap X_{i'} = \emptyset$ for each $i \neq i' \in N$, is the set of items sold to buyer i. A *price vector* is an m-tuple $\mathbf{p} = (p_1, \ldots, p_m)$, where $p_j > 0$ is the price of item j. An *outcome* of the market is a pair $(\mathbf{X}, \mathbf{p})$.

Given an outcome $(\mathbf{X}, \mathbf{p})$, we denote with $u_{ij}(\mathbf{p}) = v_{ij} - p_j$ the utility that buyer i gets when she is sold item j and with $u_i(\mathbf{X}, \mathbf{p}) = \sum_{j \in X_i} u_{ij}(\mathbf{p})$ the overall utility of buyer i in $(\mathbf{X}, \mathbf{p})$. When the outcome (or the price vector) is clear from the context, we simply write u_i and u_{ij}. An outcome $(\mathbf{X}, \mathbf{p})$ is *feasible* if $u_i \geq 0$ for each $i \in N$.

We denote with $M(\mathbf{X}) = \bigcup_{i \in N} X_i$ the set of items sold to some buyer according to the allocation vector $\mathbf{X}$. We say that a buyer i is a *winner* if $X_i \neq \emptyset$ and we denote with $W(\mathbf{X})$ the set of all the winners in $\mathbf{X}$. For an item $j \in M(\mathbf{X})$, we denote with $b_{\mathbf{X}}(j)$ the buyer $i \in W(\mathbf{X})$ such that $j \in X_i$, while, for an item $j \notin M(\mathbf{X})$, we define $b_{\mathbf{X}}(j) = 0$. Moreover, for a winner $i \in W(\mathbf{X})$, we denote with $f_{\mathbf{X}}(i) = \min\{j \in M : j \in X_i\}$ the best-quality item in X_i. Also in this case, when the allocation vector is clear from the context, we simply write $b(j)$ and $f(i)$. Finally, we denote with $\beta(\mathbf{X}) = \max\{i \in N : i \in W(\mathbf{X})\}$ the maximum index of a winner in $\mathbf{X}$. An allocation vector $\mathbf{X}$ is *monotone* if $\min_{j \in X_i}\{q_j\} \geq q_{f(i')}$ for each $i, i' \in W(\mathbf{X})$ with $v_i > v_{i'}$, that is, all the items of i are of quality greater of equal to the one of all the items of i'.

Definition 1. *A feasible outcome* $(\mathbf{X}, \mathbf{p})$ *is an envy-free outcome if, for each buyer* $i \in N$, $u_i \geq \sum_{j \in T} u_{ij}$ *for each* $T \subseteq M$ *of cardinality* d_i.

Notice that, by definition, an outcome $(\mathbf{X}, \mathbf{p})$ is envy-free if and only if the following three conditions holds: (i) $u_i \geq 0$ for each $i \in N$, (ii) $u_{ij} \geq u_{ij'}$ for each $i \in W(\mathbf{X})$, $j \in X_i$ and $j' \notin X_i$, (iii) $\sum_{j \in T} u_{ij} \leq 0$ for each $i \notin W(\mathbf{X})$ and $T \subseteq M$ of cardinality d_i. Note also that, as already remarked, envy-free solutions always exist, since the outcome $(\mathbf{X}, \mathbf{p})$ such that $X_i = \emptyset$ for each $i \in N$ and $p_j = \infty$ for each $j \in M$ is envy-free. Moreover, deciding whether an outcome is envy-free can be done in polynomial time.

By the definition of envy-freeness, if $i \in W(X)$ is a winner, then all the buyers i' with $v_{i'} > v_i$ and $d_{i'} \leq d_i$ must be winners as well, otherwise i' would envy a subset of the bundle assigned to i. This motivates the following definition, which restricts to instances not containing buyers not being a priori able to receive items in any envy-free assignment (*useless buyers*).

Definition 2. *An instance I is proper if, for each buyer $i \in N$, it holds $d_i + \sum_{i' | v_{i'} > v_i, d_{i'} \leq d_i} d_{i'} \leq m$.*

The (market) *revenue* generated by an outcome $(\mathbf{X}, \mathbf{p})$ is defined as $rev(\mathbf{X}, \mathbf{p}) = \sum_{j \in M(\mathbf{X})} p_j$. RMPSD asks for the determination of an envy-free outcome of maximum revenue. We observe that it is possible to efficiently check if an instance is proper, and if discarding useless buyers is allowed, an instance can be made proper in polynomial time, without worsening the value of its optimal solution. An instance of the RMPSD problem can be modeled as a triple $(\mathbf{V}, \mathbf{D}, \mathbf{Q})$, where $\mathbf{V} = (v_1, \ldots, v_n)$ and $\mathbf{D} = (d_1, \ldots, d_n)$ are the vectors of buyers' values and demands, while $\mathbf{Q} = (q_1, \ldots, q_m)$ is the vector of item qualities. We conclude this section with three lemmas describing some properties that need to be satisfied by any envy-free outcome.

Lemma 1 ([9]). *If an outcome $(\mathbf{X}, \mathbf{p})$ is envy-free, then $\mathbf{X}$ is monotone.*

Given an outcome $(\mathbf{X}, \mathbf{p})$, an item $j \in X_i$ is *overpriced* if $u_{ij} < 0$.

Lemma 2 ([9]). *Let $(\mathbf{X}, \mathbf{p})$ be an envy-free outcome. For each overpriced item $j' \in M(\mathbf{X})$, it holds $b(j') = \beta(\mathbf{X})$.*

3   A Pricing Scheme for Monotone Allocation Vectors

Since we are interested only in envy-free outcomes, by Lemma 1, in the following we will implicitly assume that any considered allocation vector is monotone.

We call *pricing scheme* a function which, given an allocation vector $\mathbf{X}$, returns a price vector. In this section, we propose a pricing scheme for allocation vectors which will be at the basis of our approximability and inapproximability results. For the sake of readability, in describing the following pricing function, given $\mathbf{X}$, we assume a re-ordering of the buyers in such a way that all the winners appear first, still in non-increasing order of v_i.

For an allocation vector $\mathbf{X}$, define the price vector $\widetilde{\mathbf{p}}$ such that, for each $j \in M$,

$$\widetilde{p}_j = \infty \text{ if } b(j) = 0 \text{ and } \widetilde{p}_j = v_{b(j)} q_j - \sum_{k=b(j)+1}^{\beta(\mathbf{X})} \left((v_{k-1} - v_k) q_{f(k)} \right) \text{ otherwise.}$$

Quite interestingly, such a scheme resembles one presented [20]. Next lemma shows that $\widetilde{\mathbf{p}}$ is indeed a price vector.

Lemma 3. *For each $j \in M$, it holds $\widetilde{p}_j > 0$.*

We continue by showing the following important property, closely related to the notion of envy-freeness, possessed by the outcome $(\mathbf{X}, \widetilde{\mathbf{p}})$ for each allocation vector $\mathbf{X}$.

Lemma 4. *For each allocation vector $\mathbf{X}$, the outcome $(\mathbf{X}, \widetilde{\mathbf{p}})$ is feasible and, for each winner $i \in W(\mathbf{X})$, $u_i \geq \sum_{j \in T} u_{ij}$ for each $T \subseteq M$ of cardinality d_i. Thus, the allocation is envy-free for the subset of the winners buyers.*

4   Results for Generic Instances

In this section, we show that it is hard to approximate the RMPSD to a factor $O(m^{1-\epsilon})$ for any $\epsilon > 0$, even when considering related valuations, whereas a simple m-approximation algorithm can be designed for unrelated valuations.

4.1   Inapproximability Result

For an integer $k > 0$, we denote with $[k]$ the set $\{1, \ldots, k\}$. Recall that an instance of the Partition problem is made up of k strictly positive numbers $q_1, \ldots, q_k$ such that $\sum_{i \in [k]} q_i = Q$, where $Q > 0$ is an even number. It is well-known that deciding whether there exists a subset $J \subset [k]$ such that $\sum_{i \in J} q_i = Q/2$ is an NP-complete problem. The inapproximability result that we derive in this subsection is obtained through a reduction from a specialization of the Partition problem, that we call Constrained Partition problem, which we define in the following.

An instance of the Constrained Partition problem is made up of an even number k of non-negative numbers $q_1, \ldots, q_k$ such that $\sum_{i \in [k]} q_i = Q$, where Q is an even number and $\frac{3}{2} \min_{i \in [k]} \{q_i\} \geq \max_{i \in [k]} \{q_i\}$. In this case, we are asked to decide whether there exists a subset $J \subset [k]$, with $|J| = k/2$, such that $\sum_{i \in J} q_i = Q/2$.

Lemma 5. *The* Constrained Partition *problem is* NP-*complete.*

We can now proceed to show our first inapproximability result, by means of the following reduction. Given an integer $k \geq 3$, consider an instance I of the Constrained Partition problem with $2(k-1)$ numbers $q_1, \ldots, q_{2(k-1)}$ such that $\sum_{i=1}^{2(k-1)} q_i = Q$ and define $q_{min} = \min_{i \in [2(k-1)]} \{q_i\}$. Remember that, by definition, Q is even and it holds $\frac{3}{2} q_{min} \geq \max_{i \in [2(k-1)]} \{q_i\}$. Note that, this last property, together with $Q \geq 2(k-1) q_{min}$, implies that $q_j \leq \frac{3Q}{4(k-1)} < \frac{Q}{2}$ for each $j \in [2(k-1)]$ since $k \geq 3$.

For any $\epsilon > 0$, define $\alpha = \lceil \frac{2}{\epsilon} \rceil + 1$ and $\lambda = k^{\alpha}$. Note that, by definition, $\lambda \geq k^2$. We create an instance I' of the RMPSD as follows. There are $n = 5$ buyers and $m = \lambda + k - 1$ items divided into four groups: k items of quality Q, one item of quality $Q/2$, $2(k-1)$ items of qualities q_i, with $i \in [2(k-1)]$, inherited from I, and $\lambda - 2k$ items of quality $\overline{q} := \frac{q_{min}}{100} > 0$. The five buyers are such that $v_1 = 2$ and $d_1 = k$, $v_2 = 1 + \frac{1}{\lambda} \frac{Q - 2k\overline{q} + kQ(\lambda+1)/2}{Qk + Q - 2k\overline{q} + \lambda\overline{q}}$ and $d_2 = \lambda$, $v_3 = 1 + \frac{1}{\lambda}$ and $d_3 = k$, $v_4 = 1 + \frac{1}{\lambda} \frac{Q - k\overline{q}}{Q + (\lambda - 2k)\overline{q}}$ and $d_4 = \lambda - k$, $v_5 = 1$ and $d_5 = \lambda - 2k$.

Note that it holds $v_i > v_{i+1}$ for each $i \in [4]$. In fact, $v_4 > 1 = v_5$, since $\lambda > 2k$ and $Q \geq 2(k-1) q_{min} = 200(k-1)\overline{q} > k\overline{q}$ for $k \geq 2$. Moreover, $v_4 < 1 + \frac{1}{\lambda}$, since $\lambda > k$ implies $Q - k\overline{q} < Q + (\lambda - 2k)\overline{q}$. Finally, $v_2 > 1 + \frac{1}{\lambda}$, since $\lambda > 2 = \frac{kQ}{k(Q - Q/2)} > \frac{kQ}{kQ - 2\overline{q}}$ implies $Q - 2k\overline{q} + \frac{kQ(\lambda+1)}{2} > Qk + Q - 2k\overline{q} + \lambda\overline{q}$ and $v_2 < 2 = v_1$, since $\lambda > \frac{k}{2} + 1$ implies $Q - 2k\overline{q} + \frac{kQ(\lambda+1)}{2} < \lambda(Qk + Q - 2k\overline{q} + \lambda\overline{q})$.

The basic ideas behind this reduction are the following ones: (i) although buyers 2 and 4 are useless, they do generate envy; (ii) each envy-free assignment of sufficiently high revenue has to satisfy the demand of buyer 5; (iii) in any envy-free assignment satisfying the demand of buyer 5, the set of items allocated to buyer 3 has to provide a positive answer to the ConstrainedPartition problem I, otherwise either buyer 2 or 4 become envious. In particular, we show that, if there exists a positive answer to I, then there exists an envy-free outcome for I' of revenue at least $(\lambda - 2k)\overline{q}$, while, if a positive answer to I does not exists, then no envy-free outcome of revenue greater than $6(k+3)(k-1)q_{min}$ can exist for I'.

Lemma 6. *If there exists a positive answer to I, then there exists an envy-free outcome for I' of revenue greater than $(\lambda - 2k)\overline{q}$.*

Now we stress the fact that, in any envy-free outcome $(\mathbf{X}, \mathbf{p})$ for I' such that $rev(\mathbf{X}, \mathbf{p}) > 0$, it must be $X_1 \neq \emptyset$. In fact, assume that there exists an envy-free outcome $(\mathbf{X}, \mathbf{p})$ such that $X_1 = \emptyset$ and $X_i \neq \emptyset$ for some $2 \leq i \leq 5$, then, since $d_1 \leq d_i$ and $v_1 > v_i$ for each $2 \leq i \leq 5$, it follows that there exists a subset of d_1 items T such that $u_1 > u_i \geq 0$, which contradicts the envy-freeness of $(\mathbf{X}, \mathbf{p})$. As a consequence of this fact and of the definition of the demand vector, it follows that each possible envy-free outcome $(\mathbf{X}, \mathbf{p})$ for I' can only fall into one of the following three cases:

1. $X_1 \neq \emptyset$ and $X_i = \emptyset$ for each $2 \leq i \leq 5$,
2. $X_1, X_3 \neq \emptyset$ and $X_2, X_4, X_5 = \emptyset$,
3. $X_1, X_3, X_5 \neq \emptyset$ and $X_2, X_4 = \emptyset$.

Note that, for each envy-free outcome $(\mathbf{X}, \mathbf{p})$ falling into one of the first two cases, it holds $rev(\mathbf{X}, \mathbf{p}) \leq v_1 kQ + v_3 \frac{3}{2}Q \leq Q(2k+3) \leq (2k+3)2(k-1)\frac{3}{2}q_{min} = 6(k+3)(k-1)q_{min}$. In the remaining of this proof, we will focus only on outcomes falling into case (3). First, we show that, if any such an outcome is envy-free, then the sum of the qualities of the items assigned to buyer 3 cannot exceed Q.

Lemma 7. *In any envy-free outcome $(\mathbf{X}, \mathbf{p})$ falling into case (3), it holds $\sum_{j \in X_3} q_j \leq Q$.*

On the other hand, we also show that, for any envy-free outcome $(\mathbf{X}, \mathbf{p})$ falling into case (3), the sum of the qualities of the items assigned to buyer 3 cannot be smaller than Q.

Lemma 8. *In any envy-free outcome $(\mathbf{X}, \mathbf{p})$ falling into case (3), it holds $\sum_{j \in X_3} q_j \geq Q$.*

As a consequence of Lemmas 7 and 8, it follows that there exists an envy-free outcome $(\mathbf{X}, \mathbf{p})$ falling into case (3) only if $\sum_{j \in X_3} q_j = Q$. Since, as we have already observed, in such a case the item of quality $Q/2$ has to belong to X_3, it follows that there exists an envy-free outcome $(\mathbf{X}, \mathbf{p})$ falling into case (3) only if there are $k - 1$ items inherited from I whose sum is exactly $Q/2$, that is, only if I admits a positive solution.

Any envy-free outcome not falling into case (3) can raise a revenue of at most $6(k+3)(k-1)q_{min}$. Hence, if there exists a positive answer to I, then, by Lemma 6, there exists a solution to I' of revenue greater than $(\lambda - 2k)\overline{q}$, while, if there is no positive answer to I, then there exists no solution to I' of revenue more than $6(k+3)(k-1)q_{min}$. Thus, if there exists an r-approximation algorithm for the RMPSD with $r \leq \frac{(\lambda-2k)q_{min}}{600(k+3)(k-1)q_{min}}$, it is then possible to decide in polynomial time the Constrained Partition problem, thus implying P = NP. Since, by the definition of α, $\frac{\lambda-2k}{600(k+3)(k-1)} = O\left(k^{\alpha-2}\right) = O\left(m^{1-2/\alpha}\right)$ and $m^{1-\epsilon} < m^{1-2/\alpha}$, the following theorem holds.

Theorem 1. *For any $\epsilon > 0$, the* RMPSD *cannot be approximated to a factor $O(m^{1-\epsilon})$ unless* P = NP.

We stress that this inapproximability result heavily relies on the presence of two useless buyers, namely buyers 2 and 4, who cannot be winners in any envy-free solution. This situation suggests that better approximation guarantees may be possible for proper instances, as we will show in the next section.

4.2 The Approximation Algorithm

In this subsection, we design a simple m-approximation algorithm for the generalization of the RMPSD in which the buyers have unrelated valuations. The inapproximability result given in Theorem 1 shows that, asymptotically speaking, this is the best approximation one can hope for unless P = NP.

For each $i \in N$, let $T_i = \text{argmax}_{T \subseteq M : |T| = d_i} \left\{ \sum_{j \in T} v_{ij} \right\}$ be the set of the d_i best items for buyer i and define $R_i = \left(\sum_{j \in T_i} v_{ij} \right) / d_i$. Let i^* be the index of the buyer with the highest value R_i. Consider the algorithm **best** which returns the outcome $(\mathbf{\overline{X}}, \mathbf{\overline{p}})$ such that $\overline{X}_{i^*} = T_{i^*}$, $\overline{X}_i = \emptyset$ for each $i \neq i^*$, $\overline{p}_j = R_{i^*}$ for each $j \in T_{i^*}$ and $\overline{p}_j = \infty$ for each $j \notin T_{i^*}$. It is easy to see that the computational complexity of Algorithm **best** is $O(nm)$.

Theorem 2. *Algorithm* **best** *returns an m-approximate solution for the* RMPSD *with unrelated valuations.*

5   Results for Proper Instances

Given a proper instance $I = (\mathbf{V}, \mathbf{D}, \mathbf{Q})$, denote with δ the number of different values in $\mathbf{V}$ and, for each $k \in [\delta]$, let $A_k \subseteq N$ denote the set of buyers with the kth highest value and $v(A_k)$ denote the value of all buyers in A_k. For $k \in [\delta]$, define $A_{\leq k} = \bigcup_{h=1}^{k} A_h$, $A_{\geq k} = \bigcup_{h=k}^{\delta} A_h$, $A_{>k} = A_{\geq k} \setminus A_k$ and $A_{<k} = A_{\leq k} \setminus A_k$, while, for each subset of buyers $A \subseteq N$, define $d(A) = \sum_{i \in A} d_i$. Let $\delta^* \in [\delta]$ be the minimum index such that $d(A_{\leq \delta^*}) > m$ and let $\widetilde{A} \subset A_{\delta^*}$ be a subset of buyers in A_{δ^*} such that $\widetilde{A} = \text{argmax}_{A \subset A_{\delta^*} : d(A) + d(A_{<\delta^*}) \leq m} \{d(A)\}$. In other words $\widetilde{A}$ is the subset of buyers in A_{δ^*} that feasibly extends $A_{<\delta^*}$ (i.e., such that the sum of the requested items of buyers in $A_{<\delta^*} \cup \widetilde{A}$ is at most m) and maximizes the number of allocated items.

Note that any instance I for which δ^* does not exist can be suitably extended with a dummy buyer $n + 1$, such that $v_{n+1} < v_n$ and $d_{n+1} = m + 1$, which is equivalent in the sense that it does not change the set of envy-free outcomes of I. Hence, in this section, we will always assume that δ^* is well-defined for each proper instance of the RMPSD.

For our purposes we need to break ties among values of the buyers in A_{δ^*} in such a way that each buyer in $\widetilde{A}$ comes before any buyer in $A_{\delta^*} \setminus \widetilde{A}$. In order to achieve this task, we need to explicitly compute the set of buyers $\widetilde{A}$. Such a computation can be done by reducing this problem to the knapsack problem. It is easy to see that, in this case, the well-known pseudo-polynomial time algorithm for knapsack is polynomial in the dimensions of I, as $d_i \leq m$ for every $i \in N$.

Because of the above discussion, from now on we can assume that ties among values of the buyers in A_{δ^*} are broken in such a way that each buyer in $\widetilde{A}$ comes before any buyer in $A_{\delta^*} \setminus \widetilde{A}$. For each $k \in [\delta^*]$, define

$$\alpha(k) = \begin{cases} \max\{i \in A_k\} & \text{if } k \in [\delta^* - 1], \\ \max\{i \in \widetilde{A}\} & \text{if } k = \delta^*. \end{cases}$$

By the definition of δ^* and $\widetilde{A}$ and by the tie breaking rule imposed on the buyers in A_{δ^*}, it follows that $\sum_{i=1}^{\alpha(k)} d_i \leq m$ for each $k \in [\delta^*]$.

We say that an allocation vector $\mathbf{X}$ is an h-prefix of I, with $h \in [\alpha(\delta^*)]$, if $\mathbf{X}$ is monotone and $i \in W(\mathbf{X})$ if and only if $i \in [h]$.

5.1   Computing an h-Prefix of I of Maximum Revenue

Let $\mathbf{X}$ be an h-prefix of I. We show that $(\mathbf{X}, \widetilde{\mathbf{p}})$ is an envy-free outcome.

Lemma 9. *The outcome* $(\mathbf{X}, \widetilde{\mathbf{p}})$ *is envy-free.*

Given an allocation vector $\mathbf{X}$, for each $i \in [\delta]$, denote with $M_i(\mathbf{X}) = \{j \in M(\mathbf{X}) : v_{b(j)} = v(A_i)\}$ the set of items allocated to the buyers with the ith highest value in $\mathbf{V}$. Recall that, since $\mathbf{X}$ is an h-prefix of I, it holds $\beta(\mathbf{X}) = h$. The following lemma gives a lower bound on the revenue generated by the outcome $(\mathbf{X}, \widetilde{\mathbf{p}})$.

Lemma 10. $rev(\mathbf{X}, \widetilde{\mathbf{p}}) \geq v_h \sum_{j \in M_h(\mathbf{X})} q_j$.

We now prove a very important result stating that the price vector $\widetilde{\mathbf{p}}$ is the best one can hope for when overpricing is not allowed. Such a result, of independent interest, plays a crucial role in the proof of the approximation guarantee of the algorithm we define in this section.

Lemma 11. *Let* $\mathbf{X}$ *be an* h-*prefix of* I. *Then* $(\mathbf{X}, \widetilde{\mathbf{p}})$ *is an optimal envy-free outcome when overpricing is not allowed.*

We design a polynomial time algorithm ComputePrefix (described in the full version [3]) which, given a proper instance I and a value $h \in [\alpha(\delta^*)]$, outputs the h-prefix $\mathbf{X}_h^*$ such that the outcome $(\mathbf{X}_h^*, \widetilde{\mathbf{p}})$ achieves the highest revenue among all possible h-prefixes of I in time $O(mh)$.

5.2   The Approximation Algorithm

Our approximation algorithm Prefix for proper instances generates a set of prefixes of I for which it computes the allocation of items yielding maximum revenue by exploiting the algorithm ComputePrefix as a subroutine. Then, it returns the solution with the highest revenue among them.

Prefix(input: instance I, output: allocation vector $\mathbf{X}^*$):
$opt := \emptyset$; $value := -1$;
compute $\widetilde{A}$;
reorder the buyers in such a way that each $i \in \widetilde{A}$ comes before any $i' \in A_{\delta^*} \setminus \widetilde{A}$;
for each $h = 1, \ldots, \alpha(\delta^*)$ **do**
| $\mathbf{X}_h^* :=$ ComputePrefix(I, h);
| **if** $rev(\mathbf{X}_h^*, \widetilde{\mathbf{p}}) > value$ **then**
| | $opt := \mathbf{X}_h^*$; $value := rev(\mathbf{X}_h^*, \widetilde{\mathbf{p}})$;
for each $k = 0, \ldots, \delta^* - 1$ **do**
| **for each** $i \in A_{k+1}$ **do**
| | reorder the buyers in A_{k+1} in such a way that i is the first buyer in A_{k+1};
| | **if** $d(A_{\leq k}) + d_i \leq m$ **then** $\mathbf{X}_k^* :=$ ComputePrefix$(I, |A_{\leq k}| + 1)$;
| | **if** $rev(\mathbf{X}_k^*, \widetilde{\mathbf{p}}) > value$ **then**
| | | $opt := \mathbf{X}_k^*$; $value := rev(\mathbf{X}_k^*, \widetilde{\mathbf{p}})$;
return opt;

It is easy to see that the computational complexity of Algorithm Prefix is $O(n^3 m)$. As a major positive contribution of this work, we show that it approximates the RMPSD to a factor 2 on proper instance.

Theorem 3. *The approximation ratio of Algorithm* Prefix *is 2 when applied to proper instances.*

We conclude this section by showing that the approximation ratio achieved by Algorithm Prefix is the best possible one for proper instances.

Theorem 4. *For any* $0 < \epsilon \leq 1$, *the* RMPSD *on proper instances cannot be approximated to a factor* $2 - \epsilon$ *unless* P = NP.

References

1. Aggarwal, G., Feder, T., Motwani, R., Zhu, A.: Algorithms for Multi-Product Pricing. In: Díaz, J., Karhumäki, J., Lepistö, A., Sannella, D. (eds.) ICALP 2004. LNCS, vol. 3142, pp. 72–83. Springer, Heidelberg (2004)
2. Bezjian-Avery, A., Calder, B., Iacobucci, D.: New Media Interative Advertising vs. Traditional Advertising. Journal of Advertising Research, 23–32 (1998)
3. Bilò, V., Flammini, M., Monaco, G.: Approximating the Revenue Maximization Problem with Sharp Demands. CoRR, arXiv:1312.3892 (2013)
4. Briest, P.: Uniform Budgets and the Envy-Free Pricing Problem. In: Aceto, L., Damgård, I., Goldberg, L.A., Halldórsson, M.M., Ingólfsdóttir, A., Walukiewicz, I. (eds.) ICALP 2008, Part I. LNCS, vol. 5125, pp. 808–819. Springer, Heidelberg (2008)

5. Chalermsook, P., Chuzhoy, J., Kannan, S., Khanna, S.: Improved Hardness Results for Profit Maximization Pricing Problems with Unlimited Supply. In: Gupta, A., Jansen, K., Rolim, J., Servedio, R. (eds.) APPROX 2012 and RANDOM 2012. LNCS, vol. 7408, pp. 73–84. Springer, Heidelberg (2012)
6. Chalermsook, P., Laekhanukit, B., Nanongkai, D.: Independent Set, Induced Matching, and Pricing: Connections and Tight (Subexponential Time) Approximation Hardnesses. In: Proceedings of FOCS 2013, pp. 370–379 (2013)
7. Chalermsook, P., Laekhanukit, B., Nanongkai, D.: Graph Products Revisited: Tight Approximation Hardness of Induced Matching, Poset Dimension and More. In: Proceedings of SODA 2013, pp. 1557–1576 (2013)
8. Chen, N., Deng, X.: Envy-Free Pricing in Multi-item Markets. In: Abramsky, S., Gavoille, C., Kirchner, C., Meyer auf der Heide, F., Spirakis, P.G. (eds.) ICALP 2010. LNCS, vol. 6199, pp. 418–429. Springer, Heidelberg (2010)
9. Chen, N., Deng, X., Goldberg, P.W., Zhang, J.: On Revenue Maximization with Sharp Multi-Unit Demands. In: CoRR, arXiv:1210.0203 (2012)
10. Chen, N., Ghosh, A., Vassilvitskii, S.: Optimal Envy-Free Pricing with Metric Substitutability. SIAM Journal on Computing 40(3), 623–645 (2011)
11. Demaine, E.D., Feige, U., Hajiaghayi, M., Salavatipour, M.R.: Combination Can Be Hard: Approximability of the Unique Coverage Problem. SIAM Journal on Computing 38(4), 1464–1483 (2008)
12. Deng, X., Goldberg, P., Sun, Y., Tang, B., Zhang, J.: Pricing Ad Slots with Consecutive Multi-unit Demand. In: Vöcking, B. (ed.) SAGT 2013. LNCS, vol. 8146, pp. 255–266. Springer, Heidelberg (2013)
13. Deng, X., Goldberg, P.W., Tang, B., Zhang, J.: Multi-unit Bayesian Auction with Demand or Budget Constraints. In: Proceedings of WIT-EC 2012 (2012)
14. Deng, X., Sun, Y., Yin, M., Zhou, Y.: Mechanism Design for Multi-slot Ads Auction in Sponsored Search Markets. In: Lee, D.-T., Chen, D.Z., Ying, S. (eds.) FAW 2010. LNCS, vol. 6213, pp. 11–22. Springer, Heidelberg (2010)
15. Edelman, B., Ostrovsky, M., Schwarz, M.: Internet Advertising and the Generalized Second-Price Auction. American Economic Review 97(1), 242–259 (2007)
16. Feldman, M., Fiat, A., Leonardi, S., Sankowski, P.: Revenue Maximizing Envy-Free Multi-Unit Auctions with Budgets. In: Proceedings of EC 2012, pp. 532–549 (2012)
17. Foley, D.: Resource Allocation and the Public Sector. Yale Economic Essays 7, 45–98 (1967)
18. Glynn, P., Rusmevichientong, P., Van Roy, B.: A Non-Parametric Approach to Multi-Product Pricing. Operations Research 54(1), 82–98 (2006)
19. Guruswami, V., Hartline, J.D., Karlin, A.R., Kempe, D., Kenyon, C., McSherry, F.: On Profit-Maximizing Envy-Free Pricing. In: Proceedings of SODA 2005, pp. 1164–1173 (2005)
20. Hartline, J.D., Yan, Q.: Envy, Truth, and Profit. In: Proceedings of EC 2011, pp. 243–252 (2011)
21. Nisan, N., et al.: Google's Auction for TV Ads. In: Albers, S., Marchetti-Spaccamela, A., Matias, Y., Nikoletseas, S., Thomas, W. (eds.) ICALP 2009, Part II. LNCS, vol. 5556, pp. 309–327. Springer, Heidelberg (2009)
22. Rosenkrans, G.: The Creativeness and Effectiveness of Online Interactive Rich Media Advertising. Journal of Interactive Advertising 9(2) (2009)
23. Shocker, A.D., Srinivasan, V.: Multiattribute Approaches for Product Concept Evaluation and Generation: A Critical Review. Journal of Marketing Research 16, 159–180 (1979)
24. Vickrey, W.: Counterspeculation, Auctions, and Competitive Sealed Tenders. Journal of Finance 16, 8–37 (1961)
25. Walras, L.: Elements of Pure Economics. Allen and Unwin (1954)

Reconfiguring Independent Sets in Claw-Free Graphs

Paul Bonsma[1], Marcin Kamiński[2], and Marcin Wrochna[2,*]

[1] University of Twente, Faculty of EEMCS, PO Box 217, 7500 AE Enschede,
The Netherlands
p.s.bonsma@ewi.utwente.nl
[2] Uniwersytet Warszawski, Institute of Computer Science, Warsaw, Poland
mjk@mimuw.edu.pl, mw290715@students.mimuw.edu.pl

Abstract. We present a polynomial-time algorithm that, given two independent sets in a claw-free graph G, decides whether one can be transformed into the other by a sequence of elementary steps. Each elementary
step is to remove a vertex v from the current independent set S and to
add a new vertex w (not in S) such that the result is again an independent set. We also consider the more restricted model where v and w have
to be adjacent.

1   Introduction

Reconfiguration Problems. To obtain a reconfiguration version of an algorithmic problem, one defines a *reconfiguration rule* – a (symmetric) adjacency relation
between solutions of the problem, describing small transformations one is allowed
to make. The main focus is on studying whether one given solution can be transformed into another by a sequence of such small steps. We call this a *reachability problem.* For example, in a well-studied reconfiguration version of vertex coloring [1,2,3,4,5,6], we are given two k-colorings of the vertices of a graph and we should
decide whether one can be transformed into the other by recoloring one vertex at
a time so that all intermediate solutions are also proper k-colorings.

A useful way to look at reconfiguration problems is through the concept of
the *solution graph.* Given a problem instance, the vertices of the solution graph
are all solutions to the instance, and the reconfiguration rule defines its edges.
Clearly, one solution can be transformed into another if they belong to the
same connected component of the solution graph. Other well-studied questions
in the context of reconfiguration are as follows: can one efficiently decide (for
every instance) whether the solution graph is connected? Can one efficiently
find shortest paths between two solutions? Common non-algorithmic results are
giving upper and lower bounds on the possible diameter of components of the

* The first author was supported by the European Community's Seventh Framework
Programme (FP7/2007-2013), grant agreement n° 317662. The second and third
author were supported by the Foundation for Polish Science (HOMING PLUS/2011-
4/8) and the National Science Center (SONATA 2012/07/D/ST6/02432).

R Ravi and I.L. Gørtz (Eds.): SWAT 2014, LNCS 8503, pp. 86–97, 2014.
© Springer International Publishing Switzerland 2014

solution graph, in terms of the instance size, or studying how much the solution space needs to be increased in order to guarantee connectivity.

Reconfiguration is a natural setting for real-life problems in which solutions evolve over time and an interesting theoretical framework that has been gradually attracting more attention. The theoretical interest is based on the fact that reconfiguration problems provide a new perspective and offer a deeper understanding of the solution space as well as a potential to develop heuristics to navigate that space.

The reconfiguration paradigm has recently been applied to a number of algorithmic problems: vertex coloring [1,2,3,4,5], list-edge coloring [7], clique, set cover, integer programming, matching, spanning tree, matroid bases [8], block puzzles [9], satisfiability [10], independent set [9,8,11], shortest paths [12,13,14], and dominating set [15]; recently also in the setting of parameterized complexity [16]. A recent survey [17] gives a good introduction to this area of research.

Reconfiguration of Independent Sets. The topic of this paper is reconfiguration of independent sets. An *independent set* in a graph is a set of pairwise nonadjacent vertices. We will view the elements of an independent set as tokens placed on vertices. Three different reconfiguration rules have been studied in the literature: token sliding (TS), token jumping (TJ), and token addition/removal (TAR). The reconfiguration rule in the TS model allows to slide a token along an edge. The reconfiguration rule in the TJ model allows to remove a token from a vertex and place it on another unoccupied vertex. In the TAR model, the reconfiguration rule allows to either add or remove a token as long as at least k tokens remain on the graph at any point, for a given integer k. In all three cases, the reconfiguration rule may of course only be applied if it maintains an independent set. A sequence of moves following these rules is called a *TS-sequence*, *TJ-sequence*, or *k-TAR-sequence*, respectively. Note that the TS model is more restricted than the TJ model, in the sense that any TS-sequence is also a TJ-sequence. Kamiński et al. [11] showed that the TAR model generalizes the TJ model, in the sense that there exists a TJ-sequence between two solutions I and J with $|I| = |J|$ if and only if there exists a k-TAR-sequence between them, with $k = |I| - 1$. TS seems to have been introduced by Hearn and Demaine [9], TAR was introduced by Ito et al. [8] and TJ by Kamiński et al. [11].

In all three models, the corresponding reachability problems are PSPACE-complete in general graphs [8] and even in perfect graphs [11] or in planar graphs of maximum degree 3 [9] (see also [3]). We remark that in [9], only the TS-model was explicitly considered, but since only maximum independent sets are used, this implies the result for the TJ model (see Proposition 2 below) and for the TAR model (using the aforementioned result from [11]).

Claw-Free Graphs. A *claw* is the tree with four vertices and three leaves. A graph is *claw-free* if it does not contain a claw as an induced subgraph. A claw is not a line graph of any graph and thus the class of claw-free graphs generalizes the class of line graphs. The structure of claw-free graphs is not simple but has been recently described by Chudnovsky and Seymour in the form of a decomposition theorem [18].

There is a natural one-to-one correspondence between matchings in a graph and independent sets in its line graph. In particular, a maximum matching in a graph corresponds to a maximum independent set in its line graph. Hence, Edmonds' maximum matching algorithm [19] gives a polynomial-time algorithm for finding maximum independent sets in line graphs. This results has been extended to claw-free graphs independently by Minty [20] and Sbihi [21]. Both algorithms work for the unweighted case, while the algorithm of Minty, with a correction proposed by Nakamura and Tamura in [22], applies to weighted graphs (see also [23, Section 69]). Recently Nobili and Sassano [24] improved this to give an $\mathcal{O}(n^4 \log n)$ algorithm, while Faenza et al. [25] proved a decomposition theorem that allows to solve the problem in $\mathcal{O}(n^3)$ time.

Our Results. In this paper, we study the reachability problem for independent set reconfiguration, using the TS and TJ model. Our main result is that these problems can be solved in polynomial time for the case of claw-free graphs. Along the way, we prove some results that are interesting in their own right. For instance, we show that for connected claw-free graphs, the existence of a TJ-sequence implies the existence of a TS-sequence between the same pair of solutions. This implies that for connected claw-free and even-hole-free graphs, the solution graph is always connected, answering an open question posed in [11].

Since claw-free graphs generalize line graphs, our results generalize the result by Ito et al. [8] on matching reconfiguration. Since a vertex set I of a graph G is an independent set if and only if $V(G)\backslash I$ is a vertex cover, our results also apply to the recently studied vertex cover reconfiguration problem [16]. The new techniques we introduce can be seen as an extension of the techniques introduced for finding maximum independent sets in claw-free graphs, and we expect them to be useful for addressing similar reconfiguration questions, such as efficiently deciding whether the solution graph is connected.

Because of space constraints, some proof details are omitted. Statements for which more proof details can be found in the full version of this paper [26] are marked with a star.

2   Preliminaries

For graph theoretic terminology not defined here, we refer to [27]. For a graph G and vertex set $S \subseteq V(G)$, we denote the subgraph induced by S by $G[S]$, and denote $G - S = G[V\backslash S]$. The set of neighbors of a vertex $v \in V(G)$ is denoted by $N(v)$, and the closed neighborhood of v is $N[v] = N(v) \cup \{v\}$. A *walk* from v_0 to v_k of length k is a sequence of vertices $v_0, v_1, \ldots, v_k$ such that $v_i v_{i+1} \in E(G)$ for all $i \in \{0, \ldots, k-1\}$. It is a *path* if all of its vertices are distinct, and a *cycle* if $k \geq 3$, $v_0 = v_k$ and $v_0, \ldots, v_{k-1}$ is a path. We use $V(C)$ to denote the vertex set of a path or cycle, viewed as a subgraph of G. A path or graph is called *trivial* if it contains only one vertex. Edges of a directed graph or *digraph* D are called *arcs*, and are denoted by the ordered tuple (u, v). A *directed path* in D is a sequence of distinct vertices $v_0, \ldots, v_k$ such that for all $i \in \{0, \ldots, k-1\}$, (v_i, v_{i+1}) is an arc of D.

We denote the distance of two vertices $u, v \in V(G)$ by $d_G(u,v)$. By $\text{diam}(G)$ we denote the *diameter* of a connected graph G, defined as $\max_{u,v \in V(G)} d_G(u,v)$. For a vertex set S of a graph G and integer $i \in \mathbb{N}$, we denote $N_i(S) = \{v \in V(G) \backslash S : |N(v) \cap S| = i\}$.

For a graph G, by $\text{TS}_k(G)$ we denote the graph that has as its vertex the set of all independent sets of G of size k, where two independent sets I and J are adjacent if there is an edge $uv \in E(G)$ with $I \backslash J = \{u\}$ and $J \backslash I = \{v\}$. We say that J can be obtained from I by *sliding a token from u to v*, or by the *move $u \to v$* for short. A walk in $\text{TS}_k(G)$ from I to J is called a *TS-sequence from I to J*. We write $I \leftrightarrow_{\text{TS}} J$ to indicate that there is a TS-sequence from I to J.

Analogously, by $\text{TJ}_k(G)$ we denote the graph that has as its vertex set the set of all independent sets of G of size k, where two independent sets I and J are adjacent if there is a vertex pair $u, v \in V(G)$ with $I \backslash J = \{u\}$ and $J \backslash I = \{v\}$. We say that J can be obtained from I by *jumping a token from u to v*. A walk in $\text{TS}_k(G)$ from I to J is called a *TJ-sequence from I to J*. We write $I \leftrightarrow_{\text{TJ}} J$ to indicate that there exists a TJ-sequence from I to J. Note that $\text{TS}_k(G)$ is a spanning subgraph of $\text{TJ}_k(G)$.

The *reachability problem* for token sliding (resp. token jumping) has as input a graph G and two independent sets I and J of G with $|I| = |J|$, and asks whether $I \leftrightarrow_{\text{TS}} J$ (resp. $I \leftrightarrow_{\text{TJ}} J$). These problems are called *TS-Reachability* and *TJ-Reachability*, respectively.

If H is a claw with vertex set $\{u, v, w, x\}$ such that $N(u) = \{v, w, x\}$, then H is called a *u-claw with leaves v, w, x*. Sets $I \backslash \{v\}$ and $I \cup \{v\}$ are denoted by $I - v$ and $I + v$ respectively. The symmetric difference of two sets I and J is denoted by $I \Delta J = (I \backslash J) \cup (J \backslash I)$. The following observation is used implicitly in many proofs:

Proposition 1. *Let I and J be independent sets in a claw-free graph G. Then every component of $G[I \Delta J]$ is a path or an even length cycle.*

By $\alpha(G)$ we denote the size of the largest independent set of G. An independent set I is called *maximum* if $|I| = \alpha(G)$. A vertex set $S \subseteq V(G)$ is a *dominating set* if $N[v] \cap S \neq \emptyset$ for all $v \in V(G)$. Observe that a maximum independent set is a dominating set, thus the only possible token jumps from it are between adjacent vertices, and hence all are token slides:

Proposition 2. *Let G be any graph and $k = \alpha(G)$. Then, $TS_k(G) = TJ_k(G)$. In particular, for any two maximum independent sets I and J in G, $I \leftrightarrow_{\text{TS}} J$ if and only if $I \leftrightarrow_{\text{TJ}} J$.*

3   The Equivalence of Sliding and Jumping

In our main result (Theorem 17), we will consider equal size independent sets I and J of a claw-free graph G, and show that in polynomial time, it can be verified whether $I \leftrightarrow_{\text{TS}} J$ and whether $I \leftrightarrow_{\text{TJ}} J$. In this section, we show that if G is connected and $G[I \Delta J]$ contains no cycles, then $I \leftrightarrow_{\text{TS}} J$. From this, we will subsequently conclude that for connected claw-free graphs $I \leftrightarrow_{\text{TS}} J$ holds if and only if $I \leftrightarrow_{\text{TJ}} J$ holds, even in the case of nonmaximum independent sets.

Lemma 3 (*). *Let I and J be independent sets in a connected claw-free graph G with $|I| = |J|$. If $G[I \triangle J]$ contains no cycles, then $I \leftrightarrow_{TS} J$.*

Proof sketch: We show that I or J can be modified using token slides so that the two resulting independent sets are closer to each other in the sense that either $|I \setminus J|$ is smaller, or it is unchanged and the minimum distance between vertices u, v with $u \in I \setminus J$ and $v \in J \setminus I$ is smaller. The claim then follows by induction.

Suppose first that $G[I \triangle J]$ contains at least one nontrivial component C. Since it is not a cycle by assumption, it must be a path. Choose an end vertex u of this path, and let v be its unique neighbor on the path. If $u \in J$ then $N(u) \cap I = \{v\}$, so we can obtain a new independent set $I' = I + u - v$ from I using a single token slide. The new set I' is closer to J in the sense that $|I' \setminus J| < |I \setminus J|$, so we may use induction to conclude that $I' \leftrightarrow_{TS} J$, and thus $I \leftrightarrow_{TS} J$. On the other hand, if $u \in I$ then we can obtain a new independent set $J' = J - v + u$ from J, and conclude the proof similarly by applying the induction assumption to J' and I.

In the remaining case, we may assume that $G[I \triangle J]$ consists only of isolated vertices. Choose $u \in I \setminus J$ and $v \in J \setminus I$, such that the distance $d := d_G(u, v)$ between these vertices is minimized. Starting with I, we intend to slide the token on u to v, to obtain an independent set $I' = I - u + v$ that is closer to J. To this end, we choose a shortest path $P = v_0, \ldots, v_d$ in G from $v_0 = u$ to $v_d = v$. If the token can be moved along this path while maintaining an independent set throughout, then $I \leftrightarrow_{TS} I'$, and the proof follows by induction as before.

So now suppose that this cannot be done, that is, at least one of the vertices on P is equal to or adjacent to a vertex in $I - u$. In that case, we choose i maximum such that $N(v_i) \cap I \neq \emptyset$. Using some simple observations (including the fact that G is claw-free), one can now show that $N(v_i) \cap I$ consists of a single vertex x. By choice of v_i, starting with I, the token on x can be moved along the path $x, v_i, v_{i+1}, \ldots, v_d$ while maintaining an independent set throughout. This yields an independent set $I'' = I - x + v$, with $I \leftrightarrow_{TS} I''$. It can also easily be shown that $d_G(u, x) < d_G(u, v)$ and $d_G(x, v) < d_G(u, v)$. So considering the choice of u and v, it follows that $x \in I \cap J$, and thus $|I'' \setminus J| = |I \setminus J|$. Since now the pair $u \in I'' \setminus J$ and $x \in J \setminus I''$ has a smaller distance $d_G(u, x) < d_G(u, v) = d$, we may assume by induction that $I'' \leftrightarrow_{TS} J$, and thus $I \leftrightarrow_{TS} J$.   □

Corollary 4. *Let I and J be independent sets in a connected claw-free graph G. Then $I \leftrightarrow_{TS} J$ if and only if $I \leftrightarrow_{TJ} J$.*

Proof: Let J be obtained from I by jumping a token from u to v. Then $G[I \triangle J]$ contains only two vertices and therefore no cycles. So by Lemma 3, any token jump can be replaced by a sequence of token slides.   □

We now consider implications of the above corollary for graphs that are claw- and even-hole-free. A graph is *even-hole-free* if it contains no even cycle as an induced subgraph. Kamiński et al. [11] proved the following statement.

Theorem 5 ([11]). *Let I and J be two independent sets of a graph G with $|I| = |J|$. If $G[I \triangle J]$ contains no even cycles, then there exists a TJ-sequence from I to J of length $|I \setminus J|$, which can be constructed in linear time.*

In particular, if G is even-hole-free, then $\mathrm{TJ}_k(G)$ is connected (for every k). However, $\mathrm{TS}_k(G)$ is not necessarily connected (consider a claw with two tokens). This motivated the question asked in [11] whether for connected, claw-free and even-hole-free graph G, $\mathrm{TS}_k(G)$ is connected. Combining Corollary 4 with Theorem 5 shows that the answer to this question is affirmative.

Corollary 6. *Let G be a connected claw-free and even-hole-free graph. Then $TS_k(G)$ is connected.*

4   Nonmaximum Independent Sets

We now continue studying connected claw-free graphs. By Lemma 3 it remains to consider the case that $G[I \Delta J]$ contains (even length) cycles. In this section, we show that if I and J are not maximum independent sets of G, such cycles can always be resolved. This requires various techniques developed for finding maximum independent sets in claw-free graphs, and the following definitions.

A vertex $v \in V(G)$ is *free* (with respect to an independent set I of G) if $v \notin I$ and $|N(v) \cap I| \leq 1$. Let $W = v_0, \ldots, v_k$ be a walk in G, and let $I \subseteq V(G)$. Then W is called *I-alternating* if $|\{v_i, v_{i+1}\} \cap I| = 1$ for $i = 0, \ldots, k-1$. In the case that W is a path, W is called *chordless* if $G[\{v_0, \ldots, v_k\}]$ is a path. In the case that W is a cycle (so $v_0 = v_k$), W is called *chordless* if $G[\{v_0, \ldots, v_{k-1}\}]$ is a cycle. A cycle $W = v_0, \ldots, v_k$ is called *I-bad* if it is I-alternating and chordless. A path $W = v_0, \ldots, v_k$ with $k \geq 2$ is called *I-augmenting* if it is I-alternating and chordless, and v_0 and v_k are both free vertices. This definition of I-augmenting paths differs from the usual definition, as it is used in the setting of finding *maximum independent sets*, since the chordless condition is stronger than needed in such a setting. However, we observe that in a claw-free graph G, the two definitions are equivalent, so we may apply well-known statements about I-augmenting paths proved elsewhere. In particular, we use the following two results originally proved by Minty [20] and Sbihi [21] (see also [23, Section 69.2]).

Theorem 7 ([23]). *Let I be an independent set in a claw-free graph G. In polynomial time, it can be decided whether an I-augmenting path between two given free vertices x and y exists, and if so, one can be computed.*

Proposition 8 ([23]). *Let I be a nonmaximum independent set in a claw-free graph G. Then I is not a dominating set, or there exists an I-augmenting path.*

We use Proposition 8 to handle the case of nonmaximum independent sets. The next statement is formulated for token jumping, and (by Corollary 4) implies the same result for token sliding, in the case of connected graphs.

Lemma 9 (*). *Let I be a nonmaximum independent set in a claw-free graph G. Then for any independent set J with $|J| = |I|$, $I \leftrightarrow_{\mathrm{TJ}} J$ holds.*

Proof sketch: By Theorem 5, it suffices to consider the case where $G[I \Delta J]$ contains at least one cycle C. Let $C = u_1, v_1, u_2, v_2, \ldots, v_k, u_1$, so that $u_i \in I$ and $v_i \in J$ for all i.

Suppose first that I is not a dominating set. Then we can choose a vertex w with $N[w] \cap I = \emptyset$. With a single token jump, we can obtain the independent set $I' = I + w - u_1$ from I. Next, apply the moves $u_k \to v_k$, $u_{k-1} \to v_{k-1}, \ldots$, $u_2 \to v_2$, in this order. (This is possible since C is chordless.) Finally, jump the token from w to v_1. It can be verified that this yields a token jumping sequence from I to $I' = I\Delta V(C)$. This way, all cycles can be resolved one by one, until no more cycles remain and Theorem 5 can be applied to prove the statement.

On the other hand, if I is a dominating set, then Proposition 8 shows that there exists an I-augmenting path $P = v_0, u_1, v_1, \ldots, u_d, v_d$, with $u_i \in I$ for all i. Since v_d is a free vertex, we can first apply the moves $u_d \to v_d$, $u_{d-1} \to v_{d-1}, \ldots u_1 \to v_1$, in this order (which can be done since P is chordless), to obtain an independent set I' from I, with $I \leftrightarrow_{\mathrm{TS}} I'$. Then v_0 is not dominated by I', so the previous argument shows that $I' \leftrightarrow_{\mathrm{TJ}} J$, which implies $I \leftrightarrow_{\mathrm{TJ}} J$. $\square$

5   Resolving Cycles

It now remains to study the case where $G[I\Delta J]$ contains (even) cycles and both I and J are maximum independent sets. In this case, there may not be a TS-sequence from I to J (even though we assume that G is connected and claw-free) – consider for instance the case where G itself is an even cycle. In this section, we characterize the case where $I \leftrightarrow_{\mathrm{TS}} J$ holds, by showing that this is equivalent with every cycle being resolvable in a certain sense (Theorem 11 below). Subsequently, we show that resolvable cycles fall into two cases: internally or externally resolvable cycles, which are characterized next. We first define the notion of resolving a cycle.

Cycles in $G[I\Delta J]$ are clearly both I-bad and J-bad. The I-*bipartition* of an I-bad cycle is the ordered tuple $[V(C) \cap I, V(C)\backslash I]$. We say that an I-bad cycle C with I-bipartition $[A, B]$ is *resolvable* (with respect to I) if there exists an independent set I' such that $I \leftrightarrow_{\mathrm{TS}} I'$ and $G[I' \cup B]$ contains no cycles. A corresponding TS-sequence from I to I' is called a *resolving sequence* and is said to *resolve* C. By combining such a resolving sequence with a sequence of moves similar to the previous proof, and then reversing the moves in the sequence from I' to I, except for moves of tokens on the cycle, one can show that every resolvable cycle can be 'turned':

Lemma 10 (*). *Let I be an independent set in a claw-free graph G and let C be an I-bad cycle. If C is resolvable with respect to I, then $I \leftrightarrow_{\mathrm{TS}} I\Delta V(C)$.*

We can now prove the following useful characterization: $I \leftrightarrow_{\mathrm{TS}} J$ if and only if every cycle in $G[I\Delta J]$ is resolvable. By symmetry, it does not matter whether one considers resolvability with respect to I or to J.

Theorem 11. *Let I and J be independent sets in a claw-free connected graph G. Then $I \leftrightarrow_{\mathrm{TS}} J$ if and only if every cycle in $G[I\Delta J]$ is resolvable with respect to I.*

Proof: Consider an I-bad cycle C in $G[I \Delta J]$ with I-bipartition $[A, B]$, and a TS-sequence from I to J. Since $N_2(B)$ eventually contains no tokens, this sequence must contain a move $u \to v$ with $u \in N_2(B)$ and $v \notin N_2(B)$. The first such move can be shown to resolve the cycle.

The other direction is proved by induction on the number k of cycles in $G[I \Delta J]$. If $k = 0$, then by Lemma 3, $I \leftrightarrow_{TS} J$. If $k \geq 1$, then consider an I-bad cycle C in $G[I \Delta J]$. Let $I' = I \Delta V(C)$. By Lemma 10, $I \leftrightarrow_{TS} I'$. The graph $G[I' \Delta J]$ has one cycle fewer than $G[I \Delta J]$. Every cycle in $G[I' \Delta J]$ remains resolvable with respect to I' (one can first consider a TS-sequence from I' to I, and subsequently a TS-sequence from I that resolves the cycle). So by induction, $I' \leftrightarrow_{TS} J$, and therefore, $I \leftrightarrow_{TS} J$. □

Finally, we show that if an I-bad cycle C can be resolved, it can be resolved in at least one of two very specific ways. Let $[A, B]$ be the I-bipartition of C. A move $u \to v$ is called *internal* if $\{u, v\} \subseteq N_2(B)$ and *external* if $\{u, v\} \subseteq N_0(B)$. A resolving sequence $I_0, \ldots, I_m$ for C is called *internal* (or *external*) if every move except the last is an internal (respectively, external) move. (Obviously, to resolve the cycle, the last move can neither be internal nor external, and can in fact be shown to always be a move from $N_2(B)$ to $N_1(B)$.) The I-bad cycle C is called *internally resolvable* resp. *externally resolvable* if such sequences exist.

Lemma 12 (*). *Let I be an independent set in a claw-free graph G and let C be an I-bad cycle. Then any shortest TS-sequence that resolves C is an internal or external resolving sequence.*

Proof sketch: Let $[A, B]$ be the I-bipartition of C. Since G is claw-free, it follows that there are no edges between vertices in $N_2(B)$ and $N_0(B)$. This can be used to show that informally, any resolving sequence for C remains a resolving sequence after either omitting all noninternal moves or omitting all nonexternal moves, while keeping the last move, which subsequently resolves the cycle. □

Theorem 11 and Lemma 12 show that to decide whether $I \leftrightarrow_{TS} J$, it suffices to check whether every cycle in $G[I \Delta J]$ is externally or internally resolvable. Next we give characterizations that allow polynomial-time algorithms for deciding whether an I-bad cycle is internally or externally resolvable. For the external case, we use the assumption that I is a maximum independent set to show that in a *shortest* external resolving sequence $I_0, \ldots, I_m$, every token moves at most once (that is, for every move $u \to v$, both $u \in I_0$ and $v \in I_m$ hold), so these moves outline an augmenting path in a certain auxiliary graph.

Theorem 13 (*). *Let I be a maximum independent set in a claw-free graph G and let C be an I-bad cycle with I-bipartition $[A, B]$. Then C is externally resolvable if and only if there exists an $(I \backslash A)$-augmenting path in $G - A - B$ between a pair of vertices $x \in N_0(B)$ and $y \in N_1(B)$.*

For a given I-bad cycle C with I-bipartition $[A, B]$, there is a quadratic number of vertex pairs $x \in N_0(B)$ and $y \in N_1(B)$ that need to be considered, and

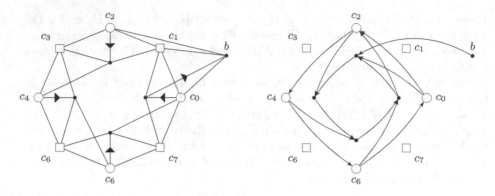

Fig. 1. An example of a claw-free graph G with an internally resolvable cycle, along with the corresponding auxiliary digraph $D(G, C)$.

for every such a pair, testing whether there is an $(I \setminus A)$-augmenting path between these in $G - A - B$ can be done in polynomial time (Theorem 7). So from Theorem 13 we conclude:

Corollary 14. *Let I be a maximum independent set in a claw-free graph G, and let C be an I-bad cycle. In polynomial time, it can be decided whether C is externally resolvable.*

Next, we characterize internally resolvable cycles. Shortest internal resolving sequences cannot be as easy to describe as external ones, since a token can move several times (see Figure 1). Nevertheless, these sequences can be shown to have a very specific structure, which can be characterized using paths in the following auxiliary digraphs.

To define these digraphs, consider an I-bad cycle $C = c_0, c_1, \ldots, c_{2n-1}, c_0$ in G, with $c_i \in I$ for even i. Let $[A, B]$ be the I-bipartition of C. For every $i \in \{0, \ldots, n-1\}$, define the corresponding *layer* as follows: $L_i = \{v \in V(G) \mid N(v) \cap B = N(c_{2i}) \cap B\}$. So when starting with I and using only internal moves, it can be seen that the token that starts on c_{2i} will stay in the layer L_i.

For such an I-bad cycle C of length at least 8, define $D(G, C)$ to be a digraph with vertex set $V(G)$, with the following arc set. For every $i \in \{0, \ldots, n-1\}$ and all pairs $u \in L_i, v \in L_{(i+1) \bmod n}$ with $uv \notin E(G)$, add an arc (u, v). For every $i \in \{0, \ldots, n-1\}$ and $b \in N_1(B)$ with $N(b) \cap B = \{c_{(2i-1) \bmod 2n}\}$, and every $v \in L_i$ with $bv \notin E(G)$, add an arc (b, v). We denote the reversed cycle by $C^{rev} = c_0, c_{2n-1}, \ldots, c_1, c_0$. This defines a similar digraph $D(G, C^{rev})$ (where arcs between layers are reversed, and arcs from $N_1(B)$ go to different layers). These graphs can be used to characterize whether C is internally resolvable.

Theorem 15 (*). *Let I be an independent set in a claw-free graph G. Let $C = c_0, c_1, \ldots, c_{2n-1}, c_0$ be an I-bad cycle ($c_0 \in I$) with I-bipartition $[A, B]$, of length at least 8. Then C is internally resolvable if and only if $D(G, C)$ or $D(G, C^{rev})$ contains a directed path from a vertex $b \in N_1(B)$ with $N(b) \cap I \subseteq A$ to a vertex in A.*

Corollary 16. *Let I be an independent set in a claw-free graph G on n vertices and let C be an I-bad cycle. It can be decided in polynomial time whether C is internally resolvable.*

Proof: If C has length at least 8, then Theorem 15 shows that it suffices to make a polynomial number of depth-first-searches in $D(G, C)$ and $D(G, C^{rev})$. Otherwise, let $[A, B]$ be the I-bipartition of C. $|A| \leq 3$, so there are only $\mathcal{O}(n^3)$ independent sets I' with $|I'| = |I|$ and $I \backslash A \subseteq I'$. So in polynomial time we can generate the subgraph of $\mathrm{TS}_k(G)$ induced by these sets, and search whether it contains a path from I to an independent set I^* with $I \backslash A \subseteq I^*$ where $G[B \cup I^*]$ contains no cycle. C is internally resolvable if and only if such a path exists. $\square$

6   Summary of the Algorithm

We now summarize how the previous lemmas yield a polynomial time algorithm for TS-Reachability and TJ-Reachability in claw-free graphs.

Theorem 17. *Let I and J be independent sets in a claw-free graph G. We can decide in polynomial time whether $I \leftrightarrow_{\mathrm{TS}} J$ and whether $I \leftrightarrow_{\mathrm{TJ}} J$.*

Proof: Assume $|I| = |J|$; otherwise, we immediately return NO. We first consider the case when G is connected. By Corollary 4, since G is connected, $I \leftrightarrow_{\mathrm{TS}} J$ if and only if $I \leftrightarrow_{\mathrm{TJ}} J$, thus we only need to consider the sliding model.

We test whether I and J are maximum independent sets of G, which can be done in polynomial time (by combining Proposition 8 and Theorem 7; see also [20,21,23]). If not, then by Lemma 9, $I \leftrightarrow_{\mathrm{TJ}} J$ holds, and thus $I \leftrightarrow_{\mathrm{TS}} J$, so we may return YES.

Now consider the case that both I and J are maximum independent sets. Theorem 11 shows that $I \leftrightarrow_{\mathrm{TS}} J$ if and only if every cycle in $G[I \triangle J]$ is resolvable with respect to I. By Lemma 12, it suffices to check for internal and external resolvability of such cycles. This can be done in polynomial time by Corollary 14 (since I is a maximum independent set of G) and Corollary 16. We return YES if and only if every cycle in C was found to be internally or externally resolvable.

Now let us consider the case when G is disconnected. Clearly tokens cannot slide between different connected components, so for deciding whether $I \leftrightarrow_{\mathrm{TS}} J$, we can apply the argument above to every component, and return YES if and only if the answer is YES for every component. If I is a not a maximum independent set then Lemma 9 shows that $I \leftrightarrow_{\mathrm{TJ}} J$ always holds. If I is maximum, then Proposition 2 shows that $I \leftrightarrow_{\mathrm{TJ}} J$ holds if and only if $I \leftrightarrow_{\mathrm{TS}} J$. $\square$

7   Discussion

The results presented here have two further implications. Firstly, combined with techniques from [28], it follows that $I \leftrightarrow_{\mathrm{TJ}} J$ can be decided for any graph G

that can be obtained from a collection of claw-free graphs using *disjoint union* and *complete join* operations. See [28] for more details.

Secondly, a closer look at constructed reconfiguration sequences shows that when G is claw-free, components of both $TS_k(G)$ and $TJ_k(G)$ have diameter bounded polynomially in $|V(G)|$. This is not surprising, since the same behavior has been observed many times. To our knowledge, the only known examples of polynomial time solvable reconfiguration problems that nevertheless require exponentially long reconfiguration sequences are on artificial instance classes, which are constructed particularly for this purpose (see e.g. [3,14]).

References

1. Bonamy, M., Bousquet, N.: Recoloring bounded treewidth graphs. Electronic Notes in Discrete Mathematics 44, 257–262 (2013)
2. Bonamy, M., Johnson, M., Lignos, I., Patel, V., Paulusma, D.: Reconfiguration graphs for vertex colourings of chordal and chordal bipartite graphs. Journal of Combinatorial Optimization 27(1), 132–143 (2014)
3. Bonsma, P., Cereceda, L.: Finding paths between graph colourings: PSPACE-completeness and superpolynomial distances. Theor. Comput. Sci. 410(50), 5215–5226 (2009)
4. Cereceda, L., van den Heuvel, J., Johnson, M.: Connectedness of the graph of vertex-colourings. Discrete Math. 308(5-6), 913–919 (2008)
5. Cereceda, L., van den Heuvel, J., Johnson, M.: Mixing 3-colourings in bipartite graphs. European J. of Combinatorics 30(7), 1593–1606 (2009)
6. Ito, T., Kawamura, K., Ono, H., Zhou, X.: Reconfiguration of list L(2, 1)-labelings in a graph. In: Chao, K.-M., Hsu, T.-S., Lee, D.-T. (eds.) ISAAC 2012. LNCS, vol. 7676, pp. 34–43. Springer, Heidelberg (2012)
7. Ito, T., Kamiński, M., Demaine, E.D.: Reconfiguration of list edge-colorings in a graph. In: Dehne, F., Gavrilova, M., Sack, J.-R., Tóth, C.D. (eds.) WADS 2009. LNCS, vol. 5664, pp. 375–386. Springer, Heidelberg (2009)
8. Ito, T., Demaine, E.D., Harvey, N.J.A., Papadimitriou, C.H., Sideri, M., Uehara, R., Uno, Y.: On the complexity of reconfiguration problems. Theoret. Comput. Sci. 412(12-14), 1054–1065 (2011)
9. Hearn, R.A., Demaine, E.D.: PSPACE-completeness of sliding-block puzzles and other problems through the nondeterministic constraint logic model of computation. Theor. Comput. Sci. 343(1-2), 72–96 (2005)
10. Gopalan, P., Kolaitis, P.G., Maneva, E.N., Papadimitriou, C.H.: The connectivity of Boolean satisfiability: Computational and structural dichotomies. SIAM J. Comput. 38(6), 2330–2355 (2009)
11. Kamiński, M., Medvedev, P., Milanič, M.: Complexity of independent set reconfigurability problems. Theor. Comput. Sci. 439, 9–15 (2012)
12. Bonsma, P.: Rerouting shortest paths in planar graphs. In: D'Souza, D., Kavitha, T., Radhakrishnan, J. (eds.) FSTTCS. LIPIcs, vol. 18, pp. 337–349. Schloss Dagstuhl - Leibniz-Zentrum für Informatik (2012)
13. Bonsma, P.: The complexity of rerouting shortest paths. Theor. Comput. Sci. 510, 1–12 (2013)
14. Kamiński, M., Medvedev, P., Milanič, M.: Shortest paths between shortest paths. Theor. Comput. Sci. 412(39), 5205–5210 (2011)

15. Suzuki, A., Mouawad, A.E., Nishimura, N.: Reconfiguration of dominating sets. CoRR abs/1401.5714 (2014)
16. Mouawad, A.E., Nishimura, N., Raman, V., Simjour, N., Suzuki, A.: On the parameterized complexity of reconfiguration problems. In: Gutin, G., Szeider, S. (eds.) IPEC 2013. LNCS, vol. 8246, pp. 281–294. Springer, Heidelberg (2013)
17. van den Heuvel, J.: The complexity of change. Surveys in Combinatorics, 127–160 (2013)
18. Chudnovsky, M., Seymour, P.D.: The structure of claw-free graphs. In: Webb, B.S. (ed.) Surveys in Combinatorics. London Mathematical Society Lecture Note Series, vol. 327, pp. 153–171. Cambridge University Press (2005)
19. Edmonds, J.: Paths, trees, and flowers. Canad. J. Math. 17, 449–467 (1965)
20. Minty, G.J.: On maximal independent sets of vertices in claw-free graphs. J. Comb. Theory, Ser. B 28(3), 284–304 (1980)
21. Sbihi, N.: Algorithme de recherche d'un stable de cardinalité maximum dans un graphe sans étoile. Discrete Mathematics 29(1), 53–76 (1980)
22. Nakamura, D., Tamura, A.: A revision of Minty's algorithm for finding a maximum weight stable set of a claw-free graph. Journal of the Operations Research Society of Japan 44(2), 194–204 (2001)
23. Schrijver, A.: Combinatorial optimization: Polyhedra and efficiency, vol. 24. Springer (2003)
24. Nobili, P., Sassano, A.: A reduction algorithm for the weighted stable set problem in claw-free graphs. Discrete Applied Mathematics (2013)
25. Faenza, Y., Oriolo, G., Stauffer, G.: An algorithmic decomposition of claw-free graphs leading to an $O(n^3)$-algorithm for the weighted stable set problem. In: SODA, pp. 630–646. SIAM (2011)
26. Bonsma, P., Kamiński, M., Wrochna, M.: Reconfiguring independent sets in claw-free graphs. CoRR abs/1403.0359 (2014)
27. Diestel, R.: Graph Theory. Electronic Edition. Springer-Verlag (2005)
28. Bonsma, P.: Independent set reconfiguration in cographs. CoRR abs/1402.1587 (2014); Extended abstract accepted for WG 2014

Competitive Online Routing on Delaunay Triangulations[*]

Prosenjit Bose[1], Jean-Lou De Carufel[1], Stephane Durocher[2], and Perouz Taslakian[3]

[1] Carleton University, Ottawa, Canada
jit@scs.carleton.ca, jdecaruf@cg.scs.carleton.ca
[2] University of Manitoba, Winnipeg, Canada
durocher@cs.umanitoba.ca
[3] American University of Armenia, Yerevan, Armenia
ptaslakian@aua.am

Abstract. The sequence of adjacent nodes (graph walk) visited by a routing algorithm on a graph G between given source and target nodes s and t is a *c-competitive* route if its length in G is at most c times the length of the shortest path from s to t in G. We present 21.766-, 17.982- and 15.479-competitive online routing algorithms on the Delaunay triangulation of an arbitrary given set of points in the plane. This improves the competitive ratio on Delaunay triangulations from the previous best of 45.749. We present a 7.621-competitive online routing algorithm for Delaunay triangulations of point sets in convex position.

1 Introduction

We study the fundamental problem of finding a route in a geometric graph from a given source vertex s to a given target vertex t. In our context, a geometric graph G is a weighted graph whose vertex set is a set P of n points in the plane, and whose edges are line segments joining pairs of points in P, where each edge is weighted by its length (the Euclidean distance between its endpoints). When full knowledge of the graph is provided, numerous algorithms exist for finding shortest paths in a weighted graph (e.g., Dijkstra's algorithm [10,12]). The problem is more challenging in the *online* setting, where a route is constructed incrementally and a partial route from s ending at a node u is extended by selecting one of u's neighbours as a function of limited information available locally at u. Without knowledge of the full graph, an online routing algorithm cannot identify a shortest path in general; the goal is to follow a path whose length is as short as possible. A path between two vertices s and t in G is a *c-spanning path* if its length is at most c times the length of the shortest path from s to t in G. An online routing algorithm is *c-competitive* on a class $\mathcal{G}$ of geometric graphs if for any graph $G \in \mathcal{G}$ and any pair of vertices $\{s, t\}$ in G,

---

[*] This work was supported in part by the Natural Sciences and Engineering Research Council of Canada (NSERC).

R Ravi and I.L. Gørtz (Eds.): SWAT 2014, LNCS 8503, pp. 98–109, 2014.
© Springer International Publishing Switzerland 2014

the algorithm constructs a c-spanning path from s to t in G. When c is a constant, we say the online routing algorithm is *competitive*. In this paper we examine the problem of designing an online routing algorithm that is c-competitive on the Delaunay triangulation for the smallest value c possible.

The Delaunay triangulation, denoted $DT(P)$, of a point set P in the plane is a triangulation of P with the property that the triangle $\triangle abc$ is a face in $DT(P)$ if and only if $\{a, b, c\} \subseteq P$ and $\bigcirc abc \cap P = \{a, b, c\}$, where $\bigcirc abc$ denotes the unique disk that has a, b, and c on its boundary. The Delaunay triangulation and its dual, the Voronoi diagram, are well studied; see [1,22] for comprehensive surveys of these structures. To simplify the presentation we assume that points in P are in general position.

An online routing algorithm sends a message m together with a header h from a source vertex s to a target vertex t in a graph G. Both the header and the message can be considered to be bit strings. Initially the algorithm only has knowledge of s, t and $N(s)$, where for each vertex v, $N(v)$ denotes the set of vertices directly adjacent to v in G (and their respective coordinates). Upon reception of a message m and its header h, a node u must select one of its neighbours to which to forward the message as a function of h and $N(u)$. This procedure repeats until the message reaches the target node t. Different routing algorithms are possible depending on the size of h and the fraction of G that is known to each node. In the setting considered in this paper, the header h stores the coordinates of the node s from which the message originated, the coordinates of the node t which is the final destination of the message, the coordinates of the neighbour of u that last forwarded the message, and possibly one additional value that is computed from distances between vertices visited by the message and may be modified by the algorithm during computation.

Online routing is also known as *local geometric routing* on geometric graphs, or simply as *local routing* when geometric information is not provided (or does not exist). Previous work in online routing includes results on triangulations [6,9,19,23], on more general planar or near-planar geometric graphs [7,9,14,15,16,17,19,21], and on arbitrary (non-geometric) graphs [3,8]. When h stores only the coordinates of the destination node t, we say an online routing algorithm is *oblivious*. That is, the forwarding decision at each node u is made as a function of only u, $N(u)$, and t. No competitive oblivious online routing algorithm exists [20], even on Delaunay triangulations [2]. In this paper we focus on *competitive* online routing algorithms. Allowing the header h to store slightly more information (some of which can be modified dynamically during routing) enables an online routing algorithm to guarantee not only that each route reaches its destination, but that it does so along a c-competitive path.

The *spanning ratio* of a graph G is the maximum ratio τ between the length of a shortest path σ on G joining any pair of nodes s and t and the Euclidean distance between s and t. That is, for any for any two vertices s and t in G there exists a path σ from s to t in G such that $|\sigma| \leq \tau|st|$, where $|\sigma|$ denotes the sum of the lengths of the edges in σ and $|st|$ denotes the Euclidean distance from s to t in G. Several previous results examine upper bounds on the spanning ratio τ of the Delaunay triangulation [11,13,18,24]. Dobkin et al. [13] proved that $\tau \leq (1 + \sqrt{5})\pi/2$ in $DT(P)$. Using this bound, Bose and Morin [6] found a $(9(1+\sqrt{5})\pi/2)$-competitive online routing algorithm for Delaunay triangulations (where $9(1 + \sqrt{5})\pi/2 \approx 45.749$). To the authors' knowledge, this was the smallest known competitive ratio for an online routing algorithm on Delaunay triangulations prior to our results.

We show that for each known upper bound τ on the spanning ratio of the Delaunay triangulation, for every set of points P and every $\{s, t\} \subseteq P$, there exists a path σ from s to t that is contained on the edges of the sequence of Delaunay triangles that intersects the line segment from s to t such that $|\sigma| \leq \tau|st|$. This property of the location of the path allows us to apply a hybrid of searching techniques developed in [5] with new techniques to define a corresponding online routing algorithm whose competitive ratio is at most 9τ for each previous upper bound on τ. The current best upper bound is $\tau \leq 1.998$, resulting in a corresponding competitive ratio of $9 \cdot 1.998 \approx 17.982$. Although this technique yields two new online routing algorithms for Delaunay triangulations, both of which improve on the previous best competitive ratio, we apply a new strategy to define a third online routing algorithm that reduces the competitive ratio further still to $\pi(5\pi + 4)/4 \approx 15.479$. Therefore, we improve the previous best competitive ratio for online routing on Delaunay triangulations by describing $(4\pi\sqrt{3})$-competitive, 17.982-competitive, and $(\pi(5\pi + 4)/4)$-competitive online routing algorithms in Sections 2.1, 2.2, and 2.3, respectively, where $4\pi\sqrt{3} \approx 21.766$ and $\pi(5\pi + 4)/4 \approx 15.479$. In Section 3 we examine Delaunay triangulations of sets of points in convex position for which we present a $(11 + 3\sqrt{2})/2$-competitive online routing algorithm using new techniques, where $(11 + 3\sqrt{2})/2 \approx 7.621$.

2   Routing on Delaunay Triangulations of Points in General Position

The problem of designing a competitive online routing algorithm on $DT(P)$ is challenging, in large part, because it seems difficult to compute a shortest path between two points in $DT(P)$ when complete knowledge of the graph is unavailable. This difficulty is related to the fact that a small perturbation in P can cause the the shortest path from s to t to change drastically. By focusing on specific local triangles in $DT(P)$ to the reduce the search space of candidate vertices to which to forward the message, and by exploiting geometric properties of the Delaunay triangulation, we can design online routing algorithms with good competitive ratios.

The search space is restricted by focusing on two specific paths that lie respectively above and below the line segment from s to t, where s and t denote the respective source and target nodes in $DT(P)$. Consider the ordered sequence of triangles that intersect the line segment st. Each triangle in this sequence has at least one edge whose interior is either completely above or completely below the line segment st. Define two ordered subsequences of triangles with one subsequence containing the triangles with an edge that lies above st, and the

other containing the triangles with an edge that lies below st. The subsequence
of edges lying above st determines a path from s to t in $DT(P)$. As is done by
Bose and Morin [5], we refer to this path as the *upper chain* from s to t and
denote it by U. The subsequence of edges lying below st forms the *lower chain*
from s to t and is denoted by L. Refer to Figure 1(a).

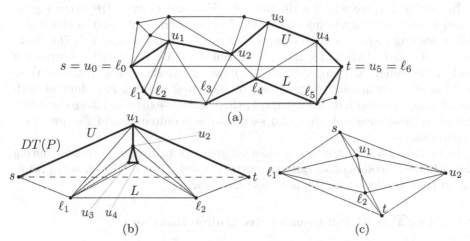

Fig. 1. (a) A Delaunay triangulation with the upper and lower chains (in bold) with
respect to s and t. (b) The upper chain U follows the sequence s, u_1, u_2, u_3, u_4, u_2,
u_1, t. (c) The vertices ℓ_1 and u_2 can be moved arbitrarily far from st, implying that
neither U nor L is a constant spanning path.

The upper chain is not necessarily a simple path since it may contain repeated edges or vertices (Figure 1(b)). Moreover, neither the upper chain nor
the lower chain is necessarily a constant spanning path (Figure 1(c)). However,
the subgraph of $DT(P)$ induced by $U \cup L$ contains a path whose length is at
most $(1+\sqrt{5})|st|\pi/2$, which is the property used to provide the only competitive
online routing algorithm [6] with competitive ratio at most $9(1 + \sqrt{5})|st|\pi/2$.

Bose and Morin [5] generalized this approach slightly to triangulated weakly
simple polygons. A polygon is *weakly simple* provided that the graph defined by
its vertices and edges is plane, the outer face is a cycle, and one bounded face
is adjacent to all vertices and edges. The weakly simple polygon is triangulated
when the bounded face is triangulated.

Theorem 1 (Bose and Morin [5]). *Given a plane geometric graph G that is
a triangulated weakly simple polygon, and two vertices s, t in G, there exists an
online competitive routing strategy that computes a path from s to t in G whose
competitive ratio is at most 9.*

Notice that the subgraph of $DT(P)$ induced by $U \cup L$ is a triangulated weakly
simple polygon since it is the ordered sequence of triangles intersecting st in
$DT(P)$. Therefore, showing the existence of a short path in this subgraph immediately gives a competitive online routing algorithm whose ratio is at most

9 times the length of this short path. This approach was used in [6], where the proof of the constant spanning ratio of the Delaunay triangulation by Dobkin et al. [13] was shown to construct a path of length at most $(1+\sqrt{5})|st|\pi/2 \approx 45.749$ in the subgraph induced by $U \cup L$. On the other hand, Xia [24] proves that there exists a path in the subgraph induced by $U \cup L$ whose length is at most $1.998|st|$, which implies an online routing algorithm whose ratio is at most 17.982.

In Section 2.1, we will use the proof by Keil and Gutwin [18] (showing an upper bound on the spanning ratio of the Delaunay triangulation) to give a new online routing algorithm with competitive ratio at most $4\pi\sqrt{3} \approx 21.766$. Note that Keil and Gutwin's [18] inductive proof does not necessarily construct a path in the subgraph induced by $U \cup L$; however, we show that whenever their proof satisfies the inductive hypothesis by including a vertex in a shortest path that lies outside the induced subgraph, there always exists an alternate vertex in the induced subgraph that also satisfies the requirements of the inductive hypothesis.

In Section 2.3 we introduce a different strategy to define an online routing algorithm with competitive ratio at most $\pi(5\pi + 4)/4 \approx 15.479$, drawing inspiration from Dobkin et al. [13] and Bose and Morin [6].

2.1 $(4\pi\sqrt{3}) \approx 21.766$-Competitive Online Routing

Keil and Gutwin [18] proved that for any two vertices s and t in $DT(P)$, there exists a path σ from s to t in $DT(P)$ such that $|\sigma| \leq \frac{4\pi\sqrt{3}}{9}|st| \leq 2.419|st|$. Although the path in the original proof may fall outside $U \cup L$, we show that the proof also implies the existence of a path of the same length among the vertices in $U \cup L$. We follow the construction given by Bose and Keil [4] (who proved the same result, but for the more general constrained Delaunay triangulations).

The proof has two main parts. The first part is a geometric property of Delaunay triangulations. The second part uses the geometric property to prove the result by induction. We begin with the former.

Consider the directed line segment st from s to t. Let $\bullet st$ be a circle through s and t such that the part of $\bullet st$ below st does not contain any points of P. We say that $\bullet st$ is a *right-empty circle* with respect to s and t. Let r denote the radius of $\bullet st$ and let $\theta(s, t)$ denote its *spanning angle*, corresponding to the reflex angle $\angle sat$, where a denotes the centre of $\bullet st$. Let $\bullet_m st$ denote the right-empty circle with respect to s and t that has the minimum spanning angle and let $\theta_m(s, t)$ denote its spanning angle. Bose and Keil [4, Lemma 2.1] proved the following lemma by induction on the rank of the minimum-spanning angles (with ties being broken arbitrarily).

Lemma 1 (Bose and Keil [4]). *For any set of points P in the plane and any $\{s, t\} \subseteq P$, if there is a right-empty circle $\bullet st$ with radius r and spanning angle $\theta(s, t)$, then there exists a path τ in $DT(P)$ from s to t whose length is at most $r \cdot \theta(s, t)$ such that every edge in τ has length at most $|st|$.*

The path τ of Lemma 1 satisfies the following property. Due to space constraints, we omit the proof.

Lemma 2. *All the vertices of the path τ are in $U \cup L$.*

We now outline the construction of the 2.419-path σ. Before doing this, we need to define a *lune*. Let p be a point on st and Γ_{sp} be the circular arc from s to p such that Γ_{sp} is above sp and the tangent to Γ_{sp} at s makes an angle of $\pi/3$ with st (refer to Figure 2(a)). Let Γ'_{sp} be the circular arc that is the reflection of

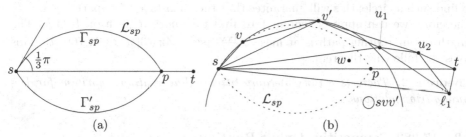

Fig. 2. (a) The lune $\mathcal{L}_{sp}$ with respect to s and p. (b) An example where the first vertex we hit by growing a lune from s is not in $U \cup L$.

Γ_{sp} across sp. The *lune* $\mathcal{L}_{sp}$ with respect to s and p is defined to be $\Gamma_{sp} \cup \Gamma'_{sp}$.

To construct the 2.419-path σ from s to t, we consider the largest empty lune $\mathcal{L}_{sp}$ that has a vertex $v \in P$ on its boundary. If there is more than one vertex on the boundary of $\mathcal{L}_{sp}$, we consider the one closest to s. We can see this as the process of growing a lune from s until it hits a vertex $v \in P$. To construct σ, we first travel from s to v using the path of Lemma 2 (by considering a specific right-empty circle $\bullet sv$; refer to the proof of Theorem 1.1 in [4]). Then, we apply induction from v to t. When we apply Lemma 2 from s to v, we need to consider a *good* right-empty circle. A right empty circle $\bullet sv$ is *good with respect to* $\mathcal{L}_{sp}$ if it is centered on so, where o is the center of Γ'_{sp}.

It is possible that the first vertex v of P we encounter by growing a lune from s is not in $U \cup L$ (refer to Figure 2(b)). In the original proof by Keil and Gutwin as well as the proof in Bose and Keil, it was not necessary for v to be in $U \cup L$ to prove the spanning ratio. However, to be able to route, we need this property to apply Theorem 1. Fortunately, we are able to show that there exists a point v' in $U \cup L$ that satisfies the same properties as v and allows the inductive argument to go through. We outline this below.

Lemma 3. *Suppose that the first vertex $v \in DT(P)$ we hit by growing a lune from s is not in $U \cup L$. Let $u_1 \in U$ and $\ell_1 \in L$ be such that $su_1 \in DT(P)$ and $s\ell_1 \in DT(P)$. If we keep growing the lune until it hits a vertex $v' \in U \cup L$, then $v' = u_1$ or $v' = \ell_1$. Moreover, if $v' = u_1$ (respectively $v' = \ell_1$), there exists a good right-empty circle $\bullet su_1$ (respectively $\bullet \ell_1 s$) with respect to $\mathcal{L}_{sp}$.*

Proof. Without loss of generality, suppose that v is above the line through st. Denote by $\mathcal{L}_{sp}$ the empty lune that has v on its boundary. Denote by $\mathcal{L}_{sp'}$ the (not necessarily empty) lune that has u_1 on its boundary. We have that v is outside of $\bigcirc su_1 \ell_1$, where $\bigcirc su_1 \ell_1$ defines $\triangle su_1 \ell_1 \in DT(P)$. Therefore, the part of $\mathcal{L}_{sp'}$ that is below su_1 is inside the empty circle $\bigcirc su_1\ell_1$. Consequently, if we keep growing $\mathcal{L}_{sp}$ until it hits a vertex $v' \in U \cup L$, then $v' = u_1$. Moreover, since the part of $\mathcal{L}_{sp'}$ that is below su_1 is empty, there exists a good right-empty circle $\bullet su_1$ with respect to $\mathcal{L}_{sp}$.   □

The proof of Theorem 1.1 in [4] is based on finding a good right-empty circle before applying induction. In our case, we can use Lemma 3 within Theorem 1.1 to find such a circle; this will guarantee that there exists a 2.419-path $\sigma \in U \cup L$. Therefore, we can apply Theorem 1 to find the shortest path on $U \cup L$. The length of our routing path is at most $9\frac{4\pi\sqrt{3}}{9}|st| = 4\pi\sqrt{3}|st| \approx 21.766|st|$. This gives the following theorem.

Theorem 2. *There is a $(4\pi\sqrt{3})$-competitive online routing algorithm for Delaunay triangulations.*

2.2   17.982-Competitive Online Routing

Xia [24] showed that the stretch factor of a Delaunay triangulation of a set of points in the plane is less than 1.998. His proof restricts the search space to the set of triangles intersecting st as outlined in the proof of Corollary 1 in [24]. Therefore, by applying Theorem 1, we obtain a competitive online routing strategy whose competitive ratio is at most 17.982.

Theorem 3. *There is a 17.892-competitive online routing algorithm for Delaunay triangulations.*

2.3 $(\pi(5\pi + 4)/4) \approx 15.479$-Competitive Online Routing

We propose an online competitive routing algorithm inspired by the work of Dobkin et al. [13] and Bose and Morin [6]. Let P denote any set of n points in general position and let s and t denote any two vertices in P. Without loss of generality, assume s and t lie on the x-axis, with s having a smaller x-coordinate than t. Let $V_0, \ldots, V_{m-1}$ be the cells of the Voronoi diagram intersected by the line segment st, with V_0 being the Voronoi cell of s and V_{m-1} being the cell of t. The path from s to t in $DT(P)$ obtained by following the sites generating the cells $V_0, \ldots, V_{m-1}$, in order, shall be referred to as the *Voronoi path* and denoted $VP(s,t)$. Label the vertices on this path $s = v_0, \ldots, v_{m-1} = t$. The Voronoi path is x-monotone and it is not necessarily a constant spanning path [13] (see Figure 3). Dobkin et al. [13] proved the following.

Lemma 4 (Dobkin et al. [13]). *Let N be the set of edges of $VP(s,t)$ that do not cross the segment st. The sum of the lengths of the edges in N is at most $|st|\pi/2$.*

If the vertices on $VP(s,t)$ all lie above the line through s and t, the Voronoi path is called *one-sided*. The above lemma implies that if $VP(s,t)$ is one-sided, then $|VP(s,t)| \leq |st|\pi/2$. Therefore, $VP(s,t)$ is a $\pi/2$-spanning path when it is one-sided. Note that $VP(s,t)$ is not necessarily a constant spanning path when

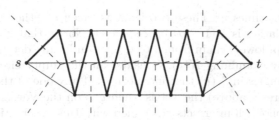

Fig. 3. This example shows that the number of times the Voronoi path (in bold) crosses st is unbounded in general. Consequently, the Voronoi path is not a constant spanning path.

it crosses st. Consider a Voronoi path from s to t that is not one-sided. Let $s = b_0, b_1, \ldots, b_q = t$ be the subsequence of vertices of the Voronoi path that lie above the x-axis. Consider two consecutive vertices in this subsequence $b_i = v_j$ and $b_{i+1} = v_k$ that are not consecutive on the Voronoi path, i.e. $k \neq j + 1$. This means that the edge $v_j v_{j+1}$ and $v_{k-1} v_k$ both cross st. (refer to Figure 4). Let

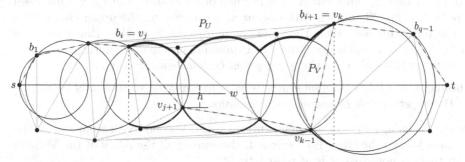

Fig. 4. The red and dashed line represents the Voronoi path P_V from $b_0 = s$ to $b_q = t$. The circles are centered on st. They are the ones that define the Voronoi path. This is an example where we would follow the Voronoi path since $h \leq \frac{1}{4}w$.

P_V be the Voronoi path $v_j, v_{j+1}, \ldots, v_k$ and let P_U be the path from v_j to v_k on the upper chain. For a point $p \in P$, let $x(p)$ and $y(p)$ be the x-coordinate and y-coordinate of p, respectively. Define $h = \min_{j < z < k} |y(v_z)|$ and $w = x(v_j) - x(v_k)$. Dobkin et al. [13] proved the following:

Lemma 5 (Dobkin et al. [13]). *If $h \leq w/4$, $|P_V|$ is at most $(1 + \sqrt{5})w\pi/2$ and the path from v_{j+1} to v_{k-1} has length at most $w\pi/2$.*

Using the construction given by Dobkin et al. [13], Bose and Morin [6] proved:

Lemma 6 (Bose and Morin [6]). *If $h > w/4$, $|P_U|$ is at most $w\pi^2/4$.*

Intuitively, the two lemmas state that when the Voronoi path from v_j to v_k comes "close" to the x-axis, then the length of the Voronoi path is at most a constant times w, otherwise the length of the upper chain from v_j to v_k is at most a constant times w. These two lemmas taken together imply that the Delaunay triangulation is a $(1 + \sqrt{5})\pi/2$-spanner. Notice that given a vertex v on the upper (resp. lower) chain from s to t, one can locally determine if v is on $VP(s,t)$ simply by examining $N(v)$. Consider all the empty circles defined by the Delaunay triangles in $N(v)$ that intersect st. If any one of these circles has its center below (resp. above) the x-axis, then v is on the Voronoi path from s to t since its Voronoi cell intersects st. Armed with this observation, Lemmas 5 and 6 seem to suggest the following competitive online routing algorithm:

When at a vertex b_i, if b_{i+1} is adjacent to b_i on the Voronoi path from s to t, follow the edge. If b_i and b_{i+1} are not adjacent on the Voronoi path, follow P_V from b_i to b_{i+1} when $h \le w/4$ and P_U when $h > w/4$. Unfortunately, the main caveat to this approach is that we do not know how to compute h or w locally from vertex b_i. It seems that knowledge of P_V is required to compute h and w, which is not necessarily available locally at b_i.

To overcome this obstacle, we slightly modify the above approach. When b_i and b_{i+1} are adjacent on the Voronoi path, we still follow the edge. However, when they are not adjacent, we take the following approach. Let $d = |v_j v_{j+1}|$. From v_j, follow P_U until either v_k is reached or a distance of at most d has been travelled on P_U. Should the latter occur at a vertex u on the upper chain, let v be the vertex furthest along the lower chain adjacent to u. Note that v must be on P_V. Move to v and continue on P_V. Proceed in this manner until t is reached. We refer to this online routing strategy as *OnlineDelaunayRoute*.

Theorem 4. *OnlineDelaunayRoute is an online routing strategy that is $(\pi(5\pi + 4)/4)$-competitive on Delaunay triangulations.*

Proof. When b_i and b_{i+1} are consecutive on the Voronoi path from s to t, the message follows the edge. By Lemma 4, the sum of all the edges of the Voronoi path that do not cross st is at most $|st|\pi/2$.

When b_i and b_{i+1} are not consecutive, the message follows two different paths depending on the length of P_U. If P_U has length at most d, then the messages travels on P_U. Otherwise, it travels on P_U for a distance of d, crosses over onto P_V and then continues travelling on P_V. Notice that by the triangle inequality, this is shorter than travelling on P_U for distance of at most d, returning to b_i and travelling on P_V. Therefore, the total distance travelled is at most $2d + |P_V|$. We bound this distance in terms of w. There are 4 cases to consider.

Case 1: $h \le w/4$ and the message travels $|P_U|$.
By Lemma 5, we have $|P_V| \le (1 + \sqrt{5})w\pi/2$. Since the edge $v_j v_{j+1} \in P_V$, we have that $d \le |P_V|$. Since the message remains on P_U, we have that $|P_U| \le d$. Therefore, $|P_U| \le (1 + \sqrt{5})w\pi/2 \le 5.09w$.

Case 2: $h \le w/4$ and the message travels $2d + |P_V|$.
By Lemma 5, we have $|P_V| \le (1 + \sqrt{5})w\pi/2$. Since the edge $v_j v_{j+1} \in P_V$, we have that $d \le |P_V|$. Therefore, $2d + |P_V| \le 3|P_V| \le 3(1 + \sqrt{5})w\pi/2 \le 15.25w$.

Case 3: $h > w/4$ and the message travels $|P_U|$.
By Lemma 6, $|P_U| \le w\pi^2/4 \le 2.47w$.

Case 4: $h > w/4$ and the message travels $2d + |P_V|$.

By Lemma 6, $|P_U| \leq w\pi^2/4$. By construction, $d \leq |P_U|$. Since the portion of P_V that lies below the x-axis is a one-sided Voronoi path, its length is at most $w\pi/2$ by Lemma 5. By the triangle inequality, $|P_V| \leq 2d + \pi w + |P_U|$. Therefore, putting it all together, we have $2d + |P_V| \leq \pi(5\pi + 4)w/4 \leq 15.479w$.

Since the cost of the path is dominated by the value obtained in Case 4, the result follows.   □

3 $(11 + 3\sqrt{2})/2 \approx 7.621$-Competitive Online Routing for Points in Convex Position

We present an online routing algorithm with a competitive ratio of at most $(11 + 3\sqrt{2})/2$ for Delaunay triangulations of sets of points in convex position, where $(11 + 3\sqrt{2})/2 \approx 7.621$. Throughout this section we assume that P is a set of points in convex position in the plane. For ease of exposition, we assume without loss of generality that the line segment st is horizontal, with s to the left of t. Let $\ominus st$ be the circle whose diameter is the line segment st. Let $S(s,t)$ be the axis-parallel square whose bisector is the line segment st. Again, let U and L denote the respective upper and lower chains of s and t in $DT(P)$. Before proving Theorem 10, we begin with a few geometric lemmas and observations used to prove the correctness of the algorithm and to bound its competitive ratio.

Lemma 7. *If a line ℓ is not parallel to any side of a convex polygon Q, then ℓ intersects the boundary of Q in at most two points.*

Lemma 8. *If vertex $v \in U$ (respectively $v \in L$) is outside of $\ominus st$ then v is adjacent to at least one vertex $v' \in L$ (respectively $v' \in U$) that is in $\ominus st$.*

Proof. Suppose that both v and v' are outside $\ominus st$. By definition, every edge between a vertex in U and a vertex in V must intersect st. Since vv' intersect st and st is the diameter of $\ominus st$, every circle with v and v' on its boundary will either contain s in its interior or t in its interior. This contradicts the fact that vv' is an edge of the Delaunay triangulation.   □

We now describe the routing algorithm. The message starts at a node s with destination t. The algorithm first forwards the message from s to one of its neighbours on $U \cup L$ that is in $S(s,t)$. Such a vertex must exist by Lemma 8. The algorithm makes a forwarding decision at each vertex v along the route, which we now describe. Without loss of generality, suppose that v is on the upper chain (an analogous symmetric case applies if v is on the lower chain). Let u be the vertex adjacent to v on the upper chain and let ℓ be the vertex adjacent to v that is furthest right on the lower chain. If u is in $S(s,t)$ then forward the message to u, otherwise forward it to ℓ. This decision can be made locally given the following information stored in the header: the source s, the destination t and $N(v)$, the set of vertices adjacent to v. Let σ be the path followed by the message.

Lemma 9. *The path σ taken by the message m crosses st at most 3 times before reaching t.*

Proof. Notice that prior to crossing the boundary of the square, the path σ crosses st. Without loss of generality, assume that σ crosses st for the first time from a vertex on the upper chain to a vertex on the lower chain. Let $x_1 y_1$ be this edge with $x_1 \in U$ and $y_1 \in L$. Since the path crosses st, x_1 must be adjacent to a vertex $x_1' \in U$ that is outside $S(s, t)$. By Lemma 8, y_1 must be in $\ominus st$ since it is also adjacent to x_1'. By Observation 7, the portion of the upper chain from x_1 to t in clockwise order and the portion of the lower chain from y_1 to t in counter-clockwise order intersects $S(s, t)$ a total of 6 times.

Suppose, for a contradiction, that σ crossed st four times with the first edge as above from x_1 to y_1. Let the other three edges be $x_2 y_2$, $x_3 y_3$, and $x_4 y_4$ with $x_i \in U$ and $y_i \in L$. This means that the upper chain intersects $S(s, t)$ twice from x_1 to x_2 since x_1' is outside $S(s, t)$ and x_2 is inside $S(s, t)$ by Lemma 8. Similarly, the lower chain between y_2 and y_3 intersects $S(s, t)$ twice. The upper chain from x_3 to x_4 intersects $S(s, t)$ twice. Finally, the edge on the lower chain adjacent to y_4 intersects $S(s, t)$ since this is what prompted the algorithm to cross to x_4. However, this is at least 7 intersections which is a contradiction. $\square$

Lemma 10. *The length of the path σ is at most $(11 + 3\sqrt{2})|st|/2$.*

Proof. Let U' be the sequence $s = u_0', u_1', \ldots u_k' = t$ of vertices followed by the message on U and L' be $s = \ell_0', \ell_1', \ldots \ell_b' = t$ be the sequence followed by the message on L. By construction, neither U' nor L' go outside $S(s, t)$. Since the union of these two sequences is a convex polygon inside $S(s, t)$, its perimeter is at most the perimeter of the square which is $4|st|$. This accounts for all of σ except for the crossing edges.

By Lemma 9, σ crosses st at most 3 times. Each of those edges has one endpoint in $S(s, t)$ and one endpoint in $\ominus st$. Therefore, its length is at most $(\sqrt{2}/2 + 1/2)|st|$ since the longest such edge has one endpoint on the corner of the square and the other diametrically opposed on the boundary of the circle. Summing the components gives an upper bound on σ of $(11 + 3\sqrt{2})|st|/2$. $\square$

Theorem 5 follows from Lemma 10:

Theorem 5. *There is a $(11 + 3\sqrt{2})/2$-competitive online routing algorithm for Delaunay triangulations of convex point sets.*

References

1. Aurenhammer, F., Klein, R., Lee, D.-T.: Voronoi Diagrams and Delaunay Triangulations. World Scientific (2013)
2. Bose, P., Brodnik, A., Carlsson, S., Demaine, E.D., Fleischer, R., López-Ortiz, A., Morin, P., Munro, I.: Online routing in convex subdivisions. Int. J. Comp. Geom. & App. 12(4), 283–295 (2002)
3. Bose, P., Carmi, P., Durocher, S.: Bounding the locality of distributed routing algorithms. Dist. Comp. 26(1), 39–58 (2013)

4. Bose, P., Keil, J.M.: On the stretch factor of the constrained Delaunay triangulation. In: ISVD, pp. 25–31 (2006)
5. Bose, P., Morin, P.: Competitive online routing in geometric graphs. Theor. Comp. Sci. 324(2-3), 273–288 (2004)
6. Bose, P., Morin, P.: Online routing in triangulations. SIAM J. Comp. 33(4), 937–951 (2004)
7. Bose, P., Morin, P., Stojmenović, I., Urrutia, J.: Routing with guaranteed delivery in ad hoc wireless networks. Wireless N. 7(6), 609–616 (2001)
8. Braverman, M.: On ad hoc routing with guaranteed delivery. In: PODC, vol. 27, p. 418. ACM (2008)
9. Chen, D., Devroye, L., Dujmović, V., Morin, P.: Memoryless routing in convex subdivisions: Random walks are optimal. In: EuroCG, pp. 109–112 (2010)
10. Cormen, T.H., Leiserson, C.E., Rivest, R.L., Stein, C.: Introduction to Algorithms, 3rd edn. MIT Press (2009)
11. Cui, S., Kanj, I.A., Xia, G.: On the stretch factor of Delaunay triangulations of points in convex position. Comp. Geom. 44(2), 104–109 (2011)
12. Dijkstra, E.W.: A note on two problems in connexion with graphs. Numer. Math. 1, 269–271 (1959)
13. Dobkin, D.P., Friedman, S.J., Supowit, K.J.: Delaunay graphs are almost as good as complete graphs. Disc. & Comp. Geom. 5, 399–407 (1990)
14. Durocher, S., Kirkpatrick, D., Narayanan, L.: On routing with guaranteed delivery in three-dimensional ad hoc wireless networks. Wireless Net. 16, 227–235 (2010)
15. Fraser, M.: Local routing on tori. Adhoc and Sensor Wireless Net. 6, 179–196 (2008)
16. Fraser, M., Kranakis, E., Urrutia, J.: Memory requirements for local geometric routing and traversal in digraphs. In: CCCG, vol. 20, pp. 195–198 (2008)
17. Guan, X.: Face Routing in Wireless Ad-Hoc Networks. PhD thesis, University of Toronto (2009)
18. Keil, J.M., Gutwin, C.A.: Classes of graphs which approximate the complete euclidean graph. Disc. & Comp. Geom. 7, 13–28 (1992)
19. Kranakis, E., Singh, H., Urrutia, J.: Compass routing on geometric networks. In: CCCG, vol. 11, pp. 51–54 (1999)
20. Kuhn, F., Wattenhofer, R., Zollinger, A.: Asymptotically optimal geometric mobile adhoc routing. In: DIALM, pp. 24–33. ACM (2002)
21. Kuhn, F., Wattenhofer, R., Zollinger, A.: Ad-hoc networks beyond unit disk graphs. In: DIALM-POMC, pp. 69–78. ACM (2003)
22. Okabe, A., Boots, B., Sugihara, K., Chiu, S.N.: Spatial Tessellations: Concepts and Applications of Voronoi Diagrams. Wiley Series in Probability and Statistics. Wiley (2009)
23. Si, W., Zomaya, A.Y.: New memoryless online routing algorithms for Delaunay triangulations. IEEE Trans. Par. & Dist. Sys. 23(8) (2012)
24. Xia, G.: The stretch factor of the Delaunay triangulation is less than 1.998. SIAM J. Comp. 42(4), 1620–1659 (2013)

Optimal Planar Orthogonal Skyline Counting Queries*

Gerth Stølting Brodal and Kasper Green Larsen

MADALGO**, Department of Computer Science, Aarhus University
{gerth,larsen}@cs.au.dk

Abstract. The skyline of a set of points in the plane is the subset of maximal points, where a point (x, y) is maximal if no other point (x', y') satisfies $x' \geq x$ and $y' \geq y$. We consider the problem of preprocessing a set P of n points into a space efficient static data structure supporting orthogonal skyline counting queries, i.e. given a query rectangle R to report the size of the skyline of $P \cap R$. We present a data structure for storing n points with integer coordinates having query time $O(\lg n / \lg \lg n)$ and space usage $O(n)$ words. The model of computation is a unit cost RAM with logarithmic word size. We prove that these bounds are the best possible by presenting a matching lower bound in the cell probe model with logarithmic word size: Space usage $n \lg^{O(1)} n$ implies worst case query time $\Omega(\lg n / \lg \lg n)$.

1 Introduction

In this paper we consider orthogonal range skyline queries for a set of points in the plane. A point $(x, y) \in \mathbb{R}^2$ *dominates* a point (x', y') if and only if $x' \leq x$ and $y' \leq y$. For a set of points P, a point $p \in P$ is *maximal* if no other point in P dominates p, and the *skyline* of P, Skyline(P), is the subset of maximal points in P.

We consider the problem of preprocessing a set P of n points in the plane with integer coordinates into a data structure to support *orthogonal range skyline counting* queries: Given an axis-aligned query rectangle $R = [x_1, x_2] \times [y_1, y_2]$ to report the size of the skyline of the subset of the points from P contained in R, i.e. report $|\text{Skyline}(P \cap R)|$. The main results of this paper are matching upper and lower bounds for data structures supporting such queries, thus completely settling the problem. Our model of computation is the standard unit cost RAM with logarithmic word size.

Previous Work. Orthogonal range searching is one of the most fundamental and well-studied topics in computational geometry, see e.g. [1] for an extensive list of previous results. For orthogonal range queries in the plane, with integer coordinates in $[n] \times [n] = \{0, \dots, n-1\} \times \{0, \dots, n-1\}$, the main results are

---

* The full version of this paper is available at arxiv.org/abs/1304.7959

** Center for Massive Data Algorithmics, a Center of the Danish National Research Foundation, grant DNRF84.

R Ravi and I.L. Gørtz (Eds.): SWAT 2014, LNCS 8503, pp. 110–121, 2014.

Table 1. Previous and new results for skyline counting queries

Space (words)	Query time	Reference
$n\frac{\lg^2 n}{\lg\lg n}$	$\frac{\lg^{3/2} n}{\lg\lg n}$	[6]
$n\lg n$	$\lg n$	[7]
$n\frac{\lg^3 n}{\lg\lg n}$	$\frac{\lg n}{\lg\lg n}$	[8]
n	$\frac{\lg n}{\lg\lg n}$	New

Table 2. Previous and new results for skyline reporting queries

Space (words)	Query time	Reference
$n\lg n$	$\lg^2 n + k$	[9] (dynamic)
$n\lg n$	$\lg n + k$	[10,7]
$n\frac{\lg n}{\lg\lg n}$	$\frac{\lg n}{\lg\lg n} + k$	[6]
$n\lg^\varepsilon n$	$(k+1)\lg\lg n$	[11]
$n\lg^\varepsilon n$	$\frac{\lg n}{\lg\lg n} + k$	New
$n\lg\lg n$	$(k+1)(\lg\lg n)^2$	[11]
$n\lg\lg n$	$\frac{\lg n}{\lg\lg n} + k\lg\lg n$	New
n	$(k+1)\lg^\varepsilon n$	[11]

the following: For the orthogonal range *counting* problem, i.e. queries report the total number of input points inside a query rectangle, optimal $O(\lg n/\lg\lg n)$ query time using $O(n)$ space was achieved in [2]. Optimality was shown in [3], where it was proved that space $n\lg^{O(1)} n$ implies query time $\Omega(\lg n/\lg\lg n)$ for range counting queries.

For range *reporting* queries it is known that space $n\lg^{O(1)} n$ implies query time $\Omega(\lg\lg n + k)$, where k is the number of points reported within the query range [4]. The best upper bounds known for range reporting are: Optimal space $O(n)$ and query time $O((k+1)\lg^\varepsilon n)$ [1], and optimal query time $O(\lg\lg n + k)$ with space $O(n\lg^\varepsilon n)$ [5]. In both cases $\varepsilon > 0$ is an arbitrarily small constant.

Orthogonal range skyline counting queries were first consider in [6], where a data structure was presented with space usage $O(n\lg^2 n/\lg\lg n)$ and query time $O(\lg^{3/2} n/\lg\lg n)$. This was subsequently improved to $O(n\lg n)$ space and $O(\lg n)$ query time [7]. Finally, a data structure achieving an even faster query time of $O(\lg n/\lg\lg n)$ was presented, however the space usage of that solution was a prohibitive $O(n\lg^3 n/\lg\lg n)$ [8]. Thus to date, no linear space solution exists with a non-trivial query time. Also, from a lower bound perspective, it is not known whether the problem is easier or harder than the standard range counting problem.

For orthogonal skyline reporting queries, the best bound is $O(n\lg n/\lg\lg n)$ space with query time $O(\lg n/\lg\lg n + k)$ [6], where k is the size of the reported skyline. Note that an $\Omega(\lg\lg n)$ search term is needed for skyline range reporting since the $\Omega(\lg\lg n)$ lower bound for standard range reporting was proved even for the case of determining whether the query rectangle is empty [4].

In [11] solutions for the sorted range reporting problem were presented, i.e. the problem of reporting the k leftmost points within a query rectangle in sorted order of increasing x-coordinate. With space $O(n)$, $O(n\lg\lg n)$ and $O(n\lg^\varepsilon n)$, respectively, query times $O((k+1)\lg^\varepsilon n)$, $O((k+1)(\lg\lg n)^2)$, and $O(k+\lg\lg n)$ were achieved, respectively. The structures of [11] support finding the rightmost (skyline) point in a query range $(k = 1)$. By recursing on the rectangle above the reported point one immediately get the bounds for skyline reporting listed in Table 2, where only the linear space solution achieves query times matching those of general orthogonal range reporting.

Our Results. In Section 3 we present a linear space data structure supporting orthogonal range skyline counting queries in $O(\lg n/\lg\lg n)$ time, thus for the first time achieving linear space and improving over all previous tradeoffs. In Section 2 we show that this is the best possible by proving a matching lower bound. More specifically, we prove a lower bound stating that the query time t must satisfy $t = \Omega(\lg n/\lg(Sw/n))$. Here $S \geq n$ is the space usage in number of words and $w = \Omega(\lg n)$ is the word size in bits. For $w = \lg^{O(1)} n$ and $S = n\lg^{O(1)} n$, this bound becomes $t = \Omega(\lg n/\lg\lg n)$. The lower bound is proved in the cell probe model of Yao [12], which is more powerful than the unit cost RAM and hence the lower bound also applies to RAM data structures.

As a side result, we can also modify our counting data structure to support reporting queries. The details are in the full version of the paper. Our reporting data structure has query time $O(\lg n/\lg\lg n+k)$ and space usage $O(n\lg^\varepsilon n)$. The best previous reporting structure with a linear term in k has $O(\lg n/\lg\lg n + k)$ query time but $O(n\lg n/\lg\lg n)$ space [6]. The reporting structure can also be modified to achieve $O(\lg n/\lg\lg n + k\lg\lg n)$ query time and $O(n\lg\lg n)$ space. See Table 2 for a comparison to previous results.

Our lower bound follows from a reduction of reachability in butterfly graphs to two-sided skyline counting queries, extending reductions by Pătraşcu [13] for two-dimensional rectangle stabbing and range counting queries. Our upper bounds are achieved by constructing a balanced search tree of degree $\Theta(\lg^\varepsilon n)$ over the points sorted by x-coordinate. At each internal node we store several space efficient rank-select data structures storing the points in the subtrees sorted by rank-reduced y-coordinates. Using a constant number of global tables, queries only need to spend $O(1)$ time at each level of the tree.

Preliminaries. If the coordinates of the input and query points are arbitrary integers fitting into a machine word, then we can map the coordinates to the range $[n]$ by using the RAM dictionary from [14], which support predecessor queries on the lexicographical orderings of the points in time $O(\sqrt{\lg n/\lg\lg n})$ using $O(n)$ space. This is less than the $O(\lg n/\lg\lg n)$ query time we are aiming for. Our solution makes extensive use of the below results from succinct data structures.

Lemma 1 ([15]). *A vector $X[1..s]$ of s zero-one values, with t values equal to one, can be stored in a data structure of size $O(t(1+\lg s/t))$ bits supporting* rank *and* select *in $O(1)$ time, where* rank(i) *returns the number of ones in $X[1..i]$, provided $X[i] = 1$, and* select(i) *returns the position of the i'th one in X.*

Lemma 2 ([16]). *Let $X[1..s]$ be a vector of s non-negative integers with total sum t. There exists a data structure of size $O(s\lg(2 + t/s))$ bits, supporting the lookup of $X[i]$ and the prefix sum $\sum_{j=1}^i X[j]$ in $O(1)$ time, for $i = 1,\ldots,s$.*

Lemma 3 ([17,18]). *Let $X[1..s]$ be a vector of integers. There exists a data structure of size $O(s)$ bits supporting range-maximum-queries in $O(1)$ time, i.e. given i and j, $1 \leq i \leq j \leq s$, reports the index k, $i \leq k \leq j$, such that $X[k] = \max(X[i..j])$. Queries only access this data structure, i.e. the vector X is not stored.*

2   Lower Bound

That an orthogonal range skyline counting data structure requires $\Omega(n \lg n)$ bits space, follows immediately since each of the $n!$ different input point sets of size n, where points have distinct x- and y-coordinates from $[n]$, can be reconstructed using query rectangles considering each possible point in $[n]^2$ independently, i.e. the space usage is at least $\lceil \lg_2(n!) \rceil = \Omega(n \lg n)$ bits.

In the remainder of this section, we prove that any data structure using $S \geq n$ words of space must have query time $t = \Omega(\lg n / \lg(Sw/n))$, where $w = \Omega(\lg n)$ denotes the word size in bits. In particular for $w = \lg^{O(1)} n$, this implies that any data structure using $n \lg^{O(1)} n$ space must have query time $t = \Omega(\lg n / \lg \lg n)$, showing that our data structure from Section 3 is optimal. Our lower bound holds even for data structures only supporting skyline counting queries inside 2-sided rectangles, i.e. query rectangles of the form $(-\infty, x] \times (-\infty, y]$. The lower bound is proved in the cell probe model of Yao [12] with word size $w = \Omega(\lg n)$. Since we derive our lower bound by reduction, we will not spend time on introducing the cell probe model, but merely note that lower bounds proved in this model applies to data structures developed in the unit cost RAM model. See e.g. [3] for a brief description of the cell probe model.

Reachability in the Butterfly Graph. We prove our lower bound by reduction from the problem known as *reachability oracles in the butterfly graph* [13]. A butterfly graph of degree B and depth d is a directed graph with $d + 1$ layers, each having B^d nodes ordered from left to right. The nodes at level 0 are the *sources* and the nodes at level d are the *sinks*. Each node, except the sinks, has out-degree B, and each node, except the sources, has in-degree B.

If we number the nodes at each level with $0, \ldots, B^d - 1$ from left to right and interpret each index $i \in [B^d]$ as a vector $v(i) = v(i)[d-1] \cdots v(i)[0] \in [B]^d$ (just write i in base B), then the node at index i at layer $k \in [d]$ has an out-going edge to each node j at layer $k + 1$ for which $v(j)$ and $v(i)$ differ only in the k'th coordinate. Here the 0'th coordinate is the coordinate corresponding to the least significant digit when thinking of $v(i)$ and $v(j)$ as numbers written in base B. Observe that there is precisely one directed path between each source-sink pair. For the s'th source and the t'th sink, this path corresponds to "morphing" one digit of $v(s)$ into the corresponding digit in $v(t)$ for each layer traversed in the butterfly graph.

The input to the problem of reachability oracles in the butterfly graph, with degree B and depth d, is a subset of the edges of the butterfly graph, i.e. we are given a subgraph G of the butterfly as input. A query is specified by a source-sink pair (s, t) and the goal is to return whether there exists a directed path from the given source s to the given sink t in G. Pătraşcu proved the following:

Theorem 1 (Pătraşcu [13], Section 5). *Any cell probe data structure answering reachability queries in subgraphs of the butterfly graph with degree B and depth d, having space usage S words of w bits, must have query time $t = \Omega(d)$, provided $B = \Omega(w^2)$ and $\lg B = \Omega(\lg Sd/N)$. Here N denotes the number of non-sink nodes in the butterfly graph.*

We derive our lower bound by showing that any cell probe data structure for skyline range counting can be used to answer reachability queries in subgraphs of the butterfly graph for any degree B and depth d.

Edges to 2-d Rectangles. Consider the butterfly graph with degree B and depth d. The first step of our reduction is inspired by the reduction Pătraşcu used to obtain a lower bound for 2-d rectangle stabbing: Consider an edge of the butterfly graph, leaving the i'th node at layer $k \in [d]$ and entering the j'th node in layer $k + 1$. We denote this edge $e_k(i,j)$. The source-sink pairs (s,t) that are connected through $e_k(i,j)$ are those for which:

1. The source has an index s satisfying $v(s)[h] = v(i)[h]$ for $h \geq k$, i.e. s and i agree on the $d - k$ most significant digits when written in base B.
2. The sink has an index t satisfying $v(t)[h] = v(j)[h]$ for $h \leq k + 1$, i.e. t and j agree on the $k + 1$ least significant digits when written in base B.

We now map each edge $e_k(i,j)$ of the butterfly graph to a rectangle in 2-d. For the edge $e_k(i,j)$, we create the rectangle $r_k(i,j) = [x_1, x_2] \times [y_1, y_2]$ where:

- $x_1 = v(i)[d-1]v(i)[d-2]\cdots v(i)[k]0\cdots 0$ in base B,
- $x_2 = v(i)[d-1]v(i)[d-2]\cdots v(i)[k](B-1)\cdots(B-1)$ in base B,
- $y_1 = v(j)[0]v(j)[1]\cdots v(j)[k+1]0\cdots 0$ in base B, and
- $y_2 = v(j)[0]v(j)[1]\cdots v(j)[k+1](B-1)\cdots(B-1)$ in base B.

The crucial observation is that for a source-sink pair, where the source is the s'th source and the sink is the t'th sink, the edges on the path from the source to the sink in the butterfly graph are precisely those edges $e_k(i,j)$ for which the corresponding rectangle $r_k(i,j)$ contains the point $(s, \mathrm{rev}_B(t))$, where $\mathrm{rev}_B(t)$ is the number obtained by writing t in base B and then reversing the digits.

We now collect the set of rectangles R, containing each rectangle $r_k(i,j)$ corresponding to an edge of the butterfly graph. Given an input subgraph G, we *mark* all rectangles $r_k(i,j) \in R$ for which the corresponding edge $e_k(i,j)$ is also in G. It follows that there is a directed path from the s'th source to the t'th sink in the subgraph G if and only if $(s, \mathrm{rev}_B(t))$ is not contained in any *unmarked* rectangle in R.

Our goal is now to transform marked and unmarked rectangles to points, such that we can use a skyline counting data structure to determine whether a given point $(s, \mathrm{rev}_B(t))$ is contained in an unmarked rectangle. Note that our reduction only works for the rectangle set R obtained from the butterfly graph, and not for any set of rectangles, i.e. we could not have reduced from the general problem of 2-d rectangle stabbing.

2-d Rectangles to Points. To avoid tedious details, we from this point on allow the input to skyline queries to have multiple points with the same x- or y-coordinate (though not two points with both coordinates identical). This assumption can easily be removed, but it would only distract the reader from the main ideas of our reduction. We still use the definition that a point (x,y) dominates a point (x',y') if and only if $x' \leq x$ and $y' \leq y$.

The next step of the reduction is to map the rectangles R to a set of points. For this, we first transform the coordinates slightly: For every rectangle $r_k(i,j) \in R$, having coordinates $[x_1, x_2] \times [y_1, y_2]$, we modify each of the coordinates in the following way: $x_1 \leftarrow dx_1 + (d-1-k)$, $x_2 \leftarrow dx_2 + d - 1$, $y_1 \leftarrow dy_1 + k$, and $y_2 \leftarrow dy_2 + d - 1$. The multiplication with d essentially corresponds to expanding each point with integer coordinates to a $d \times d$ grid of points. The purpose of adding k to y_1 and $(d-1-k)$ to x_1 is to ensure that, if two rectangles share a lower-left corner (only possible for two rectangles $r_k(i,j)$ and $r_{k'}(i',j')$ where $k \neq k'$), then those corners do not dominate each other in the transformed set of rectangles. We will see later that the particular placement of the points based on k also plays a key role. We use $\pi : [B^d]^4 \to [dB^d]^4$ to denote the above map. With this notation, the transformed set of rectangles is denoted $\pi(R)$ and each rectangle $r_k(i,j) \in R$ is mapped to $\pi(r_k(i,j)) \in \pi(R)$.

We now create the set of points P' containing the set of lower-left corner points for all rectangles $\pi(r_k(i,j)) \in \pi(R)$, i.e. for each $\pi(r_k(i,j)) = [x_1, x_2] \times [y_1, y_2]$, we add the point (x_1, y_1) to P'. The set P' has the following crucial property:

Lemma 4. *Let (x,y) be a point with coordinates in $[B^d] \times [B^d]$. Then for the two-sided query rectangle $Q = (-\infty, dx + d - 1] \times (-\infty, dy + d - 1]$, it holds that* Skyline$(Q \cap P')$ *contains precisely the points in P' corresponding to the lower-left corners of the rectangles $\pi(r_k(i,j)) \in \pi(R)$ for which $r_k(i,j)$ contains (x,y).*

Proof. First let $p = (x_1, y_1) \in P'$ be the lower-left corner of a rectangle $\pi(r_k(i,j))$ such that $r_k(i,j)$ contains the point (x,y). We want to show that $p \in$ Skyline$(Q \cap P')$. Since $r_k(i,j)$ contains the point (x,y), we have $x \geq \lfloor x_1/d \rfloor$ and $y \geq \lfloor y_1/d \rfloor$. From this, we get $dx + d - 1 \geq d\lfloor x_1/d \rfloor + (d-1-k) = x_1$ and $dy + d - 1 \geq d\lfloor y_1/d \rfloor + k = y_1$, i.e. p is inside Q. Since (x,y) is inside $r_k(i,j)$, we also have that $(dx + d - 1, dy + d - 1)$ is dominated by the upper-right corner of $\pi(r_k(i,j))$, i.e. $(dx + d - 1, dy + d - 1)$ is inside $\pi(r_k(i,j))$.

What remains to be shown is that no other point in $Q \cap P'$ dominates p. For this, assume for contradiction that some point $p' = (x_1', y_1') \in P'$ is both in Q and also dominates p. First, since p' is dominated by $(dx + d - 1, dy + d - 1)$ and also dominates p, we know that p' must be inside $\pi(r_k(i,j))$. Now let $\pi(r_{k'}(i',j')) \neq \pi(r_k(i,j))$ be the rectangle in $\pi(R)$ from which p' was generated, i.e. p' is the lower-left corner of $\pi(r_{k'}(i',j'))$. We have three cases:

1. First, if $k' = k$ we immediately get a contradiction since the rectangles $\pi(R)_k = \{\pi(r_{k'}(i',j')) \in \pi(R) \mid k' = k\}$ are pairwise disjoint and hence p' could not have been inside $\pi(r_k(i,j))$.

2. If $k' < k$, we know that $\pi(r_{k'}(i',j'))$ is shorter in x-direction and longer in y-direction than $\pi(r_k(i,j))$. From our transformation, we know that $(y_1 \bmod d) = k$ and $(y_1' \bmod d) = k' < k$. Thus since p' dominates p, we must have $\lfloor y_1'/d \rfloor > \lfloor y_1/d \rfloor$. But these two values are precisely the y-coordinates of the lower-left corners of $r_k(i,j)$ and $r_{k'}(i',j')$. By definition, we get:

$$v(j')[0]v(j')[1] \cdots v(j')[k'+1]0 \cdots 0 > v(j)[0]v(j)[1] \cdots v(j)[k+1]0 \cdots 0 \, .$$

Since $k' < k$, this furthermore gives us

$$v(j')[0]v(j')[1] \cdots v(j')[k'+1] > v(j)[0]v(j)[1] \cdots v(j)[k'+1] .$$

From this it follows that

$$v(j')[0] \cdots v(j')[k'+1]0 \cdots 0 > v(j)[0] \cdots v(j)[k+1](B-1) \cdots (B-1) ,$$

i.e. the lower-left corner of $r_{k'}(i',j')$ is outside $r_k(i,j)$, which also implies that the lower-left corner of $\pi(r_{k'}(i',j'))$ is outside $\pi(r_k(i,j))$. That is, p' is outside $\pi(r_k(i,j))$, which gives the contradiction.

3. The case for $k' > k$ is symmetric to the case $k' < k$, just using the x-coordinates instead of the y-coordinates to derive the contradiction.

The last step of the proof is to show that no point $p = (x_1, y_1) \in P'$ can be in Skyline($Q \cap P'$) but at the same time correspond to the lower-left corner of a rectangle $\pi(r_k(i,j))$ where $r_k(i,j)$ does not contains the point (x,y). First observe that $(dx + d - 1, dy + d - 1)$ is contained in precisely one rectangle $\pi(r_{k'}(i',j'))$ for each value of $k' \in [d]$. Now let $\pi(r_k(i',j')) \neq \pi(r_k(i,j))$ be the rectangle containing $(dx + d - 1, dy + d - 1)$ amongst the rectangles $\pi(R)_k$. The lower-left corner of this rectangle is dominated by $(dx + d - 1, dy + d - 1)$ but also dominates p, hence p is not in Skyline($Q \cap P'$). □

Handling Marked and Unmarked Rectangles. The above steps are all independent of the concrete input subgraph G. As discussed, we need a way to determine whether a query point is contained in an unmarked rectangle or not. This step is now very simple in light of Lemma 4: First, multiply all coordinates of points in P' by 2. This corresponds to expanding each point with integer coordinates into a 2×2 grid. Now for every point $p \in P'$, if the rectangle $\pi(r_k(i,j))$ from which p was generated is marked, then we add 1 to both the x- and y-coordinate of p, i.e. we move p to the upper-right corner of the 2×2 grid in which it is placed. If $\pi(r_k(i,j))$ is unmarked, we replace it by two points, one where we add 1 to the x-coordinate, and one where we add 1 to the y-coordinate. We denote the resulting set of points $P(G)$. It follows immediately that:

Corollary 1. *Let G be a subgraph of the butterfly graph with degree B and depth d. Also, let (x,y) be a point with coordinates in $[B^d] \times [B^d]$. Then for the two-sided query rectangle $Q = (-\infty, 2d(x+1)-1] \times (-\infty, 2d(y+1)-1]$, it holds that Skyline($Q \cap P(G)$) contains precisely one point from $P(G)$ for every marked rectangle in R that contains (x,y), two points from $P(G)$ for every unmarked rectangle in R that contains (x,y), and no other points, i.e. $|$Skyline($Q \cap P(G)$)$| - d$ equals the number of unmarked rectangles in R which contains (x,y).*

Corollary 2. *Let G be a subgraph of the butterfly graph with degree B and depth d. Let s be the index of a source and t the index of a sink. Then the s'th source can reach the t'th sink in G if and only if $|$Skyline($Q \cap P(G)$)$| = d$ for the two-sided query rectangle $Q = (-\infty, 2d(s+1)-1] \times (-\infty, 2d(\text{rev}_B(t)+1)-1]$.*

Deriving the Lower Bound. The lower bound can be derived from Corollary 2 and Theorem 1 as follows. First note that the set R contains NB rectangles, since each rectangle corresponds to an edge of the butterfly graph and each of the N non-sink nodes of the butterfly graph has B outgoing edges. Each of these rectangles gives one or two points in $P(G)$. Letting n denote $|P(G)|$, we have $NB \leq n \leq 2NB$. From $N = d \cdot B^d \leq n$ we get $d \leq \lg n$ and $d = \Theta(\lg_B N)$.

Given n, $w \geq \lg n$, and $S \geq n$, we now derive a lower bound on the query time. Setting $B = \frac{S}{n} w^2$ we have $B = \Omega(w^2)$ and $\lg B = \Omega(\lg \frac{Sd}{N})$ (as required by Theorem 1), where the last bound follows from $\lg \frac{Sd}{N} \leq \lg \frac{S \cdot \lg n}{n/2B} \leq \lg(2B \frac{S \cdot w}{n}) \leq \lg(2B^2) = O(\lg B)$. Furthermore we have $\lg \frac{Sw}{n} = \frac{1}{2} \lg(\frac{Sw}{n})^2 \geq \frac{1}{2} \lg(\frac{S}{n} w^2) = \frac{1}{2} \lg B$. From Theorem 1 we can now bound the time for a skyline counting query by $t = \Omega(d) = \Omega(\lg_B N) = \Omega(\lg n / \lg B) = \Omega(\lg n / \lg(Sw/n))$.

3   Skyline Counting Data Structure

In this section we describe a data structure using $O(n)$ space supporting orthogonal skyline counting queries in $O(\lg n / \lg \lg n)$ time. We describe the basic idea of how to support queries, and present the details of the stored data structure. The details of the query can be found in the full version of the paper.

The basic idea is to store the n points in left-to-right x-order at the leaves of a balanced tree T of degree $\Theta(\log^\varepsilon n)$, i.e. height $O(\log n / \log \log n)$, and for each internal node v have a list L_v of the points in the subtree rooted in v in sorted y-order. The *slab* of v is the narrowest infinite vertical band containing L_v. To obtain the overall linear space bound, L_v will not be stored explicitly but implicitly and rank-reduced using rank-select data structures, where navigation is performed using fractional cascading on rank-select data structures (details below). A 4-sided query R decomposes into 2-sided subqueries at $O(\log n / \log \log n)$ nodes (in Figure 1, R is decomposed into subqueries R_1-R_5, white points are nodes on the skyline within R, double circled points are the topmost points within each R_i). For skyline queries (both counting and reporting) it is important to consider the subqueries right-to-left, and the lower y-value for the subquery in R_i is raised to the maximal y-value of a point in the subqueries to the right. Since the tree T has non-constant degree, we need space efficient solutions for multislab queries at each node v. We partition L_v into blocks of size $O(\log^{2\varepsilon} n)$, and a query R_i decomposes into five subqueries (1-5), see Figure 2: (1) and (3) are on small subsets of points within a single block and can be answered by tabulation (given the *signature* of the block); (2) is a block aligned multislab query; (4) and (5) are for single slabs (at the children of v). For (2,4,5) the skyline size between points i and j (numbered bottom-up) can be computed as one plus the difference between the size of the skyline from 1 to j and 1 to k, where k is the rightmost point between i and j (see Figure 3, white and black circles and crosses are all points, crosses indicate the skyline from i to j, white circles from 1 to k, and white circles together with crosses from 1 to j). Finally, the skyline size from 1 to i can be computed from a prefix sum, if we for point i store the number of points in the skyline from 1 to $i-1$ dominated by i

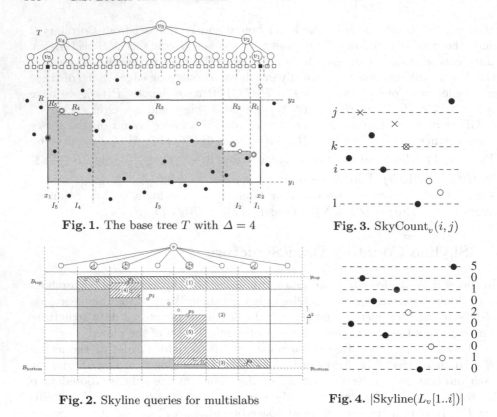

Fig. 1. The base tree T with $\Delta = 4$

Fig. 3. SkyCount$_v(i, j)$

Fig. 2. Skyline queries for multislabs

Fig. 4. $|\text{Skyline}(L_v[1..i])|$

(see Figure 4, the skyline between 1 and 6 consists of the three white nodes, and the size is $6 - (2 + 0 + 0 + 0 + 1 + 0) = 3$).

The details of the construction are as follows. We let $\Delta = \max\{2, \lceil \lg^\varepsilon n \rceil\}$ be a parameter of our construction, where $0 < \varepsilon < 1/3$ is a constant. We build a balanced *base tree* T over the set of points P, where the leafs from left-to-right store the points in P in sorted order w.r.t. x-coordinate. Each internal node of T has degree at most Δ and T has height $\lceil \lg_\Delta n \rceil + 1$. (See Figure 1)

For each internal node v of T we store a set of data structures. Before describing these we need to introduce some notation. The subtree of T rooted at a node v is denoted T_v, and the set of points stored at the leaves of T_v is denoted P_v. We let $n_v = |P_v|$ and $L_v[1..n_v]$ be the list of the points in P_v sorted in increasing y-order. We let $I_v = [\ell_v, r_v]$ denote the x-interval defined by the x-coordinates of the points stored at the leaves of T_v, and denote $I_v \times [n]$ the *slab* spanned by v. The degree of v is denoted d_v, the children of v are from left-to-right denoted $c_v^1, \ldots, c_v^{d_v}$, and the parent of node v is denoted p_v. A list L_v is partitioned into a sequence of blocks $B_v[1..\lceil n_v/\Delta^2 \rceil]$ of size Δ^2, such that $B_v[i] = L_v[(i-1)\Delta^2 + 1 .. \min\{n_v, i\Delta^2\}]$. The *signature* $\sigma_v[i]$ of a block $B_v[i]$ is a list of pairs: For each point p from $B_v[i]$ in increasing y-order we construct a pair (j, r), where j is the index of the child c_v^j of v storing p and r is the rank of p's x-coordinate among all points in $B_v[i]$ stored at the same child c_v^j as p. The total number of bits required for a signature is at most $\Delta^2(\lg \Delta + \lg \Delta^2) = O(\lg^{2\varepsilon} n \cdot \lg \lg n)$.

To achieve overall $O(n)$ space we need to encode succinctly sufficient information for performing queries. In particular we will *not* store the points in L_v explicitly at the node v, but only partial information about the points relative position will be stored.

Queries on a block $B_v[i]$ are handled using table lookups in global tables using the block signature $\sigma_v[i]$. We have tables for the below block queries, where we assume σ is the signature of a block storing points $p_1, \ldots, p_{\Delta^2}$ distributed in Δ child slabs.

Below(σ, t, i) Returns the number of points from $p_1, \ldots, p_t$ contained in slab i.

Rightmost(σ, b, t, i, j) Returns k, where p_k is the rightmost point among $p_b, \ldots, p_t$ contained in slabs $[i, j]$. If no such point exists, -1 is returned.

Topmost(σ, b, t, i, j) Returns k, where p_k is the topmost point among $p_b, \ldots, p_t$ contained in slabs $[i, j]$. If no such point exists, -1 is returned.

SkyCount(σ, b, t, i, j) Returns the size of the skyline for the subset of the points $p_b, \ldots, p_t$ contained in slabs $[i, j]$.

The arguments to each of the above lookups consists of at most $|\sigma| + 2\lg \Delta^2 + 2\lg \Delta = |\sigma| + O(\lg \lg n) = O(\lg^{2\varepsilon} n \cdot \lg \lg n)$ bits and the answer is $\lg(\Delta + 1) = O(\lg \lg n)$ bits, i.e. each query can be answered in $O(1)$ time using a table of size $O(2^{\lg^{2\varepsilon} n \cdot \lg \lg n} \cdot \lg \lg n) = o(n)$ bits, since $\varepsilon < 1/3$.

For each internal node v of T we store the following data structures, each having $O(1)$ access time.

$C_v(i)$ Compact array that for each i, where $1 \le i \le n_v$, stores the index of the child of v storing $L_v[i]$, i.e. $1 \le C_v(i) \le \Delta$. Space usage $O(n_v \lg \Delta)$ bits.

$\pi_v(i)$ For each i, $1 \le i \le n_v$, stores the index of $L_v[i]$ in L_{p_v}, i.e. $L_{p_v}[\pi_v(i)] = L_v[i]$. This can be supported by constructing the select data structure of Lemma 1 on the bit-vector X, where $X[i] = 1$ if and only if $L_{p_v}[i]$ is in L_v. A query to $\pi_v(i)$ simply becomes a select(i) query. Space usage $O(n_v \lg(n_{p_v}/n_v)) = O(n_v \lg \Delta)$ bits.

$\sigma_v(i)$ Array of signatures for the blocks $B_v[1..\lceil n_v/\Delta^2 \rceil]$. Space usage $O(n_v/\Delta^2 \cdot \Delta^2 \cdot \lg \Delta) = O(n_v \lg \Delta)$ bits.

Pred$_v(t, i)$ / Succ$_v(t, i)$ Supports finding the predecessor/successor of $L_v[t]$ in the i'th child list $L_{c_v^i}$. Returns $\max\{k \mid 1 \le k \le n_{c_v^i} \wedge \pi_{c_v^i}[k] \le t\}$ and $\min\{k \mid 1 \le k \le n_{c_v^i} \wedge \pi_{c_v^i}[k] \ge t\}$, respectively. For each child index i, we construct an array X^i of size $\lceil n/\Delta^2 \rceil$, such that $X^i[b]$ is the number of points in block $B_v[b]$ that are stored in the i'th child slab. The prefix sums of each X^i are stored using the data structure of Lemma 2 using space $O((n_v/\Delta^2) \lg(\Delta^2))$ bits. The total space for all Δ children of v becomes $O(\Delta \cdot n_v/\Delta^2 \cdot \lg \Delta) = O(n_v)$ bits. The result of a Pred$_v(t, i)$ query is $\sum_{j=1}^{\lceil t/\Delta^2 \rceil - 1} X^i[j] + \text{Below}(\sigma_v(\lceil t/\Delta^2 \rceil), 1 + (t - 1 \bmod \Delta^2), i)$, where the first term can be computed in $O(1)$ time by Lemma 2 and the second term is a constant time global table lookup. The result of Succ$_v(t, i) = \text{Pred}_v(t, i)$ if $C_v[t] = i$, otherwise Succ$_v(t, i) = \text{Pred}_v(t, i) + 1$.

Rightmost$_v(i,j)$ Returns the index k, where $i \leq k \leq j$, such that $L_v[k]$ has the maximum x-value among $L_v[i..j]$. Using Lemma 3 on the array of the x-coordinates of the points in L_v we achieve $O(1)$ time queries and space usage $O(n_v)$ bits.

SkyCount$_v(i)$ Returns $|\text{Skyline}(L_v[1..i])|$. Construct an array X, where $X[i]$ is the number of points in $\text{Skyline}(L_v[1..i-1])$ dominated by $L_v[i]$. (See Figure 4) We can now compute $|\text{Skyline}(L_v[1..i])|$ as $i - \sum_{j=1}^{i} X[j]$. Using Lemma 2 the query time becomes $O(1)$ and the space usage $O(n_v)$ bits, since $\sum_{j=1}^{n_v} X[j] \leq n_v - 1$.

SkyCount$_v(i,j)$ Returns $|\text{Skyline}(L_v[i..j])|$, computable by the following expression: SkyCount$_v(j) - $ SkyCount$_v(\text{Rightmost}_v(i,j)) + 1$ (see Figure 3).

Finally, we store for each node v and slab interval $[i,j]$ the following data structures.

Rightmost$_{v,i,j}(b,t)$ Returns k, where $L_v[k]$ is the rightmost point among the points in blocks $B_v[b..t]$ contained in slabs $[i,j]$. If no such point exists, -1 is returned. Can be solved by applying Lemma 3 to the array X, where $X[s]$ is the x-coordinate of the rightmost point in $B_v[s]$ contained in slabs $[i,j]$. A query first finds the block ℓ containing the rightmost point using this data structure, and then returns $(\ell - 1)\Delta^2 + \text{Rightmost}(\sigma_v[\ell], 1, \Delta^2, i, j)$. Space usage $O(n_v/\Delta^2)$ bits.

Topmost$_{v,i,j}(b,t)$ Returns k, where $L_v[k]$ is the topmost point among the points in blocks $B_v[b..t]$ contained in slabs $[i,j]$. If no such point exists, -1 is returned. Can be solved by first using Lemma 3 on the array X, where $X[s] = s$ if there exists a point in $B_v[s]$ contained in slabs $[i,j]$. Otherwise $X[s] = 0$. Let ℓ be the block found using Lemma 3. Return the result of $(\ell - 1)\Delta^2 + \text{Topmost}(\sigma_v[\ell], 1, \Delta^2, i, j)$. Space usage $O(n_v/\Delta^2)$ bits.

SkyCount$_{v,i,j}(b,t)$ Returns the size of the skyline for the subset of points in blocks $B_v[b..t]$ contained in slabs $[i,j]$. Can be supported by two applications of Lemma 2 on two arrays X and Y: Let $X[s] = \text{SkyCount}(\sigma_v[s], 1, \Delta^2, i, j)$, i.e. the size of the skyline of the points in block $B_v[s]$ contained in slabs $[i,j]$. Let $B_{v,i,j}[s]$ denote the points in $B_v[s]$ contained in slabs $[i,j]$. Let $Y[s] = |\text{Skyline}(B_{v,i,j}[1..s-1]) \setminus \text{Skyline}(B_{v,i,j}[1..s])|$, i.e. the number of points on $\text{Skyline}(B_{v,i,j}[1..s-1])$ dominated by points in $B_{v,i,j}[s]$. Space usage for X and Y is $O(n_v/\Delta^2 \cdot \lg \Delta^2)$ bits. We can compute SkyCount$_{v,i,j}(b,t) = \sum_{s=k}^{t} X[s] - \sum_{s=k+1}^{t} Y[s]$, where $k = \lceil \text{Rightmost}_{v,i,j}(b,t)/\Delta^2 \rceil$.

The total space of our data structure, in addition to the $o(n)$ bits for our global tables, can be bounded as follows. The total space for all $O(\Delta^2)$ multislab data structures for a node v is $O(\Delta^2 \cdot n_v/\Delta^2 \cdot \lg \Delta)$ bits. The total space for all data structures at a node v becomes $O(n_v \lg \Delta)$ bits. Since the sum of all n_v for a level of T is at most n, the total space for all nodes at a level of T is $O(n \lg \Delta)$ bits. Since T has height $O(\lg_\Delta n)$, the total space usage becomes $O(n \lg \Delta \cdot \lg_\Delta n) = O(n \lg n)$ bits, i.e. $O(n)$ words. The data structure can be constructed bottom-up in $O(n \log n)$ time.

References

1. Chan, T.M., Larsen, K.G., Pătraşcu, M.: Orthogonal range searching on the RAM, revisited. In: 27th ACM Symposium on Computational Geometry, pp. 1–10. ACM (2011)
2. JáJá, J., Mortensen, C.W., Shi, Q.: Space-efficient and fast algorithms for multidimensional dominance reporting and counting. In: Fleischer, R., Trippen, G. (eds.) ISAAC 2004. LNCS, vol. 3341, pp. 558–568. Springer, Heidelberg (2004)
3. Pătraşcu, M.: Lower bounds for 2-dimensional range counting. In: 39th Annual ACM Symposium on Theory of Computing, pp. 40–46. ACM (2007)
4. Pătraşcu, M., Thorup, M.: Time-space trade-offs for predecessor search. In: 38th Annual ACM Symposium on Theory of Computing, pp. 232–240. ACM (2006)
5. Alstrup, S., Brodal, G.S., Rauhe, T.: New data structures for orthogonal range searching. In: 41st Annual Symposium on Foundations of Computer Science, pp. 198–207. IEEE Computer Society (2000)
6. Das, A.S., Gupta, P., Kalavagattu, A.K., Agarwal, J., Srinathan, K., Kothapalli, K.: Range aggregate maximal points in the plane. In: Rahman, M. S., Nakano, S.-i. (eds.) WALCOM 2012. LNCS, vol. 7157, pp. 52–63. Springer, Heidelberg (2012)
7. Kalavagattu, A.K., Agarwal, J., Das, A.S., Kothapalli, K.: Counting range maxima points in plane. In: Arumugam, S., Smyth, W.F. (eds.) IWOCA 2012. LNCS, vol. 7643, pp. 263–273. Springer, Heidelberg (2012)
8. Das, A.S., Gupta, P., Srinathan, K.: Counting maximal points in a query orthogonal rectangle. In: Ghosh, S.K., Tokuyama, T. (eds.) WALCOM 2013. LNCS, vol. 7748, pp. 65–76. Springer, Heidelberg (2013)
9. Brodal, G.S., Tsakalidis, K.: Dynamic planar range maxima queries. In: Aceto, L., Henzinger, M., Sgall, J. (eds.) ICALP 2011, Part I. LNCS, vol. 6755, pp. 256–267. Springer, Heidelberg (2011)
10. Kalavagattu, A.K., Das, A.S., Kothapalli, K., Srinathan, K.: On finding skyline points for range queries in plane. In: 23rd Annual Canadian Conference on Computational Geometry (2011)
11. Nekrich, Y., Navarro, G.: Sorted range reporting. In: Fomin, F.V., Kaski, P. (eds.) SWAT 2012. LNCS, vol. 7357, pp. 271–282. Springer, Heidelberg (2012)
12. Yao, A.C.C.: Should tables be sorted? Journal of the ACM 28(3), 615–628 (1981)
13. Pătraşcu, M.: Unifying the landscape of cell-probe lower bounds. SIAM Journal on Computing 40(3), 827–847 (2011)
14. Beame, P., Fich, F.E.: Optimal bounds for the predecessor problem. In: 31st Annual ACM Symposium on Theory of Computing, pp. 295–304. ACM (1999)
15. Raman, R., Raman, V., Rao, S.S.: Succinct indexable dictionaries with applications to encoding k-ary trees and multisets. In: 13th Annual ACM-SIAM Symposium on Discrete Algorithms, pp. 233–242. SIAM (2002)
16. Raman, R., Raman, V., Satti, S.R.: Succinct indexable dictionaries with applications to encoding k-ary trees, prefix sums and multisets. ACM Transactions on Algorithms 3(4) (2007)
17. Sadakane, K.: Succinct data structures for flexible text retrieval systems. Journal of Discrete Algorithms 5(1), 12–22 (2007)
18. Fischer, J.: Optimal succinctness for range minimum queries. In: López-Ortiz, A. (ed.) LATIN 2010. LNCS, vol. 6034, pp. 158–169. Springer, Heidelberg (2010)

B-slack Trees: Space Efficient B-Trees

Trevor Brown

Department of Computer Science, University of Toronto, Canada
tabrown@cs.toronto.edu

Abstract. B-slack trees, a subclass of B-trees that have substantially better worst-case space complexity, are introduced. They store n keys in height $O(\log_b n)$, where b is the maximum node degree. Updates can be performed in $O(\log_{\frac{b}{2}} n)$ amortized time. A relaxed balance version, which is well suited for concurrent implementation, is also presented.

1 Introduction

B-trees are balanced trees designed for block-based storage media. Internal nodes contain between $b/2$ and b child pointers, and one less key. Leaves contain between $b/2$ and b keys. All leaves have the same depth, so access times are predictable. If memory can be allocated on a per-byte basis, nodes can simply be allocated the precise amount of space they need to store their data, and no space is wasted. However, typically, all nodes have the same, fixed capacity, and some of the capacity of nodes is wasted. As much as 50% of the capacity of each node is wasted in the worst case. This is particularly problematic when data structures are being implemented in hardware, since memory allocation and reclamation schemes are often very simplistic, allowing only a single block size to be allocated (to avoid fragmentation). Furthermore, since hardware devices must include sufficient resources to handle the worst-case, good expected behaviour is not enough to allow hardware developers to reduce the amount of memory included in their devices. To address this problem, we introduce *B-slack trees*, which are a variant of B-trees with substantially better worst-case space complexity. We also introduce relaxed B-slack trees, which are a variant of B-slack trees that are more amenable to concurrent implementation.

The development of B-slack trees was inspired by a collaboration with a manufacturer of internet routers, who wanted to build a concurrent router based on a tree. In such embedded devices, storage is limited, so it is important to use it as efficiently as possible. A suitable tree would have a simple algorithm for updates, small space complexity, fast searches, and searches that would not be blocked by concurrent updates. Updates were expected to be infrequent. One naive approach is to rebuild the entire tree after each update. Keeping an old copy of the tree while rebuilding a new copy would allow searches to proceed unhindered, but this would double the space required to store the tree.

Search trees can be either node-oriented, in which each key is stored in an internal node or leaf, or leaf-oriented, in which each key is stored at a leaf and the

R. Ravi and I.L. Gørtz (Eds.): SWAT 2014, LNCS 8503, pp. 122–133, 2014.
© Springer International Publishing Switzerland 2014

keys of internal nodes serve only to direct a search to the appropriate leaf. In a node-oriented B-tree, the leaves and internal nodes have different sizes (because internal nodes contain keys and pointers to children, and leaves contain only keys). So, if only one block size can be allocated, a significant amount of space is wasted. Moreover, deletion in a node-oriented tree sometimes requires stealing a key from a successor (or predecessor), which can be in a different part of the tree. This is a problem for concurrent implementation, since the operation involves a large number of nodes, namely, the nodes on the path between the node and its successor.

B-slack trees are leaf-oriented trees with many desirable properties. The average degree of nodes is high, exceeding $b - 1.4$ for trees of height at least three. Their space complexity is better than all of their competitors. Consider a dictionary implemented by a leaf-oriented search tree, in which, along with each key, a leaf stores a pointer to associated data. Suppose that each key and each pointer to a child or to data occupies a single word. Then, $\frac{2b}{b-2.4}n$ is an upper bound on the number of words needed to store a B-slack tree with $n > b^3$ keys. For large b, this tends to $2n$, which is optimal. B-slack trees have logarithmic height, and the number of rebalancing steps performed after a sequence of m updates to a B-slack tree of size n is amortized $O(\log(n + m))$ per update. Furthermore, the number of rebalancing steps needed to rebalance the tree can be reduced to amortized *constant* per update at the cost of slightly increased space complexity, as will be explained in the full version of the paper.

The rest of this paper is organized as follows. Section 2 surveys related work. Section 3 introduces B-slack trees and relaxed B-slack trees. Height, average degree, space complexity and rebalancing costs of relaxed B-slack trees (and, hence, of B-slack trees) are analyzed in Section 4. Finally, we conclude in Section 5.

2   Related Work

B-trees were initially proposed by Bayer and McCreight in 1970 [4]. Insertion into a full node in a B-tree causes it to split into two nodes, each half full. Deletion from a half-full node causes it to merge with a neighbour. Arnow, Tenenbaum and Wu proposed P-trees [2], which enjoy moderate improvements to average space complexity over B-trees, but waste 66% of each node in the worst case.

A number of generalizations of B-trees have been suggested that achieve much less waste if no deletions are performed. Bayer and McCreight also proposed B*-trees in [4], which improve the worst-case space complexity. At most a third of the capacity of each node in a B*-tree is wasted. This is achieved by splitting a node only when it and one of its neighbours are both full, replacing these two nodes by three nodes. Küspert [9] generalized B*-trees to trees where each node contains between $\lfloor \frac{bm}{m+1} \rfloor$ and b pointers or keys, where $m \leq b - 1$ is a design parameter. Such a tree behaves just like a B*-tree everywhere except at the leaves. An insertion into a full leaf causes keys to be shifted among the nearest $m - 1$ siblings to make room for the inserted key. If the $m - 1$ nearest siblings are also full, then these m nodes are replaced by $m + 1$ nodes which evenly share keys. Large values of m yield good worst-case space complexity.

Baeza-Yates and Per-åke Larson introduced B+trees with partial expansions [3]. Several node sizes are used, each a multiple of the block size. An insertion to a full node causes it to expand to the next larger node size. With three node sizes, at most 33% of each node can be wasted, and worst-case utilization improves with the number of block sizes used. However, this technique simply pushes the complexity of the problem onto the memory allocator. Memory allocation is relatively simple for one block size, but it quickly becomes impractical for simple hardware to support larger numbers of block sizes.

Culik, Ottmann and Wood introduced strongly dense multiway trees (SDM-trees) [7]. An SDM-tree is a node-oriented tree in which all leaves have the same depth, and the root contains at least two pointers. Apart from the root, every internal node u with fewer than b pointers has at least one sibling. Each sibling of u has b pointers if it is an internal node and b keys if it is a leaf. Insertion can be done in $O(b^3 + (\log n)^{b-2})$ time. Deletion is not supported, but the authors mention that the insertion algorithm could be modified to obtain a deletion algorithm, and the time complexity of the resulting algorithm "would be at most $O(n)$ and at least $O((\log n)^{b-1})$." Besides the long running times for each operation (and the lack of better amortized results), the insertion algorithm is very complex and involves many nodes, which makes it poorly suited for hardware implementation. Furthermore, in a concurrent setting, an extremely large section of the tree would have to be modified atomically, which would severely limit concurrency.

Srinivasan introduced a leaf-oriented B-tree variant called an *Overflow tree* [12]. For each parent of a leaf, its children are divided into one or more groups, and an overflow node is associated with each group. The tree satisfies the B-tree properties and the additional requirement that each leaf contains at least $b-1-s$ keys, where $s \geq 2$ is a design parameter and b is the maximum degree of nodes. Inserting a key into a full leaf causes the key to be inserted into the overflow node instead; if the overflow node is full, the entire group is reorganized. Deleting from a leaf is the same as in a B-tree unless it will cause the leaf to contain too few keys, in which case, a key is taken from the overflow node; if a key cannot be taken from the overflow node, the entire group is reorganized. Each search must look at an overflow node. The need to atomically modify and search two places at once makes this data structure poorly suited for concurrent implementation.

Hsuang introduced a class of node-oriented trees called *H-trees* [8], which are a subclass of B-trees parameterized by γ and δ. These parameters specify a lower bound on the number of grandchildren of each internal node (that has grandchildren), and a lower bound on the number of keys contained in each leaf, respectively. Larger values of δ and γ yield trees that use memory more efficiently. When δ and γ are as large as possible, each leaf contains at least $b - 3$ keys, and each internal node has zero or at least $\lfloor \frac{b^2+1}{2} \rfloor$ grandchildren. The paper presents $O(\log n)$ insertion and deletion algorithms for node-oriented H-trees. The algorithms are very complex and involve many cases. H-trees have a minimum average degree of approximately $b/\sqrt{2}$ for internal nodes, which is much smaller than the $b - 1.4$ of B-slack trees (for trees of height at least three).

The full version of the paper describes families of B-trees, H-trees and Overflow trees which require significantly more space than B-slack trees.

Rosenberg and Snyder introduced *compact B-trees* [11], which can be constructed from a set of keys using the minimum number of nodes possible. No compactness preserving insertion or deletion procedures are known. The authors suggested using regular B-tree updates and periodically compacting a data structure to improve efficiency. However, experiments in [1] showed that starting with a compact B-tree and adding only 1.6% more keys using standard B-tree operations reduced storage utilization from 99% to 67%.

An impressive paper by Brønnimann et al. [5] presented three ways to transform an arbitrary sequential dictionary into a more space efficient one. One of these ways will be discussed here; of the other two, one is extremely complex and poorly suited for concurrent hardware implementation, and the other pushes the complexity onto the memory allocator.

Brønnimann's transformation takes any sequential tree data structure and modifies it by replacing each key in the sequential data structure with a *chunk*, which is a group of $b - 2$, $b - 1$ or b keys, where b is the memory block size. All chunks in the data structure are also kept in a doubly linked list to facilitate iteration and movement of keys between chunks. For instance, a BST would be transformed into a tree in which each node has zero, one or two children, and $b - 2$, $b - 1$ or b keys. All keys in chunks in the left subtree of a node u would be smaller than all keys in u's chunk, and all keys in chunks in the right subtree of u would be larger than all keys in u's chunk. A search for key k behaves the same as in the sequential data structure until it reaches the only chunk that can contain k, and searches for k within the chunk. An insertion first searches for the appropriate chunk, then it inserts the key into this chunk. Inserting into a full chunk requires shifting the keys of the b nearest other chunks to make room. If the b closest neighboring chunks are full, then a key is taken from each, and a new node containing b keys is inserted using the sequential data structure's insertion algorithm. Deletion is similar. Each operation in the resulting data structure runs in $O(f(n) + b^2)$ steps, where $f(n)$ is the number of steps taken by the sequential data structure to perform the same operation.

After this transformation, a B-tree with maximum degree b requires $2n + O(n/b)$ words to store n keys and pointers to data. In the worst-case, each chunk wastes $2/b$ of its space, which is somewhat worse than in B-slack trees. Furthermore, supporting fast searches can introduce significant complexity to the hardware design. A node in the transformed B-tree contains up to $b - 1$ chunks, each of which occupies one block of memory. Therefore, hardware must be able to quickly load up to $b - 1$ blocks at once, or else deciding which child pointer to follow will be slow.

3   B-slack trees

A B-slack tree is a variant of a B-tree. Each node stores its keys in sorted order, so binary search can be used to determine which child of an internal node should be visited next by a search, or whether a leaf contains a key. Let $p_0, p_1, ..., p_m$ be the sequence of pointers contained in an internal node, and $k_1, k_2, ..., k_m$ be its sequence of keys. For each $1 \leq i \leq m$, the subtree pointed to by p_{i-1} contains keys strictly smaller than k_i, and the subtree pointed to by p_i contains keys greater than *or equal to* k_i. We say that the *degree of an internal node* is the number of non-NIL pointers it contains, and the *degree of a leaf* is the number of keys it contains. This unusual definition of degree simplifies our discussion. The degree of node v is denoted $deg(v)$. If the maximum possible degree of a node is b, and its degree is $b - x$, then we say it contains x *slack*.

A B-slack tree is a leaf-oriented search tree with maximum degree $b > 4$ in which:

P1: every leaf has the same depth,

P2: internal nodes contain between 2 and b pointers (and one less key),

P3: leaves contain between 0 and b keys, and

P4: for each internal node u, the total slack contained in the children of u is at most $b - 1$.

P4 is the key property that distinguishes B-slack trees from other variants of B-trees. It limits the aggregate space wasted by a number of nodes, as opposed to limiting the space wasted by each node. Alternatively, P4 can be thought of as a lower bound on the sum of the degrees of the children of each internal node. Formally, for each internal node with children $v_1, v_2, ..., v_l$, $deg(v_1) + deg(v_2) + ... + deg(v_l) \geq lb - (b - 1) = lb - b + 1$. This interpretation is useful to show that all nodes have large subtrees. For instance, it implies that a node u with two internal children must have at least $b + 1$ grandchildren. If these grandchildren are also internal nodes, we can conclude that u must have at least $b^2 - b + 2$ great grandchildren.

A tree that satisfies P1, and in which every node has degree $b - 1$, is an example of a B-slack tree. Another example of a B-slack tree is a tree of height two, where b is even, the root has degree two, its two children have degree $b/2$ and $b/2+1$, respectively, and the grandchildren of the root are leaves with degree b, except for two, one in the left subtree of the root, and one in the right subtree, that each have degree one. This tree contains the smallest number of keys of any B-slack tree of height two.

3.1   Relaxed B-slack trees

A relaxed balance search tree decouples updates that rebalance (or reorganize the keys of) the tree from updates that modify the set of keys stored in the tree [10]. The advantages of this decoupling are twofold. First, updates to a relaxed balance version of a search tree are smaller, so a greater degree of concurrency is possible in a multithreaded setting. Second, for some applications, it may be useful to temporarily disable rebalancing to allow a large number of updates to be performed quickly, and to gradually rebalance the tree afterwards.

A relaxed B-slack tree is a relaxed balance version of a B-slack tree that has weakened the properties. A weight of zero or one is associated with each node.

These weights serve a purpose similar to the colors red and black in a red-black tree. We define the *relaxed depth* of a node to be one less than the sum of the weights on the path from the root to this node. A relaxed B-slack tree is a leaf-oriented search tree with maximum degree $b > 4$ in which:

P0': every node with weight zero contains exactly two pointers,
P1': every leaf has the same relaxed depth,
P2': internal nodes contain between 1 and b pointers (and one less key), and
P3 : leaves contain between 0 and b keys

To clarify the difference between B-slack trees and relaxed B-slack trees, we identify several types of *violations* of the B-slack trees properties that can be present in a relaxed B-slack tree. We say that a *weight violation* occurs at a node with weight zero, a *slack violation* occurs at a node that violates P4, and a *degree violation* occurs at an internal node with only one child (violating P2). Observe that P1 is satisfied in a relaxed B-slack tree with no weight violations. Likewise, P2 is satisfied in a relaxed B-slack tree with no degree violations, and P4 is satisfied in a relaxed B-slack tree with no slack violations. Therefore, a relaxed B-slack tree that contains no violations is a B-slack tree. Rebalancing steps can be performed to eliminate violations, and gradually transform any relaxed B-slack tree into a B-slack tree.

3.2   Updates to Relaxed B-slack trees

We now describe the algorithms for inserting and deleting keys in a relaxed B-slack trees (in a way that maintains P0', P1', P2' and P3). The updates for relaxed B-slack trees are shown in Figure 1. There, weights appear to the right of nodes, and shaded regions represent slack. If u is a node that is not the root, then we let $\pi(u)$ denote the parent of u. Our insertion and deletion algorithms always ensure that all leaves have weight one.

Deletion. First, a search is performed to find the leaf u where the deletion should occur. If the leaf does not contain the key to be deleted, then the deletion terminates immediately, and the tree does not change. If the leaf contains the key to be deleted, then the key is removed from the sequence of keys stored in that leaf. Deleting this key may create a slack violation.

Insertion. To perform an insertion, a search is first performed to find the leaf u where the insertion should occur. If u contains some slack, then the key is added to the sequence of keys in u, and the insertion terminates. Otherwise, u cannot accommodate the new key, so Overflow is performed. Overflow replaces u by a subtree of height one consisting of an internal node with weight zero, and two leaves with weight one. The b keys stored in u, plus the new key, are evenly distributed between the children of the new internal node. If u was the root before the insertion, then the new internal node becomes the new root. Otherwise, u's parent $\pi(u)$ before the insertion is changed to point to the new

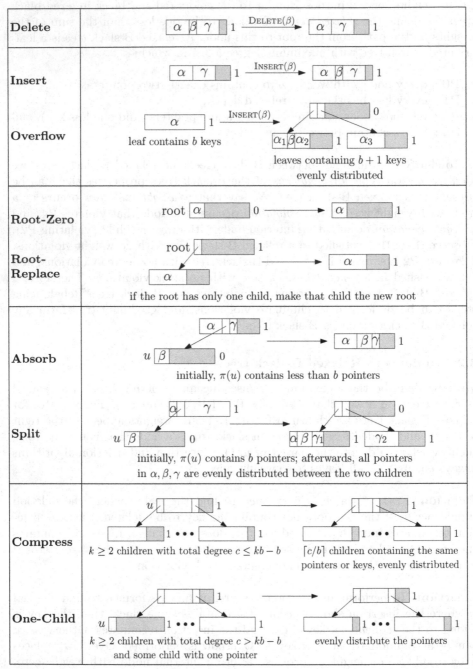

Fig. 1. Updates to B-slack trees (and relaxed B-slack trees). Nodes with weight zero contain exactly two pointers.

internal node instead of u. After Overflow, there is a weight violation at the new internal node. Additionally, since the new internal node contains $b - 2$ slack, whereas u contained no slack, there may be a slack violation at $\pi(u)$.

Delete, Insert and Overflow maintain the properties of a relaxed B-slack tree. They will also maintain the properties of a B-slack tree, provided that rebalancing steps are performed to remove any violations that are created.

3.3 Rebalancing Steps

There are six different rebalancing steps for relaxed B-slack trees: Root-Zero, Root-Replace, Absorb, Split, One-Child and Compress. If there is a degree violation at the root, then Root-Replace is performed. If not, but there is a weight violation at the root, Root-Zero is performed. If there is a weight violation at an internal node that is not the root, then Absorb or Split is performed. Suppose there are no weight violations. If there is a degree violation at a node u and no degree or slack violation at $\pi(u)$, then One-Child is performed. If there is a slack violation at a node u and no degree violation at u, then Compress is performed. Figure 1 illustrates these steps. The goal of rebalancing is to eliminate all violations, while maintaining the relaxed B-slack tree properties.

Root-Zero. Root-Zero changes the weight of the root from zero to one, eliminating a weight violation, and incrementing the relaxed depth of every node. If P1' held before Root-Zero, it holds afterwards.

Root-Replace. Root-Replace replaces the root r by its only child u, and sets u's weight to one. This eliminates a degree violation at r, and any weight violation at u. If u had weight zero before Root-Replace, then the relaxed depth of every leaf is the same before and after Root-Replace. Otherwise, the relaxed depth of every leaf is decremented by Root-Replace. In both cases, if P1' held before Root-Replace, it holds afterwards.

Absorb. Let u be a non-root node with weight zero. Absorb is performed when $\pi(u)$ contains less than b pointers. In this case, the two pointers in u are moved into $\pi(u)$, and u is removed from the tree. Since the pointer from $\pi(u)$ to u is no longer needed once u is removed, $\pi(u)$ now contains at most b pointers. The only node that was removed is u and, since it had weight zero, the relaxed depth of every leaf remains the same. Thus, if P1' held before Absorb, it also holds afterwards. Absorb eliminates a weight violation at u, but may create a slack violation at $\pi(u)$.

Split. Let u be a non-root node with weight zero. Split is performed when $\pi(u)$ contains exactly b pointers. In this case, there are too many pointers to fit in a single node. We create a new node v with weight one, and evenly distribute all of the pointers and keys of u and $\pi(u)$ (except for the pointer from $\pi(u)$ to u) between u and v. Now $\pi(u)$ has two children, u and v. The weight of u is set to one, and the weight of $\pi(u)$ is set to zero. As above, this does not change the relaxed depth of any leaf, so P1′ still holds after Split. Split moves a weight violation from u to $\pi(u)$ (closer to the root, where it can be eliminated by a Root-Zero or Root-Replace), but may create slack violations at u and v.

Compress. Compress is performed when there is a slack violation at an internal node u, there is no degree violation at u, and there are no weight violations at u or any of its $k \geq 2$ children. Let $c \leq kb - b$ be the number of pointers or keys stored in the children of u. Compress evenly distributes the pointers or keys contained in the children of u amongst the first $\lceil c/b \rceil$ children of u, and discards the other children. This will also eliminate any degree violations at the children of u if $c > 1$. After the update, u satisfies P4. Compress does not change the relaxed depth of any node, so P1′ still holds after. Compress removes at least one child of u, so it increases the slack of u by at least one, possibly creating a slack violation at $\pi(u)$. (However, it decreases the total amount of slack in the tree by at least $b - 1$.) Thus, after a Compress, it may be necessary to perform another Compress at $\pi(u)$. Furthermore, as Compress distributes keys and pointers, it may move nodes with different parents together, under the same parent. Even if two parents initially satisfied P4 (so the children of each parent contain a total of less than b slack), the children of the combined parent may contain b or more slack, creating a slack violation. Therefore, after a Compress, it may also be necessary to perform Compress at some of the children of u.

One-Child. One-Child is performed when there is a degree violation at an internal node u, there are no weight violations at u or any of its siblings, and there is no violation of any kind at $\pi(u)$. Let k be the degree of $\pi(u)$. Since there is no slack violation at $\pi(u)$, there are a total of $c > kb - b = b(k-1)$ pointers stored in u and its siblings. Since u has only one child pointer, each of its other $k-1$ siblings must contain b pointers. One-Child evenly distributes the keys and pointers of the children of $\pi(u)$. One-Child does not change the relaxed depth of any node, so P1′ still holds after. One-Child eliminates a degree violation at u, but, like Compress, it may move children with different parents together under the same parent, possibly creating slack violations at some children of $\pi(u)$. So, it may be necessary to perform Compress at some of the children of $\pi(u)$.

All of these updates maintain P0′, P2′ and P3. While rebalancing steps are being performed to eliminate the violation created by an insertion or deletion, there is at most one node with weight zero.

We prove that a rebalancing step can be applied to any relaxed B-slack tree that is not a B-slack tree.

Lemma 1. *Let T be a relaxed B-slack tree. If T is not a B-slack tree, then a rebalancing step can be performed.*

Proof. If T is not a B-slack tree, it contains a weight violation, a slack violation or a degree violation. If there is weight violation, then Root-Zero, Absorb or Split can be performed. Suppose there are no weight violations. Let u be the node at the smallest depth that has a slack or degree violation. Suppose u has a degree violation. If u is the root, then Root-Replace can be performed. Otherwise, $\pi(u)$

has no violation, so One-Child can be performed. Suppose u does not have a degree violation. Then, u must have a slack violation, and Compress can be performed.   □

4   Analysis

Due to space constraints, this section merely gives an outline of results proved about B-slack trees. In the full version of the paper, we provide a detailed analysis of B-slack trees that store n keys, by giving: an upper bound on the height of the tree, a lower bound on the average degree of nodes (and, hence, utilization), and an upper bound on the space complexity.

Arbitrary B-slack trees are difficult to analyze, so we begin by studying a class of trees called b-overslack trees. A b-overslack tree has a root with degree two, and satisfies P1, P2 and P3, but instead of P4, the children of each internal node contain a total of exactly b slack. Thus, a b-overslack tree is a relaxed B-slack tree, but not a B-slack tree. Consider a b-overslack tree T of height h that contains n keys. We prove that the total degree at depth $\delta \leq h$ in T is $d(\delta) = 2^{-\delta}(\alpha^\delta + \gamma^\delta)$, where $\alpha = b + \sqrt{b^2 - 4b}$ and $\gamma = b - \sqrt{b^2 - 4b}$. Since the total degree at the lowest depth is precisely the number of keys in the tree, every b-overslack tree of height h contains exactly $d(h)$ keys. Furthermore, when $h \geq 3$, we also have $(b - 1.4)^h < (b - \frac{\gamma}{2})^h \leq d(h) \leq b^h$. Therefore, for $n > b^3$ (which implies height at least three), h satisfies $\lceil \log_b n \rceil \leq h \leq \lceil \log_{b - \gamma/2} n \rceil < \lceil \log_{b-1.4} n \rceil$. We also prove that the average degree of nodes in T is $\frac{b \cdot d(h-1) - b + 2}{b \cdot d(h-2) - b + 3}$, which is greater than $b - 1.4$ for $h \geq 3$.

We next prove some connections between overslack trees and B-slack trees. First, we show that each b-overslack tree of height h has a smaller total degree of nodes at each depth than any B-slack tree of height h. We do this by starting with an arbitrary B-slack tree of height h, and repeatedly removing pointers and keys from the children of each internal node that satisfies P4 (taking care not to violate P1, P2 or P3), until we obtain a b-overslack tree. It follows that each b-overslack tree of height h contains fewer keys than any B-slack tree of height h. Consequently, every b-overslack tree with n keys has height at least as large as any B-slack tree with n keys. We next prove that every b-overslack tree of height h has a smaller average node degree than any B-slack tree of height h. As above, the proof starts with an arbitrary B-slack tree of height h, and removes pointers and keys from nodes until the tree becomes a b-overslack tree. However, in this proof, every time we remove a pointer, we must additionally show that the average degree of nodes in the tree decreases.

We then compute the *space complexity* of a B-slack tree containing n keys, which is the number of words needed to store it. Consider a leaf-oriented tree with maximum degree b. For simplicity, we assume that each key and each pointer to a child or data occupies one word in memory. Thus, a leaf occupies $2b$ words, and an internal node occupies $2b - 1$ words. A memory block size of $2b$ is assumed. Let $\bar{D}$ be the average degree of nodes. Then, $U = \bar{D}/b$ is the proportion of space that is utilized (which we call the *average space utilization* of the tree), and

$1 - U$ is the proportion of space that is wasted. The space complexity is $2bF$, where F is the number of nodes in the tree. Suppose the tree contains n keys. By definition, the sum of the degrees of all nodes is $F - 1 + n$, since each node, except the root, has a pointer into it and the degree of a leaf is the number of keys it contains. Additionally, $F\bar{D}$ is equal to the sum of degrees of all nodes, so $F = (n - 1)/(\bar{D} - 1)$. Therefore, the space complexity is $2b(n - 1)/(\bar{D} - 1)$. In order to compute an upper bound on the space complexity for a B-slack tree of height h, we simply need a lower bound on $\bar{D}$. Above, we saw that $\bar{D} > b - 1.4$ for B-slack trees of height at least three. It follows that a B-slack tree with $n > b^3$ keys has space complexity at most $2b(n - 1)/(b - 2.4) < \frac{2b}{b-2.4}n$.

The full version of the paper describes pathological families of B-trees, Overflow trees and H-trees, and compares the space complexity of example trees in these families with the worst-case upper bound on the space complexity of a B-slack tree. By studying these families, we obtain lower bounds on the space complexity of these trees that are above the upper bound for B-slack trees.

We also study the number of rebalancing steps necessary to maintain balance in a relaxed B-slack tree. Consider a relaxed B-slack tree obtained by starting from a B-slack tree containing n keys and performing a sequence of i insertions and d deletions. We prove that such a relaxed B-slack tree will be transformed back into a B-slack tree after at most $2i(2 + \lfloor\log_{\lfloor\frac{b}{2}\rfloor}(n + i)/2\rfloor) + d/(b - 1)$ rebalancing steps, irrespective of which rebalancing steps are performed, and in which order. Hence, insertions perform amortized $O(\log(n+i))$ rebalancing steps and deletions perform an amortized constant number of rebalancing steps.

5   Conclusion

We introduced B-slack trees, which have excellent space complexity in the worst case, and amortized logarithmic updates. The data structure is simple, requires only one block size, and is well suited for hardware implementation.

Modifying the definition of B-slack trees so that the total slack shared amongst the children of each internal node of degree k is at most $b + k - 1$, instead of $b - 1$, yields a data structure with amortized constant rebalancing (with small constants), and only a slight increase in space complexity. Specifically, such a tree containing $n > b^3$ keys occupies at most $\frac{2b}{b-3.4}n$ words. Details appear in the full version of the paper.

The recently introduced technique of Brown, Ellen and Ruppert [6] can be used to obtain a concurrent implementation of relaxed B-slack trees that tolerates process crashes and guarantees some process will always make progress. In the resulting implementation, localized updates to disjoint parts of the tree can proceed concurrently, and searches can proceed without synchronizing with updates, which makes them extremely fast. The implementation can be designed such that, in a quiescent state, when no updates are in progress, the data structure is a B-slack tree.

B-slack trees have been implemented in Java, and code is freely available from http://www.cs.utoronto.ca/~tabrown. Experiments have been performed to validate the theoretical worst-case bounds, and to better understand the level of pessimism in them. The results indicate that few rebalancing steps are performed in practice, and average degree is somewhat better than the already good worst-case bounds. For instance, for $b = 16$ and $b = 32$, over a variety of simulated random workloads with tree sizes varying between $2^5 = 32$ and $2^{20} = 1,048,576$ keys, there were at most 1.2 rebalancing steps per insertion or deletion, and average degrees for trees were approximately $b - 0.5$, which is extremely close to optimal.

Acknowledgments. This work was dramatically improved by the insightful comments of my supervisor, Faith Ellen.

References

1. Arnow, D.M., Tenenbaum, A.M.: An empirical comparison of B-trees, compact B-trees and multiway trees. ACM SIGMOD Record 14, 33–46 (1984)
2. Arnow, D.M., Tenenbaum, A.M., Wu, C.: P-trees: Storage efficient multiway trees. In: Proceedings of the 8th Annual International ACM SIGIR Conference on Research and Development in Information Retrieval, pp. 111–121. ACM (1985)
3. Baeza-Yates, R.A., Larson, P.-A.: Performance of B+-trees with partial expansions. IEEE Transactions on Knowledge and Data Eng. 1(2), 248–257 (1989)
4. Bayer, R., McCreight, E.: Organization and maintenance of large indexes. Technical Report D1-82-0989, Boeing Scientific Research Laboratories (1970)
5. Brönnimann, H., Katajainen, J., Morin, P.: Putting your data structure on a diet. CPH STL Rep, 1 (2007)
6. Brown, T., Ellen, F., Ruppert, E.: A general technique for non-blocking trees. In: Proc. of the 19th ACM SIGPLAN Symp. on Principles and Practice of Parallel Programming, PPoPP 2014, pp. 329–342. ACM, New York (2014)
7. Culik II, K., Ottmann, T., Wood, D.: Dense multiway trees. ACM Transactions on Database Systems (TODS) 6(3), 486–512 (1981)
8. Huang, S.-H.S.: Height-balanced trees of order (β, γ, δ). ACM Trans. Database Syst. 10(2), 261–284 (1985)
9. Küspert, K.: Storage utilization in B*-trees with a generalized overflow technique. Acta Informatica 19(1), 35–55 (1983)
10. Larsen, K., Soisalon-Soininen, E., Widmayer, P.: Relaxed balance through standard rotations. In: Rau-Chaplin, A., Dehne, F., Sack, J.-R., Tamassia, R. (eds.) WADS 1997. LNCS, vol. 1272, pp. 450–461. Springer, Heidelberg (1997)
11. Rosenberg, A.L., Snyder, L.: Compact B-trees. In: Proceedings of the 1979 ACM SIGMOD International Conference on Management of Data, SIGMOD 1979, pp. 43–51. ACM, New York (1979)
12. Srinivasan, B.: An adaptive overflow technique to defer splitting in B-trees. The Computer Journal 34(5), 397–405 (1991)

Approximately Minwise Independence with Twisted Tabulation[*]

Søren Dahlgaard and Mikkel Thorup

University of Copenhagen
{soerend,mthorup}@di.ku.dk

Abstract. A random hash function h is ε-minwise if for any set S, $|S| = n$, and element $x \in S$, $\Pr[h(x) = \min h(S)] = (1 \pm \varepsilon)/n$. Minwise hash functions with low bias ε have widespread applications within similarity estimation.

Hashing from a universe $[u]$, the twisted tabulation hashing of Pǎtraşcu and Thorup [SODA'13] makes $c = O(1)$ lookups in tables of size $u^{1/c}$. Twisted tabulation was invented to get good concentration for hashing based sampling. Here we show that twisted tabulation yields $\tilde{O}(1/u^{1/c})$-minwise hashing.

In the classic independence paradigm of Wegman and Carter [FOCS'79] $\tilde{O}(1/u^{1/c})$-minwise hashing requires $\Omega(\log u)$-independence [Indyk SODA'99]. Pǎtraşcu and Thorup [STOC'11] had shown that simple tabulation, using same space and lookups yields $\tilde{O}(1/n^{1/c})$-minwise independence, which is good for large sets, but useless for small sets. Our analysis uses some of the same methods, but is much cleaner bypassing a complicated induction argument.

1 Introduction

The concept of minwise hashing (or the "MinHash algorithm" according to [1]) is a basic algorithmic tool suggested by Broder et al. [1,2] for problems related to set similarity and containment. After the initial application of this algorithm in the early AltaVista search engine to detecting and clustering similar documents, the scheme has reappeared in numerous other applications[1] and is now a standard tool in data mining where it is used for estimating similarity [2,1,3], rarity [4], document duplicate detection [5,6,7,8], large-scale learning [9], etc. [10,11,12,13].

The basic motivation of minwise independence is to use hashing to select an element from a set S. With a hash function h, we simply pick the element $x \in S$ with the minimum hash value. If the hash function is fully random and no two keys get the same hash, then x is uniformly distributed in S.

A nice aspect of minwise selection is that $\min h(A \cup B) = \min\{\min h(A), \min h(B)\}$. This makes it easy, e.g., to select a random leader in many distributed settings. It also implies that that

---

[*] Research partly supported by Thorup's Advanced Grant from the Danish Council for Independent Research under the Sapere Aude research carrier programme.

[1] See http://en.wikipedia.org/wiki/MinHash

R Ravi and I.L. Gørtz (Eds.): SWAT 2014, LNCS 8503, pp. 134–145, 2014.
© Springer International Publishing Switzerland 2014

$\min h(A \cup B) \in h(A \cap B) \iff \min h(A) = \min h(B)$. Therefore, if h is fully random and collision free,

$$\Pr_h[\min h(A) = \min h(B)] = \frac{|A \cap B|}{|A \cup B|} .$$

Thus, if we, for two sets A and B, have stored $\min h(A)$ and $\min h(B)$, then we can use $[\min h(A) = \min h(B)]^2$ as an unbiased estimator for the Jaccard similarity $|A \cap B|/|A \cup B|$.

Unfortunately, we cannot realistically implement perfect minwise hash functions where each $x \in S$ has probability $1/|S|$ of being the unique minimum [2]. More precisely, to handle any subset S of a universe $\mathcal{U}$, we need a random permutation $h : \mathcal{U} \to \mathcal{U}$ represented using $\Theta(|\mathcal{U}|)$ bits.

Instead we settle for a bias ε. Formally, a random hash function $h : \mathcal{U} \to \mathcal{R}$ from some key universe $\mathcal{U}$ to some range $\mathcal{R}$ of hash values is random variable following some distribution over $\mathcal{R}^{\mathcal{U}}$. We say that h is ε-minwise or has bias ε if for every $S \subseteq \mathcal{U}$ and $x \in \mathcal{U} \setminus S$,

$$\Pr[h(x) \leqslant \min h(S)] \leqslant \frac{1 + \varepsilon}{|S| + 1} \tag{1}$$

$$\Pr[h(x) < \min h(S)] \geqslant \frac{1 - \varepsilon}{|S| + 1} \tag{2}$$

From (1) and (2), we easily get for any $A, B \subseteq \mathcal{U}$, that

$$\Pr_h[\min h(A) = \min h(B)] = (1 \pm \varepsilon) \cdot \frac{|A \cap B|}{|A \cup B|} .$$

To implement ε-minwise hashing in Wegman and Carter's [14] classic framework of k-independent hash functions $\Theta(\log \frac{1}{\varepsilon})$-independence is both sufficient [15] and necessary [16]. These results are for "worst-case" k-independent hash functions. A much more time-efficient solution is based on simple tabulation hashing of Zobrist [17]. In simple tabulation hashing, the hash value is computed by looking up $c = O(1)$ bitstrings in tables of size $|\mathcal{U}|^{1/c}$ and XORing the results. This is very fast with tables in cache. Pătraşcu and Thorup have shown [18] that simple tabulation hashing, which is not even 4-independent, has bias $\varepsilon = \tilde{O}(1/|S|^{1/c})$. Unfortunately, this bias is useless for small sets S.

In this paper, we consider the twisted tabulation of Pătraşcu and Thorup [19] which was invented to yield Chernoff-style concentration bounds, and high probability amortized performance bounds for linear probing. It is almost as fast as simple tabulation using the same number of lookups but an extra XOR and a shift. We show that with twisted tabulation, the bias is $\varepsilon = \tilde{O}(1/|U|^{1/c})$, which is independent of the set size.

It should be noted, that Thorup [20] recently introduced a double tabulation scheme yielding high independence in $O(1)$ time, hence much faster than using an

---

[2] This is the Iverson bracket notation, where $[P]$ is 1 for a predicate P if P is true and 0 otherwise.

$\omega(1)$-degree polynomial to get $\omega(1)$-independence and $o(1)$ bias. However, with table size $|U|^{1/c}$, the scheme ends up using at least $7c$ lookups [20, Theorem 1] and 12 times more space, so we expect it to be at least an order of magnitude slower than twisted tabulation[3].

When using minwise for similarity estimation, to reduce variance, we typically want to run q experiments with q independent hash functions $h_1, ..., h_q$, and save the vector of $(\min h_1(A), ..., \min h_q(A))$ as a *sketch* for the set A. We can then estimate the Jaccard similarity as $\sum_{i=1}^{q}[\min h_i(A) = \min h_i(B)]/q$. While q reduces variance, it does not reduce bias, so the bias has to be small for each h_i. This scheme is commonly referred to as $k \times$minwise. Since $\min h_1(A)$ is always compared to $\min h_1(B)$, we say that the samples of the two sketches are *aligned*. A standard alternative[1], called bottom-q, is to just use a single hash function h, and store the q smallest hash values as a set $S(A)$. Estimating the Jaccard-index is then done as $|S(A) \cap S(B) \cap \{q \text{ smallest values of } S(A) \cup S(B)\}|/q$. It turns out that a large q reduces both variance and bias [23]. However, the problem with bottom-q sketches, is that the samples lose their alignment. In applications of large-scale machine learning this alignment is needed in order to efficiently construct a dot-product for use with a linear support vector machine (SVM) [4] such as LIBLINEAR [24] or Pegasos [25]. Using the alignment of $k \times$minwise, it was shown how to construct such a dot-product in [9] based on this scheme. In such applications it is therefore important to have small bias ε. Finally, we note that when $q = 1$, both schemes reduce to basic minwise hashing with the fundamental goal of sampling a single random element from any set with only a small bias, which is exactly the problem addressed in this paper.

2   Preliminaries

Let us briefly review tabulation-based hashing. For both simple and twisted tabulation we are dealing with some universe $\mathcal{U} = \{0, 1, ..., u - 1\}$ denoted by $[u]$ and wish to hash keys from $[u]$ into some range $\mathcal{R} = [2^r]$. We view a key $x \in [u]$ as a vector of $c > 1$ *characters* from the alphabet $\Sigma = [u^{1/c}]$, i.e. $x = (x_0, ..., x_{c-1}) \in \Sigma^c$. We generally assume c to be a small constant (e.g. 4).

2.1   Simple Tabulation

In simple tabulation hashing we initialize c tables $h_0, ..., h_{c-1} : \Sigma \to \mathcal{R}$ with independent random data. The hash $h(x)$ is then computed as

$$h(x) = \bigoplus_{i \in [c]} h_i[x_i] .$$

Here $\oplus$ denotes bit-wise XOR. This is a well-known scheme dating back to [17].

---

[3] The whole area of tabulation hashing is about minimizing the number of lookups, e.g., [21] saves a factor 2 in lookups over [22] for moderate independence.

[4] See http://en.wikipedia.org/wiki/Support_vector_machine#Linear_SVM

Simple Tabulation is known to be 3-independent, but it was shown in [18] to have much more powerful properties than this would suggest. These properties include fourth moment bounds, Chernoff bounds when distribution balls into many bins and random graph properties necessary in cuckoo hashing. It was also shown that simple tabulation is ε-minwise independent with $\varepsilon = O\left(\frac{\lg^2 n}{n^{1/c}}\right)$.

We will need the following basic lemma regarding simple tabulation ([18, Lemma 2.2]):

Lemma 1. *Suppose we use simple tabulation to hash $n \leqslant m^{1-\varepsilon}$ keys into m bins for some constant $\varepsilon > 0$. For any constant γ, all bins get less than $d = \min\{((1+\gamma)/\varepsilon)^c, 2^{(1+\gamma)/\varepsilon}\}$ keys with probability $\geqslant 1 - m^{-\gamma}$.*

Specifically this implies that if we hash n keys into $m = nu^\varepsilon$ bins, then each bin has $O(1)$ elements with high probability. In this paper "with high probability" (w.h.p.) means with probability $1 - u^{-\gamma}$ for any desired constant $\gamma > 1$.

2.2   Twisted Tabulation

Twisted tabulation hashing is another tabulation-based hash function introduced in [19]. Twisted tabulation can be seen as two independent simple tabulation functions $h^\tau : \Sigma^{c-1} \to \Sigma$ and $h^S : \Sigma^c \to \mathcal{R}$. If we view a key x as the head $head(x) = x_0$ and the tail $tail(x) = (x_1, \ldots, x_{c-1})$, we can define the hash value of twisted tabulation as follows:

$$t(x) = h^\tau(tail(x))$$

$$h_{>0}(x) = \bigoplus_{i=1}^{c-1} h_i^S[x_i]$$

$$h(x) = h_{>0}(x) \oplus h_0^S[x_0 \oplus t(x)] \ .$$

We refer to the value $x_0 \oplus t(x)$ as the *twisted head* of the key x, and define the *twisted group* of a character α to be $G_\alpha = \{x \mid x_0 \oplus t(x) = \alpha\}$. For the keys in G_α, we refer to the XOR with $h_0^S[x_0 \oplus t(x)]$ as the *final (XOR)-shift*, which is common to all keys in G_α. We call $h_{>0}(x)$ the *internal hashing*.

Throughout the proofs we will rely on the independence between h^τ and h^S to fix the hash function in a specific order, i.e. fixing the twisted groups first.

One powerful property of twisted tabulation is that the keys are distributed nicely into the twisted groups. We will use the following lemma from the analysis of twisted tabulation [19, Lemma 2.1]:

Lemma 2. *Consider an arbitrary set S of keys and a constant parameter $\varepsilon > 0$. W.h.p. over the random choice of the twister hash function, h^τ, all twisted groups have size $O(1 + |S|/\Sigma^{1-\varepsilon})$.*

Twisted tabulation hashing also gives good concentration bounds in form of Chernoff-like tail bounds, which is captured by the following lemma, [19, Theorem 1.1].

Lemma 3. *Choose a random twisted tabulation hash function $h : [u] \to [u]$. For each key $x \in [u]$ in the universe, we have an arbitrary value function $v_x : [u] \to [0,1]$ assigning a value $V_x = v_x(h(x)) \in [0,1]$ to x for each possible hash value. Let $\mu_x = \mathbf{E}_{y \in [u]}[v_x(y)]$ denote the expected value of $v_x(y)$ for uniformly distributed $y \in [u]$. For a fixed set of keys $S \subseteq [u]$, define $V = \sum_{x \in S} V_x$ and $\mu = \sum_{x \in S} \mu_x$. Let γ, c, and ε be constants. Then for any $\mu < \Sigma^{1-\varepsilon}$ and $\delta > 0$ we have:*

$$\Pr[V \geq (1+\delta)\mu] \leq \left(\frac{e^\delta}{(1+\delta)^{(1+\delta)}} \right)^{\Omega(\mu)} + 1/u^\gamma \qquad (3)$$

$$\Pr[V \leq (1-\delta)\mu] \leq \left(\frac{e^{-\delta}}{(1-\delta)^{(1-\delta)}} \right)^{\Omega(\mu)} + 1/u^\gamma \qquad (4)$$

In practice, we can merge h^τ and h^S to a single simple tabulation function $h^\star : \Sigma \to \Sigma \times \mathcal{R}$, but with $h_0^\star : \Sigma \to \mathcal{R}$. This adds $\log \Sigma$ bits to each entry of the tables $h_1^\star, \ldots h_{c-1}^\star$ (in practice we want these to be 32 or 64 bits anyway). See the code in Figure 1 for an implementation of 32-bit keys in C.

```
INT32 TwistedTab32(INT32 x, INT64[4][256] H) {
    INT32 i;
    INT64 h=0;
    INT8 c;
    for (i=0;i<3;i++) {
      c=x;
      h^=H[i][c];
      x = x>> 8;
    }                          // at the end i=3
    c=x^h;                     // extra xor with h
    h^=H[i][c];
    h>>=32;                    // extra shift of h
    return ((INT32) h);
}
```

Fig. 1. C-code implementation of twisted tabulation for 32-bit keys assuming a point H to randomly fille storage

3   Minwise for Twisted Tabulation

We will now show the following theorem:

Theorem 1. *Twisted tabulation is $O\left(\frac{\log^2 u}{\Sigma} \right)$-minwise independent.*

Recall from the definition of ε-minwise, that we are given an input set S of $|S| = n$ keys and a query key $q \in \mathcal{U} \setminus S$. We will denote by Q the twisted group of the query key q. Similarly to the analysis in [18] we assume that the output range is $[0, 1)$. We pick $\ell = \gamma \log u$ and divide the output range into n/ℓ bins. Here γ is chosen such that the number of bins is a power of two and large enough that the following two properties hold.

1. The minimum bin $[0, \ell/n)$ is non-empty with probability $1 - 1/u^2$ by Lemma 3. Here $\mu = O(\log u) < \Sigma^{1-\varepsilon}$.
2. The bins are d-bounded for each twisted group (for some constant d) with probability $1 - 1/u^2$ by Lemma 1. Meaning that for any twisted group G, at most d keys land in each of the n/ℓ bins after the internal hashing is done. This holds because each twisted group has $n/\Sigma^{1-\varepsilon}$ elements w.h.p. by Lemma 2.

Similar to [18], we assume that the hash values are binary fractions of infinite precision so we can ignore collisions. The theorem holds even if we use just $\lg(n\Sigma)$ bits for the representation: Let $\tilde{h}$ be the truncation of h to $\lg(n\Sigma)$ bits. There is only a distinction when $\tilde{h}(q)$ is minimal and there exists some $x \in S$ such that $\tilde{h}(x) = \tilde{h}(q)$. Since the minimum bin is non-empty with probability $1 - 1/u^2$ we can bound the probability of this from above by

$$\Pr\left[\tilde{h}(q) \leqslant \ell/n \wedge \exists x \in S : \tilde{h}(x) = \tilde{h}(q)\right] \leqslant \frac{\ell}{n} \cdot \left(n \cdot \frac{1}{n\Sigma}\right) + 1/u^2$$

using 2-independence to conclude that $\{\tilde{h}(q) \leqslant \ell/n\}$ and $\{\tilde{h}(x) = \tilde{h}(q)\}$ are independent.

3.1   Upper Bound

To upper bound the probability that $h(q)$ is smaller than $\min h(S)$ it suffices to look at the case when q is in the minimum bin $[0, \ell/n)$, as we have

$$\Pr[h(q) < \min h(S)] \leqslant \Pr[\min h(S) \geqslant \ell/n] + \Pr[h(q) < \min(h(S) \cup \{\ell/n\})]$$
$$\leqslant 1/u^2 + \Pr[h(q) < \min(h(S) \cup \{\ell/n\})] \tag{5}$$

To bound (5) we will use the same notion of representatives as in [18]: If a non-query twisted group $G_\alpha \neq Q$ has more than one element in some bin, we pick one of these arbitrarily as the representative. Let $R(G_\alpha)$ denote the set of representatives from G_α and let R denote the union of all such sets. We trivially have that $\Pr[h(q) < \min h(S)] \leqslant \Pr[h(q) < \min h(R)]$.

The proof relies on fixing the tables associated with the hash functions h^τ and h^S in the following order:

1. Grouping into twisted groups is done by fixing h^τ. Each group has $O(1 + n/\Sigma^{1-\varepsilon})$ elements by Lemma 2 w.h.p.
2. The internal hashing of all twisted groups is done by fixing the tables $h_1^S, \ldots, h_{c-1}^S$. This determines the set of representatives R.
3. Having fixed the set R we do the final shifts of the twisted groups G_α by fixing h_0^S. We will show that the probability of q having the minimum hash value after these shifts is at most $1/(|R| + 1)$.
   Since $|R|$ is a random variable depending only on the internal hashing and twisted groups, the entire probability is bounded by $\mathbf{E}[1/(|R| + 1)]$.

To see step 3 from above we let $Rand(A)$ be a randomizing function that takes each element in a set A and replaces it with an independent uniformly random number in $[0, 1)$. We will argue that

$$\Pr[h(q) < \min h(R) \cup \{\ell/n\}] \leqslant \Pr[h(q) < \min Rand(R)] = 1/(|R| + 1) \quad (6)$$

To prove (6) fix $h(q) = p < \ell/n$ and consider some twisted group G_α. When doing the final shift of the group we note that each representative $x \in R(G_\alpha)$ is shifted randomly, so $\Pr[h(x) \leqslant p] = p$. However, since the number of bins is a power of two, and each representative in $R(G_\alpha)$ is shifted by the same value, at most one element of $R(G_\alpha)$ can land in the minimum bin. This gives $\Pr[\min h(R(G_\alpha)) \leqslant p] = |R(G_\alpha)|p$. For $Rand(R)$, a union bound gives that $\Pr[\min Rand(R(G_\alpha)) \leqslant p] \leqslant |R(G_\alpha)|p$, implying that

$$\Pr[p < \min(h(R(G_\alpha)) \cup \{\ell/n\})] \leqslant \Pr[p < \min(Rand(R(G_\alpha)) \cup \{\ell/n\})]$$

Because the shifts of different twisted groups are done independently we get

$$\begin{aligned}
\Pr[p < \min(h(R) \cup \{\ell/n\})] &= \prod_{G_\alpha \neq Q} \Pr[p < \min(h(R(G_\alpha)) \cup \{\ell/n\})] \\
&\leqslant \prod_{G_\alpha \neq Q} \Pr[p < \min(Rand(R(G_\alpha)) \cup \{\ell/n\})] \\
&= \Pr[p < \min(Rand(R) \cup \{\ell/n\})] \\
&\leqslant \Pr[p < \min Rand(R)]
\end{aligned}$$

This holds for any value $p < \ell/n$, so it also holds for our random hash value $h(q)$. Therefore

$$\Pr[h(q) < \min(h(R) \cup \{\ell/n\})] \leqslant \Pr[h(q) < \min Rand(R)] \leqslant 1/(|R| + 1)$$

This finishes the proof of (6).

All that remains is to bound the expected value $\mathbf{E}[1/(|R| + 1)]$ and thus the total probability when the internal hashing and twisted groups are random. We will do this using a convexity argument, so we need the following constraints on the random variable $|R|$: We trivially have $1 \leqslant |R| \leqslant n$. We know that the internal hashing is d-bounded with probability $1 - 1/u^2$, which gives $|R| \geqslant |S \setminus Q|/d \geqslant n/(2d)$. To bound $\mathbf{E}[|R|]$ from below, consider the probability that a key x is *not* a representative. For this to happen x must land in the query group, or another element must land in the same twisted group and bin as x. By 2-independence and a union bound the probability of this event is at most $1/\Sigma + (n - 1) \cdot 1/\Sigma \cdot \ell/n = O(\ell/\Sigma)$. The expected number of representatives is therefore

$$\begin{aligned}
\mathbf{E}[|R|] &= \sum_{x \in S} \Pr[x \in R] \\
&= \sum_{x \in S} (1 - \Pr[x \notin R]) \\
&\geqslant n \cdot (1 - O(\ell/\Sigma)) \ .
\end{aligned}$$

To bound $\mathbf{E}[1/(|R|+1)]$ we introduce a random variable r which maximizes $\mathbf{E}[1/(r+1)]$ while satisfying the constraints of $|R|$ noted above. By convexity of $1/(r+1)$ we get that $\mathbf{E}[1/(r+1)]$ is maximized when r takes the most extreme values. Hence $r = 1$ with probability $1/u^2$, $r = n/(2d)$ with the maximal probability p and $r = n$ with probability $(1-p-1/u^2)$. This gives an expected value of

$$\mathbf{E}[r] = 1/u^2 + p \cdot n/(2d) + (1-p-1/u^2) \cdot n .$$

Thus $p = O(\ell/\Sigma)$ to respect the constraints. To bound $\mathbf{E}[1/(|R|+1)]$ we have

$$\mathbf{E}[1/(|R|+1)] \leqslant \mathbf{E}[1/(r+1)]$$
$$\leqslant \frac{1}{2u^2} + \frac{p}{n/(2d)+1} + \frac{1-p-1/u^2}{n+1}$$
$$\leqslant \frac{O(p)}{n+1} + \frac{1}{n+1} + O(1/u^2)$$
$$= \frac{1}{n+1} \cdot (1 + O(\ell/\Sigma)) . \tag{7}$$

Combining (5), (6) and (7) we get

$$\Pr[h(q) < \min h(S)] \leqslant \Pr[h(q) < \min(h(S) \cup \{\ell/n\})] + O(1/u^2)$$
$$\leqslant \Pr[h(q) < \min(h(R) \cup \{\ell/n\})] + O(1/u^2)$$
$$\leqslant \mathbf{E}[1/(|R|+1)] + O(1/u^2)$$
$$= \frac{1}{n+1} \cdot \left(1 + O\left(\frac{\log u}{\Sigma}\right)\right) .$$

3.2   Lower Bound

We have two cases for the lower bound. When $n = O(\log u)$ we observe that the probability of some twisted group having more than one element is bounded from above by $n^2/\Sigma = O(\log^2 u/\Sigma)$ using 2-independence and a union bound. Since the twisted groups hash independently of each other we have in this case that all elements hash independently. The probability of q getting the smallest hash value is thus at least $1/(n+1) \cdot (1 - O(\log^2 u/\Sigma))$.

When $n = \omega(\log u)$ we again look at the case when q lands in the minimum bin $[0, \ell/n)$. We consider the query group Q separately and thus look at the expression:

$$\Pr[h(q) < \min h(S)] \geqslant \Pr[h(q) < \min(h(S) \cup \{\ell/n\})]$$
$$= \Pr[h(q) < \min(h(S \setminus Q) \cup \{\ell/n\})]$$
$$- \Pr[\min h(Q) < h(q) < \min(h(S \setminus Q) \cup \{\ell/n\})] . \tag{8}$$

Furthermore we will assume that all twisted groups have $O(1 + n/\Sigma^{1-\varepsilon})$ elements at the cost of a factor $(1 - 1/u^2)$ by Lemma 2. We will subtract this extra term later in (13). Since the twisted groups hash independently we have for a fixed $h(q) = p < \ell/n$ that

$$\Pr[p < \min h(S \setminus Q)] = \prod_{G_\alpha \neq Q} \Pr[p < \min h(G_\alpha)] \ . \tag{9}$$

We can bound this expression using [18, Lemma 5.1], which states that $1 - pk > (1 - p)^{(1+pk)k}$ for $pk \leqslant \sqrt{2} - 1$ and $p \in [0,1]$. Consider a twisted group G_α and some element $x \in G_\alpha$. We have $\Pr[h(x) < p] = p$ and a union bound gives us that $\Pr[p < \min h(G_\alpha)] \geqslant 1 - p|G_\alpha|$. Since $n = \omega(\log u)$ we have that $p|G_\alpha| \leqslant \ell/n \cdot O(1 + n/\Sigma^{1-\varepsilon}) = o(1)$, so the conditions for the lemma hold. This gives us

$$1 - p|G_\alpha| \geqslant (1 - p)^{|G_\alpha|(1+p \cdot |G_\alpha|)} \tag{10}$$

Plugging this into (9) gives

$$\Pr[p < \min h(S \setminus Q)] \geqslant \prod_{G_\alpha \neq Q} (1 - p)^{|G_\alpha|(1+p|G_\alpha|)}$$

$$\geqslant \prod_{G_\alpha \neq Q} (1 - p)^{|G_\alpha|(1+p \cdot 2(|G_\alpha|-1))}$$

$$\geqslant (1 - p)^m,$$

with

$$m = n + O(\ell/n) \cdot \sum_{G_\alpha \neq Q} (|G_\alpha| - 1)|G_\alpha| \ .$$

To bound the entire probability we thus integrate from 0 to ℓ/n:

$$\Pr[h(q) \leqslant \min(h(S \setminus Q) \cup \{\ell/n\})] = \int_0^{\ell/n} \Pr[p < \min h(S \setminus Q)] \, dp$$

$$\geqslant \int_0^{\ell/n} (1 - p)^m \, dp$$

$$\geqslant \frac{1 - (1 - \ell/n)^{m+1}}{m + 1}$$

$$> 1/(m + 1) - 1/(nu) \ . \tag{11}$$

Similar to the upper bound m only depends on the twisted groups and their internal hashing, so the entire probability is bounded by $\mathbf{E}[1/(m+1)] - 1/nu \geqslant 1/\mathbf{E}[m+1] - 1/nu$. We note that the sum $\sum_{G_\alpha \neq Q}(|G_\alpha|-1)|G_\alpha|$ counts for each key in a non-query group the number of other elements in its group, so

$$\mathbf{E}\left[\sum_{G_\alpha \neq Q} (|G_\alpha| - 1)|G_\alpha|\right] \leqslant n^2/\Sigma \ .$$

The expected value $\mathbf{E}[m + 1]$ is therefore bounded by

$$\mathbf{E}[m + 1] \leqslant (n + 1) \cdot (1 + O(\ell/\Sigma)) \ . \tag{12}$$

We can combine this with (11) and get a bound on the first part of (8). We also need to subtract the probability that the keys don't distribute nicely into twisted groups. Doing this we get the following bound:

$$
\begin{aligned}
\Pr[h(q) \leqslant \min(h(S \setminus Q) \cup \{\ell/n\})] &\geqslant \mathbf{E}[1/(m+1)] - 1/nu - 1/u^2 \\
&\geqslant 1/\mathbf{E}[m+1] - 1/nu - 1/u^2 \\
&\geqslant \frac{1}{(n+1)(1 + O(\ell/\Sigma))} - 1/nu - 1/u^2 \\
&\geqslant \frac{1}{n+1} \cdot \left(1 - O\left(\frac{\log u}{\Sigma}\right)\right)
\end{aligned} \tag{13}
$$

To finish the bound on (8) we need to give an upper bound on

$$
\Pr[\min h(Q) < h(q) < \min h(S \setminus Q) \wedge h(q) < \ell/n] . \tag{14}
$$

To do this we will again consider the set of representatives that we used in the upper bound. We start by fixing the twisted groups. Just like in the upper bound we have w.h.p. that $|R| \geqslant n/(2d)$. We can therefore bound (14) by

$$
1/u^2 + \Pr[\min h(Q) < h(q) < \min h(S \setminus Q) \wedge h(q) < \ell/n \wedge |R| \geqslant n/(2d)] .
$$

We fix $h(q) = p$ for some $p < \ell/n$. Using 2-independence between the fixed query value p and each element of Q we get $\Pr[\min h(Q) < p] \leqslant p|Q|$ and thus

$$
\Pr[\min h(Q) < p \wedge |R| \geqslant n/(2d)] \leqslant p|Q| . \tag{15}
$$

We wish to multiply this by

$$
\Pr[p < \min h(S \setminus Q) \mid \min h(Q) < p \wedge |R| \geqslant n/(2d)] .
$$

For this we use the same approach as for (6). We know that when $p < \ell/n$ we have that $\Pr[p < \min h(R)] \leqslant \Pr[p < \min Rand(R)] = (1 - p)^{|R|}$. This holds regardless of the internal hashing so our restriction of $|R| \geqslant n/(2d)$ does not change anything. We now get

$$
\Pr[p < \min h(S \setminus Q) \mid \min h(Q) < p \wedge |R| \geqslant n/(2d)] \leqslant (1 - p)^{n/(2d)} .
$$

Multiplying together with (15) we get

$$
\begin{aligned}
\Pr[\min h(Q) < p < \min h(S \setminus Q) \wedge |R| \geqslant n/(2d)] &\leqslant p|Q|(1 - p)^{n/(2d)} \\
&\leqslant p|Q|e^{-pn/(2d)}
\end{aligned}
$$

for a fixed $p < \ell/n$. To finish the bound we thus integrate from 0 to ℓ/n and get an upper bound on (14):

$$\Pr[\min h(Q) < h(q) < \min h(S \setminus Q) \wedge h(q) < \ell/n]$$
$$\leqslant 1/u^2 + \Pr[\min h(Q) < h(q) < \min h(S \setminus Q) \wedge h(q) < \ell/n \wedge |R| \geqslant n/(2d)]$$
$$\leqslant 1/u^2 + \int_0^{\ell/n} p|Q|e^{-pn/(2d)}\, dp$$
$$= 1/u^2 + O\left(\int_0^{d/n} p|Q|\, dp\right)$$
$$= 1/u^2 + O(|Q|/n^2) \ .$$

We now note that $|Q|$ is a random variable with expected value n/Σ, which gives the final bound on (14) as

$$\Pr[\min h(Q) < h(q) < \min h(S \setminus Q) \wedge h(q) < \ell/n] \leqslant \mathbf{E}\big[O(|Q|/n^2)\big]$$
$$= O(1/n\Sigma) \ . \tag{16}$$

Combining (8), (13) and (16) gives the desired bound:

$$\Pr[h(q) < \min h(S)] \geqslant \Pr[h(q) < \min h(S \setminus Q) \wedge h(q) < \ell/n]$$
$$- \Pr[\min h(Q) < h(q) < \min h(S \setminus Q) \wedge h(q) < \ell/n]$$
$$\geqslant \frac{1}{n+1} \cdot \left(1 - O\left(\frac{\log u}{\Sigma}\right)\right) - O\left(\frac{1}{n\Sigma}\right)$$
$$= \frac{1}{n+1} \cdot \left(1 - O\left(\frac{\log u}{\Sigma}\right)\right)$$

References

1. Broder, A.Z.: On the resemblance and containment of documents. In: Proc. Compression and Complexity of Sequences (SEQUENCES), pp. 21–29 (1997)
2. Broder, A.Z., Charikar, M., Frieze, A.M., Mitzenmacher, M.: Min-wise independent permutations. Journal of Computer and System Sciences 60(3), 630–659 (2000); See also STOC 1998
3. Broder, A.Z., Glassman, S.C., Manasse, M.S., Zweig, G.: Syntactic clustering of the web. Computer Networks 29, 1157–1166 (1997)
4. Datar, M., Muthukrishnan, S.M.: Estimating rarity and similarity over data stream windows. In: Möhring, R.H., Raman, R. (eds.) ESA 2002. LNCS, vol. 2461, pp. 323–334. Springer, Heidelberg (2002)
5. Broder, A.: Identifying and filtering near-duplicate documents. In: Giancarlo, R., Sankoff, D. (eds.) CPM 2000. LNCS, vol. 1848, pp. 1–10. Springer, Heidelberg (2000)
6. Manku, G.S., Jain, A., Sarma, A.D.: Detecting near-duplicates for web crawling. In: Proc. 10th WWW, pp. 141–150 (2007)
7. Yang, H., Callan, J.P.: Near-duplicate detection by instance-level constrained clustering. In: Proc. 29th SIGIR, pp. 421–428 (2006)

8. Henzinger, M.R.: Finding near-duplicate web pages: A large-scale evaluation of algorithms. In: Proc. ACM SIGIR, pp. 284–291 (2006)
9. Li, P., Shrivastava, A., Moore, J.L., König, A.C.: Hashing algorithms for large-scale learning. In: Advances in Neural Information Processing Systems, pp. 2672–2680 (2011)
10. Bachrach, Y., Herbrich, R., Porat, E.: Sketching algorithms for approximating rank correlations in collaborative filtering systems. In: Karlgren, J., Tarhio, J., Hyyrö, H. (eds.) SPIRE 2009. LNCS, vol. 5721, pp. 344–352. Springer, Heidelberg (2009)
11. Bachrach, Y., Porat, E., Rosenschein, J.S.: Sketching techniques for collaborative filtering. In: Proc. 21st IJCAI, pp. 2016–2021 (2009)
12. Cohen, E., Datar, M., Fujiwara, S., Gionis, A., Indyk, P., Motwani, R., Ullman, J.D., Yang, C.: Finding interesting associations without support pruning. IEEE Trans. Knowl. Data Eng. 13(1), 64–78 (2001)
13. Schleimer, S., Wilkerson, D.S., Aiken, A.: Winnowing: Local algorithms for document fingerprinting. In: Proc. SIGMOD, pp. 76–85 (2003)
14. Wegman, M.N., Carter, L.: New classes and applications of hash functions. Journal of Computer and System Sciences 22(3), 265–279 (1981); See also FOCS 1979
15. Indyk, P.: A small approximately min-wise independent family of hash functions. Journal of Algorithms 38(1), 84–90 (2001); See also SODA 1999
16. Pătraşcu, M., Thorup, M.: On the k-independence required by linear probing and minwise independence. In: Abramsky, S., Gavoille, C., Kirchner, C., Meyer auf der Heide, F., Spirakis, P.G. (eds.) ICALP 2010. LNCS, vol. 6198, pp. 715–726. Springer, Heidelberg (2010)
17. Zobrist, A.L.: A new hashing method with application for game playing. Technical Report 88, Computer Sciences Department, University of Wisconsin, Madison, Wisconsin (1970)
18. Pătraşcu, M., Thorup, M.: The power of simple tabulation-based hashing. Journal of the ACM 59(3) (2012); Article 14 Announced at STOC 2011
19. Pătraşcu, M., Thorup, M.: Twisted tabulation hashing. In: Proc. 24th ACM/SIAM Symposium on Discrete Algorithms (SODA), pp. 209–228 (2013)
20. Thorup, M.: Simple tabulation, fast expanders, double tabulation, and high independence. In: FOCS, pp. 90–99 (2013)
21. Klassen, T.Q., Woelfel, P.: Independence of tabulation-based hash classes. In: Proc. 10th Latin American Theoretical Informatics (LATIN), pp. 506–517 (2012)
22. Thorup, M., Zhang, Y.: Tabulation-based 5-independent hashing with applications to linear probing and second moment estimation. SIAM Journal on Computing 41(2), 293–331 (2012); Announced at SODA 2004 and ALENEX 2010
23. Thorup, M.: Bottom-k and priority sampling, set similarity and subset sums with minimal independence. In: Proc. 45th ACM Symposium on Theory of Computing, STOC (2013)
24. Fan, R.E., Chang, K.W., Hsieh, C.J., Wang, X.R., Lin, C.J.: LIBLINEAR: A library for large linear classification. Journal of Machine Learning Research 9, 1871–1874 (2008)
25. Shalev-Shwartz, S., Singer, Y., Srebro, N.: Pegasos: Primal estimated sub-gradient solver for svm. In: Proceedings of the 24th International Conference on Machine Learning, ICML 2007, pp. 807–814 (2007)

Separability of Imprecise Points

Mark de Berg[1], Ali D. Mehrabi[1,*], and Farnaz Sheikhi[2]

[1] Department of Mathematics and Computer Science,
TU Eindhoven, The Netherlands
[2] Laboratory of Algorithms and Computational Geometry, Department of
Mathematics and Computer Science, Amirkabir University of Technology

Abstract. An imprecise point is a point p with an associated imprecision region $\mathcal{I}_p$ indicating the set of possible locations of the point p. We study separability problems for a set R of red imprecise points and a set B of blue imprecise points in $\mathbb{R}^2$, where the imprecision regions are axis-aligned rectangles and each point $p \in R \cup B$ is drawn uniformly at random from $\mathcal{I}_p$. Our results include algorithms for finding *certain* separators (separating R from B with probability 1), *possible separators* (separating R from B with non-zero probability), *most likely* separators (separating R from B with maximal probability), and *maximal* separators (maximizing the expected number of correctly classified points).

1 Introduction

Separability problems are a natural class of problems arising in the analysis of categorical geometric data. In a separability problem one is given a set of n points in $\mathbb{R}^d$, each of which is categorized as either *red* or *blue*, and the goal is to decide whether the red points can be separated from the blue points by a *separator* from a given class of geometric objects. When the separator is a hyperplane the problem can be solved by linear programming in $O(n)$ time, as was observed by Megiddo [17] already 30 years ago. Since then various classes of separators have been studied, mostly for the 2-dimensional version of the problem. In particular, the separability problem in the plane has been studied for separators in the form of a circle [19], a strip and a wedge [12], and a convex [6] or simple polygon [8]. For the latter two problems the objective is not just to decide the existence of a separator but to find a minimum-complexity separator. Inspired by the reconstruction of buildings from LIDAR data, Van Kreveld *et al.* [13] recently considered arbitrarily oriented rectangles as separators, and Sheikhi *et al.* [21] studied arbitrarily oriented L-shapes.

Obviously it is not always possible to separate the given point sets by a separator of the given type. Houle [10,11] therefore introduced *weak separability*, where the goal is to maximize the number of correctly classified points. For example, for linear separability the weak separability problem asks for a line ℓ that maximizes the sum of the number of red points to the right of ℓ and the

---

* Supported by The Netherlands Organization for Scientific Research (NWO).

R Ravi and I.L. Gørtz (Eds.): SWAT 2014, LNCS 8503, pp. 146–157, 2014.

number of blue points to the left of ℓ. (A separator that correctly classifies all points is then called a *strong separator*.) Weak separability has been studied for separators in the form of a line [2,7,11] and a strip [3].

In data-analysis problems involving geometric data, the data is typically obtained by GPS, LIDAR, or some other imprecise measuring technology. Ideally, one would like to take this into account when analyzing the data. Within the computational-geometry literature, several imprecision models have been proposed [4,14,18]. The most popular models associate to each data point p an *imprecision region* $\mathcal{I}_p$, which indicates the possible locations of p. Typical choices for the imprecision regions are disks [20], axis-aligned rectangles or squares [14], and horizontal segments [14]. Horizontal segments model the situation where there is imprecision in only one of the coordinates, and rectangles or squares model the situation where the coordinates come from independent measurements. A point p with an associated imprecision region $\mathcal{I}_p$ is often called an *imprecise point*. Löffler [14] and Löffler and Van Kreveld [15,16] study classical computational-geometry problems on imprecise points. In most problems they want to find certain "extremal" structures, such as the largest possible convex hull. De Berg *et al.* [1] study the question whether a given structure is possible.

In this paper we study various separability problems for imprecise points in the plane. We extend the region-based imprecision model to include probabilistic aspects. More precisely, we assume each point p is drawn from its imprecision region $\mathcal{I}_p$ according to some distribution. In the current paper, we consider the uniform distribution. Given a set R of red points and a set B of blue points, and a class of separators, we then wish to find

- a *certain separator*, which separates R from B with probability 1;
- a *possible separator*, which separates R from B with non-zero probability;[1]
- a *most likely separator*, which separates R from B with maximal probability;
- a *maximal separator*, which is a weak separator that maximizes the expected number of correctly classified points.

Most of our results are for axis-aligned rectangles as imprecision regions. (We do not require the rectangles to have the same size or aspect ratio.) Our results are as follows. In Section 2 we observe that finding a certain separator can easily be done in $O(n)$ time, both for linear and rectangular separators. Finding possible separators is fairly easy as well: a possible linear separator can be found in $O(n \log n)$ time, while a possible rectangular separator can be found in $O(n)$ time. Most likely separators are harder. Here we study the 1-dimensional case, which already turns out to be hard to solve since it requires finding the maximum of a possibly high-degree polynomial. In Section 3 we present exact algorithms for weak separability for linear separators (running in $O(n^2)$ time), for rectangular separators (running in $O(n^3 \log n)$), and for rectangular separators when the imprecision regions are horizontal segments (running in $O(n^2\sqrt{n})$ time). We also present fast $(1 - \varepsilon)$-approximation algorithms.

---

[1] A valid separator can have zero separation probability, when the separator touches an imprecision-region boundary. Our algorithms can be adapted to this case.

2   Strong Separability

Let R be a set of red points and B be a set of blue points in the plane, with $n := |R| + |B|$. Each point $p \in R \cup B$ has an associated imprecision region $\mathcal{I}_p$, which is an axis-aligned rectangle. In this section we give algorithms to find strong separators, that is, separators that classify all points correctly.

Certain and Possible Separators. A line (or other shape) is a *certain sep-arator* if and only if the interiors of all red imprecision regions lie entirely on one side of it while the interiors of all blue imprecision regions lie entirely on the other side. Hence, deciding whether $R \cup B$ admits a certain separator is very easy: a line ℓ is a certain separator if and only if the vertices of the red and blue imprecision regions lie on opposite sides of ℓ, and so we can decide the existence of a certain separator by linear programming. Finding a rectangular certain separator is also easy: if there is an axis-aligned rectangle with, say, all red imprecision regions inside and all blue imprecision regions outside, then the bounding box of the red imprecision regions is a certain separator.

Finding possible separators is only slightly more involved than finding certain separators. First, consider linear separators. We wish to find a *possible separa-tor* ℓ (which we consider to be a directed line) that has the red points to its left and the blue points to its right. Then ℓ is a possible separator unless there is a red imprecision region lying completely to the right of ℓ or a blue imprecision lying completely to the left. Thus, we proceed as follows.

Suppose we rotate the coordinate frame over an angle ϕ in counterclockwise direction, for some $0 \leqslant \phi < 2\pi$. We call the axes in this rotated coordinate system the x_ϕ-axis and the y_ϕ-axis. For a red imprecise point $r \in R$, let $f_r(\phi)$ denote the minimum x_ϕ-coordinate of any point in $\mathcal{I}_r$. Similarly, let $g_b(\phi)$ denote the maximum x_ϕ-coordinate of any point in $\mathcal{I}_b$. Now there is a possible separator that makes an angle $\phi + \pi/2$ with the positive x-axis if and only if $\max_{r \in R} f_r(\phi) < \min_{b \in B} g_b(\phi)$. Hence, to find whether there exists an angle ϕ that admits a possible separator we compute the upper envelope $E^+(F)$ of the set $F := \{f_r : r \in R\}$ and the lower envelope $E^-(G)$ of the set $G := \{g_b : b \in B\}$, and then check whether there is an angle ϕ where $E^+(F)$ lies below $E^-(G)$. To compute $E^+(F)$ ($E^-(G)$ can be handled similarly) we proceed as follows. Note that any two functions f_r and $f_{r'}$ intersect at angles defined by a common outer tangent of $\mathcal{I}_r$ and $\mathcal{I}_{r'}$. We now split the domain $[0 : 2\pi)$ of ϕ into four sub-domains of length $\pi/2$. Within each sub-domain the vertex of a rectangle $\mathcal{I}_r$ that determines f_r is fixed. Hence, within a sub-domain any two functions f_r and $f_{r'}$ intersect at most once, namely at the angle determined by the line through the two relevant vertices of $\mathcal{I}_r$ and $\mathcal{I}_{r'}$ (if this angle lies in the sub-domain). Hence, $E^+(F)$ can be computed in $O(n \log n)$ time [9]. The same is true for $E^-(G)$.

We conclude that deciding whether a possible linear separator exists (and, if so, computing one) can be done in $O(n \log n)$ time.

We now turn our attention to possible axis-aligned rectangular separators that have all red points inside and all blue points outside. Hence, we are looking for a rectangle σ such that no red imprecision region $\mathcal{I}_r$ is completely outside σ and no blue imprecision region is completely inside σ.

Consider all right edges of the red imprecision regions, and let ℓ_{left} be the vertical line through the leftmost of these edges. Clearly, any possible separator σ must have its left edge to the left of this line. Define ℓ_{right} as the vertical line through the rightmost of the left edges of the red imprecision regions, ℓ_{bot} as the lowest horizontal line through the top edges of the red imprecision regions, and ℓ_{top} as the highest horizontal line through the bottom edges of the red imprecision regions. There are now several cases, depending on the relative positions of the lines ℓ_{left} and ℓ_{right}, and of ℓ_{bot} and ℓ_{top}.

If ℓ_{left} lies to the left of ℓ_{right} and ℓ_{bot} lies below ℓ_{top}, then any possible separator must contain the rectangular area A enclosed by these four lines. Moreover, a possible separator exists if and only if no blue imprecision region is fully contained in A. Hence, we can decide if a possible separator exists in $O(n)$ time. The other cases are even simpler, because in those cases a possible separator always exists (assuming all blue imprecision regions have non-zero area). For instance, suppose ℓ_{left} lies to the right of ℓ_{right} and ℓ_{bot} lies below ℓ_{top}. Then any vertical segment in between ℓ_{right} and ℓ_{left} and connecting ℓ_{bot} to ℓ_{top} intersects all red imprecision regions, which implies that a very thin rectangle containing such a segment is a possible separator.

Theorem 1 summarizes the results on certain and possible separators.

Theorem 1. *Let $R \cup B$ be a bichromatic set of n imprecise points in the plane, each with an imprecision region that is an axis-aligned rectangle. For linear separators, we can decide in $O(n)$ whether a certain separator exists for $R \cup B$ and in $O(n \log n)$ time whether a possible separator exists. For axis-aligned rectangular separators, both problems can be solved in $O(n)$ time.*

Most Likely Separators. Finding most likely separators is considerably harder than finding possible separators. We study the 1-dimensional version of the problem, where the imprecision regions are intervals on the real line and a linear separator is a point. Suppose we are interested in separators that have all blue points to the left and all red points to the right.

For a point $x \in \mathbb{R}$ define $F(x) := \Pr[x \text{ is a separator}]$. For a blue point b define lengthL(b, x) as the length of the part of $\mathcal{I}_b$ lying to the left of x, and for a red point r define lengthR(r, x) as the length of the part of $\mathcal{I}_r$ lying to the right of x. Obviously the probability that a blue point b lies to the correct side of x is equal to $f_b(x) := \text{lengthL}(b, x)/\text{length}(\mathcal{I}_b)$, and the probability that a red point r lies to the correct side is $f_r(x) := \text{lengthR}(r, x)/\text{length}(\mathcal{I}_r)$. Hence,

$$F(x) = \prod_{b \in B} f_b(x) \cdot \prod_{r \in R} f_r(x).$$

Let $x_{\min}$ be the rightmost left endpoint of a blue imprecision region, and let $x_{\max}$ be the leftmost right endpoint of a red imprecision region. If $x_{\min} > x_{\max}$ then no separator exists and if $x_{\min} = x_{\max}$ then this point is the only possible separator, so assume $x_{\min} < x_{\max}$. Note that $F(x)$ is non-zero exactly on the interval $[x_{\min}, x_{\max}]$, which we call the *critical domain* of F. The endpoints of the imprecision regions inside the critical domain partition it into *elementary intervals*. Over each such elementary interval, the function $F(x)$ is a polynomial whose degree is bounded by the number of imprecision regions containing that elementary interval. We now prove that $F(x)$ is unimodal over its critical domain.

Lemma 1. $F(x)$ *is unimodal over its critical domain.*

Proof. We first prove that $F(x)$ is unimodal in the interior of each elementary interval $I = [x_1, x_2]$, where we assume without loss of generality that $x_1 = 0$. For any blue point b with $I \cap \mathcal{I}_b = \emptyset$ we have $f_b(x) = 1$ for $x \in I$. (We cannot have $f_b(x) = 0$ as I is part of the critical domain.) When $I \subseteq \mathcal{I}_b$, we have $f_b(x) = (C_b + x)/\text{length}(\mathcal{I}_b)$ for a constant C_b (which is the length of the part of $\mathcal{I}_b$ lying to the left of I). Similarly, for a red point r for which $I \subseteq \mathcal{I}_r$ we have $f_r(x) = (C'_r - x)/\text{length}(\mathcal{I}_r)$ for a constant C'_r (which is the length of the part of $\mathcal{I}_r$ lying to the right of I). Note that we must have $x \leqslant C'_r$ within I. Hence, if $B(I)$ and $R(I)$ are the sets of blue and red points whose imprecision regions cover I, then for $x \in I$

$$F(x) = C \cdot \prod_{b \in B(I)} (C_b + x) \cdot \prod_{r \in R(I)} (C'_r - x), \tag{1}$$

where $C = 1/(\prod_{b \in B(I)} \text{length}(\mathcal{I}_b) \cdot \prod_{r \in R(I)} \text{length}(\mathcal{I}_r))$. Thus,

$$F'(x) = C \cdot F(x) \cdot \left(\sum_{b \in B(I)} \frac{1}{C_b + x} - \sum_{r \in R(I)} \frac{1}{C'_r - x} \right). \tag{2}$$

Note that $F(x) > 0$ for $x \in I$, all terms $1/(C_b + x)$ and $1/(C'_r - x)$ are positive, the sum $\sum_{b \in B(I)} 1/(C_b + x)$ is strictly decreasing while the sum $\sum_{r \in R(I)} 1/(C'_r - x)$ is strictly increasing, Hence, $F'(x) = 0$ at most once inside I or $F'(x) = 0$ everywhere inside I. (The latter occurs when $B(I) \cup R(I) = \emptyset$, which happens for at most one elementary interval.) Thus, $F(x)$ is unimodal inside I.

To extend the analysis to the entire critical domain, we consider two consecutive elementary intervals I_1 and I_2. Let x^* be the right endpoint of I_1 (which is also the left endpoint of I_2). Denote the left and right derivative at x^* by $(F')^-(x^*)$ and $(F')^+(x^*)$. We claim that $(F')^-(x^*) > (F')^+(x^*)$. Observe that $B(I_1) \supseteq B(I_2)$. Indeed, a blue imprecision region cannot start at x^* since then $F(x) = 0$ for $x \in I_1$. Similarly, $R(I_1) \subseteq R(I_2)$. From Equation (2) we now see that $(F')^-(x^*) > (F')^+(x^*)$. Together with the unimodality inside each elementary interval, this means $F(x)$ is unimodal over the entire critical domain. □

Lemma 1 allows us to perform a binary search over the critical domain to find the elementary interval I^* containing the most likely separator. At each step of the binary search, we need to evaluate $F(x)$ at a given x, which takes $O(n)$ time. Hence, I^* can be found in $O(n \log n)$ time in total. Unfortunately, the most likely separator X^* is not necessarily one of the endpoints of I^*. Moreover, within I^* the function $F(x)$ is a polynomial of possibly very high degree. Hence, we may have to resort to numerical methods to approximate its maximum.

Theorem 2. *Let $R \cup B$ be a bichromatic set of n imprecise points on the real line, each with an imprecision region that is an interval. Then we can locate in $O(n \log n)$ time the elementary interval that contains the most likely separator.*

3 Weak Separability

We now turn our attention to the case where we allow some of the points to be misclassified. The goal is then to find a *maximal separator*, that is, a separator that is expected to correctly classify the maximum number of points.

3.1 Weak Separability by a Line

For a line ℓ, let ℓ^- denote the halfplane to the left of ℓ and ℓ^+ the halfplane to the right of ℓ. We want to find a line ℓ that maximizes $G(\ell)$, which is defined as the expected number of red points in ℓ^- plus the expected number of blue points in ℓ^+. For a red point r we define $g_r^-(\ell)$ to be the fraction of $\mathcal{I}_r$ lying to the left of ℓ, and for a blue point b we define $g_b^+(\ell)$ to be the fraction of $\mathcal{I}_b$ to the right of ℓ. Hence, $g_r^-(\ell)$ and $g_b^+(\ell)$ give the probability that r and b are classified correctly, respectively, so

$$G(\ell) = \sum_{r \in R} g_r^-(\ell) + \sum_{b \in B} g_b^+(\ell). \tag{3}$$

To find the maximal separator, we dualize the corners of the imprecision regions, giving us a set L of $4n$ lines in dual space. With a slight abuse of notation we use $G(p)$, for a point p in dual space, to denote the value $G(\ell_p)$ of the line ℓ_p whose dual is p. Let G_C denote the function G restricted to a cell C of the arrangement $\mathcal{A}(L)$. For two neighboring cells C and C' we can obtain $G_{C'}$ from G_C by adding, subtracting or modifying one of the terms in (3). Hence, we can compute a maximal separator by constructing the arrangement $\mathcal{A}(L)$ in $O(n^2)$ time, traversing the dual graph of the arrangement while maintaining the function G, and computing the maximum value of G in each cell. We can improve the storage requirements of the algorithm by not computing the entire arrangement before we start the traversal, but by computing $\mathcal{A}(L)$ using topological sweep [5]. Besides the usual information we need to maintain for the sweep, we then also maintain the function G_C for each cell C immediately to the left of the sweep line. This way the maximal separator can be found using only $O(n)$ storage.

Theorem 3. *Let $B \cup R$ be a bichromatic set of n imprecise points in the plane, each with an axis-parallel rectangular imprecision region. We can compute a maximal line separator for $B \cup R$ in $O(n^2)$ time and using $O(n)$ storage.*

3.2 Weak Separability by a Rectangle

We now turn our attention to the problem of finding an axis-aligned rectangular separator σ that maximizes the sum of the expected number of red points inside σ and the expected number of blue points outside σ. This is equivalent to maximizing $G(\sigma) := \sum_{r \in R} g_r^-(\sigma) + \sum_{b \in B} g_b^+(\sigma)$, where $g_r^-(\sigma)$ and $g_b^+(\sigma)$ denote the fractions of $\mathcal{I}_r$ and $\mathcal{I}_b$ covered by the interior and exterior of σ, respectively.

We first observe that there must be a maximal separator all of whose edges overlap at least partially with an edge of an imprecision region. Indeed, if we keep three edges of a separator σ fixed, and move the fourth edge e, then $G(\sigma)$ changes linearly until we hit an edge of an imprecision region. Hence, there is a direction into which we can move e—either growing or shrinking σ—such that $G(\sigma)$ does not decrease until we hit an edge. This observation implies that there are only $O(n^4)$ candidates for the maximal separator. However, we can still compute a maximal separator in $O(n^3 \log n)$ time, as shown next.

Pick two vertical edges of imprecision regions. Let ℓ_{left} and ℓ_{right} be the vertical lines containing these edges, with ℓ_{left} lying to the left of ℓ_{right}. We will compute the maximal rectangular separator whose left and right edges are restricted to be contained in ℓ_{left} and ℓ_{right}, respectively, by a divide-and-conquer algorithm. Let $y_1, \ldots, y_m$ be the y-coordinates of the horizontal edges of the imprecision regions that lie at least partially inside the strip defined by ℓ_{left} and ℓ_{right}. We can assume these y-coordinates are sorted in increasing order. Let $t := \lceil m/2 \rceil$, and let ℓ_{mid} be the line $y = y_t$. The idea is to compute the best separators above and below ℓ_{mid} recursively, then compute the best separator intersecting ℓ_{mid}, and then take the best of the three separators. Computing the best separator intersecting ℓ_{mid} seems easy: We just compute the best separator whose bottom edge is contained in ℓ_{mid} by scanning the possible y-coordinates for the top edge in the order $y_{t+1}, \ldots, y_m$, do the same for the best separator whose top edge is contained in ℓ_{mid} (this time scanning downward over $y_{t-1}, \ldots, y_1$) and take the union of the two sub-rectangles found. However, in the recursive call we may have to take into account those imprecision regions whose top and bottom edges fall outside the y-range corresponding to the recursive call, and this is problematic for the running time. Hence, we refine our algorithm as follows.

In a generic call we are given a rectangular area A bounded from the left by ℓ_{left}, from the right by ℓ_{right}, from below by the line $y = y_i$ for some $1 \leqslant i < m$ (initially $i = 1$), and from the top by the line $y = y_j$, for some $i < j \leqslant m$ (initially $j = m$). We also have a sorted list of all y-coordinates $y_i, \ldots, y_j$ of the horizontal edges of the imprecision regions that intersect A, with for each y-coordinate a pointer to the imprecision region that generated it. Our goal is to compute the best separator contained in A whose left and right edges are contained in ℓ_{left} and ℓ_{right}, and whose bottom and top edges have y-coordinates chosen from the list. To this end we also need some information to deal with the imprecision regions that intersect A but do not have a horizontal edge intersecting A.

Consider such a red imprecision region $\mathcal{I}_r$, and consider a separator $\sigma(y) := [x_{\text{left}}, x_{\text{right}}] \times [y_i, y]$, where x_{left} and x_{right} are the x-coordinates of ℓ_{left} and ℓ_{right}, respectively. Define $f_r(y)$ to be the fraction of $\mathcal{I}_r$ inside $\sigma(y)$. Note that $f_r(y)$ is a linear function. Also note that the fraction of $\mathcal{I}_r$ inside a separator $[x_{\text{left}}, x_{\text{right}}] \times [y', y]$ is given by $f_r(y) - f_r(y')$. For a blue imprecision region $\mathcal{I}_b$ we define $f_b(y)$ similarly, except this time we use the fraction of $\mathcal{I}_b$ outside $\sigma(y)$.

The extra information we need in the recursive call with region A is the linear function $F_A := \sum_{r \in R} f_r + \sum_{b \in B} f_b$.

It remains to describe how to handle the recursive call with rectangle A. We split A into two rectangles A_1 and A_2 at y-coordinate y_t, where $t := \lceil (i+j)/2 \rceil$ is the median y-coordinate of the horizontal edges in A and A_1 is the lower region.

Next we compute the functions F_{A_1} and F_{A_2} that we have to pass on to the recursive calls for A_1 and A_2. The function F_{A_1} can be computed in linear time as follows. We first determine all imprecision regions that have a horizontal edge in A but not in A_1. We compute the functions f_r (resp. f_b) for all such red (resp. blue) imprecision regions. We add all these functions and then add the function F, which represents the contributions of the imprecision regions that already span A. For F_{A_2} the computations are similar, except that we should subtract $F(y_t)$, since F was defined for separators whose bottom edge has y-coordinate y_i while F_{A_2} is defined for separators whose bottom edge has y-coordinate y_t. Thus both F_{A_1} and F_{A_2} can be computed in linear time.

We now do recursive calls on A_1 with function F_{A_1} and on A_2 with function F_{A_2}, giving us two candidate separators. After the recursive calls, we have to find the best separator that intersects $y = y_t$. To this end, we first compute the best separator σ_1^* of the form $[x_{\text{left}}, x_{\text{right}}] \times [y, y_t]$ and the best separator σ_2^* of the form $[x_{\text{left}}, x_{\text{right}}] \times [y_t, y]$. Both can be computed by scanning edges of the imprecision regions in order—for the former separator we scan downwards from y_t, for the latter we scan upwards from y_t—and maintaining the expected number of correctly classified points. While we scan, we use the function F to account for the contribution of the imprecision regions without a horizontal edge inside A. This way the scans can be implemented so that they run in linear time. The best separator intersecting $y = y_t$ is now given by $\sigma_1^* \cup \sigma_2^*$.

We conclude that we need $O(n)$ time to handle A, plus the time needed for the calls on A_1 and A_2, leading to a total time of $O(n \log n)$ to find the best separator whose left and right edges are contained in the lines ℓ_{left} and ℓ_{right}. The overall time for the algorithm is therefore $O(n^3 \log n)$.

Theorem 4. *Let $B \cup R$ be a bichromatic set of n imprecise points in the plane, each with an imprecision region that is an axis-aligned rectangle. We can compute a maximal linear separator for $B \cup R$ in $O(n^3 \log n)$ time.*

Horizontal Segments as Imprecision Regions. We can improve the running time even further when the imprecision regions are horizontal unit-length segments rather than rectangles. As in the case of rectangular imprecision regions, we only have to consider separators σ whose left and right edges pass through a vertex of an imprecision region. We will first consider a special case of the problem, where the maximal rectangular separator is required to intersect a given horizontal line. The solution to this problem will be used as a subroutine in a divide-and-conquer algorithm for the general problem.

The restricted problem. Let ℓ_{hor} be a given horizontal line. We call a rectangular separator that intersects ℓ_{hor} a *restricted separator*. Our goal is to compute a restricted separator that maximizes the expected number of correctly classified points. As mentioned above, we only have to consider separators whose left edge passes through an endpoint of an imprecision region. Fix an endpoint v, and let ℓ_{vert} denote the vertical line through v. We further restrict our separator by requiring that its left edge is contained in ℓ_{vert}. We show how to compute such a maximal separator for ℓ_{vert} in $O(n\sqrt{n})$ time, leading to an algorithm for the restricted problem that runs in $O(n^2\sqrt{n})$ time.

Let $\mathcal{I}_1, \ldots, \mathcal{I}_m$ be the parts of the imprecision regions in $R \cup B$ lying to the right of ℓ_{vert}, numbered from top to bottom. (We assume for simplicity that no two imprecision regions have the same y-coordinate.) Let y_i denote the y-coordinate of $\mathcal{I}_i$. Let k be such that $\mathcal{I}_1, \ldots, \mathcal{I}_k$ lie above ℓ_{hor} and $\mathcal{I}_{k+1}, \ldots, \mathcal{I}_m$ lie below ℓ_{hor}. For each $1 \leqslant i \leqslant k$ and $x > 0$, define $\sigma_i(x)$ to be the rectangle bounded from the left by ℓ_{vert}, bounded from below by ℓ_{hor}, bounded from above by the line $y = y_i$, and bounded from the right by the vertical line at distance x from ℓ_{vert}. For $k < i \leqslant m$ we define $\sigma_i(x)$ similarly, except that now $\sigma_i(x)$ is bounded from above by ℓ_{hor} and from below by $y = y_i$. Finally, let $\sigma^*(x)$ denote the restricted separator whose left edge is contained in ℓ_{vert} and whose right edge lies at distance x from ℓ_{vert} for which the expected number of correctly classified points is maximized. Clearly $\sigma^*(x)$ is obtained by combining the best rectangle from the set $\{\sigma_i(x) : 1 \leqslant i \leqslant k\}$ with the best rectangle from $\{\sigma_i(x) : k < i \leqslant m\}$. Hence, $G(\sigma^*(x)) = \max_{1 \leqslant i \leqslant k} G(\sigma_i(x)) + \max_{k < i \leqslant m} G(\sigma_i(x))$.

To find the overall best rectangle, we need to find the maximum of $G(\sigma^*(x))$ over all $x > 0$. (In fact, we know that we only have to consider x-values corresponding to the vertical edges of the imprecision regions. However, our approach does not allow us to restrict our attention to those values only.) Thus our strategy is to compute the upper envelopes of the sets of functions $\Gamma := \{G(\sigma_i(x)) : 1 \leqslant i \leqslant k\}$ and $\overline{\Gamma} := \{G(\sigma_i(x)) : k < i \leqslant m\}$. Once we have the upper envelopes, we can add them in linear time (in the sum of their complexities) to find the best restricted rectangular separator with left edge at ℓ_{vert}. Next we describe how to compute $\mathcal{E}(\Gamma)$, the upper envelope of Γ; computing $\mathcal{E}(\overline{\Gamma})$ can done in a similar way.

To compute $\mathcal{E}(\Gamma)$ we use a divide-and-conquer algorithm. Define $\Gamma' := \{G(\sigma_i(x)) : 1 \leqslant i \leqslant t\}$ and $\Gamma'' := \{G(\sigma_i(x)) : t < i \leqslant k\}$, where $t := \lceil k/2 \rceil$. We will compute $\mathcal{E}(\Gamma')$ and $\mathcal{E}(\Gamma'')$ separately and merge the resulting envelopes to get $\mathcal{E}(\Gamma)$. Next we explain how to compute $\mathcal{E}(\Gamma')$ and $\mathcal{E}(\Gamma'')$.

Let $S := \{\mathcal{I}_1, \ldots, \mathcal{I}_k\}$ denote the parts of the imprecision regions above ℓ_{hor} and to the right of ℓ_{vert}, and define $S' := \{\mathcal{I}_1, \ldots, \mathcal{I}_t\}$ and $S'' := \{\mathcal{I}_{t+1}, \ldots, \mathcal{I}_m\}$ to be the top half and bottom half of the set S of imprecision regions, respectively. The function values $G(\sigma_i(x))$ for $t < i \leqslant k$ only depend on S'', as all imprecision regions in S' are above any rectangle $\sigma_i(x)$ for $t < i \leqslant k$. Hence, we can compute $\mathcal{E}(\Gamma'')$ recursively by only considering S''. The function values $G(\sigma_i(x))$ for $1 \leqslant i \leqslant t$ depend on both S' and S''. However, each such $G(\sigma_i(x))$ can be obtained by computing $G(\sigma_i(x))$ with respect to S' and then adding $G(\sigma_{t+1}(x))$ to take S'' into account. Hence, we recursively compute $\mathcal{E}(\Gamma')$ with respect to S' and then we add $G(\sigma_{t+1})$ to the computed envelope to obtain the true envelope. Note that $G(\sigma_{t+1})$ can easily be computed in $O(n)$ time.

As mentioned, after computing $\mathcal{E}(\Gamma')$ and $\mathcal{E}(\Gamma'')$ as just described, we merge them to obtain $\mathcal{E}(\Gamma)$. Next we analyze $|\mathcal{E}(\Gamma)|$, the complexity of $\mathcal{E}(\Gamma)$.

Lemma 2. $|\mathcal{E}(\Gamma)| = O(n\sqrt{n})$

Proof. We partition the part of the plane to the right of ℓ_v into $O(n)$ strips by drawing vertical lines through the endpoints of the imprecision regions. Now consider a function $G(\sigma_i(x))$. Since the imprecision regions are unit-length segments, the contribution of each red imprecision region $\mathcal{I}_r$ to $G(\sigma_i(x))$ is equal to the length of $\mathcal{I}_r$ inside $\sigma_i(x)$. Similarly, for a blue imprecision region $\mathcal{I}_b$, the contribution is its length outside $\sigma_i(x)$. This implies that $G(\sigma_i(x))$, which is the sum of all contributions, is linear within each strip. Moreover, the slope of $G(\sigma_i(x))$ is equal to the number of red imprecision regions inside the strip that are below y_i minus the number of such blue imprecision regions. Note that this implies that the slope of $G(\sigma_i(x))$ in adjacent strips differs by at most 1. (This assumes that all endpoints of the imprecision regions have distinct x-coordinates. When this is not the case, we can introduce a number of dummy strips of zero width at shared x-coordinates, and the argument still goes through.) Using the above observation we can now bound the complexity of the upper envelope $\mathcal{E}(\Gamma)$ using a charging scheme, as follows.

Number the strips from left to right, and consider the j-th strip. Let E_j be the collection of edges of $\mathcal{E}(\Gamma)$ that lie in the j-th strip. We charge the $\sqrt{n}$ rightmost edges of E_j to the j-th strip, and we charge the remaining edges to the function $G(\sigma_i)$ contributing this edge. Obviously the total charge to the strips is $O(n\sqrt{n})$, so it remains to bound the number of times any $G(\sigma_i)$ can be charged. To bound this number we observe that within a strip the upper envelope is a convex chain, so the slopes of the edges on the envelope strictly increases from left to right. Since all slopes are integers, this implies that the slope of any function $G(\sigma_i)$ that is charged in the j-th strip is at least $\sqrt{n}$ smaller than the slope of the function contributing the rightmost edge of the envelope in the strip. Because the slope of any function changes by at most 1 from one strip to the next, this implies that it will take at least $\sqrt{n}/2$ strips for $G(\sigma_i)$ to overtake the function contributing the rightmost edge in the j-th strip. Hence, $G(\sigma_i)$ is charged $O(\sqrt{n})$ times. $\square$

Lemma 2 implies that the running time of our algorithm for the restricted problem satisfies $T(n) = 2T(n/2) + O(n) + O(n\sqrt{n})$, when we fix the left edge of the rectangular separator to be contained in a vertical line ℓ_{vert}. This solves to $O(n\sqrt{n})$. Since ℓ_{vert} can be chosen in $O(n)$ ways, we get the following result.

Lemma 3. *Let $B \cup R$ be a bichromatic set of n imprecise points in the plane, each with an imprecision region that is a unit-length horizontal segment, and let ℓ_{hor} be a horizontal line. Then in $O(n^2\sqrt{n})$ time we can compute a maximal linear separator for $B \cup R$ that is restricted to intersect ℓ_{hor}.*

The general problem. With the solution for the restricted problem at hand we can easily obtain a divide-and-conquer algorithm for the general problem, where the separator is not required to intersect a given line. To this end we partition the plane into two half-planes by a horizontal line ℓ_{hor}, each containing half of the segments (imprecision regions) from $R \cup B$. We then recursively compute the maximal rectangular separator lying below ℓ_{hor}, and the maximal rectangular separator lying above ℓ_{hor}. Finally, we compute the maximal rectangular separator intersecting ℓ_{hor}—this can be done in $O(n^2\sqrt{n})$ by Lemma 3—and we take the best of the three separators computed. The total running time $T(n)$ of our algorithm satisfies $T(n) = 2T(n/2) + O(n^2\sqrt{n})$, which solves to $T(n) = O(n^2\sqrt{n})$.

Theorem 5. *Let $B \cup R$ be a bichromatic set of n imprecise points in the plane, each with an imprecision region that is a unit-length horizontal segment. We can compute a maximal linear separator for $B \cup R$ in $O(n^2\sqrt{n})$ time.*

3.3   Approximate Weak Separability

Our exact algorithms to compute maximal separators have at least quadratic running time both for linear and for rectangular separators. We now present a simple near-linear $(1 - \varepsilon)$-approximation algorithm for computing maximal separators. The approach works for linear as well as rectangular separators.

Define $\mathcal{R}$ to be a set of ranges corresponding to the type of separator we are interested in: for the linear separability $\mathcal{R}$ is the set of all possible halfplanes, for the rectangular separability problem $\mathcal{R}$ is the set of all possible rectangles in the plane. First we replace each imprecision region $\mathcal{I}$ with a point set $P_{\mathcal{I}}$ such that, for any range in $\mathcal{R}$, the fraction of points inside the range is a good approximation of the fraction of the area of $\mathcal{I}$ that is covered by the range. More precisely, for each imprecision region $\mathcal{I}$, we compute a point set $P_{\mathcal{I}} \subset \mathcal{I}$ whose geometric discrepancy with respect to $\mathcal{R}$ is at most $\delta_1 := \varepsilon/8$, that is, such that for any range $\rho \in \mathcal{R}$ we have $\left|\ \text{area}(\rho \cap \mathcal{I})/\text{area}(\mathcal{I}) - |\rho \cap P_{\mathcal{I}}|/|P_{\mathcal{I}}|\ \right| \leqslant \delta_1$. The points in each set $P_{\mathcal{I}}$ are assigned the same color as $\mathcal{I}$.

Let $P_R := \bigcup_{r \in R} P_{\mathcal{I}_r}$ and $P_B := \bigcup_{b \in B} P_{\mathcal{I}_b}$. We reduce the size of P_R by computing a δ_2-approximation A_R of P_R with respect to $\mathcal{R}$, that is, a subset $A_R \subset P_R$ such that $\left|\ |\rho \cap A_R|/|A_R| - |\rho \cap P_R|/|P_R|\ \right| \leqslant \delta_2$ for any range $\rho \in \mathcal{R}$, where $\delta_2 := \varepsilon/8$. The size of P_B is reduced similarly, obtaining a subset $A_B \subset P_B$.

Finally, we compute a separator σ_{ALG} from the class we are interested in—either lines or rectangles—that maximizes

$$\frac{|\sigma_{\text{ALG}}^+ \cap A_R|}{|A_R|} \cdot |P_R| + \frac{|\sigma_{\text{ALG}}^- \cap A_B|}{|A_B|} \cdot |P_B|$$

where σ_{ALG}^+ and σ_{ALG}^- denote the parts of the plane that are to the left and right (for lines) or inside and outside (for rectangles) our separator. This is done in a brute-force manner, by checking all separators on the point set $A_R \cup A_B$ (of which there are $O(|A_R \cup A_B|^2)$ for linear separators and $O(|A_R \cup A_B|^4)$ for rectangular separators). In full version of the paper we show that this gives the following theorem.

Theorem 6. *Let $B \cup R$ be a bichromatic set of n imprecise points in the plane, each with a rectangular imprecision region. We can compute a $(1 - \varepsilon)$-approximation of the maximal linear separator for $B \cup R$ in $O(\text{poly}(1/\varepsilon)n)$ time. A $(1 - \varepsilon)$-approximation of the maximal rectangular separator for $B \cup R$ can also be computed in $O(\text{poly}(1/\varepsilon)n)$ time.*

References

1. de Berg, M., Mumford, E., Roeloffzen, M.: Finding structures on imprecise points. In: 26th Europ. Workshop Comput. Geom., pp. 85–88 (2010)
2. Chan, T.M.: Low-dimensional linear programming with violations. SIAM J. Comput. 34, 879–893 (2005)
3. Cortés, C., Díaz-Báñez, J.M., Pérez-Lantero, P., Seara, C., Urrutia, J., Ventura, I.: Bichromatic separability with two boxes: A general approach. J. Alg. 64, 79–88 (2009)
4. Davoodi, M., Khanteimouri, P., Sheikhi, F., Mohades, A.: Data imprecision under λ-geometry: Finding the largest axis-aligned bounding box. In: Abstracts 27th Europ. Workshop Comput. Geom., pp. 135–138 (2011)
5. Edelsbrunner, H., Guibas, L.J.: Topologically sweeping an arrangement. J. Comput. Syst. Sci. 38(1), 165–194 (1989)
6. Edelsbrunner, H., Preparata, F.P.: Minimum polygonal separation. Inf. Comput. 77, 218–232 (1988)
7. Everett, H., Robert, J.-M., van Kreveld, M.: An optimal algorithm for computing ($\leqslant K$)-levels, with applications. Int. J. Comput. Geom. Appl. 60, 247–261 (1996)
8. Fekete, S.: On the complexity of min-link red-blue separation (1992) (manuscript)
9. Hershberger, J.: Finding the upper envelope of n line segments in $O(n \log n)$ time. Inf. Proc. Lett. 33, 169–174 (1989)
10. Houle, M.F.: Weak separability of sets. PhD thesis, McGill Univeristy (1989)
11. Houle, M.F.: Algorithms for weak and wide separation of sets. Discr. Appl. Math. 45, 139–159 (1993)
12. Hurtado, F., Noy, M., Ramos, P.A., Seara, C.: Separating objects in the plane by wedges and strips. Discr. Appl. Math. 109, 109–138 (2001)
13. van Kreveld, M., van Lankveld, T., Veltkamp, R.: Identifying well-covered minimal bounding rectangles in 2D point data. In: Abstracts 25th Europ. Workshop Comput. Geom., pp. 277–280 (2009)
14. Löffler, M.: Data Imprecision in Computational Geometry. PhD thesis, Utrecht University (2009)
15. Löffler, M., van Kreveld, M.: Largest and smallest convex hulls for imprecise points. Algorithmica 56, 235–269 (2010)
16. Löffler, M., van Kreveld, M.: Largest bounding box, smallest diameter, and related problems on imprecise points. Comput. Geom. Theory Appl. 43, 419–433 (2010)
17. Megiddo, N.: Linear-time algorithms for linear programming in $\mathbb{R}^3$ and related problems. SIAM J. Comput. 12, 759–776 (1983)
18. Myers, Y., Joskowicz, L.: Uncertain geometry with dependencies. In: Proc. 14th ACM Symp. Solid Phys. Mod., pp. 159–164 (2010)
19. O'Rourke, J., Rao Kosaraju, S., Megiddo, N.: Computing circular separability. Discr. Comput. Geom. 1, 105–113 (1986)
20. Salesin, D., Stolfi, J., Guibas, L.J.: Epsilon geometry: building robust algorithms from imprecise computations. In: Proc. 5th ACM Symp. Comput. Geom., pp. 208–217 (1989)
21. Sheikhi, F., de Berg, M., Mohades, A., Davoodi Monfared, M.: Finding monochromatic L-shapes in bichromatic point sets. In: Proc. 22nd Canadian Conf. Comput. Geom., pp. 269–272 (2010); To appear in Comput. Geom. Theory Appl.
22. Seara, C.: On Geometric Separability. PhD thesis, Universidad Politécnica de Catalunya (2002)

Line-Distortion, Bandwidth and Path-Length of a Graph

Feodor F. Dragan[1], Ekkehard Köhler[2], and Arne Leitert[1]

[1] Department of Computer Science, Kent State University, Kent, OH 44242, USA
{dragan,aleitert}@cs.kent.edu
[2] Mathematisches Institut, Brandenburgische Technische Universität Cottbus,
D-03013 Cottbus, Germany
ekoehler@math.tu-cottbus.de

Abstract. We investigate the *minimum line-distortion* and the *minimum bandwidth* problems on unweighted graphs and their relations with the *minimum length* of a Robertson-Seymour's path-decomposition. The *length* of a path-decomposition of a graph is the largest diameter of a bag in the decomposition. The *path-length* of a graph is the minimum length over all its path-decompositions. In particular, we show: (i) if a graph G can be embedded into the line with distortion k, then G admits a Robertson-Seymour's path-decomposition with bags of diameter at most k in G; (ii) for every class of graphs with path-length bounded by a constant, there exist an efficient constant-factor approximation algorithm for the minimum line-distortion problem and an efficient constant-factor approximation algorithm for the minimum bandwidth problem; (iii) there is an efficient 2-approximation algorithm for computing the path-length of an arbitrary graph; (iv) AT-free graphs and some intersection families of graphs have path-length at most 2; (v) for AT-free graphs, there exist a linear time 8-approximation algorithm for the minimum line-distortion problem and a linear time 4-approximation algorithm for the minimum bandwidth problem.

1 Introduction and Previous Work

Computing a minimum distortion embedding of a given n-vertex graph G into the line ℓ was recently identified as a fundamental algorithmic problem with important applications in various areas of computer science, like computer vision [21], as well as in computational chemistry and biology (see [15]). It asks, for a given graph $G = (V, E)$, to find a mapping f of vertices V of G into points of ℓ with minimum number k such that $d_G(x, y) \leq |f(x) - f(y)| \leq k d_G(x, y)$ for every $x, y \in V$. The parameter k is called the *minimum line-distortion* of G and denoted by $\mathsf{ld}(G)$. The embedding f is called *non-contractive* since $d_G(x, y) \leq |f(x) - f(y)|$ for every $x, y \in V$.

In [3], Bădoiu et al. showed that this problem is hard to approximate within a constant factor. They gave an exponential-time exact algorithm and a polynomial-time $\mathcal{O}(n^{1/2})$-approximation algorithm for arbitrary unweighted input graphs, along with a polynomial-time $\mathcal{O}(n^{1/3})$-approximation algorithm for

R Ravi and I.L. Gørtz (Eds.): SWAT 2014, LNCS 8503, pp. 158–169, 2014.

unweighted trees. In another paper [2] Bădoiu et al. showed that the problem is hard to approximate by a factor $\mathcal{O}(n^{1/12})$, even for weighted trees. They also gave a better polynomial-time approximation algorithm for general weighted graphs, along with a polynomial-time algorithm that approximates the minimum line-distortion k embedding of a weighted tree by a factor polynomial in k. Fast exponential-time exact algorithms for computing the line-distortion of a graph were proposed in [8,9]. Fomin et al. in [9] showed that a minimum distortion embedding of an unweighted graph into the line can be found in time $5^{n+o(n)}$. Fellows et al. in [8] gave an $\mathcal{O}(nk^4(2k + 1)^{2k})$ time algorithm that for an unweighted graph G and integer k either constructs an embedding of G into the line with distortion at most k, or concludes that no such embedding exists. They extended their approach also to weighted graphs obtaining an $\mathcal{O}(nk^{4W}(2k+1)^{2kW})$ time algorithm, where W is the largest edge weight. Thus, the problem of minimum distortion embedding of a given n-vertex graph G into the line ℓ is Fixed Parameter Tractable. Recently, Heggernes et al. in [13,14] initiated the study of minimum distortion embeddings into the line of specific graph classes. In particular, they gave polynomial-time algorithms for the problem on bipartite permutation graphs and on threshold graphs [14]. Furthermore, in [13], Heggernes et al. showed that the problem of computing a minimum distortion embedding of a given graph into the line remains NP-hard even when the input graph is restricted to a bipartite, cobipartite, or split graph, implying that it is NP-hard also on chordal, cocomparability, and AT-free graphs. They also gave polynomial-time constant-factor approximation algorithms for split and cocomparability graphs.

Minimum distortion embedding into the line may appear to be closely related to the widely known and extensively studied graph parameter *bandwidth*, denoted by $\mathsf{bw}(G)$. The only difference between the two parameters is that a minimum distortion embedding has to be *non-contractive*, whereas there is no such restriction for bandwidth. Formally, given an unweighted graph $G = (V, E)$ on n vertices, consider a 1-1 map f of the vertices V into integers in $[1, n]$; f is called a *layout* of G. The *bandwidth of layout* f is defined as the maximum stretch of any edge, i.e., $\mathsf{bw}(f) = \max_{uv \in E} |f(u) - f(v)|$. The *bandwidth* of a graph is defined as the minimum possible bandwidth achievable by any 1-1 map (layout) $V \rightarrow [1, n]$. That is, $\mathsf{bw}(G) = \min_{f:V \rightarrow [1,n]} \mathsf{bw}(f)$.

It is known that $\mathsf{bw}(G) \leq \mathsf{ld}(G)$ for every connected graph G (see, e.g., [14]). However, the bandwidth and the minimum line-distortion of a graph can be very different. For example, it is common knowledge that a cycle of length n has bandwidth 2, whereas its minimum line-distortion is exactly $n - 1$ [14]. Bandwidth is known to be one of the hardest graph problems; it is NP-hard even for very simple graphs like caterpillars of hair-length at most 3 [18], and it is hard to approximate by a constant factor even for trees [1]. Polynomial-time algorithms for the exact computation of bandwidth are known for very few graph classes, including bipartite permutation graphs [12] and interval graphs (see, e.g., [17] and papers cited therein). A constant-factor approximation algorithm is known for AT-free graphs [16]. In [10] Golovach et al. showed also that the bandwidth minimization problem is Fixed Parameter Tractable on AT-free graphs by presenting an $n2^{\mathcal{O}(k \log k)}$ time algorithm. For general (unweighted) n-vertex graphs, the minimum bandwidth can be approximated within a factor of $\mathcal{O}(\log^{3.5} n)$ [7]. For n-vertex trees and chordal graphs, the minimum bandwidth can be approximated within a factor of $\mathcal{O}(\log^{2.5} n)$ [11].

Our main tool in this paper is Robertson-Seymour's path-decomposition and its length. A *path-decomposition* ([20]) of a graph $G = (V, E)$ is a sequence of subsets $\{X_i : i \in I\}$ ($I := \{1, 2, \ldots, q\}$) of vertices of G, called *bags*, with three properties: (1) $\bigcup_{i \in I} X_i = V$; (2) For each edge $uv \in E$, there is a bag X_i such that $u, v \in X_i$; (3) For every three indices $i \leq j \leq k$, $X_i \cap X_k \subseteq X_j$ (equivalently, the subsets containing any particular vertex form a contiguous subsequence of the whole sequence). We denote a path-decomposition $\{X_i : i \in I\}$ of a graph G by $\mathcal{P}(G)$. The *width* of a path-decomposition $\mathcal{P}(G) = \{X_i : i \in I\}$ is $\max_{i \in I} |X_i| - 1$. The *path-width* of a graph G, denoted by $\mathsf{pw}(G)$, is the minimum width over all path-decompositions $\mathcal{P}(G)$ of G [20]. The caterpillars are exactly the graphs with path-width 1. Following [5] (where the notion of tree-length of a graph was introduced), we define the *length* of a path-decomposition $\mathcal{P}(G)$ of a graph G to be $\lambda := \max_{i \in I} \max_{u,v \in X_i} d_G(u, v)$ (i.e., each bag X_i has diameter at most λ in G). The *path-length* of G, denoted by $\mathsf{pl}(G)$, is the minimum length over all path-decompositions of G. Interval graphs are exactly the graphs with path-length 1; it is known that G is an interval graph if and only if G has a path-decomposition with each bag being a maximal clique of G. Following [6] (where the notion of tree-breadth of a graph was introduced), we define the *breadth* of a path-decomposition $\mathcal{P}(G)$ of a graph G to be the minimum integer r such that for every $i \in I$ there is a vertex $v_i \in V$ with $X_i \subseteq D_G(v_i, r)$ (i.e., each bag X_i can be covered by a disk $D_G(v_i, r)$ of radius at most r in G). Note that vertex v_i does not need to belong to X_i. The *path-breadth* of G, denoted by $\mathsf{pb}(G)$, is the minimum breadth over all path-decompositions of G. Evidently, for any graph G with at least one edge, $1 \leq \mathsf{pb}(G) \leq \mathsf{pl}(G) \leq 2\mathsf{pb}(G)$ holds. Hence, if one parameter is bounded by a constant for a graph G then the other parameter is bounded for G as well.

Recently, Robertson-Seymour's *tree-decompositions* with bags of bounded radius proved to be very useful in designing an efficient approximation algorithm for the problem of minimum stretch embedding of an unweighted graph into its spanning tree [6]. The decision version of the problem is the *tree t-spanner problem* which asks, for a given graph $G = (V, E)$ and an integer t, if a spanning tree T exists such that $d_T(x, y) \leq t\, d_G(x, y)$ for every $x, y \in V$. It was shown in [6] that: (a) if a graph G can be embedded to a spanning tree with stretch t, then G admits a Robertson-Seymour's tree-decomposition with bags of radius at most $\lceil t/2 \rceil$ and diameter at most t in G (i.e., the tree-breadth $\mathsf{tb}(G)$ of G is at most $\lceil t/2 \rceil$ and the tree-length $\mathsf{tl}(G)$ of G is at most t); (b) there is an efficient algorithm which constructs for an n-vertex unweighted graph G with $\mathsf{tb}(G) \leq \rho$ a spanning tree with stretch at most $2\rho \log_2 n$. As a consequence, an efficient $(\log_2 n)$-approximation algorithm for the problem of minimum stretch embedding of an unweighted graph into its spanning tree was obtained [6].

Contribution of This Paper: Motivated by [6], in this paper, we investigate possible connections between the line-distortion and the path-length (path-breadth) of a graph. We show that, for every graph G, $\mathsf{pl}(G) \leq \mathsf{ld}(G)$ and $\mathsf{pb}(G) \leq \lceil \mathsf{ld}(G)/2 \rceil$ hold. Furthermore, we demonstrate that for every class of graphs with path-length bounded by a constant, there is an efficient constant-factor approximation algorithm for the minimum line-distortion problem. As a consequence, every graph G with $\mathsf{ld}(G) = c$ can be embedded in polynomial time into the line with distortion at most $\mathcal{O}(c^2)$ (reproducing a result from [3]). Additionally, using the same technique, we show that, for every class of graphs with path-length bounded by a constant, there is an efficient constant-factor approximation algorithm for the minimum bandwidth problem. We also investigate (i) what particular graph classes have constant bounds on path-length and (ii) how fast the path-length of an arbitrary graph can be computed or sharply estimated. We present an efficient 2-approximation (3-approximation) algorithm for computing the path-length (resp., the path-breadth) of a graph. We show that AT-free graphs and some intersection families of graphs have small path-length and path-breadth. In particular, the path-length of every AT-free graph is at most 2. Using this and some additional structural properties, we give a linear time 8-approximation algorithm for the minimum line-distortion problem and a linear time 4-approximation algorithm for the minimum bandwidth problem for AT-free graphs.

2   Preliminaries

All graphs occurring in this paper are connected, finite, unweighted, undirected, loopless and without multiple edges. We call $G = (V, E)$ an *n-vertex m-edge graph* if $|V| = n$ and $|E| = m$. In this paper we consider only graphs with $n > 1$. A *clique* is a set of pairwise adjacent vertices of G. By $G[S]$ we denote a subgraph of G induced by vertices of $S \subseteq V$. For a vertex v of G, the sets $N_G(v) = \{w \in V : vw \in E\}$ and $N_G[v] = N_G(v) \cup \{v\}$ are called the *open neighborhood* and the *closed neighborhood* of v, respectively.

In a graph G the *length* of a path from a vertex v to a vertex u is the number of edges in the path. The *distance* $d_G(u, v)$ between vertices u and v is the length of a shortest path connecting u and v in G. The *diameter* in G of a set $S \subseteq V$ is $\max_{x,y \in S} d_G(x, y)$ and its *radius* in G is $\min_{x \in V} \max_{y \in S} d_G(x, y)$ (in some papers they are called the *weak diameter* and the *weak radius* to indicate that the distances are measured in G not in $G[S]$). The distance between a vertex v and a set S of G is measured as $d_G(v, S) = \min_{u \in S} d_G(v, u)$. The *disk* of G of radius k centered at vertex v is the set of all vertices at distance at most k to v: $D_G(v, k) = \{w \in V : d_G(v, w) \leq k\}$.

The following result generalizes a characteristic property of the famous class of AT-free graphs (see [4]). An independent set of three vertices such that each pair is joined by a path that avoids the neighborhood of the third is called an *asteroidal triple*. A graph G is an *AT-free graph* if it does not contain any asteroidal triples [4]. Proofs of statements in this section are omitted.

Proposition 1. *Let G be a graph with $\mathsf{pl}(G) \leq \lambda$. Then, for every three vertices u, v, w of G there is one vertex, say v, such that the disk of radius λ centered at v intercepts every path connecting u and w, i.e., the removal of disk $D_G(v, \lambda)$ from G disconnects u and w.*

We will also need the following property of graphs with $\mathsf{pl}(G) \leq \lambda$. A path P of a graph G is called k-*dominating path* of G if every vertex v of G is at distance at most k from a vertex of P, i.e., $d_G(v, P) \leq k$. A pair of vertices x, y of G is called a k-*dominating pair* if every path between x and y is a k-dominating path of G. It is known that every AT-free graph has a 1-dominating pair [4].

Corollary 1. *Every graph G with $\mathsf{pl}(G) \leq \lambda$ has a λ-dominating pair.*

The following proposition further strengthens the connections between graphs with small path-length and AT-free graphs. Recall that the k-power of a graph $G = (V, E)$ is a graph $G^k = (V, E')$ such that for every $x, y \in V$ ($x \neq y$), $xy \in E'$ if and only if $d_G(x, y) \leq k$.

Proposition 2. *For a graph G with $\mathsf{pl}(G) \leq \lambda$, $G^{2\lambda}$ is an AT-free graph.*

A subset of vertices of a graph is called *connected* if the subgraph induced by those vertices is connected. We say that two connected sets S_1, S_2 of a graph G *see each other* if they have a common vertex or there is an edge in G with one end in S_1 and the other end in S_2. A family of connected subsets of G is called a *bramble* if every two sets of the family see each other. We say that a bramble $\mathcal{F} = \{S_1, \ldots, S_h\}$ of G is k-*dominated* by a vertex v of G if in every set S_i of $\mathcal{F}$ there is a vertex $u_i \in S_i$ with $d_G(v, u_i) \leq k$.

Proposition 3. *For a graph G with $\mathsf{pb}(G) \leq \rho$, every bramble of G is ρ-dominated by a vertex.*

Corollary 2. *Let G be a graph with $\mathsf{pb}(G) \leq \rho$, S be a subset of vertices of G and $r\colon S \to \mathbb{N}$ be a radius function defined on S such that the disks of the family $\mathcal{F} = \{D_G(x, r(x)) : x \in S\}$ pairwise intersect. Then the disks $\{D_G(x, r(x) + \rho) : x \in S\}$ have a nonempty common intersection.*

3   Bandwidth of Graphs with Bounded Path-Length

In this section we show that there is an efficient algorithm that for any graph G with $\mathsf{pl}(G) = \lambda$ produces a layout f with bandwidth at most $(4\lambda + 2)\mathsf{bw}(G)$. Moreover, this statement is true even for all graphs with λ-dominating shortest paths. Recall that a shortest path P of a graph G is a k-*dominating shortest path* of G if every vertex v of G is at distance at most k from a vertex of P, i.e., $d_G(v, P) \leq k$. We will need the following auxiliary lemma.

Lemma 1 ([19]). *For each vertex $v \in V$ of an arbitrary graph G and each positive integer r, $\frac{|D_G(v,r)|-1}{2r} \leq \mathsf{bw}(G)$.*

The main result of this section is the following.

Proposition 4. *Every graph G with a k-dominating shortest path has a layout f with bandwidth at most $(4k + 2)\mathsf{bw}(G)$. If a k-dominating shortest path of G is given in advance, then such a layout f can be found in linear time.*

Proof. Let $P = (x_0, x_1, \ldots, x_i, \ldots, x_j, \ldots, x_q)$ be a k-dominating shortest path of G. Consider a Breadth-First-Search-tree T_P of G started from path P, i.e., BFS(P)-tree of G. For each vertex x_i of P, let X_i be the set of vertices of G that are located in the branch of T_P that is rooted at x_i. We have $x_i \in X_i$. Since P k-dominates G, we have $d_G(v, x_i) \leq k$ for every $i \in \{1, \ldots, q\}$ and every $v \in X_i$. Now create a layout f of G by placing vertices of X_i before all vertices of X_j, if $i < j$, and by placing vertices within each X_i in an arbitrary order.

We claim that this layout f has bandwidth at most $(4k + 2)\mathsf{bw}(G)$. Consider any edge uv of G and assume $u \in X_i$ and $v \in X_j$ ($i \leq j$). For this edge uv we have $f(v) - f(u) \leq |\bigcup_{l=i}^{j} X_l| - 1$. We know also that $d_P(x_i, x_j) = j - i \leq 2k + 1$, since P is a shortest path of G and $d_P(x_i, x_j) = d_G(x_i, x_j) \leq d_G(x_i, u) + 1 + d_G(x_j, v) \leq 2k + 1$. Consider vertex x_c of P with $c = i + \lfloor (j - i)/2 \rfloor$, i.e., a middle vertex of subpath of P between x_i and x_j. Consider an arbitrary vertex w in X_l, $i \leq l \leq j$. Since $d_G(x_c, w) \leq d_G(x_c, x_l) + d_G(x_l, w)$, $d_G(x_c, x_l) \leq \lceil 2k + 1 \rceil / 2$ and $d_G(x_l, w) \leq k$, we get $d_G(x_c, w) \leq 2k + 1$. In other words, disk $D_G(x_c, 2k + 1)$ contains all vertices of $\bigcup_{l=i}^{j} X_l$. Applying Lemma 1 to $|D_G(x_c, 2k + 1)| \geq |\bigcup_{l=i}^{j} X_l|$, we conclude $f(v) - f(u) \leq |\bigcup_{l=i}^{j} X_l| - 1 \leq |D_G(x_c, 2k + 1)| - 1 \leq 2(2k + 1)\mathsf{bw}(G) = (4k + 2)\mathsf{bw}(G)$. $\square$

Corollary 3. *For every n-vertex m-edge graph G, a layout with bandwidth at most $(4\mathsf{pl}(G) + 2)\mathsf{bw}(G)$ can be found in $\mathcal{O}(n^2 m)$ time.*

Proof. For an n-vertex m-edge graph G, a k-dominating shortest path with $k \leq \mathsf{pl}(G)$ can be found in $\mathcal{O}(n^2 m)$ time as follows. Iterate over all vertex pairs of G. For each pair pick a shortest path P connecting them and run BFS(P) to find most distant vertex v_P from P. Finally, report that path P for which $d_G(v_P, P)$ is minimum. By Corollary 1, this minimum is at most $\mathsf{pl}(G)$. $\square$

Thus, we have the following interesting conclusion.

Theorem 1. *For every class of graphs with path-length bounded by a constant, there is an efficient constant-factor approximation algorithm for the minimum bandwidth problem.*

In Section 6 we show that the path-length of every AT-free graph is at most 2. Using additional structural properties of AT-free graphs, we give for them a linear time 4-approximation algorithm for the minimum bandwidth problem. This result reproduces an approximation result from [16] with a better run-time.

4   Path-Length and Line-Distortion

In this section, we first show that the line-distortion of a graph gives an upper bound on its path-length and then demonstrate that if the path-length of a graph G is bounded by a constant then there is an efficient constant-factor approximation algorithm for the minimum line-distortion problem on G.

Proposition 5. *For an arbitrary graph G, $\mathsf{pl}(G) \leq \mathsf{ld}(G)$, $\mathsf{pw}(G) \leq \mathsf{ld}(G)$ and $\mathsf{pb}(G) \leq \lceil \mathsf{ld}(G)/2 \rceil$.*

Proof. It is known (see, e.g., [14]) that every connected graph $G = (V, E)$ has a minimum distortion embedding f into the line ℓ (called a *canonic* embedding) such that $|f(x) - f(y)| = d_G(x, y)$ for every two vertices of G that are placed next to each other in ℓ by f. Assume, in what follows, that f is such a canonic embedding and let $k := \mathsf{ld}(G)$.

Consider the following path-decomposition of G created from f. For each vertex v, form a bag B_v consisting of all vertices of G which are placed by f in the interval $[f(v), f(v)+k]$ of the line ℓ. Order these bags with respect to the left ends of the corresponding intervals. Evidently, for every vertex $v \in V$, $v \in B_v$, i.e., each vertex belongs to a bag. More generally, a vertex u belongs to a bag B_v if and only if $f(v) \leq f(u) \leq f(v) + k$. Since $\mathsf{ld}(G) = k$, for every edge uv of G, $|f(u) - f(v)| \leq k$ holds. Hence, both ends of edge uv belong either to bag B_u (if $f(u) < f(v)$) or to bag B_v (if $f(v) < f(u)$). Consider now three bags B_a, B_b, and B_c with $f(a) < f(b) < f(c)$ and a vertex v of G that belongs to B_a and B_c. We have $f(a) < f(b) < f(c) \leq f(v) \leq f(a) + k < f(b) + k$. Hence, necessarily, v belongs to B_b as well.

It remains to show that each bag B_v, $v \in V$, has in G diameter at most k, radius at most $\lceil k/2 \rceil$ and cardinality at most $k + 1$. Indeed, for any two vertices $x, y \in B_v$, we have $|f(x) - f(y)| \leq k$, i.e., $d_G(x, y) \leq |f(x) - f(y)| \leq k$. Furthermore, any interval $[f(v), f(v) + k]$ (of length k) can have at most $k + 1$ vertices of G as the distance between any two vertices placed by f to this interval is at least 1 ($|f(x) - f(y)| \geq d_G(x, y) \geq 1$). Thus, $|B_v| \leq k + 1$ for every $v \in V$.

Consider now the point $p_v := f(v) + \lfloor k/2 \rfloor$ in the interval $[f(v), f(v) + k]$ of ℓ. Assume, without loss of generality, that p_v is between $f(x)$ and $f(y)$, the images of two vertices x and y of G placed next to each other in ℓ by f. Let $f(x) \leq p_v < f(y)$. Since f is a canonic embedding, there must exist in G a vertex c on a shortest path between x and y such that $d_G(x, c) = p_v - f(x)$ and $d_G(c, y) = f(y) - p_v = d_G(x, y) - d_G(x, c)$. We claim that for every vertex $w \in B_v$, $d_G(c, w) \leq \lceil k/2 \rceil$ holds. Assume $f(w) \geq f(y)$ (the case when $f(w) \leq f(x)$ is similar). Then, we have $d_G(c, w) \leq d_G(c, y) + d_G(y, w) \leq (f(y) - p_v) + (f(w) - f(y)) = f(w) - p_v \leq f(w) - f(v) - \lfloor k/2 \rfloor \leq k - \lfloor k/2 \rfloor \leq \lceil k/2 \rceil$. $\square$

It should be noted that the difference between the path-length and the line-distortion of a graph can be very large. A complete graph K_n on n vertices has path-length 1, whereas the line-distortion of K_n is $n - 1$. Note also that the bandwidth and the path-length of a graph do not bound each other. The bandwidth of K_n is $n - 1$ while its path-length is 1. On the other hand, the path-length of cycle C_{2n} is n while its bandwidth is 2.

Now we show that there is an efficient algorithm that for any graph G with $\mathsf{pl}(G) = \lambda$ produces an embedding f of G into the line ℓ with distortion at most $(12\lambda + 7)\mathsf{ld}(G)$. Again, this statement is true even for all graphs with λ-dominating shortest paths. We will need the following auxiliary lemma from [3]. We reformulate it slightly. Recall that a subset of vertices of a graph is called *connected* if the subgraph induced by those vertices is connected.

Lemma 2 ([3]). *Any connected subset $S \subseteq V$ of a graph $G = (V, E)$ can be embedded into the line with distortion at most $2|S| - 1$ in time $\mathcal{O}(|V| + |E|)$. In particular, there is a mapping f, computable in $\mathcal{O}(|V| + |E|)$ time, of vertices from S into points of the line such that $d_G(x, y) \leq |f(x) - f(y)| \leq 2|S| - 1$ for every $x, y \in S$.*

The main result of this section is the following.

Proposition 6. *Every graph G with a k-dominating shortest path admits an embedding f of G into the line with distortion at most $(8k + 4)\mathsf{ld}(G) + (2k)^2 + 2k + 1$. If a k-dominating shortest path of G is given in advance, then such an embedding f can be found in linear time.*

Proof. Like in the proof of Proposition 4, consider a k-dominating shortest path $P = (x_0, x_1, \ldots, x_i, \ldots, x_j, \ldots, x_q)$ of G and identify by BFS(P) the sets X_i, $i \in \{1, \ldots, q\}$. We had $d_G(v, x_i) \leq k$ for every $i \in \{1, \ldots, q\}$ and every $v \in X_i$. It is clear also that each X_i is a connected subset of G. Similar to [3], we define an embedding f of G into the line ℓ by placing vertices of X_i before all vertices of X_j, if $i < j$, and by placing vertices within each X_i in accordance with the embedding mentioned in Lemma 2. Also, for each $i \in \{1, \ldots, q - 1\}$, leave a space of length $2k + 1$ between the interval of ℓ spanning the vertices of X_i and the interval spanning the vertices of X_{i+1}.

We claim that f is a (non-contractive) embedding with distortion at most $(8k+4)\mathsf{ld}(G)+(2k)^2+2k+1$. It is sufficient to show that $d_G(x,y) \leq |f(x)-f(y)|$ for every two vertices of G that are placed next to each other in ℓ by f and that $|f(v) - f(u)| \leq (8k + 4)\mathsf{ld}(G) + (2k)^2 + 2k + 1$ for every edge uv of G (see, e.g., [3,14]).

From Lemma 2, we know that $d_G(x,y) \leq |f(x) - f(y)| \leq 2|X_l| - 1$ for every $x, y \in X_l$ and $l \in \{1, 2, \ldots, q\}$. Additionally, for every $x \in X_i$ and $y \in X_{i+1}$ $(i \in \{1, 2, \ldots, q - 1\})$, we have $d_G(x,y) \leq d_G(x, x_i) + 1 + d_G(y, x_{i+1}) \leq 2k+1 \leq |f(y) - f(x)|$ (as a space of length $2k+1$ is left between the interval of ℓ spanning the vertices of X_i and the interval spanning the vertices of X_{i+1}). Hence, f is non-contractive.

Consider now an arbitrary edge uv of G and assume $u \in X_i$ and $v \in X_j$ $(i \leq j)$. For this edge uv we have $f(v) - f(u) \leq \sum_{l=i}^{j}(2|X_l| - 1 + 2k + 1) - 2k - 1 \leq 2|\bigcup_{l=i}^{j} X_l| + 2k(j - i + 1) - 2k - 1 = 2|\bigcup_{l=i}^{j} X_l| + 2k(j - i) - 1$. Recall that $d_P(x_i, x_j) = j - i \leq 2k + 1$, since P is a shortest path of G and $d_P(x_i, x_j) = d_G(x_i, x_j) \leq d_G(x_i, u)+1+d_G(x_j, v) \leq 2k+1$. Hence, $f(v)-f(u) \leq 2|\bigcup_{l=i}^{j} X_l| + 2k(2k+1) - 1$.

As in the proof of Proposition 4, $|\bigcup_{l=i}^{j} X_l| - 1 \leq (4k + 2)\mathsf{bw}(G)$. As $\mathsf{bw}(G) \leq \mathsf{ld}(G)$ for every graph G (see, e.g., [14]), we get $f(v)-f(u) \leq 2|\bigcup_{l=i}^{j} X_l|+2k(2k+1) - 1 \leq 2(4k + 2)\mathsf{bw}(G) + 2k(2k + 1) + 1 \leq (8k + 4)\mathsf{ld}(G) + 2k(2k + 1) + 1$. $\square$

Corollary 4. *For every n-vertex m-edge graph G, an embedding into the line with distortion at most $(12\mathsf{pl}(G) + 7)\mathsf{ld}(G)$ can be found in $\mathcal{O}(n^2 m)$ time.*

Proof. See the proof of Corollary 3 and note that, by Proposition 5, $\mathsf{pl}(G) \leq \mathsf{ld}(G)$. Hence, the distortion established in Proposition 6 becomes $\leq (8\mathsf{pl}(G) + 4)\mathsf{ld}(G) + 2(2\mathsf{pl}(G) + 1)\mathsf{ld}(G) + 1 \leq (12\mathsf{pl}(G) + 7)\mathsf{ld}(G)$. □

Thus, we have the following interesting conclusion.

Theorem 2. *For every class of graphs with path-length bounded by a constant, there is an efficient constant-factor approximation algorithm for the minimum line-distortion problem.*

Using inequality $\mathsf{pl}(G) \leq \mathsf{ld}(G)$ in Corollary 4 once more, we reproduce a result of [3].

Corollary 5 ([3]). *For every graph G with $\mathsf{ld}(G) = c$, an embedding into the line with distortion at most $\mathcal{O}(c^2)$ can be found in polynomial time.*

It should be noted that, since the difference between the path-length and the line-distortion of a graph can be very large (close to n), the result in Corollary 4 seems to be stronger.

Theorem 1 and Theorem 2 stress the importance of investigations of (i) what particular graph classes have constant bounds on path-length and of (ii) how fast the path-length of an arbitrary graph can be computed or sharply estimated.

5   Constant-Factor Approximation of Path-Length

Let $G = (V, E)$ be an arbitrary graph and s be its arbitrary vertex. A *layering* $\mathcal{L}(s, G)$ of G with respect to a start vertex s is the decomposition of V into the layers $L_i = \{u \in V : d_G(s, u) = i\}, i = 0, 1, \ldots, q$. We can get a path-decomposition of G by adding to each layer L_i $(i > 0)$ all vertices from layer L_{i-1} that have a neighbor in L_i. Let $L_i^+ := L_i \cup (\bigcup_{v \in L_i}(N_G(v) \cap L_{i-1}))$. Clearly, the sequence $\{L_1^+, \ldots, L_q^+\}$ is a path-decomposition of G and can be constructed in $\mathcal{O}(|E|)$ total time. We call this path-decomposition an *extended layering* of G and denote it by $\mathcal{L}^+(s, G)$. It turns out that this type of path-decomposition has length at most twice as large as the path-length of the graph.

Theorem 3. *For every graph G with $\mathsf{pl}(G) = \lambda$ there is a vertex s such that the length of the extended layering $\mathcal{L}^+(s, G)$ of G is at most 2λ. In particular, a factor 2 approximation of the path-length of an arbitrary n-vertex graph can be computed in $\mathcal{O}(n^3)$ total time.*

Proof. Consider a path-decomposition $\mathcal{P}(G) = \{X_1, \ldots, X_p\}$ of length $\mathsf{pl}(G) = \lambda$ of G. Let s be an arbitrary vertex from X_1. Consider the layering $\mathcal{L}(s, G)$ of G with respect to s where $L_i = \{u \in V : d_G(s, u) = i\}$ $(i = 0, 1, \ldots, q)$. Let x and y be two arbitrary vertices from L_i $(i \in \{1, \ldots, q\})$ and x' and y' be arbitrary vertices from L_{i-1} with $xx', yy' \in E$. We will show that $\max\{d_G(x,y), d_G(x,y'),$ $d_G(x',y)\} \leq 2\lambda$. By induction on i, we may assume that $d_G(y',x') \leq 2\lambda$ as $x',y' \in L_{i-1}$.

If there is a bag in $\mathcal{P}(G)$ containing both vertices x and y, then $d_G(x,y) \leq \lambda$ and therefore $d_G(x,y') \leq \lambda + 1 \leq 2\lambda$, $d_G(y,x') \leq \lambda + 1 \leq 2\lambda$. Assume now that all bags containing x are earlier in $\mathcal{P}(G) = \{X_1, X_2, \ldots, X_p\}$ than the bags containing y. Let B be a bag of $\mathcal{P}(G)$ containing both ends of edge xx' (such a bag necessarily exists by properties of path-decompositions). By the position of this bag B in $\mathcal{P}(G)$ and the fact that $s \in X_1$, any shortest path connecting s with y must have a vertex in B. Let w be a vertex of B that is on a shortest path of G connecting vertices s and y and containing edge yy'. Such a shortest path must exist because of the structure of the layering $\mathcal{L}(s,G)$ that starts at s and puts y' and y in consecutive layers. We have $\max\{d_G(x,w), d_G(x',w)\} \leq \lambda$. If $w = y'$ then we are done; $\max\{d_G(x,y), d_G(x,y'), d_G(x',y)\} \leq \lambda + 1 \leq 2\lambda$. So, assume that $w \neq y'$. Since $d_G(x,s) = d_G(s,y) = i$ (by the layering) and $d_G(x,w) \leq \lambda$, we must have $d_G(w,y') + 1 = d_G(w,y) = d_G(s,y) - d_G(s,w) = d_G(s,x) - d_G(s,w) \leq d_G(w,x) \leq \lambda$. Hence, $d_G(y,x) \leq d_G(y,w) + d_G(w,x) \leq 2\lambda$, $d_G(y,x') \leq d_G(y,w) + d_G(w,x') \leq 2\lambda$ and $d_G(y',x) \leq d_G(y',w) + d_G(w,x) \leq 2\lambda - 1$.

We conclude that the distance between any two vertices from L_i^+ is at most 2λ, that is, the length of tree decomposition $\mathcal{L}^+(s,G)$ of G is at most 2λ.   □

Theorem 4. *For every graph G with $\mathsf{pb}(G) = \rho$ there is a vertex s such that the breadth of the extended layering $\mathcal{L}^+(s,G)$ of G is at most 3ρ. In particular, a factor 3 approximation of the path-breadth of an arbitrary n-vertex graph can be computed in $\mathcal{O}(n^3)$ total time.*

Proof. Since $\mathsf{pl}(G) \leq 2\mathsf{pb}(G)$, by Theorem 3, there is a vertex s in G such that the length of extended layering $\mathcal{L}^+(s,G) = \{L_1^+, \ldots, L_q^+\}$ of G is at most 4ρ. Consider a bag L_i^+ of $\mathcal{L}^+(s,G)$ and a family $\mathcal{F} = \{D_G(x, 2\rho) : x \in L_i^+\}$ of disks of G. Since $d_G(u,v) \leq 4\rho$ for every pair $u, v \in L_i^+$, the disks of $\mathcal{F}$ pairwise intersect. Hence, by Corollary 2, the disks $\{D_G(x, 3\rho) : x \in L_i^+\}$ have a nonempty common intersection. A vertex w from that common intersection has all vertices of L_i^+ within distance at most 3ρ. That is, for each $i \in \{1, \ldots, q\}$ there is a vertex w_i with $L_i^+ \subseteq D_G(w_i, 3\rho)$.   □

6   Approximation of Line-Distortions of AT-Free Graphs

The path-length of every AT-free graph is bounded by 2 (proof is omitted).

Proposition 7. *If G is an AT-free graph, then $\mathsf{pb}(G) \leq \mathsf{pl}(G) \leq 2$.*

The class of AT-free graphs contains a number of intersection families of graphs, among them the permutation graphs, the trapezoid graphs and the cocomparability graphs. Theorem 2 implies already that there is an efficient constant-factor approximation algorithm for the minimum line-distortion problem on permutation graphs, trapezoid graphs, cocomparability graphs as well as AT-free graphs. Recall that for arbitrary (unweighted) graphs the minimum line-distortion problem is hard to approximate within a constant factor [3]. Furthermore, the problem remains NP-hard even when the input graph is restricted to a chordal, cocomparability, or AT-free graph [13]. Polynomial-time constant-factor approximation algorithms were known only for split and cocomparability graphs [13]. As far as we know, for AT-free graphs (the class which contains all cocomparability graphs), no prior efficient approximation algorithm was known.

In this section, using additional structural properties of AT-free graphs we give a better approximation algorithm for all AT-free graphs. It is an 8-approximation algorithm and runs in linear time. The following nice structural result from [16] will be very useful.

Lemma 3 ([16]). *Let $G = (V, E)$ be an AT-free graph. Then, there is a dominating path $\pi = (v_0, \ldots, v_k)$ and a layering $\mathcal{L} = \{L_0, \ldots, L_k\}$ with $L_i = \{u \in V : d_G(u, v_0) = i\}$ such that for all $u \in L_i$ ($i \geq 1$), $uv_i \in E$ or $uv_{i-1} \in E$. Computing π and $\mathcal{L}$ can be done in linear time.*

Theorem 5. *There is a linear time algorithm to compute an 8-approximation of the line-distortion of an AT-free graph.*

Proof. Let G be an AT-free graph. We first compute a path $\pi = (v_0, \ldots, v_k)$ and a layering $\mathcal{L} = \{L_0, \ldots, L_k\}$ as defined in Lemma 3. To define an embedding f of G into the line, we partition every layer L_i in three sets: $\{v_i\}$, $X_i = \{x : x \in L_i, v_i x \in E\}$, and $\overline{X}_i = L_i \setminus (\{v_i\} \cup X_i)$. Note that if $x \in \overline{X}_i$, then $v_{i-1}x \in E$. Since each vertex in X_i is adjacent to v_i and each vertex in $\overline{X}_i$ is adjacent to v_{i-1}, for all $x, y \in X_i$, $d_G(x, y) \leq 2$, and for all $x, y \in \overline{X}_i$, $d_G(x, y) \leq 2$. Also, for all $x \in X_i$ and $y \in \overline{X}_i$, $d_G(x, y) \leq 3$. The embedding f places vertices of G into the line in the following order: $(v_0, \ldots, v_{i-1}, \overline{X}_i, X_i, v_i, \overline{X}_{i+1}, X_{i+1}, v_{i+1}, \ldots, v_k)$. Between every two vertices x, y placed next to each other in the line, to guarantee non-contractiveness, f leaves a space of length $d_G(x, y)$ (which is either 1 or 2 or 3, where 3 occurs only when $x \in \overline{X}_i$ and $y \in X_i$ for some i).

We will now show that f approximates the minimum line-distortion of G. Since $\mathcal{L}$ is a BFS layering from v_0, i.e., it represents the distances of vertices from v_0, there is no edge xy with $x \in L_{i-1}$ and $y \in L_{i+1}$. Also note that $D_G(v_i, 2) \supseteq L_i \cup L_{i+1} \cup \{v_{i-1}\}$. By the definition of f, for all $xy \in E$ with $x, y \in L_i \cup L_{i+1}$, $|f(x) - f(y)| < |f(v_{i-1}) - f(v_{i+1})|$. Therefore, counting how many vertices are placed by f between $f(v_{i-1})$ and $f(v_{i+1})$ and the distance in G between vertices placed next to each other, we get $|f(x) - f(y)| \leq 2(|D_G(v_i, 2)| - 2) + 2 = 2(|D_G(v_i, 2)| - 1)$. Using Lemma 1 and the fact that $\mathsf{bw}(G) \leq \mathsf{ld}(G)$, we get $|f(x) - f(y)| \leq 8\,\mathsf{ld}(G)$ for all $xy \in E$.   □

It is easy to see that the order in which vertices of G placed by f into the line gives also a layout of G with bandwidth at most $4\mathsf{bw}(G)$. This reproduces an approximation result from [16] (in fact, their algorithm had complexity $\mathcal{O}(m + n \log n)$ for an n-vertex m-edge graph, since it involved a known $\mathcal{O}(n \log n)$ time algorithm to find an optimal layout for a caterpillar tree).

Corollary 6 ([16]). *There is a linear time algorithm to compute a 4-approximation of the minimum bandwidth of an AT-free graph.*

References

1. Blache, G., Karpinski, M., Wirtgen, J.: On approximation intractability of the bandwidth problem, Technical report TR98-014, University of Bonn (1997)
2. Bǎdoiu, M., Chuzhoy, J., Indyk, P., Sidiropoulos, A.: Low-distortion embeddings of general metrics into the. In: STOC 2005, pp. 225–233. ACM (2005)
3. Bǎdoiu, M., Dhamdhere, K., Gupta, A., Rabinovich, Y., Raecke, H., Ravi, R., Sidiropoulos, A.: Approximation algorithms for low-distortion embeddings into low-dimensional spaces. In: SODA 2005, pp. 119–128. ACM/SIAM (2005)
4. Corneil, D.G., Olariu, S., Stewart, L.: Asteroidal Triple-Free Graphs. SIAM Journal on Discrete Mathematics 10, 399–430 (1997)
5. Dourisboure, Y., Gavoille, C.: Tree-decompositions with bags of small diameter. Discr. Math. 307, 208–229 (2007)
6. Dragan, F.F., Köhler, E.: An Approximation Algorithm for the Tree t-Spanner Problem on Unweighted Graphs via Generalized Chordal Graphs. In: Goldberg, L.A., Jansen, K., Ravi, R., Rolim, J.D.P. (eds.) RANDOM 2011 and APPROX 2011. LNCS, vol. 6845, pp. 171–183. Springer, Heidelberg (2011)
7. Feige, U.: Approximating the bandwidth via volume respecting embedding. J. of Computer and System Science 60, 510–539 (2000)
8. Fellows, M.R., Fomin, F.V., Lokshtanov, D., Losievskaja, E., Rosamond, F.A., Saurabh, S.: Distortion Is Fixed Parameter Tractable. In: Albers, S., Marchetti-Spaccamela, A., Matias, Y., Nikoletseas, S., Thomas, W. (eds.) ICALP 2009, Part I. LNCS, vol. 5555, pp. 463–474. Springer, Heidelberg (2009)
9. Fomin, F.V., Lokshtanov, D., Saurabh, S.: An exact algorithm for minimum distortion embedding. Theor. Comput. Sci. 412, 3530–3536 (2011)
10. Golovach, P.A., Heggernes, P., Kratsch, D., Lokshtanov, D., Meister, D., Saurabh, S.: Bandwidth on AT-free graphs. Theor. Comput. Sci. 412, 7001–7008 (2011)
11. Gupta, A.: Improved Bandwidth Approximation for Trees and Chordal Graphs. J. Algorithms 40, 24–36 (2001)
12. Heggernes, P., Kratsch, D., Meister, D.: Bandwidth of bipartite permutation graphs in polynomial time. Journal of Discrete Algorithms 7, 533–544 (2009)
13. Heggernes, P., Meister, D.: Hardness and approximation of minimum distortion embeddings. Information Processing Letters 110, 312–316 (2010)
14. Heggernes, P., Meister, D., Proskurowski, A.: Computing minimum distortion embeddings into a path of bipartite permutation graphs and threshold graphs. Theoretical Computer Science 412, 1275–1297 (2011)
15. Indyk, P., Matousek, J.: Low-distortion embeddings of finite metric spaces. In: Handbook of Discrete and Computational Geometry, pp. 177–196. CRC (2004)
16. Kloks, T., Kratsch, D., Müller, H.: Approximating the Bandwidth for Asteroidal Triple-Free Graphs. J. Algorithms 32, 41–57 (1999)
17. Kratsch, D., Stewart, L.: Approximating Bandwidth by Mixing Layouts of Interval Graphs. SIAM J. Discrete Math. 15, 435–449 (2002)
18. Monien, B.: The Bandwidth-Minimization Problem for Caterpillars with Hair Length 3 is NP-Complete. SIAM J. Alg. Disc. Meth. 7, 505–512 (1986)
19. Räcke, H.: http://ttic.uchicago.edu/~harry/teaching/pdf/lecture15.pdf
20. Robertson, N., Seymour, P.: Graph minors. I. Excluding a forest. Journal of Combinatorial Theory, Series B 35, 39–61 (1983)
21. Tenenbaum, J.B., de Silva, V., Langford, J.C.: A global geometric framework for nonlinear dimensionality reduction. Science 290, 2319–2323 (2000)

Colorful Bin Packing

György Dósa[1] and Leah Epstein[2]

[1] Department of Mathematics, University of Pannonia, Veszprém, Hungary
[2] Department of Mathematics, University of Haifa, Haifa, Israel
`dosagy@almos.vein.hu`, `lea@math.haifa.ac.il`

Abstract. We study a variant of online bin packing, called colorful bin packing. In this problem, items that are presented one by one are to be packed into bins of size 1. Each item i has a size $s_i \in [0, 1]$ and a color $c_i \in \mathcal{C}$, where $\mathcal{C}$ is a set of colors (that is not necessarily known in advance). The total size of items packed into a bin cannot exceed its size, thus an item i can always be packed into a new bin, but an item cannot be packed into a non-empty bin if the previous item packed into that bin has the same color, or if the occupied space in it is larger than $1 - s_i$. This problem generalizes standard online bin packing and online black and white bin packing (where $|\mathcal{C}| = 2$). We prove that colorful bin packing is harder than black and white bin packing in the sense that an online algorithm for zero size items that packs the input into the smallest possible number of bins cannot exist for $|\mathcal{C}| \geq 3$, while it is known that such an algorithm exists for $|\mathcal{C}| = 2$. We show that natural generalizations of classic algorithms for bin packing fail to work for the case $|\mathcal{C}| \geq 3$, and moreover, algorithms that perform well for black and white bin packing do not perform well either, already for the case $|\mathcal{C}| = 3$. Our main results are a new algorithm for colorful bin packing that we design and analyze, whose absolute competitive ratio is 4, and a new lower bound of 2 on the asymptotic competitive ratio of any algorithm, that is valid even for black and white bin packing.

1 Introduction

Colorful bin packing is a packing problem where a sequence of colored items is presented to the algorithm, and the goal is to partition (or pack) the items into a minimal number of bins. The set of items is denoted by $\{1, 2, \ldots, n\}$, where $0 \leq s_i \leq 1$ is the size of item i, and $c_i \in \mathcal{C}$ is its color. The items are to be packed one by one (according to their order in the input sequence), such that the items packed into each bin have a total size of at most 1, and any two items packed consecutively into one bin have different colors. Since the input is viewed as a sequence rather than a set, the natural scenario for this problem is an online one; after an item has been packed, the next item is presented. In an online environment, the algorithm packs an item without any knowledge regarding the further items, and the set $\mathcal{C}$ (or even its cardinality) is not necessarily known to the algorithm. The number of items, n, is typically unknown to the algorithm as well. In the case that inputs are viewed as sequences and not as sets, online

R Ravi and I.L. Gørtz (Eds.): SWAT 2014, LNCS 8503, pp. 170–181, 2014.

algorithms are typically compared to optimal offline algorithms that must pack the items exactly in the same order as they appear in the input.

Consider an input for colorful bin packing with N red items of size zero, followed by N blue items of size zero. This input requires N bins, but reordering the items reduces the required number of bins to 1. Thus, distinguishing reasonable online algorithms from less successful ones cannot be done by comparison to offline algorithms that are allowed to reorder the input. The offline algorithms to which we compare our online algorithm are therefore not allowed to reorder the input. Such an optimal offline algorithm is denoted by OPT (OPT denotes a specific optimal offline algorithm, and we use OPT to denote also the number of bins that it uses for a given input). The absolute competitive ratio of an algorithm is the supremum ratio over all inputs between the number of bins that it uses and the number of bins that OPT uses (for the same input). The asymptotic competitive ratio is the limit of absolute competitive ratios R_K when K tends to infinity and R_K takes into account only inputs for which OPT uses at least K bins. Note that (by definition), for a given algorithm (for some online bin packing problem), its asymptotic competitive ratio never exceeds its absolute competitive ratio.

The special case of colorful bin packing, called black and white packing, was introduced in [1]. In this variant there are just two colors, called black and white. The motivation for black and white bin packing was in assignment to containers of items so that any two items packed consecutively into one bin can be easily distinguished later. An example for such items was articles that are printed on either white paper or recycled paper, in which case bins simply contain piles of paper, and packing articles printed on the two kinds of paper so that the two kinds alternate allows to distinguish them easily. Colorful bin packing is the generalization where there is a number of different kinds of printing paper (for example, paper of distinct colors that is used for printing advertisement flyers), and in order to distinguish between two items (two piles of flyers), they have to have different colors of printing paper.

It was shown [1] that the natural generalizations of several well-known algorithms fail to obtain finite competitive ratios. For example, Next Fit (NF) for colorful bin packing (and for black and white bin packing) packs items into a single active bin, and moves to a new active bin as soon as packing an item into the active bin is impossible. For standard bin packing, a new active bin is opened when there is no space for the new item in the previous active bin, but for colorful bin packing a new bin will be opened either in this case, or when the last item of the active bin and the new item have the same color. It was shown in [1] that this algorithm fails to achieve a finite competitive ratio (already for two colors). Harmonic algorithms [10], that partition items into sub-inputs according to sizes and pack each sub-input independently of the other sub-inputs, were also shown to have unbounded competitive ratios [1]. On the other hand, there are some basic online bin packing algorithms that can be adapted successfully for black and white bin packing. The generalizations of Any Fit (AF) algorithms, that never use a new bin unless there is not other way to pack a new item, were shown to have constant absolute competitive ratios. The generalized versions of such algorithms for colorful bin packing open a new bin only if the current item cannot be packed into an existing bin such that the color constraint is kept and the total size of items packed into the bin will remain at most 1. Three important special cases of AF are First Fit (FF), Best Fit (BF), and Worst Fit (WF). These algorithms select the bin where a new item is packed (out of the feasible options) to be the bin of minimum index, the a bin with the smallest empty space, and a bin with the largest empty space, respectively. The difference with classical bin packing is that the infeasible bins can be of two kinds, either those that do not have sufficient empty space, and those where the last packed item has the same color as the color of the new item. It was shown that all AF algorithms have absolute and asymptotic competitive ratios of at least 3 and at most 5 for black and white bin packing. Veselý [16] tightened the bound and showed an upper bound of 3 on the absolute competitive ratio of AF algorithms. The results of [1,16] in fact show that the absolute competitive ratio of WF is $2 + \frac{1}{d-1}$, if all items have sizes in $(0, \frac{1}{d}]$ (while FF and BF still have absolute and asymptotic competitive ratios of exactly 3 even in this restricted case). The positive results for AF algorithms are valid only for black and white packing but not for colorful bin packing. In contrast to these last results, we will show that AF algorithms do not have constant (absolute or asymptotic) competitive ratios for colorful bin packing with $|\mathcal{C}| \geq 3$.

Colorful bin packing is also a generalization of standard bin packing (since already black and white bin packing is such a generalization). For standard bin packing, NF has an asymptotic and an absolute competitive ratio of 2 [8]. Any Fit algorithms all have absolute competitive ratios of at most 2 [14,7,8,9,3] (some of these algorithms have smaller absolute or asymptotic competitive ratios; for example, in [3] it is shown that FF has an absolute competitive ratio of 1.7, and an asymptotic bound of 1.7 was known for FF for many years [9]). There are algorithms with smaller asymptotic competitive ratios, and the best possible asymptotic competitive ratio is known to be in [1.5403, 1.58889] [15,13,2]. Other variants of bin packing where the sequence of items must remain ordered even for offline solutions include *Packing with LIB (largest item in the bottom)* constraints, where an item can be packed into a bin with sufficient space if it is no larger than any item packed into this bin [11,6,12,5,4].

In our algorithms, we say that a bin B has color c if the last item that was packed into B has this color. Obviously, a bin changes its color as items are packed into it. For simplicity, we use names of colors as the elements of $\mathcal{C}$. Another algorithm for black and white bin packing presented in [1] is the algorithm Pseudo. This algorithm keeps a list of pseudo-bins, each being a list of (valid) bins. Each new item is assigned to a pseudo-bin and then to a bin of this pseudo-bin. The color of a (non-empty) pseudo-bin is defined to be the color of its last bin. An item is first assigned to a pseudo-bin of the opposite color (that is, a white item to a black pseudo-bin and a black item to a white pseudo-bin), opening a new pseudo-bin for the item if this assignment is impossible (there is no pseudo-bin of the other color). A pseudo-bin is split into bins in an online fashion; a new item is packed into the last bin of the pseudo-bin where it was assigned (note that this is always possible with respect to the color of the item), and a new bin (for this pseudo-bin) is opened if the empty space in the current last bin of the pseudo-bin is insufficient. In the case that there are multiple pseudo-bins that are suitable for the new item (multiple pseudo-bins have the opposite color), then in principle any one of them is chosen (that is, the analysis holds for arbitrary tie-breaking), but the algorithm was defined such that such a bin of minimum index is selected. A simple generalization of Pseudo for colorful packing is to assign a new item to a pseudo-bin of a minimum index whose color is different from the color of the new item. We show that this algorithm has an unbounded (absolute and asymptotic) competitive ratio. We show, however, that the tie-breaking rule can be modified, and a variant of this algorithm, called BALANCED-PSEUDO (BaP), has an absolute (and asymptotic) competitive ratio of 4. Roughly speaking, BaP tries to balance the colors of pseudo-bins; for a new item it finds the most frequent color of pseudo-bins (excluding the pseudo-bins having the same color as the new item), and assigns the new item to such a pseudo-bin. Interestingly, this approach is much more successful.

Finally, we design two new lower bounds. We give a lower bound of 2 on the asymptotic (and absolute) competitive ratio of any algorithm. This last lower bound is valid already for $|\mathcal{C}| = 2$ (i.e., for black and white bin packing) and it significantly improves the previous lower bound of approximately 1.7213 [1]. We also consider zero size items. It was shown in [1] that Pseudo is an optimal algorithm for zero size items (its absolute competitive ratio is 1). We show that in contrast, if $|\mathcal{C}| \geq 3$, then the asymptotic competitive ratio of any algorithm for such items is at least $\frac{3}{2}$. This implies that the two problems (colorful bin packing and black and white bin packing) are different.

In Section 2 we demonstrate that the existing algorithms have poor performance, we define algorithm BaP, analyze its competitive ratio for arbitrary items and for zero size items, and show that the analysis is tight. Lower bounds for arbitrary online algorithms are given in Section 3. Some proofs were omitted due to space constraints and can be found in http://arxiv.org/abs/1404.3990.

2 Algorithms

We start this section with examples showing that the algorithms that had a good performance for black and white bin packing (or their natural generalizations, all defined in the introduction) have a poor performance for colorful packing.

Proposition 1. *The algorithms FF, BF, WF, AF, and Pseudo have unbounded asymptotic competitive ratios for colorful bin packing.*

A New Algorithm. We define an algorithm called BALANCED-PSEUDO (BaP). The algorithm keeps a sequence of pseudo-bins denoted by $P_1, P_2, \ldots$, where each pseudo-bin is a sequence of bins. For pseudo-bin P_j, its sequence of bins is denoted by $B_1^j, B_2^j, \ldots, B_{n_j}^j$. Let k denote the number of pseudo-bins (at a given time). For any $1 \leq j \leq k$, C_j denotes the color of the last item assigned to P_j

(this will be the color of the last item of $B^j_{n_j}$), and it is called the color of the pseudo-bin P_j.

Algorithm BaP is similar to algorithm Pseudo [1], but it tries to balance the number of pseudo-bins of different colors, and it prefers to assign an item to a pseudo-bin of a color that occurs a maximum number of times (excluding pseudo-bins having the same color as the new item). For a new item i, if all pseudo-bins have the color c_i, then a new pseudo-bin P_{k+1} is opened, where it consists of one bin B^{k+1}_1. In this case, we let $k = k + 1$, $n_k = 1$. Otherwise, for any color $g \neq c_i$, let N_g be the number of pseudo-bins of color g. Let g' be a color for which $N_{g'}$ is maximal. Assign item i to a pseudo-bin P_j of color g'. If i can be packed into $B^j_{n_j}$ (with respect to the total size of items, as by definition the color of P_j is $g' \neq c_i$, so the color of i does not prevent its packing), then add it to this bin (as its last item), and otherwise, let $n_j = n_j + 1$, and pack i into $B^j_{n_j}$ as its only item. For all cases, if i was assigned to pseudo-bin P_j, then let $C_j = c_i$ (this is done no matter how j is chosen).

Analysis. The analysis separates the effect of sizes from the effect of colors. This is possible since BaP (similarly to Pseudo) already has such a separation. The number of pseudo-bins is independent of the sizes of items, while the partition of a pseudo-bin into bins is independent of the colors. The algorithm that is applied on every pseudo-bin is simply NF, and moreover, a new bin is used when there is no space for the current item in the previous bin of the same pseudo-bin. Every pair of consecutive bins of one pseudo-bin have items whose total size exceeds 1, thus the resulting bins are occupied by a total size above $\frac{1}{2}$ on average, possibly except for one bin of each pseudo-bin. We show that at each time that a new pseudo-bin is opened, an optimal solution cannot have less than half the number of bins, even if items have zero sizes. Informally, the reason is that a new pseudo-bin is opened when all pseudo-bins have the color of the new item. However, once the number of pseudo-bins of this color exceeds half the number of pseudo-bins, BaP prefers to use such bins as much as possible (in this case their number decreases), and an increase in their number can only be caused by an input where there is a large number of items of the same color arriving almost consecutively. Obviously, such inputs require large numbers of bins in any solution.

We let $LB_0 = \sum_{i=1}^n s_i$. Obviously, $OPT \geq LB_0$. Let $1 \leq i \leq j \leq n$. For any color c that appears in the subsequence of consecutive $j - i + 1$ items $i, i+1, ..., j$, let $C(i, j, c)$ be the number of times that it appears. Let

$$LB(i, j, c) = C(i, j, c) - (j - i + 1 - C(i, j, c)) = 2C(i, j, c) - j + i - 1 , \quad (1)$$

$LB(i, j) = \max_c LB(i, j, c)$, and $LB_1 = \max_{i,j} LB(i, j)$. For any non-empty input we have $LB_1 \geq 1$ since $LB(i, i, c_i) = 1$ for any i. Note that $LB(i, j, c)$ is positive only if the number of times that c appears in the subsequence $i, ..., j$ is more than $\frac{j-i+1}{2}$ (i.e., more than half the items of this subsequence are of color c), and thus for computing LB_1 it is sufficient to consider for every subsequence only a color c that appears a maximum number of times in this subsequence. The following lemma generalizes a property proved in [1].

Lemma 1. $OPT \geq LB_1$.

Consider the action of BaP, and let k be the index of the last pseudo-bin (i.e., k is the final value of the variable k). For $1 \leq m \leq k$, let LB^m denote LB_1 at the time that the first item is assigned to P_m. Let Y_m be the (index of the) first item that is assigned to P_m, and let X_m be its color (thus $Y_1 = 1$ holds by definition, i.e., the first item of the input is also the first item assigned to the first pseudo-bin). For convenience, let $Y_{k+1} = n + 1$. Let phase m be the subsequence of consecutive items $Y_m, \ldots, Y_{m+1} - 1$. In the lemmas below, when we discuss properties holding during phase m, we mean that they hold starting the time just after Y_m is packed and ending right after $Y_{m+1} - 1$ is packed.

Theorem 1. *For any $1 \leq m \leq k$, there exists $i \leq Y_m$ such that $C(i, Y_m, X_m) > \frac{m+3}{4} + \frac{Y_m - i}{2}$.*

Proof. We prove the claim by induction. For $m = 1$, $Y_m = 1$, and $C(1, 1, c_1) = 1$ as required. For $m = 2$, the items Y_2 and $Y_2 - 1$ have the same color X_2 (as $Y_2 - 1$ was assigned to P_1 and Y_2 is assigned to P_2). Thus, we find $C(Y_2 - 1, Y_2, X_2) = 2$. Next, assume that the claim holds for some $m \geq 2$. We will prove the claim for $m + 1$ by considering phase $2 < m \leq k - 1$.

Lemma 2. *If at some time in phase m (where $2 \leq m \leq k - 1$) an item i of a color that is not X_{m+1} is assigned to a pseudo-bin of a color that is not X_{m+1} (the two last items that the pseudo-bin receives are of colors different from X_{m+1}), then just before assigning i (the second item out of the two items whose colors are not X_{m+1}) there are less than $(m+1)/2$ (that is, at most $m/2$) pseudo-bins of color X_{m+1}.*

Lemma 3. *If during phase m there are always at least $(m + 1)/2$ pseudo-bins of color X_{m+1}, then $X_m = X_{m+1}$. In this case, letting t be the number of items of color X_m in phase m, phase m contains $t - 1$ items of other colors.*

If the condition of Lemma 3 holds, then let i be such that $C(i, Y_m, X_m) \geq \frac{m+3}{4} + \frac{Y_m - i}{2}$, and let t be the number of items of color $X_m = X_{m+1}$ in phase m. We have $C(i, Y_{m+1}, X_{m+1}) \geq \frac{m+3}{4} + \frac{Y_m - i}{2} + t$, and $Y_{m+1} - Y_m = 2t - 1$. Thus, $C(i, Y_{m+1}, X_{m+1}) \geq \frac{m+3}{4} + \frac{Y_m - i}{2} + \frac{Y_{m+1} - Y_m + 1}{2} > \frac{(m+1)+3}{4} + \frac{Y_{m+1} - i}{2}$ as required.

Lemma 4. *If there is a time in phase m that at most $m/2$ bins were of color X_{m+1}, then there exists an index i such that $Y_m \leq i \leq Y_{m+1} - 1$ where*

$$C(i, Y_{m+1}, X_{m+1}) \geq \frac{m + 4}{4} + \frac{Y_{m+1} - i}{2}.$$

Proof. Consider the last time during phase m that there are at most $m/2$ bins of color X_{m+1}, and let i be the first item right after this time. Since after item $Y_{m+1} - 1$ arrives, all m pseudo-bins have color X_{m+1} and $m > m/2$, the time just after $Y_{m+1} - 1$ arrives does not satisfy the condition, so the last such time must be earlier, i is well-defined, and $i \leq Y_{m+1} - 1$. We have $c_i = X_{m+1}$ as its assignment to a pseudo-bin increased the number of pseudo-bins of this color. Moreover, starting this time, there are at least $(m + 1)/2$ bins of color X_{m+1} at all times until after the arrival of Y_{m+1} (by the choice of the time, and since Y_{m+1} has the same color and causes the creation of a new pseudo-bin of this color). If m is even, then just before i is packed, there are exactly $m/2$ pseudo-bins of color X_{m+1} and $m/2$ pseudo-bins of other colors, and after item Y_{m+1} is assigned, there are $m + 1$ pseudo-bins of color X_{m+1}. Moreover, while the items $i, \ldots, Y_{m+1} - 1$ are being assigned, every item whose color is not X_{m+1} is assigned to a pseudo-bin of color X_{m+1}, so every pseudo-bin receives alternating colors (items of color X_{m+1} alternate with other colors). Thus, if there are t items whose colors are not X_{m+1} among these items, there are $t + \frac{m}{2}$ items of color X_{m+1}, and the total number of items is $Y_{m+1} - i = 2t + \frac{m}{2}$. Including Y_{m+1}, we have $C(i, Y_{m+1}, X_{m+1}) = t + \frac{m}{2} + 1 = \frac{m}{2} + 1 + \frac{Y_{m+1}-i}{2} - \frac{m}{4} = \frac{(m+1)+3}{4} + \frac{Y_{m+1}-i}{2}$ as required. If m is odd, then if there are t items whose colors are not X_{m+1} among these items, there are $t + \frac{m+1}{2}$ items of color X_{m+1}, and the total number of items is $Y_{m+1} - i = 2t + \frac{m+1}{2}$. We have $C(i, Y_{m+1}, X_{m+1}) = t + \frac{m+1}{2} + 1 = \frac{m}{2} + \frac{3}{2} + \frac{Y_{m+1}-i}{2} - \frac{m+1}{4} > \frac{m+4}{4} + \frac{Y_{m+1}-i}{2}$ as required. □

This completes the proof of the theorem.   □

The next corollary follows from choosing $j = Y_k$ and i such that $C(i, Y_k, X_k) \geq \frac{m+3}{4} + \frac{Y_m - i}{2}$, and using (1).

Corollary 1. We have $LB_1 \geq LB^k \geq LB(i, Y_k, X_k) \geq \frac{k+1}{2}$.

Corollary 2. The absolute competitive ratio of BaP is at most 4 for arbitrary items, and at most 2 for zero size items.

We can show that the analysis of BaP is tight.

Proposition 2. The asymptotic competitive ratio of BaP is at least 2 for zero size items, and at least 4 for arbitrary items.

Proof. We will use the following parameters. Let $N \geq 2$ be a large integer. Let $M = 4^{N+1}$, let $a_1 = 1$, and for $i > 1$, let $a_i = (3a_{i-1} + 2)/4$.

Lemma 5. We have $1 \leq a_i < 2$, $a_i > a_{i-1}$ for all i, and $\lim_{i \to \infty} a_i = 2$. Moreover, $a_i = 2 - (3/4)^{i-1}$ holds.

We start with an input of zero size items. In this input all items are white, red, or blue. The input consists of the following $N + 1$ phases. In phase 0, M white items arrive. In phase i (for $1 \leq i \leq N$), $a_i \cdot M/2$ red items arrive, and then $(1 - a_i/2)M$ blue items arrive. We find $a_i \cdot M/2 = (2 - (3/4)^{i-1})4^{N+1}/2 = 2(4^N - 3^{i-1} \cdot 4^{N-i+1})$, and $(1 - a_i/2)M = 2 \cdot 4^N - 2 \cdot 4^N + 2 \cdot 3^{i-1} \cdot 4^{N-i+1}$. The numbers of red and blue items are even integers in $(0, M)$, and their sum is M. Phase i ends with the arrival of M white items. We have $OPT = M$. Obviously, M bins are needed already for the first M white items. Each bin of the optimal solution receives one white item in phase 0, and in each additional phase it receives one red item or one blue item, and additionally one white item.

Lemma 6. *After i phases BaP has $a_{i+1}M$ pseudo-bins, all of which are white.*

Proof. By induction. This holds for $i = 0$. Assume that it holds after phase $i-1$. In phase i, first the red items are assigned to distinct pseudo-bins, and now there are $a_i \cdot M/2$ red pseudo-bins and $a_i \cdot M/2$ white pseudo-bins. Now the blue items are packed such that half of them join red pseudo-bins and half join white pseudo-bins. The number of white pseudo-bins is now $a_i \cdot M/2 - (1 - a_i/2)M/2 = M/4(3a_i - 2)$. The number of pseudo-bins that are either red or blue is now $a_i \cdot M/2 + (1 - a_i/2)M/2 = M(a_i + 2)/4$. Note that $(a_i + 2)/4 < 1$ since $a_i < 2$. The M white items can join $M/4(a_i + 2)$ pseudo-bins that are either red or blue, and the remaining $M - M/4(a_i + 2)$ items cause the opening of new white pseudo-bins. The total number of pseudo-bins now is $a_i \cdot M + (M - M(a_i + 2)/4)$ and they are all white. The last number is equal to $M(a_i + 1 - a_i/4 - 1/2) = M(3a_i + 2)/4 = M \cdot a_{i+1}$. $\qquad\qquad\square$

We find that after $N+1$ phases, the algorithm has $(2 - (3/4)^N) \cdot M$ pseudo-bins, each consisting of one bin, which implies the lower bound.

In order to prove that the asymptotic competitive ratio is at least 4 for arbitrary item sizes, we start with presenting the input above to BaP. At this time, all items are of three colors and have zero sizes, $OPT = M$, the algorithm has $2M - m$ pseudo-bins where $m = (\frac{3}{4})^N M$. The input continues as follows (we ensure that $OPT = M$ will hold for the complete input). There are $2M - m - 1$ items, all of different new colors (none of these colors is white or red or blue). Moreover, we reserve the color black for later, and thus we require that none of these colors is black. Each of these items has size 2ε (for some $\varepsilon < 1/(8M)$). OPT will use one bin for items of size 2ε, while BaP will assign each item to a different pseudo-bin. Now all the bins of BaP have different colors (one pseudo-bin remains white). Next, $M - 1$ black items arrive, where each item has size $1 - \varepsilon$. OPT adds them to its white bins, the algorithm assigns at most one item to a white pseudo-bin, so at least $M - 2$ items are assigned to different pseudo-bins whose color was not white, red, blue, or black (and the last item assigned to this pseudo-bin had size 2ε). Thus, there are at least $M - 1$ black pseudo-bins, and at least $M - 2$ of them consist of two bins each, as the total size of items assigned to it is above 1. Next, there are $M - 2$ items all of different and new colors and sizes of 2ε. OPT packs them into the bin that already has items of this size, while the algorithm adds them to its black pseudo-bins, and at least $M - 3$ pseudo-bins now consist of three bins. The algorithm will have at least $2M - m + (M - 2) + (M - 3) = 4M - m$ bins, while $OPT = M$. The competitive ratio approaches 4 for a sufficiently large value of N.

Note that this example does not require any assumptions regarding the behavior of BaP in cases of ties. The example requires, however, a large number of different colors. We provide a different example that is valid for a run of BaP where ties between pseudo-bins of one color are broken in favor of smaller indices, and $\mathcal{C} = \{white, red, blue\}$. Once again, the input starts with the items of zero size as above. Afterwards, there are three batches of items, consisting of M blue items, M white items, and M blue items, respectively, of sizes that we will define. Since the number of pseudo-bins is above M and all of them are white, blue items must join white pseudo-bins, and white items must join blue pseudo-bins. The three batches are packed into the first M pseudo-bins, where the jth item of a batch is packed into the pseudo-bin of index j. For $1 \leq t \leq M + 1$, let $\delta_t = \varepsilon/4^t$ (thus we have $\delta_{t+1} = \delta_t/4$). The size of the tth item in the first batch (of blue items) is δ_t ($t = 1, ..., M$). The size of the tth item in the second batch (of white items) is $1 - 3\delta_{t+1}$ ($t = 1, ..., M$). The size of the tth item in the third batch (of blue items) is δ_t ($t = 1, ..., M$). We have $\delta_t + (1 - 3 \cdot \delta_{i+1}) > 1$ since $\delta_t - 3 \cdot \delta_{t+1} = \delta_t/4$. Therefore, each pseudo-bin $t = 1, \dots, M$ consists of three bins.

We show that for this input $OPT \leq M + 2$. Given the packing into M white bins, for $t = 1, ..., M - 1$ we group the items of sizes $\delta_t, 1 - 3 \cdot \delta_t, \delta_t$ (of colors blue, white, and blue, respectively) and pack them into $M - 1$ bins. A blue item of size δ_M is added to the remaining bin, and the two items of sizes δ_M and $1 - 3 \cdot \delta_{M+1}$ are packed into new bins.   □

3   Lower Bounds

The (absolute or asymptotic) competitive ratio cannot decrease if the cardinality of $\mathcal{C}$ grows. Thus, when we claim a negative result for $|\mathcal{C}| \geq \ell$, it is sufficient to prove it for $|\mathcal{C}| = \ell$. Thus, the lower bound for arbitrary items is proved for $|\mathcal{C}| = 2$, and the lower bound for zero size items is proved for $|\mathcal{C}| = 3$.

3.1   An Asymptotic Lower Bound of 2

We will consider an algorithm, and construct an input consisting of black and white items based on its behavior. The construction is carried out in phases, where in each phase the algorithm has to pack a black item after a white item. If they are packed together, it turns out that it would have been better to pack this last black item separately, since another smaller black item arrives, and a large white item that should have been combined with the first black item of this phase. Since no other combination is possible, the algorithm has two new bins instead of just one. If the algorithm uses a new bin for the first black item, it turns out that the phase ends, and the algorithm used a new bin when this was not necessary. The first situation is slightly better for the algorithm, and a ratio of 2 will follow from that. The precise construction is presented in the proof of the following theorem.

Theorem 2. *The asymptotic competitive ratio of any algorithm for colorful bin packing is at least 2.*

Proof. Consider an online algorithm A. Let $N > 3$ be a large integer. Let $\varepsilon = \frac{1}{N^3}$, and $\delta_i = \frac{1}{5^i \cdot N^3}$ for $1 \leq i \leq N^2$. Let $\mathcal{C} = \{$ black, white$\}$. The list of items will consist of white items called *regular white items*, each of size ε, white items called *huge white items*, whose sizes are either of the form $1 - 2\delta_i$ (for some $1 \leq i \leq N^2$) or 1, black items called *special black items*, whose sizes are of the form $3\delta_i$, and black items called *regular black items* whose sizes are of the form δ_i.

The list is created as follows. An index i is used for the number of regular white items that have arrived so far (each such item is followed by a regular black item). An index j is used for the number of huge white items that have arrived so far (each such item is preceded by a black item and followed by a black item). The input stops when one of $i = N^2$ and $j = N$ happens (even if the second event did not happen). Let $i = 0$ and $j = 0$.

1. If $j = N$, then stop. Else, if $i = N^2$, then $N - j$ huge white items of size 1 each arrive; stop.
2. Let $i = i + 1$; a regular white item arrives; a regular black item of size δ_i arrives.
3. If the last black item is packed into a new bin, the phase ends. Go to step 1 to start a new phase.
4. Else, it must be the case that the last black item is packed into a bin where the last item is white. Let $j = j + 1$, a special black item of size $3\delta_i$ arrives, then a huge white item of size $1 - 2\delta_i$ arrives, and finally, a regular black item of size δ_i arrives, and the phase ends. Go to step 1 to start a new phase.

Lemma 7. *Any huge white item is strictly larger than $1 - \varepsilon$. Any black item is strictly smaller than ε. The total size of a huge white item of phase i and a black item of an earlier phase is above 1.*

Lemma 8. $N \leq OPT \leq N + 1$.

Proof. There are N huge white items, each of size above $\frac{1}{2}$, thus, since a pair of such items cannot be packed into a bin together even with a black item, $OPT \geq N$. We create a packing with $N + 1$ bins as follows. If there are huge white items of size 1, each such item is packed into a separate bin. We show how the remaining items can be packed into j bins (where j is the final value of the variable j). Every remaining huge white item is packed in a bin with the last regular black item that arrived before it, and the regular black item that arrived after it. The total size of such three items of phase i is 1. This leaves a sequence of items of alternating colors, where some of the black items are special. The white items in the remaining input are regular, and the black item of phase i has a size of either δ_i or $3\delta_i$. In this sequence, every item is no larger than ε, and there are $2i \leq 2N^2$ items (where i is the final value of this variable). Thus, the total size of these items is below 1, and they are all packed into a single bin. □

Lemma 9. *The number of bins used by the algorithm up to a time when $i = i'$ is at least i'. The number of black bins at a time when $j = j'$ is at least $2j' + 1$.*

For a fixed value of N, if the input was terminated since $i = N^2$ but $j < N$, then the cost of the algorithm is at least $N^2 + N - j \geq N^2 + 1$. As $OPT \leq N + 1$, we find a competitive above $N - 1 > 2$. If $j = N$, then the cost of the algorithm is at least $2N + 1$ (as this is a lower bound on the number of black bins), while $OPT \leq N + 1$, and we find a ratio of at least $2 - \frac{1}{N+1}$. We found that for any $N > 3$, there is an input where $OPT \geq N$, and the competitive ratio for this input is at least $2 - \frac{1}{N+1}$. This implies the claim. □

3.2 A Lower Bound for Zero Size Items

It was shown in [1] that if all items have zero sizes, then the algorithm Pseudo finds an optimal solution (that is, its absolute competitive ratio is 1). Our analysis of BaP implies that its absolute and asymptotic competitive ratios for zero size items are equal to 2. Here, we show that there cannot be an online algorithm for colorful bin packing with at least three colors and zero size items that produces an optimal solution (a solution that uses the minimum number of bins).

Theorem 3. *Any algorithm for zero size items with $|\mathcal{C}| \geq 3$ has an asymptotic competitive ratio of at least $\frac{3}{2}$.*

Proof. We will use $\mathcal{C} = \{\text{white, red, blue}\}$. Recall that all items have zero sizes, thus for every presented item we only specify its color. Let $M \geq 2$ be a large integer. We construct an input for which $M \leq OPT \leq M + 3$. The input starts with phase 0 that consists of M white items. Thus, $OPT \geq M$. The remainder of the input is presented in phases. In parallel to presenting the input, we will create a packing π for the complete input. This packing will consist of $M + 3$ bins. The M items of phase 0 are packed in π into M bins called regular bins. In addition to the M regular bins of π, there will be a special bin of each color in π (this bin is empty after phase 0). The regular bins of π (M bins in total), will always be of one color (this color can be any of the three colors). Each phase i will have a color $G(i)$ associated with it. This is the color of the M regular bins of π. The color associated with phase 0 is white.

Phase i is defined as follows. Let c_i and c_i' be the two colors that are not the color associated with phase $i - 1$ (i.e., $c_i, c_{i'} \in \mathcal{C} \setminus \{G(i-1)\}$, $c_i \neq c_{i'}$). There are $2M$ items of alternating colors; the items of odd indices are of color c_i, and the items of even indices are of color c_i'. Let W_i, R_i, and B_i, be the numbers of white, red, and blue bins, that the algorithm has after the last $2M$ items have arrived. Phase i ends with M items of the color for which the number of bins of the algorithm is maximal after the $2M$ first items of phase i have been packed by the algorithm (that is, letting $X = \max\{W_i, R_i, B_i\}$, the last M items are white if $X = W_i$, otherwise, if $X = R_i$, then they are red, and otherwise they are blue). Let $G(i)$ be the color of the last M items of phase i.

Let N_i be the number of bins of the algorithm after phase i. We have $N_0 = M$. In phase $i \geq 1$ the algorithm obviously has at least N_{i-1} bins after the first $2M$ items of phase i have arrived, and there are at least $\frac{N_{i-1}}{3}$ bins of color $G(i)$. Therefore, after M items of color $G(i)$ arrive, the algorithm has M additional bins of color $G(i)$, and there are at least $\frac{N_{i-1}}{3} + M$ bins of color $G(i)$. We get $N_i \geq \frac{N_{i-1}}{3} + M$. Thus, $N_i \geq M \cdot \frac{3^{i+1}-1}{2 \cdot 3^i}$. This holds for $i = 0$ as $N_0 = M$, and $\frac{3^1-1}{2 \cdot 3^0} = 1$, and using the recurrence, $N_{i+1} \geq (\frac{3^{i+1}-1}{2 \cdot 3^i})M/3 + M = (\frac{3^{i+2}-1}{2 \cdot 3^{i+1}})M$.

Due to symmetry, we describe the packing π for the case that the color associated with phase $i - 1$ is white, and the first $2M$ items of phase i alternate between red and blue (starting with red). If the last M items of phase i are blue or red, then the first $2M$ items are packed into the blue special bin (which remains blue), and the last M items are packed into the M regular bins. If the last M items are white, each bin receives a red item and an blue item. Now all regular bins are blue, and the last M white items can be packed into them. The color associated with phase i is indeed $G(i)$.

We find that the competitive ratio of the algorithm is at least $\frac{M}{M+3} \cdot \frac{3^{i+1}-1}{2\cdot 3^i}$. Letting M and i grow without bound we find a lower bound of $\frac{3}{2}$ on the asymptotic competitive ratio. □

References

1. Balogh, J., Békési, J., Dósa, G., Epstein, L., Kellerer, H., Tuza, Z.: Online results for black and white bin packing. Theory of Computing Systems (to appear)
2. Balogh, J., Békési, J., Galambos, G.: New lower bounds for certain classes of bin packing algorithms. Theoretical Computer Science 440-441, 1–13 (2012)
3. Dósa, G., Sgall, J.: First fit bin packing: A tight analysis. In: Proc. of the 30th International Symposium on Theoretical Aspects of Computer Science (STACS 2013), pp. 538–549 (2013)
4. Dósa, G., Tuza, Z., Ye, D.: Bin packing with "largest in bottom" constraint: tighter bounds and generalizations. Journal of Combinatorial Optimization 26(3), 416–436 (2013)
5. Epstein, L.: On online bin packing with LIB constraints. Naval Research Logistics 56(8), 780–786 (2009)
6. Finlay, L., Manyem, P.: Online LIB problems: Heuristics for bin covering and lower bounds for bin packing. RAIRO Operetions Research 39(3), 163–183 (2005)
7. Johnson, D.S.: Near-optimal bin packing algorithms. PhD thesis, MIT, Cambridge, MA (1973)
8. Johnson, D.S.: Fast algorithms for bin packing. Journal of Computer and System Sciences 8(3), 272–314 (1974)
9. Johnson, D.S., Demers, A., Ullman, J.D., Garey, M.R., Graham, R.L.: Worst-case performance bounds for simple one-dimensional packing algorithms. SIAM Journal on Computing 3, 256–278 (1974)
10. Lee, C.C., Lee, D.T.: A simple online bin packing algorithm. Journal of the ACM 32(3), 562–572 (1985)
11. Manyem, P.: Bin packing and covering with longest items at the bottom: Online version. The ANZIAM Journal 43(E), E186–E232 (2002)
12. Manyem, P., Salt, R.L., Visser, M.S.: Approximation lower bounds in online LIB bin packing and covering. Journal of Automata, Languages and Combinatorics 8(4), 663–674 (2003)
13. Seiden, S.S.: On the online bin packing problem. Journal of the ACM 49(5), 640–671 (2002)
14. Ullman, J.D.: The performance of a memory allocation algorithm. Technical Report 100, Princeton University, Princeton, NJ (1971)
15. van Vliet, A.: An improved lower bound for on-line bin packing algorithms. Information Processing Letters 43(5), 277–284 (1992)
16. Veselý, P.: Competitiveness of fit algorithms for black and white packing. Manuscript. presented in MATCOS 2013 (2013)

Algorithms Parameterized by Vertex Cover and Modular Width, through Potential Maximal Cliques*

Fedor V. Fomin[1], Mathieu Liedloff[2], Pedro Montealegre[2], and Ioan Todinca[2]

[1] Department of Informatics, University of Bergen, N-5020 Bergen, Norway
fomin@ii.uib.no
[2] Univ. Orléans, INSA Centre Val de Loire, LIFO EA 4022, BP 6759, F-45067
Orléans Cedex 2, France
{mathieu.liedloff,ioan.todinca,pedro.montealegre}@univ-orleans.fr

Abstract. In this paper we give upper bounds on the number of *minimal separators* and *potential maximal cliques of graphs* w.r.t. two graph parameters, namely *vertex cover* (vc) and *modular width* (mw). We prove that for any graph, the number of minimal separators is $\mathcal{O}^*(3^{vc})$ and $\mathcal{O}^*(1.6181^{mw})$, the number of potential maximal cliques is $\mathcal{O}^*(4^{vc})$ and $\mathcal{O}^*(1.7347^{mw})$, and these objects can be listed within the same running times. (The $\mathcal{O}^*$ notation suppresses polynomial factors in the size of the input.) Combined with known results [3,12], we deduce that a large family of problems, e.g., TREEWIDTH, MINIMUM FILL-IN, LONGEST INDUCED PATH, FEEDBACK VERTEX SET and many others, can be solved in time $\mathcal{O}^*(4^{vc})$ or $\mathcal{O}^*(1.7347^{mw})$.

1 Introduction

The *vertex cover* of a graph G, denoted by $vc(G)$, is the minimum number of vertices that cover all edges of the graph. The *modular width* $mw(G)$ can be defined as the maximum degree of a prime node in the modular decomposition of G (see [20] and Section 4 for definitions). The main results of this paper are of combinatorial nature: we show that the number of *minimal separators* and the number of *potential maximal cliques* of a graph (see Section 2 and also [3] for definitions) are upper bounded by a function in each of these parameters. More specifically, we prove the number of minimal separators is at most 3^{vc} and $\mathcal{O}^*(1.6181^{mw})$, and the number of potential maximal cliques is $\mathcal{O}^*(4^{vc})$ and $\mathcal{O}^*(1.7347^{mw})$, and these objects can be listed within the same running time bounds. Recall that the $\mathcal{O}^*$ notation suppresses polynomial factors in the size of the input, i.e., $\mathcal{O}^*(f(k))$ should be read as $f(k) \cdot n^{\mathcal{O}(1)}$ where n is the number of vertices of the input graph. Minimal separators and potential maximal cliques have been used for solving several classical optimization problems, e.g., TREEWIDTH, MINIMUM FILL-IN [10], LONGEST INDUCED PATH, FEEDBACK

* Supported by the European Research Council under the European Union's Seventh Framework Programme (FP/2007-2013) / ERC Grant Agreement n. 267959.

R Ravi and I.L. Gørtz (Eds.): SWAT 2014, LNCS 8503, pp. 182–193, 2014.

Vertex Set or Independent Cycle Packing [12]. Pipelined with our combinatorial bounds, we obtain a series of algorithmic consequences in the area of FPT algorithms parameterized by the vertex cover and the modular width of the input graph. In particular, the problems mentioned above can be solved in time $\mathcal{O}^*(4^{vc})$ and $\mathcal{O}^*(1.7347^{mw})$. These results are complementary in the sense that graphs with small vertex cover are sparse, while graphs with small modular width may be dense.

Vertex cover and modular width are strongly related to treewidth (tw) and cliquewidth (cw) parameters, since for any graph G we have $\text{tw}(G) \leq \text{vc}(G)$ and $\text{cw}(G) \leq \text{mw}(G) + 2$. The celebrated theorem of Courcelle [6] states that all problems expressible in Counting Monadic Second Order Logic (CMSO$_2$) can be solved in time $f(\text{tw}) \cdot n$ for some function f depending on the problem. A similar result for cliquewidth [7] shows that all CMSO$_1$ problems can be solved in time $f(\text{cw}) \cdot n$, if the clique-decomposition is also given as part of the input. (See the full version [11] for definitions of different types of logic. Informally, CMSO$_2$ allows logic formulae with quantifiers over vertices, edges, edge sets and vertex sets, and counting modulo constants. The CMSO$_1$ formulae are more restricted, we are not allowed to quantify over edge sets.)

Typically function f is a tower of exponentials, and the height of the tower depends on the formula. Moreover Frick and Grohe [15] proved that this dependency on treewidth or cliquewidth cannot be significantly improved in general. Lampis [18] shows that the running time for CMSO$_2$ problems can be improved $2^{2^{\mathcal{O}(vc)}} \cdot n$ when parametrized by vertex cover, but he also shows that this cannot be improved to $\mathcal{O}^*(2^{2^{o(vc)}})$ (under the exponential time hypothesis). We are not aware of similar improvements for parameter modular width, but we refer to [16] for discussions on problems parameterized by modular width.

Most of our algorithmic applications concern a restricted, though still large subset of CMSO$_2$ problems, but we guarantee algorithms that are single exponential in the vertex cover: $\mathcal{O}^*(4^{vc})$ and in the modular width: $\mathcal{O}^*(1.7347^{mw})$. We point out that our result for modular width extends the result of [13,12], who show a similar bound of $\mathcal{O}^*(1.7347^n)$ for the number of potential maximal cliques and for the running times for these problems, but parameterized by the number of vertices of the input graph.

We use the following generic problem proposed by [12], that encompasses many classical optimization problems. Fix an integer $t \geq 0$ and a CMSO$_2$ formula φ. Consider the problem of finding, in the input graph G, an induced subgraph $G[F]$ together with a vertex subset $X \subseteq F$, such that the treewidth of $G[F]$ is at most t, the graph $G[F]$ together with the vertex subset X satisfy formula φ, and X is of maximum size under this conditions. This optimization problem is called Max Induced Subgraph of tw $\leq t$ Satisfiying φ:

$$\begin{array}{ll} \text{Max} & |X| \\ \text{subject to} & \text{There is a set } F \subseteq V \text{ such that } X \subseteq F; \\ & \text{The treewidth of } G[F] \text{ is at most } t; \\ & (G[F], X) \models \varphi. \end{array} \tag{1}$$

Note that our formula φ has a free variable corresponding to the vertex subset X. For several examples, in formula φ the vertex set X is actually equal to F. E.g., even when φ only states that $X = F$, for $t = 0$ we obtain the MAXIMUM INDEPENDENT SET PROBLEM, and for $t = 1$ we obtain the MAXIMUM INDUCED FOREST. If $t = 1$ and φ states that $X = F$ and $G[F]$ is a path we obtain the LONGEST INDUCED PATH problem. Still under the assumption that $X = F$, we can express the problem of finding the largest induced subgraph $G[F]$ excluding a fixed planar graph H as a minor, or the largest induced subgraph with no cycles of length $0 \mod l$. But X can correspond to other parameters, e.g. we can choose the formula φ such that $|X|$ is the number of connected components of $G[F]$. Based on this we can express problems like INDEPENDENT CYCLE PACKING, where the goal is to find an induced subgraph with a maximum number of components, and such that each component induces a cycle.

The result of [12] states that problem MAX INDUCED SUBGRAPH OF tw $\leq t$ SATISFIYING φ can be solved in a running time of the type $\# \mathrm{pmc} \cdot n^{t+4} \cdot f(\varphi, t)$ where $\# \mathrm{pmc}$ is the number of potential maximal cliques of the graph, assuming that the set of all potential maximal cliques is also part of the input. Thanks to our combinatorial bounds we deduce that the problem MAX INDUCED SUBGRAPH OF tw $\leq t$ SATISFIYING φ can be solved in time $\mathcal{O}(4^{\mathrm{vc}} n^{t+c})$ and $\mathcal{O}(1.7347^{\mathrm{mw}} n^{t+c})$, for some small constant c.

There are several other graph parameters that can be computed in time $\mathcal{O}^*(\# \mathrm{pmc})$ if the input graph is given together with the set of its potential maximal cliques. E.g.,TREEWIDTH, MINIMUM FILL-IN [10], their weighted versions [1,17] and several problems related to phylogeny [17], or TREELENGTH [19]. Pipelined with our main combinatorial result, we deduce that all these problems can be solved in time $\mathcal{O}^*(4^{\mathrm{vc}})$ or $\mathcal{O}^*(1.7347^{\mathrm{mw}})$. Recently Chapelle et al. [5] provided an algorithm solving TREEWIDTH and PATHWIDTH in $\mathcal{O}^*(3^{\mathrm{vc}})$, but those completely different techniques do not seem to work for MINIMUM FILL-IN or TREELENGTH. The interested reader may also refer., e.g., to [8,9] for more (layout) problems parameterized by vertex cover.

2   Minimal Separators and Potential Maximal Clique

Let $G = (V, E)$ be an undirected, simple graph. We denote by n its number of vertices and by m its number of edges. The *neighborhood* of a vertex v is $N(v) = \{u \in V : \{u, v\} \in E\}$. We say that a vertex x *sees* a vertex subset S (or vice-versa) if $N(x)$ intersects S. For a vertex set $S \subseteq V$ we denote by $N(S)$ the set $\bigcup_{v \in S} N(v) \setminus S$. We write $N[S]$ (resp. $N[x]$) for $N(S) \cup S$ (resp. $N(x) \cup \{x\}$). Also $G[S]$ denotes the subgraph of G induced by S, and $G - S$ is the graph $G[V \setminus S]$.

A *connected component* of graph G is the vertex set of a maximal induced connected subgraph of G. Consider a vertex subset S of graph G. Given two vertices u and v, we say that S is a u, v-separator if u and v are in different connected components of $G - S$. Moreover, if S is inclusion-minimal among all u, v-separators, we say that S is a *minimal u, v-separator*. A vertex subset S is called a *minimal separator* of G if S is a u, v-minimal separator for some pair of vertices u and v.

Let C be a component of $G - S$. If $N(C) = S$, we say that C is a *full component* associated to S.

Proposition 1 (folklore). *A vertex subset S of G is a minimal separator if and only if $G - S$ has at least two full components associated to S. Moreover, S is a minimal x, y-separator if and only if x and y are in different full components associated to S.*

A graph H is *chordal* or *triangulated* if every cycle with four or more vertices has a chord, i.e., an edge between two non-consecutive vertices of the cycle. A *triangulation* of a graph $G = (V, E)$ is a chordal graph $H = (V, E')$ such that $E \subseteq E'$. Graph H is a *minimal triangulation* of G if for every edge set E'' with $E \subseteq E'' \subset E'$, the graph $F = (V, E'')$ is not chordal.

A set of vertices $\Omega \subseteq V$ of a graph G is called a *potential maximal clique* if there is a minimal triangulation H of G such that Ω is a maximal clique of H.

The following statement due to Bouchitté and Todinca [3] provides a characterization of potential maximal cliques, and in particular allows to test in polynomial time if a vertex subset Ω is a potential maximal clique of G:

Proposition 2 ([3]). *Let $\Omega \subseteq V$ be a set of vertices of the graph $G = (V, E)$ and $\{C_1, \ldots, C_p\}$ be the set of connected components of $G - \Omega$. We denote $\mathcal{S}(\Omega) = \{S_1, S_2, \ldots, S_p\}$, where $S_i - N(C_i)$ for all $i \subset \{1, \ldots, p\}$. Then Ω is a potential maximal clique of G if and only if*

1. *each $S_i \in \mathcal{S}(\Omega)$ is strictly contained in Ω;*
2. *the graph on the vertex set Ω obtained from $G[\Omega]$ by completing each $S_i \in \mathcal{S}(\Omega)$ into a clique is a complete graph.*

Moreover, if Ω is a potential maximal clique, then $\mathcal{S}(\Omega)$ is the set of minimal separators of G contained in Ω.

Another way of stating the second condition is that for any pair of vertices $u, v \in \Omega$, if they are not adjacent in G then there is a component C of $G - \Omega$ seeing both x and y.

To illustrate Proposition 2, consider, e.g., the cube graph depicted in Figure 2. The set $\Omega_1 = \{a, e, g, c, h\}$ is a potential maximal clique and the minimal separators contained in Ω_1 are $\{a, e, g, c\}$ and $\{a, h, c\}$. Another potential maximal clique of the cube graph is $\Omega_2 = \{a, c, f, h\}$ containing the minimal separators $\{a, c, f\}$, $\{a, c, h\}$, $\{a, f, h\}$ and $\{c, f, h\}$.

Based on Propositions 1 and 2, one can easily deduce:

Corollary 1 (see e.g., [3]). *There is an $O(m)$ time algorithm testing if a given vertex subset S is a minimal separator of G, and $O(nm)$ time algorithm testing if a given vertex subset Ω is a potential maximal clique of G.*

We also need the following observation.

Proposition 3 ([3]). *Let Ω be a potential maximal clique of G and let $S \subset \Omega$ be a minimal separator. Then $\Omega \setminus S$ is contained in a unique component C of $G - S$, and moreover C is a full component associated to S.*

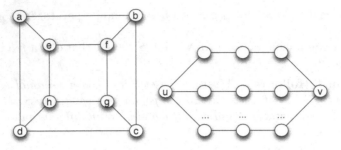

Fig. 1. Cube graph (left) and watermelon graph (right)

3   Relations to Vertex Cover

A vertex subset W is a *vertex cover* of G if each edge has at least one endpoint in W. Note that if W is a vertex cover, that $V \setminus W$ induces an *independent set* in G, i.e. $G - W$ contains no edges. We denote by $\text{vc}(G)$ the size of a minimum vertex cover of G. The parameter $\text{vc}(G)$ is called the *vertex cover number* or simply (by a slight abuse of language) the *vertex cover* of G. There is a well-known (folklore) branching algorithm computing the vertex cover of the input graph in time $\mathcal{O}^*(2^{vc})$.

Let us show that any graph G has at most $3^{\text{vc}(G)}$ minimal separators.

Lemma 1. *Let $G = (V, E)$ be a graph, W be a vertex cover and $S \subseteq V$ be a minimal separator of G. Consider a three-partition (D_1, S, D_2) of V such that both D_1 and D_2 are formed by a union of components of $G - S$, and both D_1 and D_2 contain some full component associated to S. Denote $D_1^W = D_1 \cap W$ and $D_2^W = D_2 \cap W$.*

Then $S \setminus W = \{x \in V \setminus W \mid N(x) \text{ intersects both } D_1^W \text{ and } D_2^W\}$.

Proof. Let $C_1 \subseteq D_1$ and $C_2 \subseteq D_2$ be two full components associated to S. Let $x \in S \setminus W$. Vertex x must have neighbors both in C_1 and C_2, hence both in D_1 and D_2. Since $x \notin W$ and W is a vertex cover, we have $N(x) \subseteq W$. Consequently x has neighbors both in D_1^W and D_2^W.

Conversely, let $x \in V \setminus W$ s.t. $N(x)$ intersects both D_1^W and D_2^W. We prove that $x \in S$. By contradiction, assume that $x \notin S$, thus x is in some component C of $G - S$. Suppose w.l.o.g. that $C \subseteq D_1$. Since $N(x) \subseteq C \cup N(C)$, we must have $N(x) \subseteq D_1 \cup S$. Thus $N(x)$ cannot intersect D_2—a contradiction.   □

Theorem 1. *Any graph G has at most $3^{\text{vc}(G)}$ minimal separators. Moreover the set of its minimal separators can be listed in $\mathcal{O}^*(3^{\text{vc}(G)})$ time.*

Proof. Let W be a minimum size vertex cover of G. For each three-partition (D_1^W, S^W, D_2^W) of W, let $S = S^W \cup \{x \in V \setminus W \mid N(x) \text{ intersects } D_1^W \text{ and } D_2^W\}$. According to Lemma 1, each minimal separator of G will be generated this way, by an appropriate partition (D_1^W, S^W, D_2^W) of W. Thus the number of minimal separators is at most $3^{\text{vc}(G)}$, the number of three-partitions of W.

These arguments can be easily turned into an enumeration algorithm, we simply need to compute an optimum vertex cover (recall this can be done in $\mathcal{O}^*(2^{(vc(G))})$ time) then test, for each set S generated from a three-partition, if S is indeed a minimal separator. The latter takes $\mathcal{O}(m)$ time for each set S using Corollary 1.   □

Observe that the bound of Theorem 1 is tight up to a constant factor. Indeed consider the watermelon graph $W_{k,3}$ formed by k disjoint paths of three vertices plus two vertices u and v adjacent to the left, respectively right ends of the paths (see Figure 2). Note that this graph has vertex cover $k + 2$ (the minimum vertex cover contains the middle of each path and vertices u and v) and it also has 3^k minimal u, v-separators, obtained by choosing arbitrarily one of the three vertices on each of the k paths.

We now extend Theorem 1 to a similar result on potential maximal cliques. Let us distinguish a particular family of potential maximal cliques, which have *active* separators. They have a particular structure which makes them easier to handle.

Definition 1 ([4]). *Let $\Omega \subseteq V$ be a potential maximal clique of graph $G = (V, E)$, let $\{C_1, \ldots, C_p\}$ be the set of connected components of $G - \Omega$ and let $S_i = N(C_i)$, for $1 \leq i \leq p$.*

Consider now the graph G^+ obtained from G by completing into a clique all minimal separators S_j, $2 \leq i \leq p$, such that $S_j \nsubseteq S_1$.

We say that S_1 is an active *separator for Ω if Ω is not a clique in this graph G^+. A pair of vertices $x, y \in \Omega$ that are not adjacent in G^+ is called an* active *pair. Note that, by Proposition 2, we must have $x, y \in S_1$.*

The following statement characterizes potential maximal cliques with active separators.

Proposition 4. *Let Ω be a potential maximal clique having an active separator $S \subset \Omega$, with an active pair $x, y \in S$. Denote by C the unique component of $G - S$ containing $\Omega \setminus S$. Then $\Omega \setminus S$ is a minimal x, y-separator in the graph $G[C \cup \{x, y\}]$.*

Again on the cube graph of Figure 2, for the potential maximal clique $\Omega_1 = \{a, e, g, c, h\}$, both minimal separators are active. E.g., for the minimal separator $S = \{a, e, g, c\}$ the pair $\{e, g\}$ is active. Not all potential maximal cliques have active separators, as illustrated by the potential maximal clique $\Omega_2 = \{a, c, f, h\}$ of the same graph.

Let us first focus on potential maximal cliques having an active separator. We give a result similar to Lemma 1, showing that such a potential maximal clique can be determined by a certain partition of the vertex cover W of G.

Lemma 2. *Let $G = (V, E)$ be a graph and W be a vertex cover of G. Consider a potential maximal clique Ω of G having an active separator $S \subseteq \Omega$ and an active pair $x, y \in S$. Let C be the unique connected component of $G - S$ intersecting Ω and let D_S be the union of all other connected components of $G - S$. Denote* by D_x the union of components of $G - \Omega$ contained in C, seeing x, by D_y the union of components of $G - \Omega$ contained in C not seeing x.

Now let $D_S^W = D_S \cap W$, $D_x^W = D_s \cap W$ and $D_y^W = D_y \cap W$.

Then one of the following holds:

1. There is a vertex $t \in \Omega$ such that $\Omega \setminus S = N(t) \cap C$.
2. There is a vertex $t \in \Omega$ such that $\Omega = N[t]$.
3. A vertex $z \notin W$ is in Ω if and only if
   (a) z sees D_S^W and $D_x^W \cup D_y^W$, or
   (b) z does not see D_S^W but is sees $D_x^W \cup \{x\}$, $D_y^W \cup \{y\}$ and $D_x^W \cup D_y^W$.

Proof. Note that D_x, D_y, D_S and Ω form a partition of the vertex set V.

We first prove that any vertex $z \notin W$ satisfying conditions 3a or 3b must be in Ω.

Consider first the case 3a when z sees D_S^W and $D_x^W \cup D_y^W$. So z sees D_S and C; we can apply Lemma 1 to partition (D_S, S, C) thus $z \in S$. Consider now the case 3b when z sees $D_x^W \cup D_y^W$, $D_x \cup \{x\}$ and $D_y \cup \{y\}$ but not D_S^W. Again by Lemma 1 applied to partition (D_S, S, C), vertex z cannot be in S. Since z has a neighbor in $D_x \cup D_y$, we have $z \in C$. Let $H = G[C \cup \{x, y\}]$ and $T = \Omega \cap C$ (thus we also have $T = \Omega \setminus S$). Recall that T is an x, y-minimal separator in H by Proposition 4. By definition of set D_x, we have that $D_x \cup \{x\}$ is exactly the component of $H - T$ containing x. Note that $D_y \cup \{y\}$ is the union of the component of $H - T$ containing y and of all other components of $H - T$ (that no not see x nor y). By applying Lemma 1 on graph H, with vertex cover $(W \cap C) \cup \{x, y\}$ and with partition $(D_x \cup \{x\}, T, D_y \cup \{y\})$ we deduce that $z \in T$.

Conversely, let $z \in \Omega \setminus W$. We must prove that either z satisfies conditions 3a or 3b, or we are in one of the first two cases of the Lemma. We distinguish the cases $z \in S$ and $z \in T$. When $z \in S$, by Lemma 1 applied to partition (D_S, S, C), z must see D_S and C. If z sees some vertex in $C \setminus \Omega$, we are done because z sees $D_x^W \cup D_y^W$ so we are in case 3a. Assume now that $N(z) \cap C \subseteq \Omega$, we prove that actually $N(z) \cap C = T = \Omega \setminus S$, so we are in case 1. Assume there is $u \in T \setminus N(z)$. By Proposition 2, there must be a connected component D of $G - \Omega$ such that $z, u \in N(D)$. Since $u \in C$, this component D must be a subset of C, so $D \subseteq C \setminus \Omega$. Together with $z \in N(D)$, this contradicts the assumption $N(z) \cap C \subseteq \Omega$.

It remains to treat the case $z \in T$. Clearly $z \in C$ cannot see D_S because S separates C from D_S. We again take graph H, with vertex cover $(W \cap C) \cup \{x, y\}$, and apply Lemma 1 with partition $(D_x \cup \{x\}, T, D_y \cup \{y\})$. We deduce that z sees both $D_x^W \cup \{x\}$ and $D_y^W \cup \{y\}$. Assume that z does not see $D_x^W \cup D_y^W$. So $N(z) \cap C \setminus \Omega = \emptyset$ thus $N[z] \subseteq \Omega$. If Ω contains some vertex $u \notin N[z]$, no component of $G - \Omega$ can see both z and u (because $N(z) \subseteq \Omega$), contradicting Proposition 2. We conclude that either z sees $D_x^W \cup D_y^W$ (so satisfies condition 3b) or $\Omega = N[z]$ (thus we are in the second case of the Lemma).   □

Theorem 2. *Any graph G has $\mathcal{O}^*(4^{vc(G)})$ potential maximal cliques. Moreover the set of its potential maximal cliques can be listed in $\mathcal{O}^*(4^{vc(G)})$ time.*

Proof. Let us first give the upper bound and the enumeration algorithm for potential maximal cliques with active separators.

The number of potential maximal cliques with active separators satisfying the second condition of Lemma 2 is at most n, and they can all be listed in polynomial time by checking, for each vertex t, if $N[t]$ is a potential maximal clique.

For enumerating the potential maximal cliques with active separators satisfying the first condition of Lemma 2, we enumerate all minimal separators S using Theorem 1, then for each $t \in S$ and each of the at most n components C of $G - S$ we check if $S \cup (C \cap N(t))$ is a potential maximal clique. Recall that testing if a vertex set is a potential maximal clique can be done in polynomial time by Corollary 1. Thus the whole process takes $\mathcal{O}^*(3^{vc(G)})$ time, and this is also an upper bound on the number of listed objects.

It remains to enumerate the potential maximal cliques with active separators satisfying the third condition of Lemma 2. For this purpose, we "guess" the sets D_S^W D_x^W, D_y^W as in the Lemma and then we compute Ω. More formally, for each four-partition $(D_S^W, D_x^W, D_y^W, \Omega^W)$ of W, we let $\Omega^{\overline{W}}$ be the set of vertices $z \notin W$ satisfying conditions 3a or 3b of Lemma 2, and we test using Corollary 1 if $\Omega = \Omega^W \cup \Omega^{\overline{W}}$ is indeed a potential maximal clique. By Lemma 2, this enumerates in $\mathcal{O}^*(4^{vc(G)})$ all potential maximal cliques of this type.

We have proven that G has $\mathcal{O}^*(4^{vc(G)})$ potential maximal cliques with active separators and these objects can be listed within the same running time. Due to space restrictions, the extension to all potential maximal cliques, including the ones with no active separators, is given in the full version [11]. □

4   Relations to Modular Width

A *module* of graph $G = (V, E)$ is a set of vertices W such that, for any vertex $x \in V \setminus W$, either $W \subseteq N(x)$ or W does not intersect $N(x)$. For the reader familiar with the modular decompositions of graphs, the modular width mw(G) of a graph G is the maximum size of a prime node in the modular decomposition tree. Equivalently, graph G is of modular width at most k if:

1. G has at most one vertex (the base case).
2. G is a disjoint union of graphs of modular width at most k.
3. G is a *join* of graphs of modular width at most k. I.e., G is obtained from a family of disjoint graphs of modular width at most k by taking the disjoint union and then adding all possible edges between these graphs.
4. The vertex set of G can be partitioned into $p \leq k$ modules $V_1, \ldots, V_p$ such that $G[V_i]$ is of modular width at most k, for all $i, 1 \leq i \leq p$.

The modular width of a graph can be computed in linear time, using e.g. [20]. Moreover, this algorithm outputs the algebraic expression of G corresponding to this grammar.

Let $G = (V, E)$ be a graph with vertex set $V = \{v_1, \ldots, v_k\}$ and let $M_i = (V_i, E_i)$ be a family of pairwise disjoint graphs, for all $i, 1 \leq i \leq k$. Denote by H the graph obtained from G by replacing each vertex v_i by the module M_i. I.e.,

$H = (V_1 \cup \cdots \cup V_k, E_1 \cup \cdots \cup E_k \cup \{ab \mid a \in V_i, b \in V_j \text{ s.t. } v_i v_j \in E\})$. We say that graph H has been obtained from G by *expanding* each vertex v_i by the module M_i.

A vertex subset W of H is an *expansion* of vertex subset W_G of G if $W = \cup_{v_i \in W_G} V_i$. Given a vertex subset W of H, the *contraction* of W is $\{v_i \mid V_i \text{ intersects } W\}$.

We prove in Lemma 3 (resp. Lemma 4) that each minimal separator (resp. each potential maximal clique of H) actually corresponds to a minimal separator (resp. potential maximal clique) of G or to a minimal separator (resp. potential maximal clique) of one of the modules M_i. Due to space restrictions, the proofs of these statements are given [11].

Lemma 3. *Let S be a minimal separator of H. One of the following holds :*

1. *S is the expansion of a minimal separator S_G of G.*
2. *There is $i \in \{1, \ldots, k\}$ such that $S \cap V_i$ is a minimal separator of M_i and $S \setminus V_i = N_H(V_i)$.*

Lemma 4. *Let Ω be a potential maximal clique of H. One of the following holds :*

1. *Ω is the expansion of a potential maximal clique Ω_G of G.*
2. *There is some $i \in \{1, \ldots, k\}$ such that $\Omega \cap V_i$ is a potential maximal clique of M_i and $\Omega \setminus V_i = N_H(V_i)$.*

Lemma 3 (resp. Lemma 4) provide an injective mapping from the set of minimal separators (resp. the set of potential maximal cliques) of H to the union of the sets of minimal separators (resp. of potential maximal cliques) of G and of the graphs M_i. Therefore we have:

Corollary 2. *The number of minimal separators (resp. of potential maximal cliques) of graph H is at most the number of minimal separators (resp. of potential maximal cliques) of G plus the number of minimal separators (resp. of potential maximal cliques) of each M_i.*

The following proposition bounds the number of minimal separators and potential maximal cliques of arbitrary graphs with respect to n.

Proposition 5 ([13,14]). *Every n-vertex graph has $\mathcal{O}(1.6181^n)$ minimal separators and $\mathcal{O}(1.7347^n)$ potential maximal cliques. Moreover, these objects can be enumerated within the same running times.*

We can now prove the main result of this section.

Theorem 3. *For any graph $G = (V, E)$, the number of its minimal separators is $\mathcal{O}(n \cdot 1.6181^{\mathrm{mw}(G)})$ and the number of its potential maximal cliques is $\mathcal{O}(n \cdot 1.7347^{\mathrm{mw}(G)})$. Moreover, the minimal separators and the potential maximal cliques can be enumerated in $\mathcal{O}^*(1.6181^{\mathrm{mw}(G)})$ and $\mathcal{O}^*(1.7347^{\mathrm{mw}(G)})$ time respectively.*

Proof. Let $k = \mathrm{mw}(G)$. By definition of modular width, there is a decomposition tree of graph G, each node corresponding to a leaf, a disjoint union, a join or a decomposition into at most k modules. The leaves of the decomposition tree are disjoint graphs with a single vertex, thus these vertices form a partition of V. There are at most n leaves and, since each internal node is of degree at least two, there are $O(n)$ nodes in the decomposition tree. For each node N, let $G(N)$ be the graph associated to the subtree rooted in N. We prove that $G(N)$ has $O(n(N) \cdot 1.6181^k)$ minimal separators and $O(n(N) \cdot 1.7347^k)$ potential maximal cliques, where $n(N)$ is the number of nodes of the subtree rooted in N. We proceed by induction from bottom to top. The statement is clear for leaves.

Let N be an internal node $N_1, N_2, \ldots, N_p$ be its sons in the tree. Graph $G(N)$ is the expansion of some graph $G'(N)$ by replacing the i-th vertex with module $G(N_i)$. If N is a *join* node, then $G'(N)$ is a clique. When N is a *disjoint union* node, graph $G'(N)$ is an independent set, and in the last case $G'(N)$ is a graph of at most k vertices. In all cases, by Proposition 5 graph $G'(N)$ has $O(1.6181^k)$ minimal separators. Thus $G(N)$ has at most $O(1.6181^k)$ more minimal separators than all its sons taken together, which completes our proof for minimal separators.

Concerning potential maximal cliques, when $G'(N)$ is a clique it has exactly one potential maximal clique, and when $G'(N)$ is of size at most k is has $O(1.7347^k)$ potential maximal cliques. We must be more careful in the case when $G'(N)$ is an independent set (i.e., N is a disjoint union node), since in this case it has p potential maximal cliques, one for each vertex, and p can be as large as n. Consider a potential maximal clique Ω of $G(N)$ corresponding to an expansion of vertices of $G'(N)$ (see Lemma 4). It follows that this potential maximal clique is exactly the vertex set of some $G(N_i)$, for a child N_i of N. By construction this vertex set is disconnected from the rest of $G(N)$, and by Proposition 2 the only possibility is that this vertex set induces a clique in $G(N)$. But in this case Ω is also a potential maximal clique of $G(N_i)$. This proves that, when N is of type disjoint union, $G(N)$ has no more potential maximal cliques than the sum of the numbers of potential maximal cliques of all its sons. Hence the whole graph G has $O(n \cdot 1.7347^k)$ potential maximal cliques. All arguments are constructive and can be turned into enumeration algorithms for these objects.   □

5 Applications

The *treewidth* of graph $G = (V, E)$, denoted tw(G), is the minimum number k such that G has a triangulation $H = (V, E')$ of clique size at most $k + 1$. The *minimum fill in* of G is the minimum size of F, over all (minimal) triangulations $H = (V, E \cup F)$ of G. The *treelength* of G is the minimum k such that there exists a minimal triangulation H, with the property that any two vertices adjacent in H are at distance at most k in graph G.

Proposition 6. *Let Π_G denote the set of potential maximal cliques of graph G. The following problems are solvable in $O^*(|\Pi_G|)$ time, when Π_G is given in the input : (WEIGHTED) TREEWIDTH [10,2], (WEIGHTED) MINIMUM FILL-IN [10,17], TREELENGTH [19].*

Recall the MAX INDUCED SUBGRAPH OF tw $\leq t$ SATISFIYING φ problem where, for a fixed integer t and a fixed CMSO$_2$ formula φ, the goal is to find a pair of vertex subsets $X \subseteq F \subseteq V$ such that tw$(G[F]) \leq t$, $(G[F], X)$ models φ and X is of maximum size.

Proposition 7 ([12]). *For any fixed integer $t > 0$ and any fixed CMSO$_2$ formula φ, problem* MAX INDUCED SUBGRAPH OF tw $\leq t$ SATISFIYING φ *is solvable in $\mathcal{O}(|\Pi_G| \cdot n^{t+4})$ time, when Π_G is given in the input.*

Problem MAX INDUCED SUBGRAPH OF tw $\leq t$ SATISFIYING φ generalizes many classical problems, for example MAXIMUM INDUCED FOREST, LONGEST INDUCED PATH, MAXIMUM INDUCED MATCHING, INDEPENDENT CYCLE PACKING, k-IN-A-PATH, k-IN-A-TREE, MAXIMUM INDUCED SUBGRAPH WITH A FORBIDDEN PLANAR MINOR. More examples of particular cases are given in the full version [11], see also [12]. From Theorems 2 and 3, we deduce:

Theorem 4. *Problems* MAX INDUCED SUBGRAPH OF tw $\leq t$ SATISFIYING φ, (WEIGHTED) TREEWIDTH, (WEIGHTED) MINIMUM FILL-IN *and* TREELENGTH *can be solved in time* $\mathcal{O}^*(4^{vc})$ *and in time* $\mathcal{O}^*(1.7347^{mw})$.

6   Conclusion

We have provided single exponential upper bounds for the number of minimal separators and the number of potential maximal cliques of graphs, with respect to parameters vertex cover and modular width.

A natural question is whether these results can be extended to other natural graph parameters. We point out that for parameters like clique-width or maximum leaf spanning tree, one cannot obtain upper bounds of type $\mathcal{O}^*(f(k))$ for any function f. A counterexample is provided by the graph $W_{p,q}$, formed by p disjoint paths of q vertices plus two vertices u and v seeing the left, respectively right ends of the paths (similar to the watermelon graph of Figure 2). Indeed this graph has a maximum leaf spanning tree with p leaves and a cliquewidth of no more than $2p + 1$, but it has roughly $(n/p)^p$ minimal u, v-separators.

Finally, we point out that our bounds on the number of potential maximal cliques w.r.t. vertex cover and to modular width do not seem to be tight. Any improvement on these bounds, together with faster enumeration algorithms for the potential maximal cliques, will immediately provide improved algorithms for the problems mentioned in Section 5.

References

1. Bodlaender, H.L., Fomin, F.V.: Tree decompositions with small cost. Discrete Applied Mathematics 145(2), 143–154 (2005)
2. Bodlaender, H.L., Rotics, U.: Computing the treewidth and the minimum fill-in with the modular decomposition. Algorithmica 36(4), 375–408 (2003)

3. Bouchitté, V., Todinca, I.: Treewidth and minimum fill-in: Grouping the minimal separators. SIAM J. Comput. 31(1), 212–232 (2001)
4. Bouchitté, V., Todinca, I.: Listing all potential maximal cliques of a graph. Theor. Comput. Sci. 276(1-2), 17–32 (2002)
5. Chapelle, M., Liedloff, M., Todinca, I., Villanger, Y.: Treewidth and pathwidth parameterized by the vertex cover number. In: Dehne, F., Solis-Oba, R., Sack, J.-R. (eds.) WADS 2013. LNCS, vol. 8037, pp. 232–243. Springer, Heidelberg (2013)
6. Courcelle, B.: The monadic second-order logic of graphs. I. Recognizable sets of finite graphs. Inf. Comput. 85(1), 12–75 (1990)
7. Courcelle, B., Makowsky, J.A., Rotics, U.: Linear time solvable optimization problems on graphs of bounded clique-width. Theory Comput. Syst. 33(2), 125–150 (2000)
8. Cygan, M., Lokshtanov, D., Pilipczuk, M., Pilipczuk, M., Saurabh, S.: On cutwidth parameterized by vertex cover. Algorithmica 68(4), 940–953 (2014)
9. Fellows, M.R., Lokshtanov, D., Misra, N., Rosamond, F.A., Saurabh, S.: Graph layout problems parameterized by vertex cover. In: Hong, S.-H., Nagamochi, H., Fukunaga, T. (eds.) ISAAC 2008. LNCS, vol. 5369, pp. 294–305. Springer, Heidelberg (2008)
10. Fomin, F.V., Kratsch, D., Todinca, I., Villanger, Y.: Exact algorithms for treewidth and minimum fill-in. SIAM J. Comput. 38(3), 1058–1079 (2008)
11. Fomin, F.V., Liedloff, M., Montealegre, P., Todinca, I.: Algorithms parameterized by vertex cover and modular width, through potential maximal cliques (2014), http://arxiv.org/abs/1404.3882
12. Fomin, F.V., Todinca, I., Villanger, Y.: Large induced subgraphs via triangulations and cmso. In: Chekuri, C. (ed.) SODA, pp. 582–583. SIAM (2014), http://arxiv.org/abs/1309.1559
13. Fomin, F.V., Villanger, Y.: Finding induced subgraphs via minimal triangulations. In: Marion, J.Y., Schwentick, T. (eds.) STACS. LIPIcs, vol. 5, pp. 383–394. Schloss Dagstuhl - Leibniz-Zentrum fuer Informatik (2010)
14. Fomin, F.V., Villanger, Y.: Treewidth computation and extremal combinatorics. Combinatorica 32(3), 289–308 (2012)
15. Frick, M., Grohe, M.: The complexity of first-order and monadic second-order logic revisited. Ann. Pure Appl. Logic 130(1-3), 3–31 (2004)
16. Gajarský, J., Lampis, M., Ordyniak, S.: Parameterized algorithms for modular-width. In: Gutin, G., Szeider, S. (eds.) IPEC 2013. LNCS, vol. 8246, pp. 163–176. Springer, Heidelberg (2013)
17. Gysel, R.: Potential maximal clique algorithms for perfect phylogeny problems. CoRR, abs/1303.3931 (2013)
18. Lampis, M.: Algorithmic meta-theorems for restrictions of treewidth. Algorithmica 64(1), 19–37 (2012)
19. Lokshtanov, D.: On the complexity of computing treelength. Discrete Applied Mathematics 158(7), 820–827 (2010)
20. Tedder, M., Corneil, D.G., Habib, M., Paul, C.: Simpler linear-time modular decomposition via recursive factorizing permutations. In: Aceto, L., Damgård, I., Goldberg, L.A., Halldórsson, M.M., Ingólfsdóttir, A., Walukiewicz, I. (eds.) ICALP 2008, Part I. LNCS, vol. 5125, pp. 634–645. Springer, Heidelberg (2008)

Win-Win Kernelization for Degree Sequence Completion Problems

Vincent Froese*, André Nichterlein, and Rolf Niedermeier

Institut für Softwaretechnik und Theoretische Informatik, TU Berlin, Germany
{vincent.froese,andre.nichterlein,rolf.niedermeier}@tu-berlin.de

Abstract. We study the kernelizability of a class of NP-hard graph modification problems based on vertex degree properties. Our main positive results refer to NP-hard graph completion (that is, edge addition) cases while we show that there is no hope to achieve analogous results for the corresponding vertex or edge deletion versions. Our algorithms are based on a method that transforms graph completion problems into efficiently solvable number problems and exploits f-factor computations for translating the results back into the graph setting. Indeed, our core observation is that we encounter a win-win situation in the sense that either the number of edge additions is small (and thus faster to find) or the problem is polynomial-time solvable. This approach helps in answering an open question by Mathieson and Szeider [JCSS 2012].

1 Introduction

In this work, we propose a general approach for achieving polynomial-size problem kernels for a class of graph completion problems where the goal graph has to fulfill certain degree properties. Thus, we explore and enlarge results on provably effective polynomial-time preprocessing for these NP-hard graph problems. To a large extent, the initial motivation for our work comes from studying the NP-hard graph modification problem DEGREE CONSTRAINT EDITING(S) for non-empty subsets $S \subseteq \{v^-, e^+, e^-\}$ of editing operations (v^-: "vertex deletion", e^+: "edge addition", e^-: "edge deletion") as introduced by Mathieson and Szeider [22].[1] The definition reads as follows.

> DEGREE CONSTRAINT EDITING(S) (DCE(S))
> **Input:** An undirected graph $G = (V, E)$, two integers $k, r > 0$, and a "degree list function" $\tau \colon V \to 2^{\{0,\dots,r\}}$.
> **Question:** Is it possible to obtain a graph $G' = (V', E')$ from G using at most k editing operations of type(s) as specified by S such that $\deg_{G'}(v) \in \tau(v)$ for all $v \in V'$?

---

* Supported by DFG, project DAMM (NI 369/13).

[1] Mathieson and Szeider [22] originally introduced a weighted version of the problem, where the vertices and edges can have positive integer weights incurring a cost for each editing operation. Here, we focus on the unweighted version.

R Ravi and I.L. Gørtz (Eds.): SWAT 2014, LNCS 8503, pp. 194–205, 2014.
© Springer International Publishing Switzerland 2014

In our work, the set S always consists of a single editing operation. Our studies focus on the two most natural parameters: the number k of editing operations and the maximum allowed degree r. We will show that, although all three variants are NP-hard, $DCE(e^+)$ is amenable to a generic kernelization method we propose. This method is based on dynamic programming solving a corresponding number problem and f-factor computations. For $DCE(e^-)$ and $DCE(v^-)$, however, we show that there is little hope to achieve analogous results.

Previous Work. There are basically two fundamental starting points for our work. First, there is our previous theoretical work on degree anonymization in social networks [15] motivated and strongly inspired by a preceding heuristic approach due to Liu and Terzi [19]. Indeed, our previous work for degree anonymization very recently inspired empirical work with encouraging experimental results [16]. A fundamental contribution of this work now is to systematically reveal what the problem-specific parts (tailored towards degree anonymization) and what the "general" parts of that approach are. In this way, we develop this approach into a general method of significantly wider applicability for a large number of graph completion problems based on degree properties. The second fundamental starting point is Mathieson and Szeider's work [22] on $DCE(S)$. They showed several parameterized preprocessing (also known as kernelization) results and left open whether it is possible to reduce $DCE(e^+)$ in polynomial time to a problem kernel of size polynomial in r —we will affirmatively answer this question. Finally, Golovach [13] achieved a number of kernelization results for closely related graph editing problems; his methods, however, significantly differ from ours.

From a more general perspective, all these considerations fall into the category of "graph editing to fulfill degree constraints", which recently received increased interest in terms of parameterized complexity analysis [10, 13, 23].

Our Contributions. Answering an open question of Mathieson and Szeider [22], we present an $O(kr^2)$-vertex kernel for $DCE(e^+)$ which we then transfer into an $O(r^5)$-vertex kernel using a strategy rooted in previous work [15, 19]. A further main contribution of our work in the spirit of meta kernelization [2] is to clearly separate problem-specific from problem-independent aspects of this strategy, thus making it accessible to a wider class of degree sequence completion problems. We observe that in case that the goal graph shall have "small" maximum degree r, then the actual graph structure is in a sense negligible and thus allows for a lot of freedom that can be algorithmically exploited. This paves the way to a *win-win situation* of either having guaranteed a small number of edge additions or the overall problem being solvable in polynomial-time anyway.

Besides our positive kernelization results, we exclude polynomial-size problem kernels for $DCE(e^-)$ and $DCE(v^-)$ subject to the assumption that NP $\not\subseteq$ coNP/poly, thereby showing that the exponential-size kernel results by Mathieson and Szeider [22] are essentially tight. In other words, this demonstrates that in our context edge completion is much more amenable to kernelization than edge deletion or vertex deletion are. We also prove NP-hardness of $DCE(v^-)$ and $DCE(e^+)$ for graphs of maximum degree three, implying that the maximum degree is not a useful parameter for kernelization purposes. Last but not least,

we develop a general preprocessing approach for DEGREE SEQUENCE COMPLE-
TION problems which yields a search space size that is polynomially bounded
in the parameter. While this per se does not give polynomial kernels, we derive
fixed-parameter tractability with respect to the combined parameter maximum
degree and solution size. The usefulness of our method is illustrated by further
example degree sequence completion problems.

Notation. All graphs in this paper are undirected, loopless, and simple (that is,
without multiple edges). For a graph $G = (V, E)$, we set $n := |V|$ and $m := |E|$.
The degree of a vertex $v \in V$ is denoted by $\deg_G(v)$, the maximum vertex degree
by Δ_G, and the minimum vertex degree by δ_G. For a finite set U, we denote
with $\binom{U}{2}$ the set of all size-two subsets of U. We denote by $\overline{G} := (V, \binom{V}{2} \setminus E)$
the complement graph of G. For a vertex subset $V' \subseteq V$, the subgraph induced
by V' is denoted by $G[V']$. For an edge subset $E' \subseteq \binom{V}{2}$, $V(E')$ denotes the set
of all endpoints of edges in E' and $G[E'] := (V(E'), E')$. For a set E' of edges
with endpoints in a graph G, we denote by $G + E' := (V, E \cup E')$ the graph
that results by inserting all edges in E' into G. Similarly, we define for a vertex
set $V' \subseteq V$, the graph $G - V' := G[V \setminus V']$. For each vertex $v \in V$, we denote
by $N_G(v)$ the open neighborhood of v in G and by $N_G[v] := N_G(v) \cup \{v\}$ the
closed neighborhood. We omit subscripts if the corresponding graph is clear from
the context. A vertex $v \in V$ with $\deg(v) \in \tau(v)$ is called *satisfied* (otherwise
unsatisfied). We denote by $U \subseteq V$ the set of all unsatisfied vertices, formally
$U := \{v \in V \mid \deg_G(v) \notin \tau(v)\}$.

Parameterized Complexity. This is a two-dimensional framework for studying
computational complexity [8, 11, 24]. One dimension of a parameterized problem
is the input size s, and the other one is the *parameter* (usually a positive integer).
A parameterized problem is called *fixed-parameter tractable* (fpt) with respect to
a parameter ℓ if it can be solved in $f(\ell) \cdot s^{O(1)}$ time, where f is a computable function only depending on ℓ. This definition also extends to *combined parameters*.
Here, the parameter usually consists of a tuple of positive integers $(\ell_1, \ell_2, \ldots)$
and a parameterized problem is called fpt with respect to $(\ell_1, \ell_2, \ldots)$ if it can be
solved in $f(\ell_1, \ell_2, \ldots) \cdot s^{O(1)}$ time.

A core tool in the development of fixed-parameter algorithms is polynomial-
time preprocessing by *data reduction* [1, 14, 20]. Here, the goal is to transform a
given problem instance I with parameter ℓ in polynomial time into an equivalent
instance I' with parameter $\ell' \leq \ell$ such that the size of I' is upper-bounded by
some function g only depending on ℓ. If this is the case, we call I' a (problem)
kernel of size $g(\ell)$. If g is a polynomial, then we speak of a *polynomial kernel*.
Usually, this is achieved by applying polynomial-time executable data reduction
rules. We call a data reduction rule $\mathcal{R}$ *correct* if the new instance I' that results
from applying $\mathcal{R}$ to I is a yes-instance if and only if I is a yes-instance. The whole
process is called *kernelization*. It is well known that a parameterized problem is
fixed-parameter tractable if and only if it has a problem kernel.

Due to a lack of space several proofs are deferred to a full version.[2]

---

[2] Available on arXiv:1404.5432.

2   Degree Constraint Editing

Mathieson and Szeider [22] showed fixed-parameter tractability of DCE(S) for all non-empty subsets $S \subseteq \{v^-, e^-, e^+\}$ with respect to the combined parameter (k, r) and W[1]-hardness with respect to the single parameter k. The fixed-parameter tractability is in a sense tight as Cornuéjols [7] proved that DCE(e^-) is NP-hard on planar graphs with maximum degree three and with $r = 3$ and thus presumably not fixed-parameter tractable with respect to r. We complement his result by showing that DCE(v^-) is NP-hard on cubic (that is three-regular) planar graphs, even if $r = 0$, and that DCE(e^+) is NP-hard on graphs with maximum degree three.

Theorem 1. DCE(v^-) *is NP-hard on cubic planar graphs, even if $r = 0$.*

Proof (Sketch). We provide a polynomial-time many-one reduction from the NP-hard VERTEX COVER on cubic planar graphs [12]. Let $(G = (V, E), h)$ be a VERTEX COVER instance with the cubic planar graph G. It is not hard to see that (G, h) is a yes-instance of VERTEX COVER if and only if $(G, h, 0, \tau)$ with $\tau(v) = \{0\}$ for all $v \in V$ is a yes-instance of DCE(v^-).   □

Theorem 2. DCE(e^+) *is NP-hard on planar graphs with maximum degree three.*

In contrast to DCE(e^-) and DCE(v^-), unless P = NP, DCE(e^+) cannot be NP-hard for constant values of r since we later show fixed-parameter tractability for DCE(e^+) with respect to the parameter r.

Excluding Polynomial Kernels. Mathieson and Szeider [22] gave exponential-size problem kernels for DCE(v^-) and DCE($\{v^-, e^-\}$) with respect to the combined parameter (k, r). We prove that these results are tight in the sense that, under standard complexity-theoretic assumptions, neither DCE(e^-) nor DCE(v^-) admits a polynomial-size problem kernel when parameterized by (k, r).

Theorem 3. DCE(e^-) *does not admit a polynomial-size problem kernel with respect to (k, r) unless NP $\subseteq$ coNP/poly.*

Theorem 4. DCE(v^-) *does not admit a polynomial-size problem kernel with respect to (k, r) unless NP $\subseteq$ coNP/poly.*

Having established these computational lower bounds, we now show that in contrast to DCE(e^-) and DCE(v^-), DCE(e^+) admits a polynomial kernel.

2.1   A Polynomial Kernel for DCE(e^+) with Respect to (k, r)

In order to describe the kernelization, we need some further notation: For $i \in \{0, \ldots, r\}$, a vertex $v \in V$ is of *type i* if and only if $\deg(v) + i \in \tau(v)$, that is, v can be satisfied by adding i edges to it. The set of all vertices of type i is denoted by T_i. Observe that a vertex can be of multiple types, implying that for $i \neq j$ the vertex sets T_i and T_j are not necessarily disjoint. Furthermore, notice that the

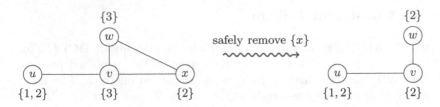

Fig. 1. An example for safely removing a vertex from a graph. The sets next to the vertices denote the degree lists defined by τ. Observe that in both graphs u is of type zero and of type one, v is of type zero, and w is of type one.

type-0 vertices are exactly the satisfied ones. We remark that there are instances for $\mathrm{DCE}(e^+)$ where we might have to add edges between two satisfied vertices (though this may seem counter-intuitive): Consider, for example, a three-vertex graph without any edges, the degree list function values are $\{2\}, \{0, 2\}, \{0, 2\}$, and $k = 3$. The two vertices with degree list $\{0, 2\}$ are satisfied. However, the only solution for this instance is to add *all* edges.

Now, we can describe our kernelization algorithm: The basic strategy is to keep the unsatisfied vertices U and "enough" arbitrary vertices of each type (from the satisfied vertices) and delete all other vertices. The idea behind the correctness is that the vertices in a solution are somehow "interchangeable". If an unsatisfied vertex needs an edge to a satisfied vertex of type i, then it is not important which satisfied type-i vertex is used. We only have to take care not to "reuse" the satisfied vertices to avoid the creation of multiple edges.

Next, we specify what we mean by "enough" vertices: The "magic number" is $\alpha := k(\Delta_G + 2)$. This leads to the definition of α-*type sets*: An α-type set $C \subseteq V$ is a vertex subset containing all unsatisfied vertices U and $\min\{\alpha, |T_i \setminus U|\}$ type-i vertices from $T_i \setminus U$ for each $i \in \{1, \dots, r\}$. We will soon show that for any fixed α-type set C, deleting all vertices in $V \setminus C$ results in an equivalent instance. However, deleting a vertex changes the degrees of its neighbors. Thus, we also have to adjust their degree lists. Formally, for a vertex subset $V' \subseteq V$, we define $\tau_{V'} : (V \setminus V') \to 2^{\{0, \dots, r\}}$, where for each $u \in V \setminus V'$, we set $\tau_{V'}(u) := \{d \in \mathbb{N} \mid d + |N_G(u) \cap V'| \in \tau(u)\}$. Then, *safely removing* a vertex set $V' \subseteq V$ from the instance (G, k, r, τ) means to replace the instance with $(G - V', k, r, \tau_{V'})$, see Figure 1 for an example. With these definitions we can provide our reduction rules leading to a polynomial-size problem kernel.

Reduction Rule 1. Let $(G = (V, E), k, r, \tau)$ be an instance of $\mathrm{DCE}(e^+)$ and let $C \subseteq V$ be an α-type set in G. Then, safely remove all vertices in $V \setminus C$.

Lemma 1. *Reduction Rule 1 is correct and can be applied in linear time.*

As each α-type set contains at most α satisfied vertices of each vertex type, it follows that after one application of Reduction Rule 1 the graph contains at most $|C| = |U| + r\alpha$ vertices. The number of unsatisfied vertices in an α-type set can always be bounded by $|U| \leq 2k$ since we can increase the degrees of at most $2k$ vertices by adding k edges. If there are more unsatisfied vertices, then we return a trivial no-instance. Thus, we end up with $|C| \leq 2k + rk(\Delta_G + 2)$. To obtain a polynomial-size problem kernel with respect to the combined parameter (k, r), we need to bound Δ_G. However, this can easily be achieved: Since we only allow edge additions, for each vertex $v \in V$, we have $\deg(v) \leq \max \tau(v) \leq r$. Formalized as a data reduction rule, this reads as follows:

Reduction Rule 2. Let $(G = (V, E), k, r, \tau)$ be an instance of $\mathrm{DCE}(e^+)$. If G contains more than $2k$ unsatisfied vertices or if there exists a vertex $v \in V$ with $\deg(v) > \max \tau(v)$, then return a trivial no-instance.

Having applied Reduction Rules 1 and 2 once, it holds that $\Delta_G \leq r$ and thus the graph contains at most $2k + rk(r+2)$ vertices. Lemma 1 ensures that we can apply Reduction Rule 1 in linear time. Note that linear time means $O(m + |\tau|)$ time, where $|\tau| \geq n$ denotes the encoding size of τ. Clearly, Reduction Rule 2 can be applied in linear time too. This leads to the following.

Theorem 5. $\mathrm{DCE}(e^+)$ *admits a problem kernel containing* $O(kr^2)$ *vertices computable in* $O(m + |\tau|)$ *time.*

2.2 A Polynomial Kernel for $\mathrm{DCE}(e^+)$ with Respect to r

In this subsection, we show how to extend the polynomial-size problem kernel provided in Theorem 5 to a polynomial-size problem kernel for the single parameter r. To this end, among other things, we adapt some ideas of Hartung et al. [15] to show how to bound k in a polynomial of r. The general strategy, inspired by a heuristic of Liu and Terzi [19], will be as follows: First, remove the graph structure and solve the problem on the degree sequence of the input graph by using dynamic programming. The solution to this number problem will indicate the *demand* for each vertex, that is, the number of added edges incident to that vertex. Then, using a result of Katerinis and Tsikopoulos [17], we prove that either $k \leq r(r + 1)^2$ or we can find a set of edges satisfying the specified demands in polynomial time.

We start by formally defining the corresponding number problem and showing its polynomial-time solvability.

> NUMBER CONSTRAINT EDITING (NCE)
> **Input:** A function $\phi : \{1, \ldots, n\} \to 2^{\{0, \ldots, r\}}$ and positive integers $d_1, \ldots, d_n, k, r$.
> **Question:** Are there n positive integers $d'_1, \ldots, d'_n$ such that $\sum_{i=1}^{n} (d'_i - d_i) = k$ and for all $i \in \{1, \ldots, n\}$ it holds that $d'_i \geq d_i$ and $d'_i \in \phi(i)$?

Lemma 2. NCE *is solvable in* $O(n \cdot k \cdot r)$ *time.*

Lemma 2 can be proved with a dynamic program that specifies the demand for each vertex, that is, the number of added edges incident to each vertex. Given these demands, the remaining problem is to decide whether there exists a set of edges that satisfy these demands and are not contained in the input graph G. This problem is closely related to the polynomial-time solvable f-FACTOR problem [21], a special case of $DCE(e^-)$ where $|\tau(v)| = 1$ for all $v \in V$; it is formally defined as follows:

> f-FACTOR
> **Input:**   A graph $G = (V, E)$ and a function $f: V \to \mathbb{N}_0$.
> **Question:** Is there an f-factor, that is, a subgraph $G' = (V, E')$ of G
>   such that $\deg_{G'}(v) = f(v)$ for all $v \in V$?

Observe that our problem of satisfying the demands of the vertices in G is essentially the question whether there is an f-factor in the complement graph $\overline{G}$ where the function f stores the demand of each vertex. Using a result of Katerinis and Tsikopoulos [17], we can show the following lemma about the existence of an f-factor:

Lemma 3. *Let $G = (V, E)$ be a graph with n vertices, $\delta_G \geq n - r - 1, r \geq 1$, and let $f: V \to \{1, \ldots, r\}$ be a function such that $\sum_{v \in V} f(v)$ is even. If $n \geq (r+1)^2$, then G has an f-factor.*

We now have all ingredients to show that we can upper-bound k by $r(r + 1)^2$ or solve the given instance of $DCE(e^+)$ in polynomial time. The main technical statement towards this is the following.

Lemma 4. *Let $I := (G = (V, E), k, r, \tau)$ be an instance of $DCE(e^+)$ with $k \geq r(r + 1)^2$ and $V = \{v_1, \ldots, v_n\}$. If there exists a $k' \in \{r(r + 1)^2, \ldots, k\}$ such that $(\deg(v_1), \ldots, \deg(v_n), 2k', r, \phi)$ with $\phi(i) := \tau(v_i)$ is a yes-instance of NCE, then I is a yes-instance of $DCE(e^+)$.*

Proof. Assume that $(\deg(v_1), \ldots, \deg(v_n), 2k', r, \phi)$ is a yes-instance of NCE. Let $d'_1, \ldots, d'_n$ be integers such that $d'_i \in \tau(v_i), \sum_{i=1}^{n} d'_i - \deg(v_i) = 2k'$, and $d'_i \geq d_i$. Hence, we know that the degree constraints can numerically be satisfied, giving rise to a new target degree d'_i for each vertex v_i. Let $A := \{v_i \in V \mid d'_i > \deg(v_i)\}$ denote the set of *affected* vertices containing all vertices which require addition of some edges. It remains to show that the degree sequence of the affected vertices can in fact be realized by adding k' edges to $G[A]$. To this end, it is sufficient to prove the existence of an f-factor in the complement graph $\overline{G[A]}$ with $f(v_i) := d'_i - \deg(v_i) \in \{1, \ldots, r\}$ for all $v_i \in A$ since such an f-factor contains exactly the k' edges we want to add to G. Thus, it remains to check that all conditions of Lemma 3 are indeed satisfied to conclude the existence of the sought f-factor. First, note that $\delta_{\overline{G[A]}} \geq |A| - r - 1$ since $\Delta_{G[A]} \leq r$. Moreover, $\sum_{v_i \in A}(d'_i - \deg(v_i)) = 2k' \leq |A|r$, and thus $|A| \geq 2k'/r \geq 2(r+1)^2$. Finally, $\sum_{v_i \in A} f(v_i) = 2k'$ is even and thus Lemma 3 applies. $\qquad \square$

As NCE is polynomial-time solvable, Lemma 4 states a win-win situation: either the solution is bounded in size or can be found in polynomial time. From this and Theorem 5, we obtain the polynomial-size problem kernel.

Theorem 6. *$DCE(e^+)$ admits a problem kernel containing $O(r^5)$ vertices computable in $O(k^2 \cdot r \cdot n + m + |\tau|)$ time.*

3   A General Approach for Degree Sequence Completion

In the previous section, we dealt with the problem $DCE(e^+)$, where one only has to *locally* satisfy the degree of each vertex. In this section, we show how the presented ideas for $DCE(e^+)$ can also be used to solve more *globally* defined problems where the degree sequence of the solution graph G' has to fulfill a given property. For example, consider the problem of adding a minimum number of edges to obtain a regular graph, that is, a graph where all vertices have the same degree. In this case the degree of a vertex in the solution is a priori not known but depends on the degrees of the other vertices.

The *degree sequence* of a graph G with n vertices is an n-tuple containing the vertex degrees. Then, for some tuple property Π, we consider the following problem:

> Π-DEGREE SEQUENCE COMPLETION (Π-DSC)
> **Input:**   A graph $G = (V, E)$, an integer $k \in \mathbb{N}$.
> **Question:** Is there a set of edges $E' \subseteq \binom{V}{2} \setminus E$ with $|E'| \leq k$ such that the degree sequence of $G + E'$ fulfills Π?

Note that Π-DSC is not a generalization of $DCE(e^+)$ since in $DCE(e^+)$ one can require for two vertices u and v of the same degree that u gets two more incident edges and v not. This cannot be expressed in Π-DSC. We remark that the results stated in this section can be extended to hold for a generalized version of Π-DSC where a "degree list function" τ is given as additional input and the vertices in the solution graph G' also have to satisfy τ, thus generalizing $DCE(e^+)$. For simplicity, however, we stick to the easier problem definition as stated above.

3.1   Fixed-Parameter Tractability of Π-DSC

In this subsection, we first generalize the ideas behind Theorem 5 to show fixed-parameter tractability of Π-DSC with respect to the combined parameter (k, Δ_G). Then, we present an adjusted version of Lemma 4 and apply it to show fixed-parameter tractability for Π-DSC with respect to the parameter $\Delta_{G'}$. Clearly, a prerequisite for both these results is that the following problem has to be fixed-parameter tractable with respect to the parameter $\Delta_T :=$ $\max\{d_1, \ldots, d_n\}$.

> Π-DECISION
> **Input:**   An integer tuple $T = (d_1, \ldots, d_n)$.
> **Question:** Does T fulfill Π?

For the next result, we need some definitions. For $0 \leq d \leq \Delta_G$, let $D_G(d) :=$ $\{v \in V \mid \deg_G(v) = d\}$ be the *block* of degree d, that is, the set of all vertices with degree d in G. A subset $V' \subseteq V$ is an *α-block set* if V' contains for every $d \in \{1, \ldots, \Delta_G\}$ exactly $\min\{\alpha, |D_G(d)|\}$ vertices. Recall that $\alpha = (\Delta_G + 2)k$, see Section 2.1, and notice the similarity of α-block sets and α-type sets. This similarity is not a coincidence for we use ideas of Reduction Rule 1 and Lemma 1 to obtain the following lemma.

Lemma 5. *Let $I := (G = (V, E), k)$ be a yes-instance of Π-DSC and let $C \subseteq V$ be an α-block set. Then, there exists a set of edges $E' \subseteq \binom{C}{2} \setminus E$ with $|E'| \leq k$ such that the degree sequence of $G + E'$ fulfills Π.*

In the context of $\mathrm{DCE}(S)$, we introduced the notion of safely removing a vertex subset to obtain a problem kernel. On the contrary, in the context of Π-DSC, it seems impossible to remove vertices in general without further knowledge about the tuple property Π. Thus, Lemma 5 does not lead to a problem kernel but only to a reduced search space for a solution, namely any α-block set. Clearly, an α-block set C can be computed in polynomial time. Then, one can simply try out all possibilities to add edges with endpoints in C and check whether in one of the cases the degree sequence of the resulting graph satisfies Π. As $|C| \leq (\Delta_G + 2)k\Delta_G$, there are at most $O(2^{((\Delta_G+2)k\Delta_G)^2})$ possible subsets of edges to add. Overall, this leads to the following theorem.

Theorem 7. *Let Π be some tuple property. If Π-DECISION is fixed-parameter tractable with respect to Δ_T, then Π-DSC is fixed-parameter tractable with respect to (k, Δ_G).*

Bounding the Solution Size k in $\Delta_{G'}$. We now show how to extend the ideas of Section 2.2 to the context of Π-DSC in order to bound the solution size k by a polynomial in $\Delta_{G'}$. The general procedure still is the one inspired by Liu and Terzi [19]: Solve the number problem corresponding to Π-DSC on the degree sequence of the input graph and then try to "realize" the solution. To this end, we define the corresponding number problem as follows:

Π-NUMBER SEQUENCE COMPLETION (Π-NSC)
Input: Positive integers $d_1, \ldots, d_n, k, \Delta$.
Question: Are there n nonnegative integers $x_1, \ldots, x_n$ with $\sum_{i=1}^n x_i = k$ such that $(d_1 + x_1, \ldots, d_n + x_n)$ fulfills Π and $d_i + x_i \leq \Delta$?

With these problem definitions, we can now generalize Lemma 4.

Lemma 6. *Let $I := (G, k)$ be an instance of Π-DSC with $V = \{v_1, \ldots, v_n\}$ and $k \geq \Delta_{G'}(\Delta_{G'} + 1)^2$. If there exists a $k' \in \{\Delta_{G'}(\Delta_{G'} + 1)^2, \ldots, k\}$ such that the corresponding Π-NSC instance $I' := (\deg(v_1), \ldots, \deg(v_n), 2k', \Delta_{G'})$ is a yes-instance, then I is a yes-instance.*

Let function $g(|I|)$ denote the running time for solving the Π-NSC instance I. Clearly, if there is a solution for an instance of Π-DSC, then there also exists a solution for the corresponding Π-NSC instance. It follows that we can decide whether there is a large solution for Π-DSC (with at least $\Delta_{G'}(\Delta_{G'} + 1)^2$ edges) in $k \cdot g(n \log(n))$ time. Hence, we arrive at the following win-win situation:

Corollary 1. *Let $I := (G, k)$ be an instance of Π-DSC. Then, either one can decide in $k \cdot g(n \log(n))$ time that I is a yes-instance, or I is a yes-instance if and only if $(G, \min\{k, \Delta_{G'}(\Delta_{G'} + 1)^2\})$ is a yes-instance.*

Using Corollary 1, we can transfer fixed-parameter tractability of Π-NSC with respect to Δ to fixed-parameter tractability of Π-DSC with respect to $\Delta_{G'}$. Notice that $\Delta_{G'} \leq k + \Delta_G$, that is, $\Delta_{G'}$ is a smaller and thus "stronger" parameter [18]. Also, showing Π-NSC to be fixed-parameter tractable with respect to Δ is a significantly easier task than proving fixed-parameter tractability for Π-DSC with respect to $\Delta_{G'}$ directly since the graph structure can be completely ignored.

Theorem 8. *If Π-NSC is fixed-parameter tractable with respect to Δ, then Π-DSC is fixed-parameter tractable with respect to $\Delta_{G'}$.*

If Π-NSC can be solved in polynomial time, then Corollary 1 shows that we can assume that $k \leq \Delta_{G'}(\Delta_{G'} + 1)^2$. Thus, as in the DCE(e$^+$) setting (Theorem 6), polynomial kernels with respect to (k, Δ_G) transfer to the parameter $\Delta_{G'}$, leading to the following.

Theorem 9. *If Π-NSC is polynomial-time solvable and Π-DSC admits a polynomial kernel with respect to (k, Δ_G), then Π-DSC also admits a polynomial kernel with respect to $\Delta_{G'}$.*

3.2 Applications

As our general approach is inspired by ideas of Hartung et al. [15], it is not surprising that it can be applied to "their" DEGREE ANONYMITY problem, where given an undirected graph $G = (V, E)$ and two positive integers k and s, one seeks an edge set E' over V of size at most s such that $G' := G + E'$ is k-*anonymous*, that is, for each vertex $v \in V$, there are at least $k - 1$ other vertices in G' having the same degree. The property Π of being k-anonymous clearly can be decided in polynomial time for a given degree sequence, and thus, by Theorem 7, we immediately get fixed-parameter tractability with respect to (s, Δ_G). Theorem 9 then basically yields the kernel results obtained by Hartung et al. [15]. There are more general versions of DEGREE ANONYMITY as proposed by Chester et al. [6]. For example, just a given subset of the vertices has to be anonymized or the vertices can have labels. As in each of these generalizations one can decide in polynomial time whether a given graph satisfies the particular anonymity requirement, Theorem 7 applies also in these scenarios. However, checking in which of these more general settings the conditions of Theorem 8 or Theorem 9 are fulfilled is future work.

Besides the graph anonymization setting, one could think of further, more general constraints on the degree sequence. For example, if $p_i(\mathcal{D})$ denotes how often degree i appears in a degree sequence $\mathcal{D}$, then being k-anonymous translates into $p_i(\mathcal{D}_{G'}) \geq k$ for all degrees i occurring in the degree sequence $\mathcal{D}_{G'}$ of the modified graph G'. Now, it is natural to consider not only a lower bound $k \leq p_i(\mathcal{D})$, but also an upper bound $p_i(\mathcal{D}) \leq u$ or maybe even a set of allowed frequencies $p_i(\mathcal{D}) \in F_i \subseteq \mathbb{N}$. Constraints like this allow to express properties not of individual degrees but of the whole distribution of the degrees in the resulting sequence. For example, in order to have some "balancedness" one can require that each occurring degree occurs exactly ℓ times for some $\ell \in \mathbb{N}$ [5]. To obtain some sort of "robustness" it might be useful to ask for an h-index of ℓ, that is, in the solution graph there are at least ℓ vertices with degree at least ℓ [9].

Another range of problems which fit naturally into our framework involves completion problems to a graph class that is completely characterized by degree sequences. For example, a graph is a *unigraph* if it is determined by its degree sequence up to isomorphism [4]. Given a degree sequence $\mathcal{D} = (d_1, \ldots, d_n)$, one can decide in linear time whether $\mathcal{D}$ defines a unigraph [3]. Thus, by Theorem 8, we conclude fixed-parameter tractability for the unigraph completion problem with respect to the parameter $\Delta_{G'}$.

4   Conclusion

We proposed a method for deriving efficient preprocessing algorithms for degree sequence completion problems. DCE(e^+) served as our main illustrating example. Roughly speaking, the core of the approach (as basically already used in previous work [15, 19]) consists of extracting the degree sequence from the input graph, efficiently solving a simpler number editing problem, and translating the obtained solution back into a solution for the graph problem using f-factors. While previous work [15, 19] was specifically tailored towards an application for graph anonymization, we generalized the approach by filtering out problem-specific parts and "universal" parts. Thus, whenever one can solve these problem-specific parts efficiently, we can automatically obtain efficient preprocessing and fixed-parameter tractability results.

Our approach seems promising for future empirical investigations concerning its practical usefulness, a very recent experimental work has been performed for DEGREE ANONYMITY [16]. Another line of future research could be to study polynomial-time approximation algorithms for the considered degree sequence completion problems. Perhaps parts of our preprocessing approach might find use here as well. A more specific open question concerning our work would be how to deal with additional connectivity requirements for the generated graphs.

References

[1] Bodlaender, H.L.: Kernelization: New upper and lower bound techniques. In: Chen, J., Fomin, F.V. (eds.) IWPEC 2009. LNCS, vol. 5917, pp. 17–37. Springer, Heidelberg (2009)

[2] Bodlaender, H.L., Fomin, F.V., Lokshtanov, D., Penninkx, E., Saurabh, S., Thilikos, D.M.: (Meta) kernelization. In: Proc. 50th FOCS, pp. 629–638. IEEE (2009)

[3] Borri, A., Calamoneri, T., Petreschi, R.: Recognition of unigraphs through superposition of graphs. J. Graph Algorithms Appl. 15(3), 323–343 (2011)

[4] Brandstädt, A., Le, V.B., Spinrad, J.P.: SIAM Monographs on Discrete Mathematics and Applications, vol. 3. SIAM (1999)

[5] Chartrand, G., Lesniak, L., Mynhardt, C.M., Oellermann, O.R.: Degree uniform graphs. Ann. N. Y. Acad. Sci. 555(1), 122–132 (1989)

[6] Chester, S., Kapron, B., Srivastava, G., Venkatesh, S.: Complexity of social network anonymization. Social Netw. Analys. Mining 3(2), 151–166 (2013)

[7] Cornuéjols, G.: General factors of graphs. J. Combin. Theory Ser. B 45(2), 185–198 (1988)

[8] Downey, R.G., Fellows, M.R.: Fundamentals of Parameterized Complexity. Springer (2013)

[9] Eppstein, D., Spiro, E.S.: The h-index of a graph and its application to dynamic subgraph statistics. J. Graph Algorithms Appl. 16(2), 543–567 (2012)

[10] Fellows, M.R., Guo, J., Moser, H., Niedermeier, R.: A generalization of Nemhauser and Trotter's local optimization theorem. J. Comput. System Sci. 77(6), 1141–1158 (2011)

[11] Flum, J., Grohe, M.: Parameterized Complexity Theory. Springer (2006)

[12] Garey, M.R., Johnson, D.S.: Computers and Intractability: A Guide to the Theory of NP-Completeness. Freeman (1979)

[13] Golovach, P.A.: Editing to a connected graph of given degrees. CoRR, abs/1308.1802 (2013)

[14] Guo, J., Niedermeier, R.: Invitation to data reduction and problem kernelization. SIGACT News 38(1), 31–45 (2007)

[15] Hartung, S., Nichterlein, A., Niedermeier, R., Suchý, O.: A refined complexity analysis of degree anonymization on graphs. In: Fomin, F.V., Freivalds, R., Kwiatkowska, M., Peleg, D. (eds.) ICALP 2013, Part II. LNCS, vol. 7966, pp. 594–606. Springer, Heidelberg (2013)

[16] Hartung, S., Hoffman, C., Nichterlein, A.: Improved upper and lower bound heuristics for degree anonymization in social networks. CoRR, abs/1402.6239 (2014)

[17] Katerinis, P., Tsikopoulos, N.: Minimum degree and f-factors in graphs. New Zealand J. Math. 29(1), 33–40 (2000)

[18] Komusiewicz, C., Niedermeier, R.: New races in parameterized algorithmics. In: Rovan, B., Sassone, V., Widmayer, P. (eds.) MFCS 2012. LNCS, vol. 7464, pp. 19–30. Springer, Heidelberg (2012)

[19] Liu, K., Terzi, E.: Towards identity anonymization on graphs. In: ACM SIGMOD Conference, SIGMOD 2008, pp. 93–106. ACM (2008)

[20] Lokshtanov, D., Misra, N., Saurabh, S.: Kernelization - preprocessing with a guarantee. In: Bodlaender, H.L., Downey, R., Fomin, F.V., Marx, D. (eds.) Fellows Festschrift 2012. LNCS, vol. 7370, pp. 129–161. Springer, Heidelberg (2012)

[21] Lovász, L., Plummer, M.D.: Matching Theory. Annals of Discrete Mathematics, vol. 29. North-Holland (1986)

[22] Mathieson, L., Szeider, S.: Editing graphs to satisfy degree constraints: A parameterized approach. J. Comput. System Sci. 78(1), 179–191 (2012)

[23] Moser, H., Thilikos, D.M.: Parameterized complexity of finding regular induced subgraphs. J. Discrete Algorithms 7(2), 181–190 (2009)

[24] Niedermeier, R.: Invitation to Fixed-Parameter Algorithms. Oxford University Press (2006)

On Matchings and *b*-Edge Dominating Sets: A 2-Approximation Algorithm for the 3-Edge Dominating Set Problem

Toshihiro Fujito[*]

Department of Computer Science and Engineering
Toyohashi University of Technology
Toyohashi 441-8580 Japan
fujito@cs.tut.ac.jp

Abstract. We consider a multiple domination version of the edge dominating set problem, called the *b-EDS* problem, where an edge set $D \subseteq E$ of minimum cardinality is sought in a given graph $G = (V, E)$ with a demand vector $b \in \mathbb{Z}^E$ such that each edge $e \in E$ is required to be dominated by $b(e)$ edges of D. When a solution D is not allowed to be a multi-set, it is called the *simple b*-EDS problem. We present 2-approximation algorithms for the simple *b*-EDS problem for the cases of $\max_{e \in E} b(e) = 2$ and $\max_{e \in E} b(e) = 3$. The best approximation guarantee previously known for these problems is 8/3 due to Berger et al. [2] who showed the same guarantee to hold even for the minimum cost case and for arbitrarily large *b*. Our algorithms are designed based on an LP relaxation of the *b*-EDS problem and locally optimal matchings, and the optimum of *b*-EDS is related to either the size of such a matching or to the optimal LP value.

1 Introduction

In an undirected graph an edge is said to *dominate* itself and all the edges adjacent to it, and a set of edges is an *edge dominating set* (abbreviated to *eds*) if the edges in it collectively dominate all the edges in a graph. The *edge dominating set problem (EDS)* asks to find an eds of minimum cardinality (cardinality case) or of minimum total cost (cost case). It was shown by Yannakakis and Gavril that, although EDS has important applications in areas such as telephone switching networking, it is NP-complete even when graphs are planar or bipartite of maximum degree 3 [12]. The classes of graphs for which its NP-completeness holds were later refined and extended by Horton and Kilakos to planar bipartite graphs, line and total graphs, perfect claw-free graphs, and planar cubic graphs [7], although EDS admits a PTAS (polynomial time approximation scheme) for planar [1] or λ-precision unit disk graphs [8]. Meanwhile, some polynomially solvable special cases have been also discovered for trees [9], claw-free chordal graphs, locally connected claw-free graphs, the line graphs of total graphs, the line

---

[*] Supported in part by the Kayamori Foundation of Informational Science Advancement and a Grant in Aid for Scientific Research of the Ministry of Education, Science, Sports and Culture of Japan.

R Ravi and I.L. Gørtz (Eds.): SWAT 2014, LNCS 8503, pp. 206–216, 2014.

graphs of chordal graphs [7], bipartite permutation graphs, cotriangulated graphs [11], and so on.

There are various variants of the basic EDS problem, and the most general one among them was introduced by Berger et al. [2] in the form of the *capacitated b-edge dominating set* problem (b, c)-EDS, where an instance consists of a graph $G = (V, E)$, a demand vector $b \in \mathbb{Z}_+^E$, a capacity vector $c \in \mathbb{Z}_+^E$ and a cost vector $w \in \mathbb{Q}_+^E$. A set D of edges in G is called a (b, c)-eds if each $e \in E$ is adjacent to at least $b(e)$ edges in D, where we allow D to contain at most $c(e)$ multiple copies of edge e. The problem asks to find a minimum cost (b, c)-eds. The (b, c)-EDS problem generalizes the EDS problem in much the same way that the set multicover problem generalizes the set cover problem. In the special case when all the capacities c are set to $+\infty$, we call the resulting problem the *uncapacitated b-EDS* problem and its feasible solutions *uncapacitated b-eds's*, whereas it is called the *simple b-EDS* problem when $c(e) = 1$ for all $e \in E$. If $b(e)$'s are set to a same value for all $e \in E$, it is called *uniform* (b, c)-EDS.

Let $b_{\max}$ denote $\max_{e \in E} b_e$. We mainly focus on the simple b-EDS problem (i.e., $(b, 1)$-EDS), and b-EDS (or b-eds) in what follows usually means the simple one unless otherwise stated explicitly. It should be noted, however, that this does not impose serious restrictions in problem solving, as long as $b_{\max}$ is bounded by some constant, since general (b, c)-EDS can be reduced to $(b, 1)$-EDS by introducing $\min\{b_{\max}, c(e)\} - 1$ many copies of e, each of them parallel to e with $b = 0$, for all the edges $e \in E$.

1.1 Previous Work

It was shown by Yannakakis and Gavril that the minimum EDS can be efficiently approximated to within a factor of 2 by computing any maximal matching [12]. They used the theorem of Harary [6] to lower bound the cardinality of a minimum eds by that of a smallest maximal matching. More recently, a 2.1-approximation algorithm first [4], and then 2-approximation algorithms [5,10] have been successfully obtained for the cost case of EDS problem via polyhedral approaches.

Among the approximation results obtained in [2] those relevant to ours are summarized in the following list (note: their results hold even for the cost cases of (b, c)-EDS problems):

- The (b, c)-EDS problem can be approximated within a factor of 8/3.
- The uniform and uncapacitated b-EDS problem can be approximated within a factor of 2.1 if $b = 1$ or a factor of 2 if $b \geq 2$.
- The integrality gap of the LP relaxation they used for (b, c)-EDS, much more complex one than ours with additional valid inequalities, is at most 8/3 and it is tight even when $b(e) \in \{0, 1\}, \forall e \in E$.

A linear-time 2-approximation algorithm for uncapacitated b-EDS was obtained by Berger and Parekh [3].

1.2 Our Work

One of the main subjects studied in the current paper is the *approximate* min-max relationships between simple b-EDS and locally optimal matchings. A most well-known example of such is perhaps the one between the vertex cover number and the size of a maximal matching. Let $\tau(G)$ denote the vertex cover number of G (i.e., the cardinality of any smallest vertex cover for G) and M be any maximal matching in G. Then, $|M| \le \tau(G) \le 2|M|$.

Let us simplify matters in the following discussion by restricting ourselves to the case of uniform and simple b-EDS, and let $\gamma_b(G)$ denote the cardinality of any smallest $(b, 1)$-eds for $G = (V, E)$ where $b(e) \equiv b, \forall e \in E$. To introduce lower bounds on $\gamma_b(G)$, we start with the following integer program, the most natural IP formulation for simple b-EDS:

$$\min \{x(E) \mid x(\delta(e)) \ge b(e) \text{ and } x_e \in \{0, 1\}, \forall e \in E\},$$

where $x(F) = \sum_{e \in F} x_e$ for $F \subseteq E$, and $\delta(e) = \{e\} \cup \{e' \in E \mid e' \text{ is adjacent to } e\}$ for $e \in E$. Replacing the integrality constraints by linear constraints $0 \le x_e \le 1$ would result in the following LP:

$$\min \{x(E) \mid x(\delta(e)) \ge b(e) \text{ and } 0 \le x_e \le 1, \forall e \in E\}.$$

Relaxing the LP above further by dropping the upper bound constraint on each x_e, we obtain an LP and its dual in the following forms:

LP: (P) $\min z_P(x) = x(E)$ LP: (D) $\max z_D(y) = \sum_{e \in E} b(e) y_e$

subject to: $x(\delta(e)) \ge b(e), \quad \forall e \in E$ subject to: $y(\delta(e)) \le 1, \quad \forall e \in E$

$x_e \ge 0, \quad \forall e \in E$ $y_e \ge 0, \quad \forall e \in E$

Notice here that (P) coincides with the LP relaxation of uncapacitated b-EDS rather than simple one.

For any matching M in $G = (V, E)$, let $y^M \in \mathbb{R}^E$ be a vector of dual variables such that

$$y_e^M = \begin{cases} \frac{1}{2} & \text{if } e \in M \\ 0 & \text{otherwise} \end{cases}$$

As $\delta(e)$ contains at most two edges of M for any $e \in E$, y^M is always feasible for (D), and its value $z_D(y)$ equals to $b|M|/2$. So, $b|M|/2$ can serve as a lower bound on $\gamma_b(G)$ for any matching M in G.

Following the way locally optimal solutions are often termed in the local search optimization, we say that a matching M in G is k-opt if, for any matching N in G larger than M, $|M \setminus N| \ge k$ (So, a *maximal* matching is 1-opt). As for upper bounds on $\gamma_b(G)$, $|M_1|$ provides itself as such a bound on $\gamma_1(G)$ for any 1-opt matching M_1. It is also the case, as will be shown later (Corollary 1), that $\gamma_2(G) \le 2|M_2|$ for any 2-opt matching M_2 in G. So it would be extremely pleasing if the following min-max relationships hold for all $b \in \mathbb{N}$:

$$\frac{b|M_b|}{2} \le \gamma_b(G) \le b|M_b|,$$

where M_b is any b-opt matching in G. It is, however, too good to be true, and it will be shown (in Section 4) that $\gamma_b(G)$ cannot be bounded above by $b|M_b|$ in general when $b \ge 3$.

Nevertheless, $\gamma_3(G)$ can be related to a stronger bound as follows. Letting dual(G) denote the optimal value of LP:(D) above for graph G, it will be seen (in Corollary 2) that the following min-max relation to hold:

$$\text{dual}(G) \leq \gamma_3(G) \leq 2 \cdot \text{dual}(G)$$

for any G (recall that dual(G) is the optimal value of the LP relaxation for *uncapacitated* b-EDS).

These upper bounds are obtained algorithmically; our algorithms, building solutions upon b-opt matchings, approximate the (*unweighted*) simple b-EDS problems within a factor of 2, where b is not assumed to be uniform, for $b_{max} = 2$ and for $b_{max} = 3$. Unlike the polyhedral approaches explored in [2], ours is more graph theoretic and our algorithms are purely combinatorial.

2   Preliminaries

In this paper only graphs with no loops are considered. For an edge set $F \subseteq E$, $V[F]$ denotes the set of vertices induced by the edges in F (i.e., the set of all the endvertices of the edges of F). For a vertex set $S \subseteq V$ let $\delta(S)$ denote the set of edges incident to a vertex in S. When S is an edge set, we let $\delta(S) = \delta(\cup_{e \in S} e)$ where edge e is a set of two vertices; then, $\delta(S)$ also denotes the set of edges dominated by S. When S is a singleton set $\{s\}$, $\delta(\{s\})$ is abbreviated to $\delta(s)$. For a vertex set $U \subseteq V$, $N(U)$ denotes the set of neighboring vertices of those in U (i.e., $N(U) = \{v \in V \mid \{u, v\} \in E$ for some $u \in U\}$), and $N(u)$ means $N(\{u\})$. The degree of a vertex u is denoted by $d(u)$. When $\delta(S), N(U)$, and $d(u)$ are considered only within a subgraph H of G (or when restricted to within a vertex subset or edge subset T), they are denoted by $\delta_H(S), N_H(U)$, and $d_H(u)$ (or $\delta_T(S), N_T(U)$, and $d_T(u)$), respectively.

When an edge e is dominated by up to $b(e)$ edges, it is said to be *fully dominated*.

3   A 2-Opt Algorithm for 2-EDS

Here a 2-approximation algorithm for the simple 2-EDS problem is presented. The algorithm is quite simple: Compute a 2-opt matching M_2 so that no augmenting path of length 3 or shorter occurs. Then, for each matched edge $e \in M_2$, if one of its endvertices is a neighbor of an exposed vertex via edge e', add e' besides e itself to a solution set while, if neither has, add any edge adjacent to e.

Divide the edge set E of an instance graph G according to demands into E_1 and E_2, where $E_i = \{e \in E \mid b(e) = i\}$. Let $G_i = (V_i, E_i)$ denote the subgraph of G induced by E_i.

1. Compute a 2-opt matching M_2 in $G_2 = (V_2, E_2)$; so no augmenting paths of length 3 or shorter occurs.
2. $D_2 \leftarrow M_2$.

Let $X \subseteq V_2$ denote the set of vertices in G_2 exposed by M_2, and consider $N_{G_2}(u)$ and $N_{G_2}(v)$ for each $e = \{u, v\} \in M_2$. If both of them contain vertices exposed by M_2, they must be same and unique, i.e., $N_{G_2}(u) \cap X = N_{G_2}(v) \cap X = \{x\}$ for some $x \in X$; otherwise, an augmenting path of length 3 having e in the middle is found. One edge adjacent to e is added to D_2, and which one to add is determined according to which of $N_{G_2}(u)$ and $N_{G_2}(v)$ contains an exposed vertex.

3. For each $e = \{u, v\} \in M_2$,
   (a) if $N_{G_2}(u)$ contains an exposed vertex $x \in X$, then add edge $\{u, x\}$ into D_2,
   (b) else if $N_{G_2}(v)$ contains an exposed vertex, then add any edge in $\delta_G(v)$ (other than e) into D_2,
   (c) else add any edge in $\delta_G(e)$ (other than e) into D_2.

Note: If it is only to dominate twice the edges in $\delta_{G_2}(u)$, we may choose any one of them in Step 3(a). It could be the case, however, that both of $N_{G_2}(u)$ and $N_{G_2}(v)$ contain the exposed vertex x as a unique exposed vertex in common, and it is then necessary to choose $\{u, x\}$ in this step to fully dominate $\{v, x\}$.

By this time all the edges in E_2 are fully dominated. It remains only to dominate those in E_1 that are not yet dominated even once.

4. Set $E_1' \leftarrow E_1 \setminus \delta_G(D_2)$.
5. Compute a 1-opt matching M_1 in $G[E_1']$ and output $D_2 \cup M_1$.

Theorem 1. *The 2-opt algorithm given above is a 2-approximation algorithm for the* $(b, 1)$*-EDS problem when* $b_{\max} = 2$.

Proof. Consider an edge $e \in E_2$. It becomes fully dominated, if $e \in M_2$, when an edge adjacent to e is added to D_2 in step 3. For $e \notin M_2$, if both of its endvertices are matched by M_2, it is made dominated fully by M_2 (in step 2). If $e = \{u, x\} \notin M_2$ is incident to an exposed vertex x, another unmatched edge incident to either u or x must be chosen into D_2 in step 3. Therefore, all the edges in E_2 become fully dominated after step 3. As not-yet-dominated edges in E_1 are taken care of in step 5 when a maximal matching M_1 is entirely chosen into a solution, the algorithm computes a simple 2-eds for G.

The performance analysis of this algorithm is omitted here as it can be subsumed by the one for the 3-opt algorithm for 3-EDS presented in Section 5. $\square$

In case when $b(e)$ is uniformly equal to 2 in the above, $G_2 = G$ and $|D_2| \leq 2|M_2|$. Thus,

Corollary 1. *For any* 2*-opt matching* M_2 *in* G, $\gamma_2(G) \leq 2|M_2|$.

4 b-Opt Matchings and $\gamma_b(G)$

4.1   Case of 3-EDS

This subsection shows that the ratio of $\gamma_b(G)$ to $b|M_b|$ for a b-opt matching M_b in G can be larger than 1 even for $b = 3$.

Let $P_4 = \{e_{i,1}e_{i,2}e_{i,3}e_{i,4} \mid 1 \leq i \leq k\}$ be a collection of simple paths of length 4, starting and ending at the common vertices u_1 and u_2 respectively, and being mutually vertex disjoint except at these two vertices. Construct a graph G by attaching two edges $e_{0,1}$ and $e_{0,2}$ at u_1 and u_2, respectively, but disjointly at the other endvertices of $e_{0,1}$ and $e_{0,2}$ from other vertices in G. Let M be a matching in G such that $M = \{e_{0,1}, e_{0,2}, e_{i,3} \mid 1 \le i \le k\}$. Then, M is a maximum matching in G with $|M| = k+2$ as there exists no augmenting path w.r.t. M. Meanwhile, $\gamma_3(G) = 4k$ since all the edges in all the paths of P_4 must be used to constitute a 3-eds for G, and hence, $\frac{\gamma_3(G)}{3|M|} = \frac{4k}{3(k+2)} > 1$ for $k > 6$ even if M is a maximum matching in G.

4.2   Case of b-EDS

This subsection shows that the ratio of $\gamma_b(G)$ to $b|M_b|$ for a b-opt matching M_b in G can be arbitrarily large as b grows.

Let S_i be a star graph centered at vertex s_i with $b/2$ edges, for $1 \le i \le b/2$ ("small" stars). Let L_i also be a star graph centered at vertex l_i with $(b/2)^2$ edges, for $1 \le i \le b/2$ ("large" stars). Construct a bipartite graph G with all the center vertices in S_i's and L_i's on one side, and a set U of $(b/2)^2$ vertices on the other side, by attaching leaves of S_i's and L_i's at the vertices in U. Each leaf of L_i is attached to a distinct vertex of U for $1 \le i \le b/2$. There are $(b/2)^2$ leaves of S_i's in total, and they are also attached to the vertices of U distinctively.

Observe now that $d(s_i) = b/2, d(l_i) = (b/2)^2$ for $1 \le i \le b/2, d(u) = b/2 + 1$ for $u \in U$, and $|\delta(e)| = b/2 + b/2 = b$ for any edge e in $\delta(s_i)$. Therefore, any b-eds for $G = (V, E)$ must contain all of $\delta(e)$'s for all $e \in \delta(s_i)$, covering all the edges of G, and meaning that $\gamma_b(G) = |E| = (b/2)^2(b/2+1)$. On the other hand, there exist $b/2+b/2 = b$ vertices on the other side of U, and hence, $|M| \le b$ for any matching M in G. It thus follows that

$$\frac{\gamma_b(G)}{b|M|} \ge \frac{(b/2)^2(b/2+1)}{b^2} = \frac{b+2}{8}$$

even if M is a maximum matching in G.

5   A 3-Opt Algorithm for 3-EDS

Here a 2-approximation algorithm for the simple 3-EDS problem is presented. In the beginning the algorithm dominates all the edges with demands of 3 using a 3-opt matching M_3. As was observed in the previous section, however, it is not good enough to choose the M_3-edges along with some edges adjacent to them. Moreover, as we treat the case when edges with demands of 2 or less are allowed to coexist, a part of the solution dominating those demand-3 edges may interfere with another part dominating those with smaller demands, and it makes the task of designing an algorithm more complicated than otherwise.

Divide the edge set E of an instance graph G according to demands into E_1, E_2, and E_3, where $E_i = \{e \in E \mid b(e) = i\}$. Let $G_i = (V_i, E_i)$ denote the subgraph of G induced by E_i.

1. Compute a 3-opt matching M_3 in $G_3 = (V_3, E_3)$; so no augmenting paths of length 5 or shorter occurs.
2. $D_3 \leftarrow M_3$.

Note: At this point an edge in M_3 is dominated once, and one in $E_3 \setminus M_3$ is also dominated once if it is incident to an exposed vertex but otherwise, it is dominated twice by $D_3 = M_3$.

We need to exercise special care in handling those edges incident to the vertices exposed by M_3, and a bipartite subgraph of G_3 induced by those edges is constructed for that purpose as follows; this will be the main body of algorithmic operations and analysis provided later.

- Let $X \subseteq V_3$ be the set of vertices in G_3 exposed by M_3. Notice that X is an independent set in G_3 since M_3 does not allow an augmenting path of length 1 to exist.
- Let $A \subseteq V_3$ be the set of neighboring vertices of X in G_3, i.e., $A = N_{G_3}(X)$.
- Let $B = (X \cup A, E_B)$ denote the bipartite subgraph graph of G_3, consisting of the vertex partition (X, A), and the set E_B of E_3-edges lying between them.
- Let $M' \subseteq M_3$ be the set of matched edges having an endvertex in A, i.e., $M' = \{e \in M_3 \mid e \cap A \neq \emptyset\}$.
- Let $M_c = \{e \in M' \mid e \subseteq A\}$. Then, each edge in $M' \setminus M_c$ has exactly one of its endvertices in A; denote it by $a(e)$ and the other endvertex of e by $\bar{a}(e)$, for each $e \in M' \setminus M_c$.
- Divide $M' \setminus M_c$ further, according to the G-degree of $a(e)$, into $M_s = \{e \in M' \setminus M_c \mid d_G(a(e)) = 2\}$, and $M_d = \{e \in M' \setminus M_c \mid d_G(a(e)) \geq 3\}$.
- Accordingly divide A into $A_c = \{$both endvertices of $e \mid e \in M_c\}$, $A_s = \{a(e) \mid e \in M_s\}$, and $A_d = \{a(e) \mid e \in M_d\}$, and E_B into $E_c = \delta_B(A_c)$, $E_s = \delta_B(A_s)$, and $E_d = \delta_B(A_d)$.

Clearly, $d_B(a) \geq 1$ for all $a \in A$. Observe that $d_B(a(e)) = 1$ for all $a(e) \in A_s$ since only two edges are incident to $a(e)$ in G and they are $\{x, a(e)\}$ for some $x \in X$ and $\{a(e), \bar{a}(e)\} \in M_s$ where $\bar{a}(e) \notin A$. It is also the case that $d_B(a) = 1$ for all $a \in A_c$: Consider a pair of vertices, a_1, a_2, in A_c such that they are the endvertices of one edge in M_c. Then, they must be adjacent to a unique and same vertex in X as otherwise, an augmenting path of length 3 would result.

Consider the subgraph $B_d = (X_d \cup A_d, E_d)$ of B induced by E_d, where $X_d = N_B(A_d) \subseteq X$.

3. Compute a *maximal* edge subset N of E_d in B_d such that $|\delta_B(x) \cap N| \leq 2$ for each $x \in X_d$ and $|\delta_B(a) \cap N| \leq 1$ at the same time for each $a \in A_d$.
4. Set $D_3 \leftarrow D_3 \cup E_c \cup E_s \cup N$.

At this point, every edge in E_c (and those in M_c) is fully dominated by $E_c \cup M_c \subseteq D_3$, and there could be such edges also in $E_s \cup E_d$. Let $\tilde{E}_s$ and $\tilde{E}_d$ denote the subsets of E_s and E_d, respectively, consisting of edges (of E_s and E_d) not-yet fully dominated by $M_3 \cup E_c \cup E_s \cup N$. As each edge in E_s is dominated at least twice by $M_3 \cup E_s$, if it is adjacent to any other in $E_c \cup E_s \cup N$, it must be fully dominated. Therefore, $\tilde{E}_s$ forms a matching in G, and no edge in $\tilde{E}_s$ is adjacent to any in $N \cup E_c$.

In the next two steps, all the edges in $E_s \cup M_s$ will be made fully dominated.

5. For $e \in \tilde{E}_s$, add one edge from E incident to the exposed endvertex of e occurring in X, into D_3; such an edge must exist as otherwise, e cannot be fully dominated.

6. For each $e \in M_s$, add any edge in $\delta_G(\bar{a}(e))$ (such an edge must exist in $\delta_G(\bar{a}(e))$ as, otherwise, e cannot be fully dominated).

Suppose, for some $a \in A_d$, $\delta_B(a) \cap N = \emptyset$. Then, every edge in $\delta_B(a)$ must be fully dominated by $N \cup M_d$; if $e \in \delta_B(a)$ is not, it can be added to N contradicting the maximality of N. Therefore, every edge in $\tilde{E}_d$ must be either in N or adjacent to an N-edge, implying that it is dominated at least twice by $N \cup M_3$.

In the next two steps, the algorithm adds edges to D_3 so that all the edges in $E_d \cup M_d$ become fully dominated. Let $M_N \subseteq M_d$ denote the set of M_d-edges adjacent to an edge in N.

7. For each $e \in M_N$, if $\delta_B(a(e))$ contains an edge in $\tilde{E}_d$, add one more edge from $\delta_G(a(e)) \setminus N$ into D_3, which must exist as $d_G(a(e)) \geq 3$. If $\delta_B(a(e))$ contains no edge in $\tilde{E}_d$, add any edge in $\delta_G(\bar{a}(e))$ if it exists (if it doesn't, add instead any edge in $\delta_G(a(e))$), into D_3.

8. For each $e \in M_d \setminus M_N$, add two edges into D_3, one from $\delta_G(a(e))$ and another from $\delta_G(\bar{a}(e))$ if it exists (if it doesn't, add instead any edge in $\delta_G(a(e))$).

We also need to take care of the edges in $M_3 \setminus M'$ and those around them, and two adjacent edges are added in a simple way for each of these matched edges.

9. For each $e \in M_3 \setminus M'$, add any edge in E incident to the endvertices of e, one each, into D_3; in case when either of them does not exist take two edges, instead of one, from the other end of e.

Finally, all the remaining edges in $E \setminus E_3$ are taken care of by simply running the 2-opt algorithm for 2-EDS on G after the demands are appropriately adjusted. Let E_i' denote the set of edges with demands of i adjusted right after step 9. Then, $E_3' = \emptyset$ and $E_3 \subseteq E_0'$ since any edge in E_3 has been fully dominated by now (Lemma 1). Moreover, $E_2' = E_2 \setminus \delta_{E_2}(D_3)$, $E_1' \subseteq (E_1 \setminus \delta_{E_1}(D_3)) \cup (E_2 \cap \delta_{E_2}(D_3))$ and $E_0' \subseteq (E_1 \cap \delta_{E_1}(D_3)) \cup (E_2 \cap \delta_{E_2}(D_3)) \cup E_3$.

10. Run the 2-opt algorithm for 2-EDS on G after setting $b'(e) \leftarrow \max\{0, b(e) - |\delta_G(e) \cap D_3|\}$ for $e \in E$, and compute a 2-eds $D_2 \cup M_1$ for (G, b').

11. Output $D_3 \cup D_2 \cup M_1$.

Lemma 1. *Every edge in E_3 becomes fully dominated after step 9.*

Proof. 1. Consider $e \in M' \cup E_B$. Since $M' \cup E_c \cup E_s \subseteq D_3$, $e \in M_c \cup E_c$ is fully dominated whereas $e \in M_s \cup E_s$ is at least twice dominated by the end of step 4, and if not yet fully dominated, e is made so in steps 5 and 6. Any edge in E_d becomes fully dominated by the end of step 7, while any edge in M_d does by the end of step 8.

2. Any edge $e \in M_3 \setminus M'$ is made fully dominated in step 9.

3. Consider $e \in E_3 \setminus (M_3 \cup E_B)$. Both endvertices of e are matched by M_3 ensuring that e is twice dominated by M_3. Observe now that for any matched vertex $u \in V_3 \setminus X$,

$\delta_G(u)$ contains only one edge in D_3 (namely, the matched edge incident to u) only if $d_G(u) = 1$ or $u = \bar{a}(e)$ for some $e \in M_d$ with an N-edge incident to $a(e)$. For any other matched vertex $u \in V_3 \setminus X$, $\delta_G(u) \setminus M_3$ contains at least one D_3-edge, and hence, if either endvertex of e is such a vertex, e is fully dominated.

The case that $d_G(u) = 1$ at an endvertex of e is excluded since e is unmatched. What remains is the case when $u = \bar{a}(e_1)$ and $v = \bar{a}(e_2)$ for $e = \{u, v\}$ such that both e_1 and e_2 are in M_d and each of $a(e_1)$ and $a(e_2)$ has an N-edge incident to it. Since no augmenting path of length 5 exists in G_3, $\{e, e_1, e_2\}$ together with those two N-edges incident to $a(e_1)$ and $a(e_2)$ must form a blossom (of length 5). There cannot exist another edge in E_B incident to either $a(e_1)$ or $a(e_2)$ as it would imply an augmenting path of length 5. So, each of $a(e_1)$ and $a(e_2)$ has only one incident edge in B, and both of them are N-edges having a common exposed vertex at their endvertices. It means, however, that those N-edges are fully dominated even *before* step 7, and hence, $\delta_B(a(e))$ contains no edge in $\tilde{E}_d$ when step 7 is executed. Therefore, an unmatched edge in $\delta_G(u)$ or $\delta_G(v)$ is added to D_3 in step 7, ensuring e being fully dominated.   □

Thus, the correctness of the algorithm above follows from this lemma and the correctness of the 2-opt algorithm for 2-EDS:

Theorem 2. *The 3-opt algorithm for the $(b, 1)$-EDS problem given above computes a feasible 3-eds for G when $b_{\max} = 3$.*

5.1 Performance Analysis of 3-opt Algorithm for 3-EDS

Recall the dual of our LP relaxation for b-EDS:

$$\text{LP: (D)} \quad \max z_D(y) = \sum_{e \in E} b(e) y_e$$

$$\text{subject to:} \quad y(\delta(e)) \leq 1, \quad \forall e \in E$$
$$y_e \geq 0, \quad \forall e \in E$$

Suppose $M_3 \subseteq E_3$ is a matching computed in step 1 of the 3-opt algorithm. Recall that $\tilde{E}_s \subseteq E_s$ forms a matching in G, and no edge in $\tilde{E}_s$ is adjacent to any in $N \cup E_c$. Set the value of y_e for $e \in E_3$ as follows:

$$y_e = \begin{cases} \frac{1}{2} & \text{if } e \in M_3 \setminus M_N \\ \frac{1}{4} & \text{if } e \in N \cup M_N \\ \frac{1}{6} & \text{if } e \in \tilde{E}_s \\ 0 & \text{otherwise} \end{cases}$$

Let M_2 and M_1 denote the matchings computed, within the run of the 2-opt algorithm for 2-EDS, in step 10 of the 3-opt algorithm. Recall E_i', the set of edges with demands of i adjusted right after step 9. Note that $M_2 \subseteq E_2' = E_2 \setminus \delta_{E_2}(D_3)$ and hence, M_2 contains edges with $b(e) = 2$ only, and no edge in M_2 can be adjacent to any in D_3. The set E_1' on the other hand may contain E_2-edges e as $b(e)$ could have been lowered to 1 if it is dominated *once* by D_3, and so may $M_1 \subseteq E_1' \setminus \delta_G(D_2)$.

Set the value of y_e for $e \in E_2 \cup E_1$ as follows:

$$y_e = \begin{cases} \frac{1}{2} & \text{if } e \in M_2 \\ \frac{1}{2} & \text{if } e \in M_1 \cap E_1 \\ \frac{1}{4} & \text{if } e \in M_1 \cap E_2 \\ 0 & \text{otherwise} \end{cases}$$

Lemma 2. *The vector* $y \in \mathbb{R}^E$ *of dual variables with its values assigned as above is feasible in LP:(D).*

Proof. The dual feasibility of y follows easily if $y(\delta_G(u)) \leq 1/2$ for all $u \in V$. Although this does not hold for all the vertices in G, we will check how large $y(\delta_G(u))$ could be depending on where u is located, and will consider the cases when it exceeds $1/2$ in what follows.

As stated above, $V[D_3] \cap V[M_2] = \emptyset$ and $V[D_2] \cap V[M_1] = \emptyset$, but $V[D_3]$ and $V[M_1]$ are not necessarily disjoint. So, if $u \in V[D_2]$, $y(\delta_G(u)) = y(\delta_{D_2}(u)) = y(\delta_{M_2}(u))$, and hence, $y(\delta_G(u)) \leq 1/2$ in this case.

Suppose $u \in V[D_3] \cap V[M_1]$. Then, the M_1-edge e in $\delta_G(u)$ must come from E_2 and it has to be dominated exactly once by D_3. Consider now for which vertex u of $V[D_3]$ we may have 1) exactly one edge of D_3 is incident to u, 2) the edge in 1) carries a positive dual, and 2) $d_G(u) \geq 2$. It can be verified that such u must be either $\bar{a}(e)$ for $e \in M_N$, or the exposed endvertex of an N edge. In either case the positive dual carried by a D_3-edge is $1/4$, while the one carried by an M_1-edge is also $1/4$; hence, $y(\delta_G(u)) \leq 1/4 + 1/4 = 1/2$ in this case.

If $u \in V[M_1] \setminus V[D_3]$, the M_1-edge contained in $\delta_G(u)$ must come from E_1, and hence, $y(\delta_G(u)) = 1/2$.

What remains is the case when $u \in V[D_3] \setminus V[M_1]$. As observed in passing within the algorithm description, $\tilde{E}_s$ forms a matching and no edge in it is adjacent to any in N. It can be verified from such observations that $\delta_{E_3}(e)$ contains at most two edges with positive dual values for any $e \in D_3$, and those two are either one each from $\tilde{E}_s$ and M_s, one each from M_N and N, or both from N. Among these $y(\delta_{E_3}(u)) = 1/2 + 1/6 > 1/2$ in the first case only, and $y(\delta_{E_3}(u)) \leq 1/2$ in the remaining cases. In the first case, however, $d_G(u) = 2$ and there is no edges incident to u other than those two edges, $e_1 \in \tilde{E}_s$ and $e_2 \in M_s$. Moreover, letting u_1 and u_2 be the other endvertices of e_1 and e_2, respectively, the algorithm adds an edge, with no positive dual, incident to each of u_1 and u_2 into D_3 resulting in $d_{D_3}(u_1) = d_{D_3}(u_2) = 2$. Hence, no more edge can be added to either of u_1 or u_2 in step 10, and each of $y(\delta_G(e_1))$ and $y(\delta_G(e_2))$ remains no larger than 1 in the end.

Therefore, we may conclude that $y(\delta_G(e)) \leq 1$ for all $e \in E$.   □

Lemma 3. *For* $y \in \mathbb{R}^E$ *of dual variables with its values assigned as above, the 3-opt algorithm computes an output of size no larger than twice the objective value of y in LP:(D), i.e.,*

$$|D_3 \cup D_2 \cup M_1| \leq 2z_D(y) = 2 \sum_{e \in E} b(e)y_e.$$

Proof. The term in the objective function of LP:(D) corresponding to y_e is $b(e)y_e$. So if at most $2b(e)y_e$ edges are used per e in dominating all the edges, the claimed inequality holds. For $e \in M_2 \cup (M_1 \cap E_1)$, y_e is set to $1/2$, and 2 edges per $e \in M_2$ and 1 edge per $e \in M_1 \cap E_1$ are used. On the other hand, 1 edge per $e \in M_1 \cap E_2$ is used with $y_e = 1/4$, and it suffices because $2b(e)y_e = 2 \cdot 2 \cdot (1/4) = 1$.

As for D_3, 3 edges are used per $e \in M_3$ where $y_e = 1/2$ if $e \in M_3 \setminus M_N$ but $y_e = 1/4$ if $e \in M_N$. For each $e_1 \in M_N$, however, there exists a mate $e_2 \in N$ of its own, carrying $y_{e_2} = 1/4$, and hence, together with e_2, e_1 can pay $1/2$ that is sufficient for 3 edges.

Besides e and two edges adjacent to e per $e \in M_3$, D_3 uses one more edge per $e' \in \tilde{E}_s$, and it can be paid for by $y_{e'}$ as $2b(e')y_{e'} = 2 \cdot 3 \cdot (1/6) = 1$.   □

It follows immediately from these preceding two lemmas that the 3-opt algorithm computes a feasible eds of size no larger than twice the optimum:

Theorem 3. *The* 3-*opt algorithm is a* 2-*approximation algorithm for the* $(b, 1)$-*EDS problem when* $b_{\max} = 3$.

Corollary 2. $\gamma_3(G) \leq 2 \cdot dual(G)$.

References

1. Baker, B.S.: Approximation algorithms for NP-complete problems on planar graphs. J. ACM 41, 153–180 (1994)
2. Berger, A., Fukunaga, T., Nagamochi, H., Parekh, O.: Approximability of the capacitated b-edge dominating set problem. Theoret. Comput. Sci. 385(1-3), 202–213 (2007)
3. Berger, A., Parekh, O.: Linear time algorithms for generalized edge dominating set problems. Algorithmica 50(2), 244–254 (2008)
4. Carr, R., Fujito, T., Konjevod, G., Parekh, O.: A $2\frac{1}{10}$-approximation algorithm for a generalization of the weighted edge-dominating set problem. Journal of Combinatorial Optimization 5(3), 317–326 (2001)
5. Fujito, T., Nagamochi, H.: A 2-approximation algorithm for the minimum weight edge dominating set problem. Discrete Appl. Math. 118, 199–207 (2002)
6. Harary, F.: Graph Theory. Addison-Wesley, Reading (1969)
7. Horton, J.D., Kilakos, K.: Minimum edge dominating sets. SIAM J. Discrete Math. 6(3), 375–387 (1993)
8. Hunt III, H.B., Marathe, M.V., Radhakrishnan, V., Ravi, S.S., Rosenkrantz, D.J., Stearns, R.E.: A unified approach to approximation schemes for NP- and PSPACE-hard problems for geometric graphs. In: van Leeuwen, J. (ed.) ESA 1994. LNCS, vol. 855, pp. 424–435. Springer, Heidelberg (1994)
9. Mitchell, S., Hedetniemi, S.: Edge domination in trees. In: Proc. 8th Southeastern Conf. on Combinatorics, Graph Theory, and Computing, pp. 489–509 (1977)
10. Parekh, O.: Edge dominating and hypomatchable sets. In: Proc. 13th SODA, pp. 287–291 (2002)
11. Srinivasan, A., Madhukar, K., Nagavamsi, P., Pandu Rangan, C., Chang, M.-S.: Edge domination on bipartite permutation graphs and cotriangulated graphs. Inform. Process. Lett. 56, 165–171 (1995)
12. Yannakakis, M., Gavril, F.: Edge dominating sets in graphs. SIAM J. Appl. Math. 38(3), 364–372 (1980)

Covering Problems
in Edge- and Node-Weighted Graphs

Takuro Fukunaga*

National Institute of Informatics, Tokyo, Japan
JST, ERATO, Kawarabayashi Large Graph Project, Japan
takuro@nii.ac.jp

Abstract. This paper discusses the graph covering problem in which
a set of edges in an edge- and node-weighted graph is chosen to satisfy some covering constraints while minimizing the sum of the weights.
In this problem, because of the large integrality gap of a natural linear programming (LP) relaxation, LP rounding algorithms based on the
relaxation yield poor performance. Here we propose a stronger LP relaxation for the graph covering problem. The proposed relaxation is applied
to designing primal-dual algorithms for two fundamental graph covering problems: the prize-collecting edge dominating set problem and the
multicut problem in trees. Our algorithms are an exact polynomial-time
algorithm for the former problem, and a 2-approximation algorithm for
the latter problem, respectively. These results match the currently known
best results for purely edge-weighted graphs.

1 Introduction

1.1 Motivation

Choosing a set of edges in a graph that optimizes some objective function under
constraints on the chosen edges constitutes a typical combinatorial optimization
problem and has been investigated in many varieties. For example, the spanning
tree problem seeks an acyclic edge set that spans all nodes in a graph, the edge
cover problem finds an edge set such that each node is incident to at least one
edge in the set, and the shortest path problem selects an edge set that connects
two specified nodes. All these problems seek to minimize the sum of the weights
assigned to edges.

This paper discusses several graph covering problems. Formally, the *graph
covering problem* is defined as follows in this paper. Given a graph $G = (V, E)$
and family $\mathcal{E} \subseteq 2^E$, find a subset F of E that satisfies $F \cap C \neq \emptyset$ for each
$C \in \mathcal{E}$, while optimizing some function depending on F. As indicated above,
the popular approaches assume an edge weight function $w\colon E \to \mathbb{R}_+$ is given,
where $\mathbb{R}_+$ denotes the set of non-negative real numbers, and seeks to minimize

---

* This work was partially supported by Japan Society for the Promotion of Science
(JSPS), Grants-in-Aid for Young Scientists (B) 25730008.

R Ravi and I.L. Gørtz (Eds.): SWAT 2014, LNCS 8503, pp. 217–228, 2014.

$\sum_{e \in F} w(e)$. On the other hand, we aspire to simultaneously minimize edge and node weights. Formally, we let $V(F)$ denote the set of end nodes of edges in F. Given a graph $G = (V, E)$ and weight function $w \colon E \cup V \to \mathbb{R}_+$, we seek a subset F of E that minimizes $\sum_{e \in F} w(e) + \sum_{v \in V(F)} w(v)$ under the constraints on F. Hereafter, we denote $\sum_{e \in F} w(e)$ and $\sum_{v \in V(F)} w(v)$ by $w(F)$ and $w(V(F))$, respectively.

Most previous investigations of the graph covering problem have focused on edge weights. By contrast, node weights have been largely neglected, except in the problems of choosing node sets, such as the vertex cover and dominating set problems. To our knowledge, when node weights have been considered in graph covering problems for choosing edge sets, they have been restricted to the Steiner tree problem or its generalizations, possibly because the inclusion of node weights greatly complicates the problem. For example, the Steiner tree problem in edge-weighted graphs can be approximated within a constant factor (the best currently known approximation factor is 1.39 [5,15]). Conversely, the Steiner tree problem in node-weighted graphs is known to extend the set cover problem (see [19]), indicating that achieving an approximation factor of $o(\log |V|)$ is NP-hard. The literature is reviewed in Section 2. As revealed later, the inclusion of node weights generalizes the set cover problem in numerous fundamental problems.

However, from another perspective, node weights can introduce rich structure into the above problems. In fact, node weights provide useful optimization problems. The objective function counts the weight of a node only once, even if the node is shared by multiple edges. Hence, the objective function defined from node weights includes a certain subadditivity, which cannot be captured by edge weights.

The aim of the present paper is to give algorithms for fundamental graph covering problems in edge- and node-weighted graphs. In solving the problems, we adopt a basic linear programming (LP) technique. Algorithms for combinatorial optimization problems are typically designed using LP relaxations. However, in problems with node-weighted graphs, the integrality gap of natural relaxations may be excessively large. Therefore, we propose tighter LP relaxations that preclude unnecessary integrality gaps. We then discuss upper bounds on the integrality gap of these relaxations in two fundamental graph covering problems: the edge dominating set (EDS) problem and multicut problem in trees. We prove upper bounds by designing primal-dual algorithms for both problems. The approximation factors of our proposed algorithms match the current best approximations in purely edge-weighted graphs.

1.2   Problem Definitions

The EDS problem covers edges by choosing adjacent edges in undirected graphs. For any edge e, let $\delta(e)$ denote the set of edges that share end nodes with e, including e itself. We say that an edge e *dominates* another edge f if $f \in \delta(e)$, and a set F of edges dominates an edge f if F contains an edge that dominates f. Given an undirected graph $G = (V, E)$, a set of edges is called an EDS if it dominates each edge in E. The EDS problem seeks to minimize the weight of the EDS. In other words, the EDS problem is the graph covering problem with $\mathcal{E} = \{\delta(e) \colon e \in E\}$.

The multicut problem specifies an undirected graph $G = (V, E)$ and demand pairs $(s_1, t_1), \ldots, (s_k, t_k) \in V \times V$. A *multicut* is an edge set C whose removal from G disconnects the nodes in each demand pair. This problem seeks a multicut of minimum weight. Let $\mathcal{P}_i$ denote the set of paths connecting s_i and t_j. The multicut problem is equivalent to the graph covering problem with $\mathcal{E} = \bigcup_{i=1}^{k} \mathcal{P}_i$.

Our proposed algorithms for solving these problems assume that the given graph G is a tree. In fact, our algorithms are applicable to the prize-collecting versions of these problems, which additionally specifies a penalty function $\pi \colon \mathcal{E} \to \mathbb{R}_+$. In this scenario, an edge set F is a feasible solution even if $F \cap C = \emptyset$ for some $C \in \mathcal{E}$, but imposes a penalty $\pi(C)$. The objective is to minimize the sum of $w(F)$, $w(V(F))$, and the penalty $\sum_{C \in \mathcal{E} : F \cap C = \emptyset} \pi(C)$. The prize-collecting versions of the EDS and multicut problems are referred to as the *prize-collecting EDS problem* and the *prize-collecting multicut problem*, respectively.

1.3 Our Results

Thus far, the EDS problem has been applied only to edge-weighted graphs. The vertex cover problem can be reduced to the EDS problem while preserving the approximation factors [6]. The vertex cover problem is solvable by a 2-approximation algorithm, which is widely regarded as the best possible approximation. Indeed, assuming the unique game conjecture, Khot and Regev [18] proved that the vertex cover problem cannot be approximated within a factor better than 2. Fujito and Nagamochi [10] showed that a 2-approximation algorithm is admitted by the EDS problem, which matches the approximation hardness known for the vertex cover problem. In the Appendix, we show that the EDS problem in bipartite graphs generalizes the set cover problem if assigned node weights and generalizes the non-metric facility location problem if assigned edge and node weights. This implies that including node weights increases difficulty of the problem even in bipartite graphs.

On the other hand, Kamiyama [17] proved that the prize-collecting EDS problem in an edge-weighted graph admits an exact polynomial-time algorithm if the graph is a tree. As one of our main results, we show that this idea is extendible to problems in edge- and node-weighted trees.

Theorem 1. *The prize-collecting EDS problem admits a polynomial-time exact algorithm for edge- and node-weighted trees.*

The proof of Theorem 1 will be sketched in Section 4. We can also show that the prize-collecting EDS problem in general edge- and node-weighted graphs admits an $O(\log |V|)$-approximation, which matches the approximation hardness on the set cover problem and the non-metric facility location problem.

The multicut problem is hard even in edge-weighted graphs; the best reported approximation factor is $O(\log k)$ [13]. The multicut problem is known to be both NP-hard and MAX SNP-hard [9], and admits no constant factor approximation algorithm under the unique game conjecture [7]. However, Garg, Vazirani, and Yannakakis [14] developed a 2-approximation algorithm for the multicut problem with edge-weighted trees. They also mentioned that, although the graphs are restricted to trees, the structure of the problem is sufficiently rich. They showed that the tree multicut problem includes the set cover problem with tree-representable set systems. They also showed that the vertex cover problem in general graphs is simply reducible to the multicut problem in star graphs, while preserving the approximation factor. This implies that the 2-approximation seems to be tight for the multicut problem in trees. As a second main result, we extended this 2-approximation to edge- and node-weighted trees, as stated in the following theorem.

Theorem 2. *The prize-collecting multicut problem admits a 2-approximation algorithm for edge- and node-weighted trees.*

Both algorithms claimed in Theorems 1 and 2 are primal-dual algorithms, that use the LP relaxations we propose. These algorithms fall into the same frameworks as those proposed in [14,17] for edge-weighted graphs. However, they need several new ideas to achieve the claimed performance because our LP relaxations are much more complicated than those used in [14,17].

The remainder of this paper is organized as follows. After surveying related work in Section 2, we define our LP relaxation for the prize-collecting graph covering problem in Section 3. In Sections 4, we sketch the proof of Theorem 1 that uses our proposed LP relaxation. The paper concludes with Section 5. We omit the proof of Theorem 2, and discussion on the prize-collecting EDS problem in general graphs with edge- and node-weights, for which we recommend referring to the full version [12] of the current paper.

2   Related Work

As mentioned in Section 1, the graph covering problem in node-weighted graphs has thus far been applied to the Steiner tree problem and its generalizations. Klein and Ravi [19] proposed an $O(\log |V|)$-approximation algorithm for the Steiner tree problem with node weights. Nutov [23,24] extended this algorithm to the survivable network design problem with higher connectivity requirements. An $O(\log |V|)$-approximation algorithm for the prize-collecting Steiner tree problem with node weights was provided by Moss and Rabani [21]; however, as noted by Könemann, Sadeghian, and Sanità [20], the proof of this algorithm contains a technical error. This error was corrected in [20]. Bateni, Hajiaghayi, and Liaghat [1] proposed an $O(\log |V|)$-approximation algorithm for the prize-collecting Steiner forest problem and applied it to the budgeted Steiner tree problem. Chekuri, Ene, and Vakilian [8] gave an $O(k^2 \log |V|)$-approximation algorithm for the prize-collecting survivable network design problem with edge-connectivity requirements of maximum value k. Later, they improved their approximation factor to $O(k \log |V|)$, and also extended it to node-connectivity requirements (see [28]). Naor, Panigrahi, and Singh [22] established an online algorithm for the Steiner tree problem with node weights which was extended to the Steiner forest problem by Hajiaghayi, Liaghat, and Panigrahi [16]. The survivable network design problem with node weights has also been extended to a problem called the network activation problem [26,25,11].

The prize-collecting EDS problem generalizes the $\{0,1\}$-EDS problem, in which given demand edges require being dominated by a solution edge set. The $\{0,1\}$-EDS problem in general edge-weighted graphs admits a $8/3$-approximation, which was proven by Berger et al. [2]. This $8/3$-approximation was extended to the prize-collecting EDS problem by Parekh [27]. Berger and Parekh [3] designed an exact algorithm for the $\{0,1\}$-EDS problem in edge-weighted trees, but their result contains an error [4]. Since the prize-collecting EDS problem embodies the $\{0,1\}$-EDS problem, the latter problem could be alternatively solved by an algorithm developed for the prize-collecting EDS problem in edge-weighted trees, proposed by Kamiyama [17].

3 LP Relaxations

This section discusses LP relaxations for the prize-collecting graph covering problem in edge and node-weighted graphs.

In a natural integer programming (IP) formulation of the graph covering problem, each edge e is associated with a variable $x(e) \in \{0,1\}$, and each node v is associated with a variable $x(v) \in \{0,1\}$. $x(e) = 1$ denotes that e is selected as part of the solution set, while $x(v) = 1$ indicates the selection of an edge incident to v. In the prize-collecting version, each demand set $C \in \mathcal{E}$ is also associated with a variable $z(C) \in \{0,1\}$, where $z(C) = 1$ indicates that the covering constraint corresponding to C is not satisfied. For $F \subseteq E$, we let $\delta_F(v)$ denote the set of edges incident to v in F. The subscript may be removed when $F = E$. An IP of the prize-collecting graph covering problem is then formulated as follows.

$$\text{minimize} \quad \sum_{e \in E} w(e)x(e) + \sum_{v \in V} w(v)x(v) + \sum_{C \in \mathcal{E}} \pi(C)z(C)$$

$$\text{subject to} \quad \sum_{e \in C} x(e) \geq 1 - z(C) \qquad \text{for } C \in \mathcal{E},$$

$$x(v) \geq x(e) \qquad \text{for } v \in V, e \in \delta(v),$$

$$x(e) \geq 0 \qquad \text{for } e \in E,$$

$$x(v) \geq 0 \qquad \text{for } v \in V,$$

$$z(C) \geq 0 \qquad \text{for } C \in \mathcal{E}.$$

In the above formulation, the first constraints specify the covering constraints, while the second constraints indicate that if the solution contains an edge e incident to v, then $x(v) = 1$. In the graph covering problem (without penalties), z is fixed at 0.

To obtain an LP relaxation, we relax the definitions of x and z in the above IP to $x \in \mathbb{R}_+^{E \cup V}$ and $z \in \mathbb{R}_+^C$. However, this relaxation may introduce a large integrality gap into the graph covering problem with node-weighted graphs, as shown in the following example. Suppose that $\mathcal{E}$ comprises a single edge set C, and each edge in C is incident to a node v. Let the weights of all edges and nodes other than v be 0. In this scenario, the optimal value of the graph covering problem is $w(v)$. On the other hand, the LP relaxation admits a feasible solution x such that $x(v) = 1/|C|$ and $x(e) = 1/|C|$ for each edge $e \in C$. The weight of this solution is $w(v)/|C|$, and the integrality gap of the relaxation for this instance is $|C|$. This phenomenon occurs even in the EDS problem and multicut problem in trees.

The above poor example can be excluded if the second constraints in the relaxation are replaced by $x(v) \geq \sum_{e \in \delta(v)} x(e)$ for $v \in V$. However, the LP obtained by this modification does not relax the graph covering problem if the optimal solutions contain high-degree nodes. Thus, we introduce a new variable $y(C, e)$ for each pair of $C \in \mathcal{E}$ and $e \in C$, and replace the second constraints by $x(v) \geq \sum_{e \in \delta(v)} y(C, e)$, where $v \in V$ and $C \in \mathcal{E}$. $y(C, e) = 1$ indicates that e is chosen to satisfy the covering constraint of C, and $y(C, e) = 0$ implies the opposite. Roughly speaking, $y(C, \cdot)$ represents a minimal fractional solution for covering a single demand set C. If a single covering constraint is imposed, the degree of each node is at most one in any minimal integral solution. Then the graph covering problem is relaxed by the LP even after modification. Summing up, we formulate our LP relaxation for an instance $I = (G, \mathcal{E}, w, \pi)$ of the prize-collecting graph covering problem as follows.

$$P(I) =$$

$$\text{minimize} \quad \sum_{e \in E} w(e)x(e) + \sum_{v \in V} w(v)x(v) + \sum_{C \in \mathcal{E}} \pi(C)z(C)$$

$$\text{subject to} \quad \sum_{e \in C} y(C, e) \geq 1 - z(C) \qquad \text{for } C \in \mathcal{E},$$

$$x(v) \geq \sum_{e \in \delta_C(v)} y(C, e) \qquad \text{for } v \in V, C \in \mathcal{E},$$

$$x(e) \geq y(C, e) \qquad \text{for } C \in \mathcal{E}, e \in C,$$

$$x(e) \geq 0 \qquad \text{for } e \in E,$$

$$x(v) \geq 0 \qquad \text{for } v \in V,$$

$$y(C, e) \geq 0 \qquad \text{for } C \in \mathcal{E}, e \in C,$$

$$z(C) \geq 0 \qquad \text{for } C \in \mathcal{E}.$$

Theorem 3. *Let I be an instance of the prize-collecting graph covering problem in edge- and node-weighted graphs. $P(I)$ is at most the optimal value of I.*

Proof. Let F be an optimal solution of I. We define a solution (x, y, z) of $P(I)$ from F. For each $C \in \mathcal{E}$, we set $z(C)$ to 0 if $F \cap C \neq \emptyset$, and 1 otherwise. If $F \cap C \neq \emptyset$, we choose an arbitrary edge $e \in F \cap C$, and let $y(C, e) = 1$. For the remaining edges e', we assign $y(C, e') = 0$. In this way, the values of variables in y are defined for each $C \in \mathcal{E}$. $x(e)$ is set to 1 if $e \in F$, and 0 otherwise. $x(v)$ is set to 1 if F contains an edge incident to v, and 0 otherwise. (x, y, z) is feasible, and its objective value in $P(I)$ is the optimal value of I.   □

In some graph covering problems, $\mathcal{E}$ is not explicitly given, and $|\mathcal{E}|$ is not bounded by a polynomial on the input size of the problem. In such cases, the above LP may not be solved in polynomial time because it cannot be written compactly. However, in this scenario, we may define a tighter LP than the natural relaxation if we can find $\mathcal{E}_1, \ldots, \mathcal{E}_t \subseteq \mathcal{E}$ such that $\cup_{i=1}^{t} \mathcal{E}_i = \mathcal{E}$, t is bounded by a polynomial of input size, and the degree of each node is small in any minimal edge set covering all demand sets in $\mathcal{E}_i$ for each $i \in \{1, \ldots, t\}$. Applying these conditions, the present author obtained a new approximation algorithm for solving a problem generalizing some prize-collecting graph covering problems [11].

4   Prize-Collecting EDS Problem in Trees

In this section, we prove Theorem 1. We regard the input graph G as a rooted tree, with an arbitrary node r selected as the root. The *depth* of a node v is the number of edges on the path between r and v. When v lies on the path between r and another node u, we say that v is an *ancestor* of u and u is a *descendant* of v. If the depth of node v is the maximum among all ancestors of u, then v is defined as the *parent* of u. If v is the parent of u, then u is a *child* of v. The upper and lower end nodes of an edge e are denoted by u_e and l_e, respectively. We say that an edge e is an ancestor of a node v and v is a descendant of e when $l_e = v$ or l_e is an ancestor of v. Similarly, an edge e is a descendant of a node v and v is an ancestor of e if $v = u_e$ or v is an ancestor of u_e. An edge e is defined as an ancestor of another edge f if e is an ancestor of u_f.

Recall that $\mathcal{E} = \{\delta(e) : e \in E\}$ in the EDS problem. Let $I = (G, w, \pi)$ be an instance of the prize-collecting EDS problem. We denote $\bigcup_{e \in \delta(v)} \delta(e)$ by $\delta'(v)$ for each $v \in V$. Then the dual of $P(I)$ is formulated as follows.

$D(I) =$

maximize $\displaystyle\sum_{e \in E} \xi(e)$

subject to $\displaystyle\sum_{e \in \delta(e')} \nu(e', e) \leq w(e')$ for $e' \in E$,   (1)

$\displaystyle\sum_{e \in \delta'(v)} \mu(v, e) \leq w(v)$ for $v \in V$,   (2)

$\xi(e) \leq \mu(u, e) + \mu(v, e) + \nu(e', e)$ for $e \in E, e' = uv \in \delta(e)$,   (3)

$\xi(e) \leq \pi(e)$ for $e \in E$,   (4)

$\xi(e) \geq 0$ for $e \in E$,

$\nu(e', e) \geq 0$ for $e' \in E, e \in \delta(e')$,

$\mu(v, e) \geq 0$ for $v \in V, e \in \delta'(v)$.

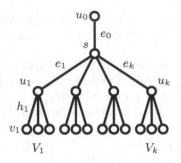

Fig. 1. Edges and nodes in Case B

For an edge set $F \subseteq E$, let $\tilde{F}$ denote $\{e \in E: \delta_F(e) = \emptyset\}$, and let $\pi(\tilde{F})$ denote $\sum_{e \in \tilde{F}} \pi(e)$. For the instance I, our algorithm yields a solution $F \subseteq E$ and a feasible solution (ξ, ν, μ) to $D(I)$, both satisfying

$$w(F) + w(V(F)) + \pi(\tilde{F}) \le \sum_{e \in E} \xi(e). \tag{5}$$

Since the right-hand side of (5) is at most $P(I)$, F is an optimal solution of I. We note that the dual solution (ξ, ν, μ) is required only for proving the optimality of the solution and need not be computed.

The algorithm operates by induction on the number of nodes of depth exceeding one. In the base case, all nodes are of depth one, indicating that G is a star centered at r. The alternative case is divided into two sub-cases: Case A, in which a leaf edge e of maximum depth satisfies $\pi(e) > 0$; and Case B, which contains no such leaf edge. In this paper, we discuss only Case B due to the space limination.

Case B

In this case, $\pi(e) = 0$ holds for all leaf edges e of maximum depth. Let s be the grandparent of a leaf node of maximum depth. Also, let $u_1, \ldots, u_k$ be the children of s, and e_i be the edge joining s and u_i for $i \in [k]$. In the following discussion, we assume that s has a parent, and that each node u_i has at least one child. This discussion is easily modified to cases in which s has no parent or some node u_i has no child. We denote the parent of s by u_0, and the edge between u_0 and s by e_0. For each $i \in [k]$, let V_i be the set of children of u_i, and H_i be the set of edges joining u_i to its child nodes in V_i. Also define $h_i = u_i v_i$ as an edge that attains $\min_{u_i v \in H_i}(w(u_i v) + w(v))$. The relationships between these nodes and edges are illustrated in Fig. 1.

Now define $\theta_1 = \min_{i=0}^{k}(w(e_i) + w(u_i) + w(s))$, $\theta_2 = \sum_{i=1}^{k} \min\{w(u_i) + w(v_i) + w(h_i), \pi(e_i)\}$, and let $\theta = \min\{\theta_1, \theta_2\}$. We denote the index $i \in [k]$ of an edge e_i that attains $\theta_1 = w(e_i) + w(u_i) + w(s)$ by i^*, and specify $K = \{i \in [k]: w(u_i) + w(v_i) + w(h_i) \le \pi(e_i)\}$. For a real number ψ, we let $(\psi)_+$ denote $\max\{0, \psi\}$.

We define $I' = (G', w', \pi')$ as follows. If $\theta_1 \geq \theta_2$, then G' is the tree obtained by removing all edges in $\bigcup_{i \in [k]} H_i$ and all nodes in $\bigcup_{i \in [k]} V_i$ from G, and $\pi' \colon E' \to \mathbb{R}_+$ is defined such that

$$\pi'(e) = \begin{cases} 0 & \text{if } e \in \{e_1, \ldots, e_k\}, \\ \pi(e) & \text{otherwise} \end{cases}$$

for $e \in E'$. In this case, $w' \colon V' \cup E' \to \mathbb{R}_+$ is defined by

$$w'(v) = \begin{cases} (w(s) - \theta)_+ & \text{if } v = s, \\ w(u_i) - (\theta - w(s) - w(e_i))_+ & \text{if } v = u_i, i \in [k]^*, \\ w(v) & \text{otherwise} \end{cases}$$

for $v \in V'$, and

$$w'(e) = \begin{cases} (w(e_i) - (\theta - w(s))_+)_+ & \text{if } e = e_i, i \in [k]^*, \\ w(e) & \text{otherwise,} \end{cases}$$

for $e \in E'$. If $\theta_1 < \theta_2$, then $e_1, \ldots, e_k$, and their descendants are removed from G to obtain G', and π' is defined by

$$\pi'(e) = \begin{cases} 0 & \text{if } e = e_0, \\ \pi(e) & \text{otherwise.} \end{cases}$$

Moreover, w' for E' and V' is defined as in the case $\theta_1 \geq \theta_2$, disregarding the weights of edges and nodes removed from G'.

Since G' has fewer nodes of depth exceeding one than G, the algorithm inductively finds a solution F' to I', and a feasible solution (ξ', ν', μ') to $D(I')$ satisfying (5). F is constructed from F' as follows.

$$F = \begin{cases} F' \cup \{e_0\} & \text{if } \delta_{F'}(u_0) \neq \emptyset, \theta > w(s) + w(e_0), \\ F' & \text{if } \delta_{F'}(u_0) = \emptyset \text{ or } \theta \leq w(s) + w(e_0), \delta_{F'}(s) \neq \emptyset, \\ F' \cup \{h_i : i \in K\} & \text{if } \delta_{F'}(u_0) = \emptyset \text{ or } \theta \leq w(s) + w(e_0), \delta_{F'}(s) = \emptyset, \theta_1 \geq \theta_2, \\ F' \cup \{e_{i^*}\} & \text{if } \delta_{F'}(u_0) = \emptyset \text{ or } \theta \leq w(s) + w(e_0), \delta_{F'}(s) = \emptyset, \theta_1 < \theta_2. \end{cases}$$

We define $\xi(e_1), \ldots, \xi(e_k)$ such that $\xi(e_i) \leq \min\{w(u_i) + w(v_i) + w(h_i), \pi(e_i)\}$ for $i \in [k]$ and $\sum_{i=1}^{k} \xi(e_i) = \theta$, which is possible because $\sum_{i=1}^{k} \min\{w(u_i) + w(v_i) + w(h_i), \pi(e_i)\} = \theta_2 \geq \theta$. We also define $\xi(e) = 0$ for each $e \in \bigcup_{i=1}^{k} H_i$. The other variables in ξ are set to their values in ξ'. The following lemma states that this ξ can form a feasible solution to $D(I)$.

Lemma 1. *Suppose that $\xi(e_1), \ldots, \xi(e_k)$ satisfy $\xi(e_i) \leq \min\{w(u_i) + w(v_i) + w(h_i), \pi(e_i)\}$ for each $i \in [k]$ and $\sum_{i=1}^{k} \xi(e_i) = \theta$. Further, suppose that $\xi(e) = 0$ holds for each $e \in \bigcup_{i=1}^{k} H_i$, and the other variables in ξ are set to their values in ξ'. Then there exist ν and μ such that (ξ, ν, μ) is feasible to $D(I)$.*

Proof. For $i \in [k]$ and $v \in V_i$, we define $\mu(v, e_i)$ and $\nu(u_iv, e_i)$ such that $\mu(v, e_i) + \nu(u_iv, e_i) = \min\{w(v_i) + w(h_i), \xi(e_i)\}$. This may be achieved without violating the constraints, because $w(v) + w(u_iv) \geq w(v_i) + w(h_i)$. We also define $\nu(u_iv, e_i)$ as $(\xi(e_i) - w(u_i) - w(h_i))_+$. These variables satisfy (1) for u_iv, (2) for v and u_i, and (3) for (e_i, u_iv). $\nu(e_j, e_i)$ for $i \in [k]$ and $j \in [k]^*$, and $\mu(v, e_i)$ for $i \in [k]$ and $v \in \{s\} \cup \{u_j \colon j \in [k]^*, j \neq i\}$ are set to 0. The other variables in ν and μ are set to their values in ν' and μ'. To advance the proof, we introduce an algorithm that increases $\nu(e_j, e_i)$ for $i \in [k]$ and $j \in [k]^*$, and $\mu(v, e_i)$ for $i \in [k]$ and $v \in \{s, u_0, \ldots, u_k\}$. At the completion of the algorithm, (ξ, ν, μ) is a feasible solution to $D(I)$.

The algorithm performs k iterations, and the i-th iteration increases the variables to satisfy (3) for each pair of e_i and e_j, where $j \in [k]^*$. The algorithm retains a set Var of variables to be increased. We introduce a notion of time: Over one unit of time, the algorithm simultaneously increases all variables in Var by one. The time consumed by the i-th iteration is $\xi(e_i)$.

At the beginning of the i-th iteration, Var is initialized to $\{\mu(u_j, e_i) \colon j \in [k]^*\}$. The algorithm updates Var during the i-th iteration as follows:

- At time $(\xi(e_i) - w(v_i) - w(h_i))_+$, $\mu(u_i, e_i)$ is added to Var if Var $\neq \{\mu(s, e_i)\}$;
- If (2) becomes tight for u_j under the increase of $\mu(u_j, e_i) \in$ Var, then $\mu(u_j, e_i)$ is replaced by $\nu(e_j, e_i)$ for each $j \in [k]^*$;
- If (1) becomes tight for e_j under the increase of $\nu(e_j, e_i) \in$ Var with some $j \in [k]^*$, then Var is reset to $\{\mu(s, e_i)\}$.

We note that the time spent between two consecutive updates may be zero.

Var always contains a variable that appears in the right-hand side of (3) for (e_i, e_j) with $j \in [k]^* \setminus \{i\}$, and for (e_i, e_i) after time $(\xi(e_i) - w(v_i) - w(h_i))_+$. The algorithm updates Var so that (1) and (2) hold for all variables except s. Hence, to show that (ξ, ν, μ) is a feasible solution to $D(I)$, it suffices to show that (2) for s does not become tight before the algorithm is completed.

We complete the proof by contradiction. Suppose that (2) for s tightens at time $\tau < \xi(e_i)$ in the i-th iteration. Since Var $= \{\mu(s, e_i)\}$ at this moment, there exists $j \in [k]^*$ such that (1) for e_j and (2) for u_j are tight. The variables in the left-hand sides of (1) for e_j and (2) for u_j and s are not simultaneously increased. Nor are these variables increased over time $(\xi(e_j) - w(v_j) - w(h_j))_+$ in the j-th iteration, and $\mu(u_j, e_j)$ is initialized to $(\xi(e_j) - w(v_j) - w(h_j))_+$. From this argument, it follows that $w(s) + w(u_j) + w(e_j) < \sum_{i'=1}^{k} \xi(e_{i'}) \leq \theta$. However, this result is contradicted by the definition of θ, which implies that $\theta \leq \theta_1 \leq w(s) + w(u_j) + w(e_j)$. Thus, the claim is proven. □

Lemma 2. F and ξ satisfy (5).

Proof. For each $i \in [k]$, either $e_i \notin E'$ holds, or $\xi'(e_i) = 0$ holds (because $\pi'(e_i) = 0$). Hence, $\sum_{e \in E} \xi(e) = \sum_{i=1}^{k} \xi(e_i) + \sum_{e \in E'} \xi'(e) = \theta + \sum_{e \in E'} \xi'(e)$. Therefore, it suffices to prove that $\sum_{e \in F} w(e) \leq \theta + \sum_{e \in F'} w'(e)$.

Without loss of generality, we can assume $|\delta_{F'}(e_0)| \leq 1$ (if false, we can remove edges e_i, $i \in [k]^*$ from F' until $|\delta_{F'}(e_0)| = 1$). In the sequel, we discuss only the case of $\delta_{F'}(u_0) \neq \emptyset$ and $\theta > w(s) + w(e_0)$. In the alternative case, the claim immediately follows from the definitions of F and w'. $\delta_{F'}(u_0) \neq \emptyset$ implies that $w'(u_0)$ is counted in the objective value of F'. Moreover, $w'(s) = w'(e_0) = 0$ follows from $\theta > w(s) + w(e_0)$. Thus, the objective values increase from F' to F by $w(u_0) - w'(u_0) + w(e_0) + w(s)$, which equals θ. □

5   Conclusion

In this paper, we emphasized a large integrality gap when the natural LP relaxation is applied to the graph covering problem that minimizes node weights. We then formulated an alternative LP relaxation for graph covering problems in edge- and node-weighted graphs that is stronger than the natural relaxation. This relaxation was incorporated into an exact algorithm for the prize-collecting EDS problem in trees, and a 2-approximation algorithm for the multicut problem in trees. The approximation guarantees for these algorithms match the previously known best results for purely edge-weighted graphs. In many other graph covering problems, the integrality gap in the proposed relaxation would increase if node weights were introduced, because the problems in node-weighted graphs admit stronger hardness results. Nonetheless, the proposed relaxation is a potentially useful tool for designing heuristics or using IP solvers to solve the above problems.

References

1. Bateni, M., Hajiaghayi, M., Liaghat, V.: Improved approximation algorithms for (budgeted) node-weighted Steiner problems. In: Fomin, F.V., Freivalds, R., Kwiatkowska, M., Peleg, D. (eds.) ICALP 2013, Part I. LNCS, vol. 7965, pp. 81–92. Springer, Heidelberg (2013)
2. Berger, A., Fukunaga, T., Nagamochi, H., Parekh, O.: Approximability of the capacitated b-edge dominating set problem. Theoretical Computer Science 385(1-3), 202–213 (2007)
3. Berger, A., Parekh, O.: Linear time algorithms for generalized edge dominating set problems. Algorithmica 50(2), 244–254 (2008)
4. Berger, A., Parekh, O.: Erratum to: Linear time algorithms for generalized edge dominating set problems. Algorithmica 62(1-2), 633–634 (2012)
5. Byrka, J., Grandoni, F., Rothvoß, T., Sanità, L.: Steiner tree approximation via iterative randomized rounding. Journal of the ACM 60(1), 6 (2013)
6. Carr, R.D., Fujito, T., Konjevod, G., Parekh, O.: A $2\frac{1}{10}$-approximation algorithm for a generalization of the weighted edge-dominating set problem. Journal of Combinatorial Optimization 5(3), 317–326 (2001)
7. Chawla, S., Krauthgamer, R., Kumar, R., Rabani, Y., Sivakumar, D.: On the hardness of approximating multicut and sparsest-cut. Computational Complexity 15(2), 94–114 (2006)
8. Chekuri, C., Ene, A., Vakilian, A.: Prize-collecting survivable network design in node-weighted graphs. In: Gupta, A., Jansen, K., Rolim, J., Servedio, R. (eds.) APPROX/RANDOM 2012. LNCS, vol. 7408, pp. 98–109. Springer, Heidelberg (2012)

9. Dahlhaus, E., Johnson, D.S., Papadimitriou, C.H., Seymour, P.D., Yannakakis, M.: The complexity of multiterminal cuts. SIAM Journal on Computing 23(4), 864–894 (1994)
10. Fujito, T., Nagamochi, H.: A 2-approximation algorithm for the minimum weight edge dominating set problem. Discrete Applied Mathematics 118(3), 199–207 (2002)
11. Fukunaga, T.: Spider covers for prize-collecting network activation problem. CoRR abs/1310.5422 (2013)
12. Fukunaga, T.: Covering problems in edge- and node-weighted graphs. CoRR abs/1404.4123 (2014)
13. Garg, N., Vazirani, V.V., Yannakakis, M.: Approximate max-flow min-(multi)cut theorems and their applications. SIAM Journal on Computing 25(2), 235–251 (1996)
14. Garg, N., Vazirani, V.V., Yannakakis, M.: Primal-dual approximation algorithms for integral flow and multicut in trees. Algorithmica 18(1), 3–20 (1997)
15. Goemans, M.X., Olver, N., Rothvoß, T., Zenklusen, R.: Matroids and integrality gaps for hypergraphic steiner tree relaxations. In: STOC, pp. 1161–1176 (2012)
16. Hajiaghayi, M., Liaghat, V., Panigrahi, D.: Online node-weighted Steiner forest and extensions via disk paintings. In: FOCS, pp. 558–567 (2013)
17. Kamiyama, N.: The prize-collecting edge dominating set problem in trees. In: Hliněný, P., Kučera, A. (eds.) MFCS 2010. LNCS, vol. 6281, pp. 465–476. Springer, Heidelberg (2010)
18. Khot, S., Regev, O.: Vertex cover might be hard to approximate to within $2 - \epsilon$. Journal of Computer and System Sciences 74(3), 335–349 (2008)
19. Klein, P.N., Ravi, R.: A nearly best-possible approximation algorithm for node-weighted Steiner trees. Journal of Algorithms 19(1), 104–115 (1995)
20. Könemann, J., Sadeghabad, S.S., Sanità, L.: An LMP $O(\log n)$-approximation algorithm for node weighted prize collecting Steiner tree. In: FOCS, pp. 568–577 (2013)
21. Moss, A., Rabani, Y.: Approximation algorithms for constrained node weighted Steiner tree problems. SIAM Journal on Computing 37(2), 460–481 (2007)
22. Naor, J., Panigrahi, D., Singh, M.: Online node-weighted Steiner tree and related problems. In: FOCS, pp. 210–219 (2011)
23. Nutov, Z.: Approximating Steiner networks with node-weights. SIAM Journal on Computing 39(7), 3001–3022 (2010)
24. Nutov, Z.: Approximating minimum-cost connectivity problems via uncrossable bifamilies. ACM Transactions on Algorithms 9(1), 1 (2012)
25. Nutov, Z.: Survivable network activation problems. In: Fernández-Baca, D. (ed.) LATIN 2012. LNCS, vol. 7256, pp. 594–605. Springer, Heidelberg (2012)
26. Panigrahi, D.: Survivable network design problems in wireless networks. In: SODA, pp. 1014–1027 (2011)
27. Parekh, O.: Approximation algorithms for partially covering with edges. Theoretical Computer Science 400(1-3), 159–168 (2008)
28. Vakilian, A.: Node-weighted prize-collecting survivable network design problems. Master's thesis, University of Illinois at Urbana-Champaign (2013)

Colored Range Searching in Linear Space

Roberto Grossi[1,*] and Søren Vind[2,**]

[1] Università di Pisa, Dipartimento di Informatica
grossi@di.unipi.it
[2] Technical University of Denmark, DTU Compute
sovi@dtu.dk

Abstract. In *colored range searching*, we are given a set of n colored points in $d \geq 2$ dimensions to store, and want to support orthogonal range queries taking colors into account. In the *colored range counting* problem, a query must report the number of distinct colors found in the query range, while an answer to the *colored range reporting* problem must report the distinct colors in the query range.

We give the first linear space data structure for both problems in two dimensions ($d = 2$) with $o(n)$ worst case query time. We also give the first data structure obtaining almost-linear space usage and $o(n)$ worst case query time for points in $d > 2$ dimensions. Finally, we present the first dynamic solution to both counting and reporting with $o(n)$ query time for $d \geq 2$ and $d \geq 3$ dimensions, respectively.

1 Introduction

In standard range searching a set of points must be stored to support queries for any given orthogonal d-dimensional query box Q (see [3] for an overview). Two of the classic problems are standard range reporting, asking for the points in Q, and standard range counting, asking for the number of points in Q. In 1993, Janardan and Lopez [10] introduced one natural generalisation of standard range searching, requiring each point to have a color. This variation is known as *colored range searching*[1]. The two most studied colored problems are *colored range reporting*, where a query answer must list the distinct colors in Q, and *colored range counting*, where the number of distinct colors in Q must be reported.

As shown in Tables 1–2, there has been a renewed interest for these problems due to their applications. For example, in database analysis and information retrieval the d-dimensional points represent entities and the colors are *classes* (or categories) of entities. Due to the large amount of entities, statistics about their classes is the way modern data processing is performed. A colored range query works in this scenario and reports some *statistical analysis* for the classes using

---

* Partially supported by Italian MIUR PRIN project AMANDA.

** Supported by a grant from the Danish National Advanced Technology Foundation. Research carried out while the author was visiting Università di Pisa.

[1] Also known in the literature as *categorical* or *generalized* range searching.

R Ravi and I.L. Gørtz (Eds.): SWAT 2014, LNCS 8503, pp. 229–240, 2014.

Table 1. Known and new solutions to *colored range counting*, ordered according to decreasing space use in each dimension group. Dyn. column shows if solution is dynamic.

Dim.	Query Time	Space Usage	Dyn.	Ref.
$d = 1$	$O(\lg n / \lg \lg n)$	$O(n)$	$\times$	[9]
$d = 2$	$O(\lg^2 n)$	$O(n^2 \lg^2 n)$		[7]
	$O(\lg^2 n)$	$O(n^2 \lg^2 n)$		[11]
	$O(X \lg^7 n)$	$O((n/X)^2 \lg^6 n + n \lg^4 n)$		[11]
	$O\left(\left(\frac{\sigma}{\lg n} + \frac{n}{\lg^c n}\right)(\lg \lg^c n)^{2+\epsilon}\right)$	$O(n)$		New
$d > 2$	$O(\lg^{2(d-1)} n)$	$O(n^d \lg^{2(d-1)} n)$		[11]
	$O(X \lg^{d-1} n)$	$O((n/X)^{2d} + n \lg^{d-1} n)$		[11]
	$O\left(\left(\frac{\sigma}{\lg n} + \frac{n}{\lg^c n}\right)(\lg \lg^c n)^{d-1} \lg \lg \lg^c n\right)$	$O(n(\lg \lg^c n)^{d-1})$		New
$d \geq 2$	$O\left(\left(\frac{\sigma}{\lg n} + \frac{n}{\lg^c n}\right)(\lg \lg^c n)^d\right)$	$O(n(\lg \lg^c n)^{d-1})$	$\times$	New

the range on the entities as a filtering method: "which kinds of university degrees do European workers with age between 30 and 40 years and salary between 30,000 and 50,000 euros have?". Here the university degrees are the colors and the filter is specified by the range of workers that are European with the given age and salary. The large amount of entities involved in such applications calls for nearly *linear space* data structures, which is the focus of this paper.

Curiously, counting is considered harder than reporting among the colored range queries as it is not a decomposable problem: knowing the number of colors in two halves of Q does not give the query answer. This is opposed to reporting where the query answer can be obtained by merging the list of colors in two halves of Q and removing duplicates. For the standard problems, both types of queries are decomposable and solutions to counting are generally most efficient.

In the following, we denote the set of d-dimensional points by P and let $n = |P|$. The set of distinct colors is Σ and $\sigma = |\Sigma| \leq n$, with $k \leq \sigma$ being the number of colors in the output. We use the notation $\lg^a b = (\lg b)^a$ and adopt the RAM with word size $w = \Theta(\lg n)$ bits, and the size of a data structure is the number of occupied words.

Observe that for both colored problems, the trivial solution takes linear time and space, storing the points and looking through all of them to answer the query. Another standard solution is to store one data structure for all points of each color that supports standard range emptiness queries ("is there any point inside Q?"). In two dimensions, this approach can answer queries in $O(\sigma \lg n)$ time and linear space using a range emptiness data structure by Nekrich [17]. However, since $\sigma = n$ in the worst case, this does not guarantee a query time better than trivial. Due to the extensive history and the number of problems considered, we will present our results before reviewing existing work.

Table 2. Known and new solutions to *colored range reporting*, ordered according to decreasing space use in each dimension group. Dyn. column shows if solution is dynamic.

Dim.	Query Time	Space Usage	Dyn.	Ref.
$d = 1$	$O(1 + k)$	$O(n)$	$\times$	[20]
$d = 2$	$O(\lg n + k)$	$O(n \lg n)$		[23]
	$O(\lg^2 n + k \lg n)$	$O(n \lg n)$	$\times$	[4,7]
	$O\left(\left(\frac{\sigma}{\lg n} + \frac{n}{\lg^c n}\right)(\lg \lg^c n)^{2+\epsilon} + k\right)$	$O(n)$		New
$d \geq 2$	$O\left(\left(\frac{\sigma}{\lg n} + \frac{n}{\lg^c n}\right)(\lg \lg^c n)^d + k\right)$	$O(n(\lg \lg^c n)^{d-1})$	$\times$	New
$d = 3$	$O(\lg^2 n + k)$	$O(n \lg^4 n)$		[7]
$d > 3$	$O(\lg n + k)$	$O(n^{1+\epsilon})$		[8,12]
$d \geq 3$	$O\left(\left(\frac{\sigma}{\lg n} + \frac{n}{\lg^c n}\right)(\lg \lg^c n)^{d-1} \lg \lg \lg^c n + k\right)$	$O(n(\lg \lg^c n)^{d-1})$		New

1.1 Our Results

We observe (see Section 1.2) that there are no known solutions to any of the colored problems in two dimensions that uses $O(n)$ words of space and answer queries in $o(n)$ worst case time. Furthermore, for colored range reporting there are no known solutions in $d > 3$ dimensions using $o(n^{1+\epsilon})$ words of space and answering queries in $o(n)$ worst case time. For colored range counting, no solutions with $o(n \operatorname{polylg} n)$ words of space and $o(n)$ worst case time exist.

We present the first data structures for colored range searching achieving these bounds, improving almost logarithmically over previously known solutions in the worst case (see Section 1.2 and Tables 1–2). Specifically, we obtain

- $o(n)$ query time and $O(n)$ space in two dimensions,
- $o(n)$ query time and $o(n \operatorname{polylg} n)$ space for counting in $d \geq 2$ dimensions,
- $o(n)$ query time and $o(n^{1+\epsilon})$ space for reporting in $d > 3$ dimensions,
- $o(n)$ query time supporting $O(\operatorname{polylg} n)$ updates in $d \geq 2$ and $d \geq 3$ dimensions for counting and reporting, respectively.

We note that while our bounds have an exponential dimensionality dependency (as most previous results), it only applies to $\lg \lg n$ factors in the bounds. Our solutions can be easily implemented and parallelized, so they are well-suited for the distributed processing of large scale data sets. Our main results can be summarised in the following theorems, noting that $c > 1$ is an arbitrarily chosen integer, and $\sigma \leq n$ is the number of distinct colors.

Theorem 1. *There is a linear space data structure for two-dimensional colored range counting and reporting storing n points, each assigned one of σ colors. The data structure answers queries in time $O((\sigma/\lg n + n/\lg^c n)(\lg \lg^c n)^{2+\epsilon})$, with reporting requiring an additive term $O(k)$.*

Theorem 2. *There is a $O(n(\lg \lg^c n)^{d-1})$ space data structure storing n d-dimensional colored points each assigned one of σ colors. Each* *colored range counting and reporting query takes time $O((\sigma/\lg n + n/\lg^c n)(\lg\lg^c n)^{d-1}\lg\lg\lg^c n)$. Reporting requires an additive $O(k)$ time.*

To obtain these results, we partition points into groups depending on their color. Each group stores all the points for at most $\lg n$ specific colors. Because the colors are partitioned across the groups, we can obtain the final result to a query by merging query results for each group (and we have thus obtained a decomposition of the problem along the color dimension). A similar approach was previously used in [11].

In order to reduce the space usage of our data structure, we partition the points in each group into a number of buckets of at most $\lg^c n$ points each. The number of buckets is $O(\sigma/\lg n + n/\lg^c n)$, with the first term counting all underfull buckets and the second counting all full buckets. Each bucket stores $m \le \lg^c n$ points colored with $f \le \lg n$ different colors. To avoid counting a color several times across different buckets, we use a solution to the d-dimensional colored range reporting problem in each bucket for which answers to queries are given as bitstrings. Answers to the queries in buckets can be merged efficiently using bitwise operations on words. We finally use an $o(n)$ space lookup table to obtain the count or list of colors present in the merged answer.

The solution to d-dimensional colored range reporting for each bucket is obtained by building a d-dimensional range tree for the m points, which uses a new linear space and $O(\lg\lg m)$ time solution to restricted one-dimensional colored range reporting as the last auxiliary data structure. In total, each bucket requires $O(m\lg^{d-1}m)$ space and $O(\lg^{d-1}m\lg\lg m)$ query time. In two dimensions, we reduce the space to linear by only storing representative rectangles in the range tree covering $O(\lg m)$ points each. Using the linear space range reporting data structure by Nekrich [17], we enumerate and check the underlying points for each of the $O(\lg m)$ range tree leaves intersecting the query range, costing us a small penalty of $O(\lg^\epsilon n)$ per point. We thus obtain a query time of $O(\lg^{2+\epsilon} m)$ for two-dimensional buckets in linear space.

Using classic results on dynamisation of range trees, we can dynamise the data structure with a little additional cost in query time. Previously, there was no known dynamic data structures with $o(n)$ query time for colored range counting in $d \ge 2$ dimensions and colored range reporting in $d \ge 3$ dimensions. Consequently, this is the first such dynamic data structure.

Theorem 3. *There is a dynamic $O(n(\lg\lg^c n)^{d-1})$ space data structure storing n d-dimensional colored points each assigned one of σ colors. The data structure answers colored range counting and reporting queries in time $O((\sigma/\lg n + n/\lg^c n)(\lg\lg^c n)^d)$. Reporting requires an additive $O(k)$ time and updates are supported in $O((\lg\lg^c n)^d)$ amortised time.*

Finally, if paying a little extra space, we can get a solution to the problem where the query time is bounded by the number of distinct colors instead of the number of points. This is simply done by not splitting color groups into buckets, giving the following result.

Corollary 1. *There is a $O(n \lg^{d-1} n)$ space data structure for n d-dimensional colored points each assigned one of σ colors. The data structure answers colored range counting and reporting queries in time $O(\sigma \lg^{d-2} n \lg \lg n)$. Reporting requires a additive $O(k)$ time.*

In two dimensions this is a logarithmic improvement over the solution where a range emptiness data structure is stored for each color at the expense of a $\lg n$ factor additional space. The above approach can be combined with the range emptiness data structure by Nekrich [17] to obtain an output-sensitive result where a penalty is paid per color reported:

Corollary 2. *There is a $O(n \lg n)$ space data structure storing n two-dimensional colored points each assigned one of σ colors. The data structure answers colored range counting and reporting queries in time $O(\sigma + k \lg n \lg \lg n)$.*

1.2   Previous Results

Colored range counting. Colored range counting is challenging, with a large gap in the known bounds compared to standard range counting, especially in two or more dimensions. For example, a classic range tree solves two-dimensional standard range counting in logarithmic time and $O(n \lg n)$ space, but no poly-logarithmic time solutions in $o(n^2)$ space are known for colored range counting.

Larsen and van Walderveen [7, 14] showed that colored range counting in one dimension is equivalent to two-dimensional standard range counting. Thus, the optimal $O(\lg n / \lg \lg n)$ upper bound for two-dimensional standard range counting by JáJá et al. [9] which matches a lower bound by Patrascu [22] is also optimal for one-dimensional colored range counting.

In two dimensions, Gupta et al. [7] show a solution using $O(n^2 \lg^2 n)$ space that answers queries in $O(\lg^2 n)$ time. They obtain their result by storing n copies of a data structure which is capable of answering three-sided queries. The same bound was matched by Kaplan et al. [11] with a completely different approach in which they reduce the problem to standard orthogonal range counting in higher dimensions. Kaplan et al. also present a tradeoff solution with $O(X \lg^7 n)$ query time and $O((\frac{n}{X})^2 \lg^6 n + n \lg^4 n)$ space for $1 \leq X \leq n$. Observe that the minimal space use for the tradeoff solution is $O(n \lg^4 n)$.

In $d > 2$ dimensions, the only known non-trivial solutions are by Kaplan et al. [11]. One of their solutions answers queries in $O(\lg^{2(d-1)} n)$ time and $O(n^d \lg^{2(d-1)} n)$ space, and they also show a number of tradeoffs, the best one having $O(X \lg^{d-1} n)$ query time and using $O((\frac{n}{X})^{2d} + n \lg^{d-1} n)$ space for $1 \leq X \leq n$. In this case, the minimal space required by the tradeoff is $O(n \lg^{d-1} n)$.

Kaplan et al. [11] showed that answering n two dimensional colored range counting queries in $O(n^{p/2})$ time (including all preprocessing time) yields an $O(n^p)$ time algorithm for multiplying two $n \times n$ matrices. For $p < 2.373$, this would improve the best known upper bound for matrix multiplication [25]. Thus, solving two dimensional colored range counting in polylogarithmic time per query and $O(n \operatorname{polylg} n)$ space would be a major breakthrough. This suggest that even in two dimensions, no polylogarithmic time solution may exist.

Colored range reporting. The colored range reporting problem is relatively well-studied [4, 6–8, 10, 12, 13, 16, 18–20, 23], with output-sensitive solutions almost matching the time and space bounds obtained for standard range reporting in one and two dimensions. In particular, Nekrich and Vitter recently gave a dynamic solution to one dimensional colored range reporting with optimal query time $O(1 + k)$ and linear space [20], while Gagie et al. earlier presented a succinct solution with query time logarithmic in the length of the query interval [6].

In two dimensions, Shi and JáJá obtain a bound of $O(\lg n + k)$ time and $O(n \lg n)$ space [23] by querying an efficient static data structure for three-sided queries, storing each point $O(\lg n)$ times. Solutions for the dynamic two-dimensional case were developed in [4, 7], answering queries with a logarithmic penalty per answer. If the points are located on an $N \times N$ grid, Agarwal et al. [1] present a solution with query time $O(\lg \lg N + k)$ and space use $O(n \lg^2 N)$. Gupta et al. achieve a static data structure using $O(n \lg^4 n)$ space and answering queries in $O(\lg^2 n + k)$ [7] in the three-dimensional case. To the best of our knowledge, the only known non-trivial data structures for $d > 3$ dimensions are by van Kreveld and Gupta et al., answering queries in $O(\lg n + k)$ time and using $O(n^{1+\epsilon})$ space [8, 12]. Other recent work on the problem include external memory model solutions when the points lie on a grid [13, 18, 19].

2   Colored Range Searching in Almost-Linear Space

We present here the basic approach that is modified to obtain our theorems. We first show how to partition the points into $O(\sigma/\lg n + n/\lg^c n)$ buckets each storing $m = O(\lg^c n)$ points of $f = O(\lg n)$ distinct colors, for which the results can be easily combined. We then show how to answer queries in each bucket in time $O(\lg^{d-1} m \lg \lg m)$ and space $O(m \lg^{d-1} m)$, thus obtaining Theorem 2.

2.1   Color Grouping and Bucketing

We partition the points of P into a number of groups P_i, where $i = 1, \ldots, \frac{\sigma}{\lg n}$, depending on their color. Each group stores all points having $f = \lg n$ distinct colors (except for the last group which may store points with less distinct colors). For each group P_i we store an ordered color list L_i of the f colors in the group. That is, a group may contain $O(n)$ points but the points have at most f distinct colors. Since colors are partitioned among groups, we can clearly answer a colored query by summing or merging the results to the same query in each group.

Each group is further partitioned into a number of buckets containing $m = \lg^c n$ points each (except for the last bucket, which may contain fewer points). Since the buckets partition the points and there cannot be more than one bucket with fewer than $\lg^c n$ points in each group, the total number of buckets is $O(\sigma/\lg n + n/\lg^c n)$. See Figure 1 for an example of the grouping and bucketing.

We require that each bucket in a group P_i supports answering *restricted colored range reporting queries* with an f-bit long bitstring, where the jth bit

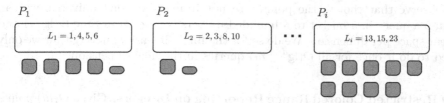

Fig. 1. Grouping and bucketing of some point set with $f = 4$. The white boxes are the groups and the grey boxes below each group are buckets each storing $O(\lg^c n)$ points.

indicates if color $L_i[j]$ is present in the query area Q in that bucket. Clearly, we can obtain the whole answer for Q and P_i by using bitwise OR operations to merge answers to the restricted colored range reporting query Q for all buckets in P_i. We call the resulting bitstring $F_{i,Q}$, which indicates the colors present in the query range Q across the entire group P_i.

Finally, we store a lookup table T of size $O(\sqrt{n} \lg n) = o(n)$ for all possible bitstrings of length $\frac{f}{2}$. For each bitstring, the table stores the number of 1s present in the bitstring and the indices where the 1s are present. Using two table lookups in T with the two halves of $F_{i,Q}$, we can obtain both the number of colors present in P_i for Q, and their indices into L_i in $O(1)$ time per index.

Summarising, we can merge answers to the restricted colored range reporting queries in $O(1)$ time per bucket and obtain the full query results for each group P_i. Using a constant number of table lookups per group, we can count the number of colors present in Q. There is $O(1)$ additional cost per reported color.

2.2 Restricted Colored Range Reporting for Buckets

Each bucket in a group P_i stores up to $m = \lg^c n$ points colored with up to $f = \lg n$ distinct colors, and must support restricted colored range reporting queries, reporting the colors in query range Q using an f-bit long bitstring. A simple solution is to use a classic d-dimensional range tree R, augmented with an f-bit long bitstring for each node on the last level of R (using the L_i ordering of the f colors). The colors within the range can thus be reported by taking the bitwise OR of all the bitstrings stored at the $O(\lg^d m)$ summary nodes of R spanning the range in the last level. This solution takes total time $O(\frac{f}{w} \lg^d m) = O(\lg^d m)$ and space $O(m \lg^{d-1} m \frac{f}{w}) = O(m \lg^{d-1} m)$, and it can be constructed in time $O(m \lg^{d-1} m)$ by building the node bitstrings from the leaves and up (recall that $w = \Theta(\lg n)$ is the word size).

The above solution is enough to obtain some of our results, but we can improve it by replacing the last level in R with a new data structure for restricted one-dimensional colored range reporting over integers that answer queries in time $O(\lg \lg m)$ and linear space. A query may perform $O(\lg^{d-1} m)$ one-dimensional queries on the last level of the range tree, so the query time is reduced to $O(\lg^{d-1} m \lg \lg m)$ per bucket. The new data structure used at the last level is given in the next section.

Observe that though the points are not from a bounded universe, we can remap a query in a bucket to a bounded universe of size m in time $O(\lg m)$ and linear space per dimension. We do so for the final dimension, noting that we only need to do it once for all $O(\lg^{d-1} m)$ queries in the final dimension.

1D Restricted Colored Range Reporting on Integers. Given $O(m)$ points in one dimension from a universe of size m, each colored with one of $f = \lg n$ distinct colors, we now show how to report the colors contained in a query range in $O(\lg \lg m)$ time and linear space, encoded as an f-bit long bitstring. First, partition the points into $j = O(m/\lg m)$ intervals spanning $\Theta(\lg m)$ consecutive points each. Each interval is stored as a balanced binary search tree of height $O(\lg \lg m)$, with each node storing a f-bit long bitstring indicating the colors that are present in its subtree. Clearly, storing all these trees take linear space.

We call the first point stored in each interval a *representative* and store a predecessor data structure containing all of the $O(m/\lg m)$ representatives of the intervals. Also, each representative stores $O(\lg m)$ f-bit long bitstrings, which are summaries of the colors kept in the $1, 2, \ldots, 2^{\lg j}$ neighboring intervals. We store these bitstrings both towards the left and the right from the representative, in total linear space.

A query $[a, b]$ is answered as follows. We decompose the query into two parts, first finding the answer for all intervals fully contained in $[a, b]$, and then finding the answer for the two intervals that only intersect $[a, b]$. The first part is done by finding the two outermost representatives inside the interval (called a', b', where $a \le a' \le b' \le b$) by using predecessor queries with a and b on the representatives. Since we store summaries for all power-of-2 neighboring intervals of the representatives, there are two bitstrings stored with a' and b' which summarises the colors in all fully contained intervals.

To find the answer for the two intervals that contain a or b, we find $O(\lg \lg m)$ nodes of the balanced binary tree for the interval and take the bitwise OR of the bitstrings stored at those nodes in $O(\lg \lg m)$ total time. Using one of the classic predecessor data structures [5, 15, 24] for the representatives, we thus obtain a query time of $O(\lg \lg m)$ and linear space.

3   2D Colored Range Searching in Linear Space

To obtain linear space in two dimensions and the proof of Theorem 1, we use the same grouping and bucketing approach as in Section 2. For each group $P_{i'}$, we only change the solution of each bucket B_i in $P_{i'}$, recalling that B_i contains up to $m = \lg^c n$ points with $f = \lg n$ distinct colors, so as to use linear space instead of $O(m \lg m)$ words of space.

We store a linear space 2D standard range reporting data structure A_i by Nekrich [17] for all points in the bucket B_i. As shown in [17], A_i supports orthogonal standard range reporting queries in $O(\lg m + r \lg^\epsilon m)$ time and updates in $O(\lg^{3+\epsilon} m)$ time, where r is the reported number of points and $\epsilon > 0$.

We also store a simple 2D range tree R_i augmented with f-bit long bitstrings on the last level as previously described in Section 2.2, but instead of storing points in R_i, we reduce its space usage by only storing areas covering $O(\lg m)$ points of B_i each. This can be done by first building R_i taking $O(m \lg m)$ space and time, and then cutting off subtrees at nodes at maximal height (called cutpoint nodes) such that at most $c' \lg m$ points are covered by each cutpoint node, for a given constant $c' > 0$. In this way, each cutpoint node is implicitly associated with $O(\lg m)$ points, which can be succinctly represented with $O(1)$ words as they all belong to a distinct 2D range. Note that the parent of a cutpoint node has $\Omega(\lg m)$ descending points, hence there are $O(m/\lg m)$ cutpoint nodes.

A query is answered by finding $O(\lg^2 m)$ summary nodes in R_i that span the entire query range Q. Combining bitstrings as described in Section 2.2, the colors for all fully contained ranges that are not stored in the leaves can thus be found. Consider now one such leaf ℓ covering an area intersecting Q: since the $O(\lg m)$ points spanned by ℓ may not be all contained in Q, we must check those points individually. Recall that the points associated with ℓ are those spanning a certain range Q', so they can be succinctly represented by Q'. To actually retrieve them, we issue a query Q' to A_i, check which ones belong to $Q' \cap Q$, and build a bitstring for the colors in $Q' \cap Q$. We finally merge the bitstrings for all summary nodes and intersecting leaves in constant time per bitstring to obtain the final result.

The time spent answering a query is $O(\lg^2 m)$ to find all bitstrings in non-leaf nodes of R_i and to combine all the bitstrings. The time spent finding the bitstring in leaves is $O(\lg^{1+\epsilon} m)$ per intersecting leaf as we use Nekrich's data structure A_i with $r = O(\lg m)$. Observe that only two leaves spanning a range of $O(\lg m)$ points may be visited in each of the $O(\lg m)$ second level data structures visited, so the time spent in all leaves is $O(\lg^{2+\epsilon} m)$, which is also the total time. Finally, since we reduced the size of the range tree by a factor $\Theta(\lg m)$, the total space usage is linear. This concludes the proof of Theorem 1.

4   Dynamic Data Structures

We now prove Theorem 3 by discussing how to support operations INSERT(p, c) and DELETE(p), inserting and deleting a point p with color c, respectively. Note that the color c may be previously unused. We still use parameters f and m to denote the number of colors in groups and points in buckets, respectively. We first give bounds on how to update a bucket, and then show how to support updates in the color grouping and point bucketing.

4.1   Updating a Bucket

Updating a bucket with a point corresponds to updating a d-dimensional range tree using known techniques in dynamic data structures. Partial rebuilding [2,21] requires amortised time $O(\lg^d m)$, including updating the bitstrings in the partially rebuilt trees and in each node of the last level trees (which takes constant time). Specifically, the bitstrings for the $O(\lg^{d-1} m)$ trees on the last level where a point was updated may need to have the bitstrings fixed on the path to the root on that level. This takes time $O(\lg m)$ per tree, giving a total amortised update time of $O(\lg^d m)$.

4.2   Updating Color Grouping and Point Bucketing

When supporting INSERT(p, c), we first need to find the group to which c belongs. If the color is new and there is a group P_i with less than f colors, we must update the color list L_i. Otherwise, we can create a new group P_i for the new color. In the group P_i, we must find a bucket to put p in. If possible, we put p in a bucket with less than m points, or otherwise we create a new bucket for p. Keeping track of sizes of groups and buckets can be done using priority queues in time $O(\lg \lg n)$. Note that we never split groups or buckets on insertions.

As for supporting DELETE(p), we risk making both groups and buckets underfull, thus requiring a merge of either. A bucket is underfull when it contains less than $m/2$ points. We allow at most one underfull bucket in a group. If there are two underfull buckets in a group, we merge them in time $O(m \lg^d m)$. Since merging buckets can only happen after $\Omega(m)$ deletions, the amortized time for a deletion in this case is $O(\lg^d m)$. A group is underfull if it contains less than $f/2$ colors and, as for buckets, if there are any two underfull groups P_i, P_j, we merge them. When merging P_i, P_j into a new group P_r, we concatenate their color lists L_i, L_j into L_r, removing the colors that are no more present while keeping the relative ordering of the surviving colors from L_i, L_j. In this way, a group merge does not require us to merge the underlying buckets, as points are partitioned arbitrarily into the buckets. However, a drawback arises: as the color list L_r for the merged group is different from the color lists L_i, L_j used for answering bucket queries, this may introduce errors in bucket query answers. Recall that an answer to a bucket query is an f-bit long bitstring which marks with 1s the colors in L_i that are in the range Q. So we have a bitstring for L_i, and one for L_j, for the buckets previously belonging to P_i, P_j, but we should instead output a bitstring for L_r in time proportional to the number of buckets in P_r. We handle this situation efficiently as discussed in Section 4.3.

4.3   Fixing Bucket Answers during a Query

As mentioned in Section 4.2, we do not change the buckets when two or more groups are merged into P_r. Consider the f-bit long bitstring b_i that is the answer for one merged group, say P_i', relative to its color list, say L_i. However, after the merge, only a sublist $L_i' \subseteq L_i$ of colors survives as a portion of the color list L_r for P_r. We show how to use L_i and L_i' to contribute to the f-bit long bitstring a that is the answer to query Q for the color list L_r in P_r. The time constraint is that we can spend time proportional to the number, say g, of buckets in P_r.

We need some additional information. For each merged group P_i', we create an f-bit long bitstring v_i with bit j set to 1 if and only if color $L_i[j]$ survives in L_r (i.e. some point in P_r has color $L_i[j]$). We call v_i the *possible answer bitstring* and let o_i be the number of 1s in v_i: in other words, L'_i is the sublist built from L_i by choosing the colors $L_i[j]$ such that $v_i[j] = 1$, and $o_i = |L'_i|$.

Consider now the current group P_r that is the outcome of $h \leq f$ old merged groups, say $P'_1, P'_2, \ldots, P'_h$ in the order of the concatenation of their color lists, namely, $L_r = L'_1 \cdot L'_2 \cdots L'_h$. Since the number of buckets in P_r is $g \geq h$, we can spend $O(g)$ time to obtain the f-bit long bitstrings $b_1, b_2, \ldots, b_h$, which are the answers for the old merged groups $P'_1, P'_2, \ldots, P'_h$, and combine them to obtain the answer a for P_r.

Here is how. The idea is that the bits in a from position $1 + \sum_{l=1}^{i-1} o_l$ to $\sum_{l=1}^{i} o_l$ are reserved for the colors in L'_i, using 1 to indicate which color in L'_i is in the query Q and 0 which is not. Let us call b'_i this o_i-bit long bitstring. Recall that we have b_i, which is the f-bit long bitstring that is the answer for P'_i and refers to L_i, and also v_i, the possible answer bitstring as mentioned before.

To obtain b'_i from b_i, v_i and o_i in constant time, we would like to employ a lookup table $S[b, v]$ for all possible f-bitstrings b and v, precomputing all the outcomes (in the same fashion as the Four Russians trick). However, the size of S would be $2^f \times 2^f \times f$ bits, which is too much (remember $f = \lg n$). We therefore build S for all possible $(f/3)$-bitstrings b and v, so that S uses $o(n)$ words of memory. This table is periodically rebuilt when n doubles or becomes one fourth, following a standard rebuilding rule. We therefore compute b'_i from b_i, v_i by dividing each of them in three parts, looking up S three times for each part, and combining the resulting three short bitstrings, still in $O(1)$ total time.

Once we have found $b'_1, b'_2, \ldots, b'_h$ in $O(h)$ time as shown above, we can easily concatenate them with bitwise shifts and ORs, in $O(h)$ time, so as to produce the wanted answer $a = b'_1 \cdot b'_2 \cdots b'_h$ as a f-bit long bitstring for P_r and its color list L_r. Recall that P_r consists of h buckets where $h \leq g \leq f$. Indeed, if it were $h > f$, there would be some groups with no colors. Since $\Omega(f)$ deletions must happen before two groups are merged, we can clean and remove the groups that have no more colors, i.e, with $o_i = 0$, and maintain the invariant that $h \leq g \leq f$.

5   Open Problems

There are a lot of loose ends in colored range searching that deserve to be investigated, and we will shortly outline a few of them. The hardness reduction by Kaplan et al. [11] gives hope that colored range counting can be proven hard, and we have indeed assumed that this is the case here. If taking instead an upper bound approach as this paper, improved time bounds obtainable in little space, or with some restriction on the number of colors, would be very interesting motivated by the large scale applications of the problem.

References

1. Agarwal, P.K., Govindarajan, S., Muthukrishnan, S.M.: Range searching in categorical data: Colored range searching on grid. In: Möhring, R.H., Raman, R. (eds.) ESA 2002. LNCS, vol. 2461, pp. 17–28. Springer, Heidelberg (2002)

2. Andersson, A.: General balanced trees. J. Algorithms 30(1), 1–18 (1999)
3. de Berg, M., Cheong, O., van Kreveld, M., Overmars, M.: Computational Geometry: Algorithms and Applications, 3rd edn. (2008)
4. Bozanis, P., Kitsios, N., Makris, C., Tsakalidis, A.K.: New upper bounds for generalized intersection searching problems. In: Fülöp, Z. (ed.) ICALP 1995. LNCS, vol. 944, pp. 464–474. Springer, Heidelberg (1995)
5. van Emde Boas, P., Kaas, R., Zijlstra, E.: Design and implementation of an efficient priority queue. Theory Comput. Syst. 10(1), 99–127 (1976)
6. Gagie, T., Kärkkäinen, J., Navarro, G., Puglisi, S.J.: Colored range queries and document retrieval. TCS (2012)
7. Gupta, P., Janardan, R., Smid, M.: Further Results on Generalized Intersection Searching Problems: Counting, Reporting, and Dynamization. J. Algorithms 19(2), 282–317 (1995)
8. Gupta, P., Janardan, R., Smid, M.: A technique for adding range restrictions to generalized searching problems. Inform. Process. Lett. 64(5), 263–269 (1997)
9. JáJá, J., Mortensen, C.W., Shi, Q.: Space-efficient and fast algorithms for multidimensional dominance reporting and counting. In: Fleischer, R., Trippen, G. (eds.) ISAAC 2004. LNCS, vol. 3341, pp. 558–568. Springer, Heidelberg (2004)
10. Janardan, R., Lopez, M.: Generalized intersection searching problems. IJCGA 3(01), 39–69 (1993)
11. Kaplan, H., Rubin, N., Sharir, M., Verbin, E.: Counting colors in boxes. In: Proc. 18th SODA. pp. 785–794 (2007)
12. van Kreveld, M.: New results on data structures in computational geometry. PhD thesis, Department of Computer Science, University of Utrecht, Netherlands (1992)
13. Larsen, K.G., Pagh, R.: I/O-efficient data structures for colored range and prefix reporting. In: Proc. 23rd SODA. pp. 583–592 (2012)
14. Larsen, K.G., van Walderveen, F.: Near-Optimal Range Reporting Structures for Categorical Data. In: Proc. 24th SODA. pp. 265–276 (2013)
15. Mehlhorn, K., Näher, S.: Bounded ordered dictionaries in $O(\lg \lg N)$ time and $O(n)$ space. Inform. Process. Lett. 35(4), 183–189 (1990)
16. Mortensen, C.W.: Generalized static orthogonal range searching in less space. Tech. rep., TR-2003-22, The IT University of Copenhagen (2003)
17. Nekrich, Y.: Orthogonal Range Searching in Linear and Almost-linear Space. Comput. Geom. Theory Appl. 42(4), 342–351 (2009)
18. Nekrich, Y.: Space-efficient range reporting for categorical data. In: Proc. 31st PODS. pp. 113–120 (2012)
19. Nekrich, Y.: Efficient range searching for categorical and plain data. ACM TODS 39(1), 9 (2014)
20. Nekrich, Y., Vitter, J.S.: Optimal color range reporting in one dimension. In: Bodlaender, H.L., Italiano, G.F. (eds.) ESA 2013. LNCS, vol. 8125, pp. 743–754. Springer, Heidelberg (2013)
21. Overmars, M.H.: Design of Dynamic Data Structures (1987)
22. Patrascu, M.: Lower bounds for 2-dimensional range counting. In: Proc. 39th STOC, pp. 40–46 (2007)
23. Shi, Q., JáJá, J.: Optimal and near-optimal algorithms for generalized intersection reporting on pointer machines. Inform. Process. Lett. 95(3), 382–388 (2005)
24. Willard, D.: Log-logarithmic worst-case range queries are possible in space $\Theta(N)$. Inform. Process. Lett. 17(2), 81–84 (1983)
25. Williams, V.V.: Multiplying matrices faster than Coppersmith-Winograd. In: Proc. 44th STOC. pp. 887–898 (2012)

Fast Dynamic Graph Algorithms
for Parameterized Problems

Yoichi Iwata and Keigo Oka

Department of Computer Science
Graduate School of Information Science and Technology
The University of Tokyo
{y.iwata,ogiekako}@is.s.u-tokyo.ac.jp

Abstract. Fully dynamic graph is a data structure that (1) supports edge insertions and deletions and (2) answers problem specific queries. The time complexity of (1) and (2) are referred to as the update time and the query time respectively. There are many researches on dynamic graphs whose update time and query time are $o(|G|)$, that is, sublinear in the graph size. However, almost all such researches are for problems in P. In this paper, we investigate dynamic graphs for NP-hard problems exploiting the notion of fixed parameter tractability (FPT).

We give dynamic graphs for Vertex Cover and Cluster Vertex Deletion parameterized by the solution size k. These dynamic graphs achieve almost the best possible update time $O(\text{poly}(k) \log n)$ and the query time $O(f(\text{poly}(k), k))$, where $f(n, k)$ is the time complexity of any static graph algorithm for the problems. We obtain these results by dynamically maintaining an approximate solution which can be used to construct a small problem kernel. Exploiting the dynamic graph for Cluster Vertex Deletion, as a corollary, we obtain a quasilinear-time (polynomial) kernelization algorithm for Cluster Vertex Deletion. Until now, only quadratic time kernelization algorithms are known for this problem.

1 Introduction

1.1 Background

Parameterized Algorithms. Assuming P $\neq$ NP, there are no polynomial-time algorithms solving NP-hard problems. On the other hand, some problems are efficiently solvable when a certain parameter, e.g. the size of a solution, is small. *Fixed parameter tractability* is one of the ways to capture such a phenomenon.

A problem is in the class *fixed parameter tractable (FPT)* with respect to a parameter k if there is an algorithm that solves any problem instance of size n with parameter k in $O(n^d f(k))$ time (*FPT time*), where d is a constant and f is some computable function.

Dynamic Graphs. (Fully) dynamic graph is a data structure that supports edge insertions, edge deletions, and answers certain problem specific queries.

R Ravi and I.L. Gørtz (Eds.): SWAT 2014, LNCS 8503, pp. 241–252, 2014.

Table 1. The time complexities of the dynamic graphs in this paper. d is the degree bound, and n is the number of the vertices. The parameter for Chromatic Number is cvd number (the size of a minimum cluster vertex deletion), and parameters for the other problems are its solution size.

Problem	Update Time	Query Time	Section
Vertex Cover	$O(k^2)$	$f_{VC}(k^2, k)$	3
Cluster Vertex Deletion	$O(k^8 + k^2 \log n)$	$f_{CVD}(k^5, k)$	4
Cluster Vertex Deletion	$O(8^k k^6)$	$O(1)$	[14]
Chromatic Number	$O(2^{2^k} \log n)$	$O(1)$	[14]
Feedback Vertex Set	$O(7.66^k k^3 + 2^k k^3 d^3 \log n)$	$O(1)$	[14]

There are a lot of theoretical research on dynamic graphs for problems that belong to P, such as Connectivity [10,11,24,27,8,15], k-Connectivity [11,8], Minimum Spanning Forest [11,8], Bipartiteness [11,8], Planarity Testing [13,8,16], All-pairs Shortest Path [25,4,26,22,20] and Directed Connectivity [3,18,19,21,23], and races for faster algorithms are going on.

On the contrary there have been few research on dynamic graphs related to FPT algorithms. To the best of our knowledge, a dynamic data structure for counting subgraphs in sparse graphs proposed by Zdeněk Dvořák and Vojtěch Tůma [7] and a dynamic data structure for tree-depth decomposition proposed by Zdeněk Dvořák, Martin Kupec and Vojtěch Tůma [6] are only such dynamic graphs. Both data structures support insertions and deletions of edges, and compute the solution of the problems in time depending only on k, where k is the parameter of the problem. For a fixed property expressed in monadic second-order logic, the dynamic graph in [6] also can answer whether the current graph has the property. For both algorithms, hidden constants are (huge) exponential in k. In particular, update time of both algorithms become super-linear in graph size n even if k is very small, say $O(\log \log n)$.

1.2   Our Contribution

In this paper, we investigate dynamic data structures for basic graph problems in FPT. Table 1 shows the problems we deal with and the time complexities of the algorithms. In Section 3 and 4, we present fully dynamic graphs for Vertex Cover and Cluster Vertex Deletion, respectively. Both dynamic data structures support additions and deletions of edges, and can answer the solution of the problem in time depending only on the solution size k. Due to the space limitation, the details of the latter three results are omitted. Please see the full version which will appear soon [14].

For the dynamic graph for Vertex Cover, the time complexity of an edge addition or deletion is $O(k^2)$ and the one of a query is $f_{VC}(k^2, k)$ where $f_{VC}(n, k)$ is the time complexity of any static algorithm for Vertex Cover on a graph of size n.

For the dynamic graph for Cluster Vertex Deletion, the time complexity of an update is $O(k^8 \log n)$ and the one of a query is $f_{CVD}(k^5, k)$ where $f_{CVD}(n, k)$ is the time complexity of any static algorithm for Cluster Vertex Deletion on a graph of size n. The extra $\log n$ factor arises because we use persistent data structures to represent some vertex sets. This enables us to copy a set in constant time.

Note that the time complexity of an update is $\mathrm{poly}(k)\mathrm{polylog}(n)$ for both algorithms, instead of an exponential function in k. As for the time complexity of a query, its exponential term in k is no more than any static algorithms.

Let us briefly explain how the algorithms work. Throughout the algorithm, we keep an approximate solution. When the graph is updated, we efficiently construct a $\mathrm{poly}(k)$ size kernel by exploiting the approximate solution, and then compute a new approximate solution on this kernel. Here, we compute not an exact solution but an approximate solution to achieve the update time polynomial in k. To answer a query, we apply a static exact algorithm to the kernel.

To see goodness of these algorithms, consider the situation such that a query is applied for every r updates. A trivial algorithm answers a query by running a static algorithm. Let the time complexity of the static algorithm be $O(f(n,k))$. In this situation, to deal with consecutive r updates and one query, our algorithm takes $O(r\mathrm{poly}(k)\mathrm{polylog}(n)+f(\mathrm{poly}(k),k))$ time, and the trivial algorithm takes $O(f(n,k))$ time. For example, let $f(n,k) = c^k + kn$ be the time complexity of the static algorithm. (The time complexity of the current best FPT algorithm for Vertex Cover is $O(1.2738^k + kn)$ [2].) Then if $r = \sqrt{n}$ and $c^k = \sqrt{n}$, the time complexity for the dynamic graph algorithm is $\sqrt{n}\mathrm{polylog}(n) = o(n)$, sublinear in n. That is, our algorithm works well even if the number of queries is fewer than the number of updates. This is an advantage of the polynomial-time update. If $r = 1$, our algorithm is faster than the trivial algorithm whenever $c^k < n$. Even if c^k is the dominant term, our algorithm is never slower than the trivial algorithm.

Let us consider the relation between our results and the result by Dvořák, Kupec and Tůma [6]. The size of a solution of Vertex Cover is called *vertex cover number*, and the size of a solution of Cluster Vertex Deletion is called *cluster vertex deletion number (cvd number)*. It is easy to show that tree-depth can be arbitrarily large even if cvd number is fixed and vice versa. Thus our result for Cluster Vertex Deletion is not included in their result. On the other hand, tree-depth is bounded by vertex cover number + 1. Thus their result indicates that Vertex Cover can be dynamically computed in $O(1)$ time if vertex cover number is a constant. However, if it is not a constant, say $O(\log \log n)$, the time complexity of their algorithm becomes no longer sublinear in n. The time complexity of our algorithm for Vertex Cover is further moderate as noted above.

As an application of the dynamic graph for Cluster Vertex Deletion, we can obtain a quasilinear-time kernelization algorithm for Cluster Vertex Deletion. To compute a problem kernel of a graph $G = (V, E)$, starting from an empty graph, we iteratively add the edges one by one while updating an approximate solution. Finally, we compute a kernel from the approximate solution. As shown in Section 4, the size of the problem kernel is $O(k^5)$ and the time for an update is $O(k^8 \log |V|)$. Thus, we obtain a polynomial kernel in $O(k^8|E| \log |V|)$ time.

Protti, Silva and Szwarcfiter [17] proposed a linear-time kernelization algorithm for Cluster Editing applying modular decomposition techniques. On the other hand, to the best of our knowledge, for Cluster Vertex Deletion, only quadratic time kernelization algorithms [12] are known (until now). Though Cluster Vertex Deletion and Cluster Editing are similar problems, it seems that their techniques cannot be directly applied to obtain a linear-time kernelization algorithm for Cluster Vertex Deletion.

2   Notations

Let $G = (V, E)$ be a simple undirected graph with vertices V and edges E. We consider that each edge in E is a set of vertices of size two. Let $|G|$ denote the *size* of the graph $|V| + |E|$. The *neighborhood* $N_G(v)$ of a vertex v is $\{u \in V \mid \{u, v\} \in E\}$, and the neighborhood $N_G(S)$ of a vertex set $S \subseteq V$ is $\bigcup_{v \in S} N_G(v) \setminus S$. The *closed neighborhood* $N_G[v]$ of a vertex v is $N_G(v) \cup \{v\}$, and the closed neighborhood $N_G[S]$ of a vertex set $S \subseteq V$ is $N_G(S) \cup S$. We denote the degree of a vertex v by $d_G(v)$. We omit the subscript if the graph is apparent from the context. The *induced subgraph* $G[S]$ of a vertex set S is the graph $(S, \{e \in E \mid e \subseteq S\})$.

By default, we use $k(G)$ or k to denote the parameter value of the current graph G. When an algorithm updates a graph G to G', we use $k = \max\{k(G), k(G')\}$ as a parameter. Note that, $k(G)$ and $k(G')$ are not greatly different in most problems. In particular, it is easy to prove that for all problems we deal with in this paper, $k(G')$ is at most $k(G) + 1$.

3   Dynamic Graph for Vertex Cover

Let $G = (V, E)$ be a graph. Vertex Cover is the problem of finding a minimum set of vertices that covers all edges. Let $k = k(G)$ be the size of a minimum vertex cover of G. The current known FPT algorithm solving Vertex Cover whose exponential function in k is smallest is by Chen, Kanj and Xia [2], and its running time is $O(|G|k + 1.2738^k)$. Let us now state the main result of this section.

Theorem 1. *There is a data structure representing a graph G which supports the following three operations.*

1. *Answers the solution for* Vertex Cover *of the current graph G.*
2. *Add an edge to G.*
3. *Remove an edge from G.*

Let k be the size of a minimum vertex cover of G. Then the time complexity of an edge addition or removal is $O(k^2)$, and of a query is $O(f(k^2, k))$, where $f(|G|, k)$ is the time complexity of any static algorithm for Vertex Cover *on a graph of size $|G|$.*

Note that the update time is polynomial in k, and the exponential term in k of the query time is same to the one of the static algorithm.

Our dynamic data structure is simply represented as a pair of the graph $G = (V, E)$ itself and a 2-approximate solution $X \subseteq V$ for Vertex Cover of G,

---

Algorithm 1. compute a 2-approximate solution

1: $X_0 := \emptyset$
2: $V' := \emptyset$
3: **for all** x in X **do**
4:     **if** $d(x) > |X|$ **then** $X_0 := X_0 \cup \{x\}$
5:     **else** $V' := V' \cup N[x]$
6: $V' := V' \setminus X_0$
7: $Y :=$ 2-approximate solution for Vertex Cover of $G[V']$.
8: $X' := X_0 \cup Y$

---

that is, we maintain a vertex set X such that X is a vertex cover of G and $|X| \leq 2k(G)$.

For both query and update, we compute a problem kernel. To do this, we exploit the fact that we already know rather small vertex cover X. When an edge $\{u, v\}$ is added to G, we add u to X making X a vertex cover and use Algorithm 1 to compute a new 2-approximate solution X' of G. When an edge is removed from G, we also use Algorithm 1 to compute a new 2-approximate solution.

Lemma 1. *Algorithm 1 computes a 2-approximate solution X' in $O(k^2)$ time, where $k = k(G)$.*

Proof. Let X^* be a minimum vertex cover of the updated graph. We have $|X^*| \leq |X|$. If $x \notin X^*$ for some vertex $x \in X_0$, $N(x)$ must be contained in X^*. Thus it holds that $|X^*| \geq |N(x)| = d(x) > |X|$, which is a contradiction. Therefore, it holds that $X_0 \subseteq X^*$. At line 7 of Algorithm 1, V' equals to $N[X \setminus X_0] \setminus X_0$. Thus we have:

(1) $X^* \setminus X_0$ is a vertex cover of $G[V']$ because $X_0 \cap V' = \emptyset$ and X^* is a vertex cover of G, and

(2) any vertex cover of $G[V']$ together with X_0 covers all edges in G because all edges not in $G[V']$ are covered by X_0.

Putting (1) and (2) together, we can prove that $X^* \setminus X_0$ is a minimum vertex cover of $G[V']$.

Since Y is a 2-approximate solution on $G[V']$ and $X^* \setminus X_0$ is a minimum vertex cover of $G[V']$, we have $|Y| \leq 2|X^* \setminus X_0|$. From (2), $X' = X_0 \cup Y$ is a vertex cover of G. Thus X' is a 2-approximate solution because $|X'| = |X_0| + |Y| \leq |X_0| + 2|X^* \setminus X_0| \leq 2|X^*|$.

The size of X is at most $2k + 1$, and thus the size of V' at line 7 is $O(k^2)$. Moreover, the number of edges in $G[V']$ is $O(k^2)$ because for each edge in $G[V']$, at least one endpoint lies on $X \setminus X_0$ and the degree of any vertex x in $X \setminus X_0$ is at most $|X|$. A 2-approximate solution can be computed in linear time using a simple greedy algorithm [9]. Thus the total time complexity is $O(k^2)$. $\square$

To answer a query, we use almost the same algorithm as Algorithm 1, but compute an exact solution at line 7 instead of an approximate solution. The validity of the algorithm can be proved by almost the same argument. The bottleneck part is to compute an exact vertex cover of the graph $G[V']$. Since the size of the solution is at most k, we can obtain the solution in $O(f(k^2, k))$ time where $f(|G|, k(G))$ is the time complexity of any algorithm solving Vertex Cover for a graph G. For example, using the algorithm in [2], we can compute the solution in $O(k^3 + 1.2738^k)$ time. We have finished the proof of Theorem 1.

4   Dynamic Graph for Cluster Vertex Deletion

4.1   Problem Definition and Time Complexity

A graph is called *cluster graph* if every its connected component is a clique, or equivalently, it contains no *induced* path with three vertices (P_3). Each maximal clique in a cluster graph is called a *cluster*. Given a graph, a subset of its vertices is called a *cluster vertex deletion* if its removal makes the graph a cluster graph. Cluster Vertex Deletion is the problem to find a minimum cluster vertex deletion. We call the size of a minimum cluster vertex deletion as a *cluster vertex deletion number* or a *cvd number* in short.

There is a trivial algorithm to find a 3-approximate solution for Cluster Vertex Deletion with time complexity $O(|E||V|)$ [12]. The algorithm greedily finds P_3 and adds all the vertices on the path to the solution until we obtain a cluster graph. According to [12], it is still open whether it is possible to improve the trivial algorithm or not.

Let us now state the main result of this section.

Theorem 2. *There is a data structure representing a graph G which supports the following three operations.*

1. *Answers the solution for Cluster Vertex Deletion of the current graph G.*
2. *Add an edge to G.*
3. *Remove an edge from G.*

Let k be the cvd number of G. Then the time complexity of an edge addition or removal is $O(k^8 + k^2 \log |V|)$, and of a query is $O(f(k^5, k))$, where $f(|G|, k)$ is the time complexity of any static algorithm for Cluster Vertex Deletion on a graph of size $|G|$.

As the static algorithm, we can use an $O(2^k k^9 + |V||E|)$-time algorithm by Hüffner, Komusiewicz, Moser, and Niedermeier [12] or an $O(1.9102^k |G|)$-time algorithm by Boral, Cygan, Kociumaka, and Pilipczuk [1].

4.2   Data Structure

We dynamically maintain the variables listed in Table 2. We always keep a 3-approximate solution X. Each cluster in $G[V \setminus X]$ is assigned a distinct *cluster label*. For each cluster label l, C_l is the set of vertices on the cluster having the label l. We keep the vertex set C_l by using a persistent data structure that

Table 2. Variables maintained in the algorithm

X	3-approximate solution
C_l for each cluster label l	the vertices in the cluster labeled l
l_u for each $u \in V \setminus X$	label of the cluster that u belongs to
L_x for each $x \in X$	$\{l_u \mid u \in N(x) \setminus X\}$
$P_{x,l}^+$ for each $x \in X$ and $l \in L_x$	$C_l \cap N(x)$
$P_{x,l}^-$ for each $x \in X$ and $l \in L_x$	$C_l \setminus N(x)$

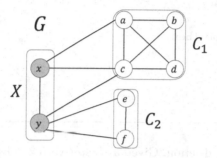

Fig. 1. An example of a graph and a 3-approximate solution

supports an update in $O(\log |C_l|)$ time. One of such data structures is a persistent red-black tree developed by Driscoll, Sarnak, Sleator and Tarjan [5]. The reason why the persistent data structure is employed is that it enables us to copy the set in constant time. For each $u \in V \setminus X$, l_u is a label of the cluster u belongs to. For a vertex x and a cluster, we say that x is incident to the cluster if at least one vertex in the cluster is incident to x. For each $x \in X$, $L_x = \{l_u \mid u \in N(x) \setminus X\}$ is the labels of the clusters that x is incident to. For each $x \in X$ and $l \in L_x$, $P_{x,l}^+ = C_l \cap N(x)$ is the set of the neighbors of x in C_l and $P_{x,l}^- = C_l \setminus N(x)$ is the set of the non-neighbors of x in C_l. Note that all variables are uniquely determined when G, X and the labels for all clusters are fixed.

For example, look at the graph depicted in Fig. 1. $X = \{x, y\}$ is a 3-approximate solution, and C_1 and C_2 are clusters. Here, the set of cluster labels is $\{1, 2\}$, $l_a = l_b = l_c = l_d = 1$ and $l_e = l_f = 2$. $L_x = \{1\}$ and $L_y = \{1, 2\}$. $P_{x,1}^+ = \{a, c\}$, $P_{x,1}^- = \{b, d\}$, $P_{y,1}^+ = \{c\}$, $P_{y,1}^- = \{a, b, d\}$, $P_{y,2}^+ = \{e, f\}$ and $P_{y,2}^- = \{\}$.

4.3 Algorithm

Update Let us explain how to update the data structure when an edge is added or removed. Before describing the whole algorithm, let us explain subroutines used in the algorithm. Algorithm 2 is used to add a vertex u in $V \setminus X$ to X, and Algorithm 3 is to remove a vertex y from X under the condition that $X \setminus \{y\}$

Algorithm 2. add $u \in V \setminus X$ to X

1: $l := l_u$
2: remove u from C_l
3: **for all** any $x \in X$ such that $l \in L_x$ **do**
4:     **if** $\{x, u\} \in E$ **then**
5:         remove u from $P^+_{x,l}$
6:         If $P^+_{x,l}$ becomes empty, remove l from L_x
7:     **else**
8:         remove u from $P^-_{x,l}$
9: add u to X
10: **if** C_l is still not empty **then**
11:     $L_u := \{l\}$
12:     copy C_l into $P^+_{u,l}$
13:     $P^-_{u,l} := \emptyset$
14: **else**
15:     $L_u := \emptyset$

---

is still a cluster vertex deletion. Given a cluster vertex deletion X, Algorithm 4 computes a 3-approximate solution X'.

Lemma 2. *Algorithm 2 adds a vertex u to X and updates the data structure correctly in $O(|X| \log n)$ time.*

Lemma 3. *If $G[V \setminus (X \setminus \{y\})]$ is a cluster graph, Algorithm 3 removes a vertex y from X and updates the data structure correctly in $O(|X| \log n)$ time.*

Due to the space limitation, the proofs of these two Lemmas are omitted. Please see the full version [14].

Lemma 4. *Algorithm 4 computes a 3-approximate solution in $O(|X|^8)$ time.*

In order to prove Lemma 4, let us prove Lemma 5 and 6.

Lemma 5. *Let V' and X_0 be the sets computed by Algorithm 4. If $S \subseteq V'$ is a cluster vertex deletion of $G[V']$ such that $|S| \leq |X \setminus X_0|$, then $S \cup X_0$ is a cluster vertex deletion of G.*

Proof. Assume that S is not a cluster vertex deletion of $G[V \setminus X_0]$. This implies that there is an induced P_3 in $G[(V \setminus X_0) \setminus S]$. Let x, y be vertices in $X \setminus X_0$ and u, v be vertices in $V \setminus (X \cup S)$. There are four possible types of induced paths: (1) xuy, (2) xyu, (3) xuv, and (4) uxv. We will rule out all these cases by a case analysis (see Fig. 2).

(1) Let $A = \{w \in V' \cap C_{l_u} \mid xw \in E \wedge yw \notin E\}$, $B = \{w \in V' \cap C_{l_u} \mid xw \notin E \wedge yw \in E\}$ and $C = \{w \in V' \cap C_{l_u} \mid xw \in E \wedge yw \in E\}$. By the construction of V', $|A| + |C| \geq |X| + 1$ and $|B| + |C| \geq |X| + 1$. Thus $\min\{|A|, |B|\} \geq |X| - |C| + 1$. Since $x, y \notin S$ and $\{x, y\} \notin E$, S must contain $C \cup B$ or $C \cup A$. Thus $|S| \geq |X| + 1$, which is a contradiction.

Algorithm 3. remove $y \in X$ from X assuming $G[V \setminus (X \setminus \{y\})]$ is a cluster graph

---

1: remove y from X
2: **if** $L_y = \emptyset$ **then**
3:      $l_y :=$ new label
4:      $C_{l_y} := \{y\}$
5: **else**
6:      $|L_y|$ must be one. Let l_y be the unique element in L_y.
7:      add y to C_{l_y}

8: $l := l_y$
9: **for all** $x \in X$ such that $l \in L_x$ **do**
10:      **if** $\{x,y\} \in E$ **then** add y to $P_{x,l}^+$
11:      **else** add y to $P_{x,l}^-$

12: **for all** $x \in X$ such that $y \in N(x)$ and $l \notin L_x$ **do**
13:      add l to L_x
14:      $P_{x,l}^+ := \{y\}$
15:      copy C_l into $P_{x,l}^-$ and remove y from $P_{x,l}^-$

Algorithm 4. compute a new 3-approximate solution X'

1: $V' := \emptyset$
2: $X_0 := \emptyset$
3: **for all** $x \in X$ **do**
4:      **if** $|L_x| > |X| + 1$ **then**
5:           add x to X_0
6:      **else**
7:           add x to V'
8:           **for all** $l \in L_x$ **do**
9:                take $\min(|P_{x,l}^+|, |X| + 1)$ vertices from $P_{x,l}^+$, and add them to V'
10:               take $\min(|P_{x,l}^-|, |X| + 1)$ vertices from $P_{x,l}^-$, and add them to V'
11: $Y :=$ 3-approximate cluster vertex deletion of $G[V']$
12: **if** $|Y| > |X \setminus X_0|$ **then** $X' := X$
13: **else** $X' := X_0 \cup Y$

---

(2) Let $A = \{w \in V' \cap C_{l_u} \mid xw \notin E \wedge yw \notin E\}$, $B = \{w \in V' \cap C_{l_u} \mid xw \in E \wedge yw \in E\}$ and $C = \{w \in V' \cap C_{l_u} \mid xw \notin E \wedge yw \in E\}$. By the construction of V', $|A| + |C| \geq |X| + 1$ and $|B| + |C| \geq |X| + 1$. Thus $\min\{|A|, |B|\} \geq |X| - |C| + 1$. Since $x, y \notin S$ and $\{x, y\} \in E$, S must contain $C \cup B$ or $C \cup A$. Thus $|S| \geq |X| + 1$, which is a contradiction.

(3) Since $|S| \leq |X|$, there is a vertex $u' \in (V' \cap C_{l_u}) \setminus S$ such that $\{x, u'\} \in E$ and a vertex $v' \in (V' \cap C_{l_u}) \setminus S$ such that $\{x, v'\} \notin E$. However it contradicts the fact that $G[V' \setminus S]$ contains no induced P_3.

(4) Since $|S| \leq |X|$, there is a vertex $u' \in (V' \cap C_{l_u}) \setminus S$ such that $\{x, u'\} \in E$ and a vertex $v' \in (V' \cap C_{l_v}) \setminus S$ such that $\{x, v'\} \in E$. However it contradicts the fact that $G[V' \setminus S]$ contains no induced P_3. $\square$

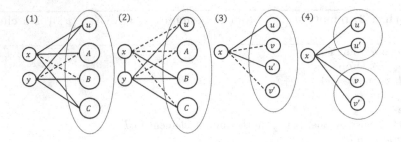

Fig. 2. Case analysis in the proof of Lemma 5. A dotted line denotes there is no edge(s)

Lemma 6. *Let V' and X_0 be the sets computed by Algorithm 4. For any cluster vertex deletion T of G such that $|T| \le |X|$, the following hold:*

1. T contains X_0,
2. $T \cap V'$ is a cluster vertex deletion of $G[V']$.

Proof. First, let us prove that T contains X_0. Assume there exists $x \in X_0 \setminus T$. Since $|L_x|$, the number of adjacent clusters of x, is more than $|X|+1$, in order to avoid induced P_3, T must contain at least $|L_x| - 1 > |X|$ vertices from adjacent clusters. It contradicts the fact that $|T| \le |X|$. Thus, T contains X_0, and so $T \setminus X_0$ is a cluster vertex deletion of $G[V \setminus X_0]$.

Since $G[V \setminus T]$ is a cluster graph, its induced subgraph $G[V' \setminus T]$ is also a cluster graph. Thus $T \cap V'$ is a cluster vertex deletion of $G[V']$. $\square$

Proof (of Lemma 4). Let X^* be a minimum cluster vertex deletion. Since X is a cluster vertex deletion, we have $|X^*| \le |X|$. By Lemma 6, it holds that $X_0 \subseteq X^*$, and $X^* \setminus X_0$ is a cluster vertex deletion of $G[V']$. $X^* \setminus X_0$ is actually a minimum cluster vertex deletion of $G[V']$, because otherwise there is a cluster vertex deletion S of $G[V']$ such that $|S| < |X^* \setminus X_0| \le |X \setminus X_0|$, but then by Lemma 5, $S \cup X_0$ becomes a cluster vertex deletion of G of size less than $|X^*|$, which is a contradiction.

If the size of the set Y computed at line 11 is larger than $|X \setminus X_0|$, the set X remains a 3-approximate solution. Otherwise, from Lemma 5, $Y \cup X_0$ is a cluster vertex deletion of G. Since Y is a 3-approximate solution and $X^* \setminus X_0$ is a minimum cluster vertex deletion of $G[V']$, we have

$$|Y \cup X_0| \le 3|X^* \setminus X_0| + |X_0| \le 3|X^*|. \tag{1}$$

Thus, $X' = Y \cup X_0$ is a 3-approximate solution on G.

The claimed time complexity is obtained as follows. The size of V' at line 11 is at most $2|X|(|X|+1)^2 = O(|X|^3)$. The number of edges in the graph $G[V']$ is maximized when $G[V' \setminus X]$ is composed of $|X|+1$ cliques with size $|X|(|X|+1)$. Thus the number of edges is at most $|X|^2(|X|+1)^3 = O(|X|^5)$. Thus, a 3-approximate solution can be computed in $O(|X|^8)$ time using the trivial algorithm described in Section 4.1, and thus the claimed time complexity holds. $\square$

Now we are ready to describe how to update the data structure when an edge is modified. To add (remove) an edge $\{u, v\}$ to (from) a graph G, before modifying G, we add u and v to X one by one using Algorithm 2 unless the vertex is already in X. After the operation, we add (remove) the edge $\{u, v\} \subseteq X$ to (from) G. Note that this operation does not affect any variables in our data structure. Now X is a cluster vertex deletion but may no longer be a 3-approximate solution. Then we compute a new 3-approximate solution X' using Algorithm 4.

Finally we replace X by X' as follows. Let R be $X \setminus X'$ and R' be $X' \setminus X$. We begin with adding every vertex in R' to X one by one using Algorithm 2. Then we remove every vertex in R from X one by one using Algorithm 3, and finish the replacement. During the process, X is always a cluster vertex deletion of the graph, and thus the assumption of Algorithm 3 is satisfied.

Let k be the maximum of the cvd numbers before and after the edge modification. During the above process, the size of X is increased to at most $6k$. Algorithm 4 is called only once, and Algorithm 2 and 3 are called $O(k)$ times. Thus together with Lemma 2, 3 and 4, the update time is $O(k^8 + k^2 \log n)$.

Query Let us explain how to answer a query. To compute a minimum cluster vertex deletion X', we use almost the same algorithm as Algorithm 4, but compute an exact solution Y at line 11 instead of an approximate solution. The validity of the algorithm can be proved by almost the same argument. The bottleneck of the algorithm is to compute a minimum cluster vertex deletion of the graph $G[V']$. Since the number of edges in $G[V']$ is $O(k^5)$ as noted in the proof of Lemma 4, using an $O(f(|G|, k))$-time static algorithm for Cluster Vertex Deletion, we can obtain the solution in $O(f(k^5, k))$ time. For example, using the algorithm in [2], we can compute the solution in $O(1.9102^k k^5)$ time.

Acknowledgement. Yoichi Iwata is supported by Grant-in-Aid for JSPS Fellows (256487). Keigo Oka is supported by JST, ERATO, Kawarabayashi Large Graph Project.

References

1. Boral, A., Cygan, M., Kociumaka, T., Pilipczuk, M.: Fast branching algorithm for cluster vertex deletion. CoRR, abs/1306.3877 (2013)
2. Chen, J., Kanj, I.A., Xia, G.: Improved upper bounds for vertex cover. Theor. Comput. Sci. 411(40-42), 3736–3756 (2010)
3. Demetrescu, C., Italiano, G.F.: Fully dynamic transitive closure: Breaking through the o(n²) barrier. In: FOCS, pp. 381–389 (2000)
4. Demetrescu, C., Italiano, G.F.: A new approach to dynamic all pairs shortest paths. In: STOC, pp. 159–166 (2003)
5. Driscoll, J.R., Sarnak, N., Sleator, D.D., Tarjan, R.E.: Making data structures persistent. J. Comput. Syst. Sci. 38(1), 86–124 (1989)
6. Dvorak, Z., Kupec, M., Tuma, V.: Dynamic data structure for tree-depth decomposition. CoRR, abs/1307.2863 (2013)

7. Dvořák, Z., Tůma, V.: A dynamic data structure for counting subgraphs in sparse graphs. In: Dehne, F., Solis-Oba, R., Sack, J.-R. (eds.) WADS 2013. LNCS, vol. 8037, pp. 304–315. Springer, Heidelberg (2013)

8. Eppstein, D., Galil, Z., Italiano, G.F., Nissenzweig, A.: Sparsification-a technique for speeding up dynamic graph algorithms (extended abstract). In: FOCS, pp. 60–69 (1992)

9. Gary, M.R., Johnson, D.S.: Computers and intractability: A guide to the theory of np-completeness (1979)

10. Henzinger, M.R., King, V.: Randomized dynamic graph algorithms with polylogarithmic time per operation. In: STOC, pp. 519–527 (1995)

11. Holm, J., de Lichtenberg, K., Thorup, M.: Poly-logarithmic deterministic fully-dynamic algorithms for connectivity, minimum spanning tree, 2-edge, and biconnectivity. J. ACM 48(4), 723–760 (2001)

12. Hüffner, F., Komusiewicz, C., Moser, H., Niedermeier, R.: Fixed-parameter algorithms for cluster vertex deletion. Theory Comput. Syst. 47(1), 196–217 (2010)

13. Italiano, G.F., Poutré, J.A.L., Rauch, M.H.: Fully dynamic planarity testing in planar embedded graphs (extended abstract). In: Lengauer, T. (ed.) ESA 1993. LNCS, vol. 726, pp. 212–223. Springer, Heidelberg (1993)

14. Iwata, Y., Oka, K.: Fast dynamic graph algorithms for parameterized problems (2014) (manuscript)

15. Patrascu, M., Demaine, E.D.: Lower bounds for dynamic connectivity. In: STOC, pp. 546–553 (2004)

16. Poutré, J.A.L.: Alpha-algorithms for incremental planarity testing (preliminary version). In: STOC, pp. 706–715 (1994)

17. Protti, F., da Silva, M.D., Szwarcfiter, J.L.: Applying modular decomposition to parameterized cluster editing problems. Theory Comput. Syst. 44(1):91–104 (2009)

18. Roditty, L.: A faster and simpler fully dynamic transitive closure. In: SODA, pp. 404–412 (2003)

19. Roditty, L., Zwick, U.: Improved dynamic reachability algorithms for directed graphs. In: FOCS, pp. 679– (2002)

20. Roditty, L., Zwick, U.: Dynamic approximate all-pairs shortest paths in undirected graphs. In: FOCS, pp. 499–508 (2004)

21. Roditty, L., Zwick, U.: A fully dynamic reachability algorithm for directed graphs with an almost linear update time. In: STOC, pp. 184–191 (2004)

22. Roditty, L., Zwick, U.: On dynamic shortest paths problems. In: Albers, S., Radzik, T. (eds.) ESA 2004. LNCS, vol. 3221, pp. 580–591. Springer, Heidelberg (2004)

23. Sankowski, P.: Dynamic transitive closure via dynamic matrix inverse (extended abstract). In: FOCS, pp. 509–517 (2004)

24. Thorup, M.: Near-optimal fully-dynamic graph connectivity. In: STOC, pp. 343–350 (2000)

25. Thorup, M.: Fully-dynamic all-pairs shortest paths: Faster and allowing negative cycles. In: Hagerup, T., Katajainen, J. (eds.) SWAT 2004. LNCS, vol. 3111, pp. 384–396. Springer, Heidelberg (2004)

26. Thorup, M.: Worst-case update times for fully-dynamic all-pairs shortest paths. In: STOC, pp. 112–119 (2005)

27. Wulff-Nilsen, C.: Faster deterministic fully-dynamic graph connectivity. In: SODA, pp. 1757–1769 (2013)

Extending Partial Representations
of Proper and Unit Interval Graphs*

Pavel Klavík[1], Jan Kratochvíl[2], Yota Otachi[3], Ignaz Rutter[4,2],
Toshiki Saitoh[5], Maria Saumell[6], and Tomáš Vyskočil[2]

[1] Computer Science Institute, Faculty of Mathematics and Physics,
Charles University in Prague, Czech Republic
`klavik@iuuk.mff.cuni.cz`
[2] Department of Applied Mathematics, Faculty of Mathematics and Physics,
Charles University in Prague, Czech Republic
`{honza,whisky}@kam.mff.cuni.cz`
[3] School of Information Science,
Japan Advanced Institute of Science and Technology, Japan
`otachi@jaist.ac.jp`
[4] Institute of Theoretical Informatics, Faculty of Informatics,
Karlsruhe Institute of Technology (KIT), Germany
`rutter@kit.edu`
[5] Graduate School of Engineering, Kobe University, Kobe, Japan
`saitoh@eedept.kobe-u.ac.jp`
[6] Department of Mathematics, University of West Bohemia,
Plzeň, Czech Republic
`saumell@kma.zcu.cz`

Abstract. The recently introduced problem of extending partial interval representations asks, for an interval graph with some intervals predrawn by the input, whether the partial representation can be extended to a representation of the entire graph. In this paper, we give a linear-time algorithm for extending proper interval representations and an almost quadratic-time algorithm for extending unit interval representations.

We also introduce the more general problem of *bounded representations* of unit interval graphs, where the input constrains the positions of intervals by lower and upper bounds. We show that this problem is NP-complete for disconnected input graphs and give a polynomial-time algorithm for a special class of instances, where the ordering of the connected components of the input graph along the real line is fixed. This includes the case of partial representation extension.

The hardness result sharply contrasts the recent polynomial-time algorithm for bounded representations of proper interval graphs [Balko et al. ISAAC'13]. So unless P = NP, proper and unit interval representations have very different structure. This explains why partial representation extension problems for these different types of representations require substantially different techniques.

---

* For the full version of this paper, see `arXiv:1207.6960`.

R Ravi and I.L. Gørtz (Eds.): SWAT 2014, LNCS 8503, pp. 253–264, 2014.
© Springer International Publishing Switzerland 2014

1   Introduction

Geometric intersection graphs, and in particular intersection graphs of objects in the plane, have gained a lot of interest for their practical motivations, algorithmic applications, and interesting theoretical properties. Undoubtedly the oldest and the most studied among them are *interval graphs* (INT), i.e., intersection graphs of intervals on the real line. They were introduced by Hájos [13] in the 1950's and the first polynomial-time recognition algorithm appeared already in the early 1960's [12]. Several linear-time algorithms are known, see [4,9].

Only recently, the following natural generalization of the recognition problem has been considered [18]. The input of the *partial representation extension* problem consists of a graph and a part of the representation and it asks whether it is possible to extend this partial representation to a representation of the entire graph. Klavík et al. [18] give a quadratic-time algorithm for interval graphs and a cubic-time algorithm for proper interval graphs. (For interval graphs the problem can be solved in linear time [3,17].) There are also polynomial-time algorithms for function and permutation graphs [15] as well as for circle graphs [6]. Chordal graph representations (intersection graphs of subtrees in a tree) [16] and intersection representations of planar graphs [5] are mostly hard to extend. For related simultaneous representation problems see [3,14].

In this paper, we extend the line of research on partial representation extension problems of proper interval graphs and unit interval graphs. Although it is well known that these graph classes are identical [20], the representation extension problems differ substantially. This is due to the fact that for proper interval graphs, in whose representations no interval is a proper subset of another interval, the extension problem is essentially topological and can be treated in a purely combinatorial manner. On the other hand, unit interval representations, where all intervals have length one, are inherently geometric, and the corresponding algorithms have to take geometric constraints into account.

It has been observed in other contexts that geometric problems are sometimes more difficult than the corresponding topological problems. For example, the partial drawing extension of planar graphs is linear-time solvable [1] for topological drawing but NP-hard for straight-line drawings [19]. Together with the result of Balko et al. [2] our results show that a generalization of partial representation extension exhibits this behavior already in 1-dimensional geometry. The bounded representation problem is polynomial-time solvable for proper interval graphs [2] and NP-complete for unit interval graphs.

Intersection Representations and Interval Graphs. For a graph G, an *intersection representation* $\mathcal{R}$ is a collection of sets $\{R_u : u \in V(G)\}$ such that $R_u \cap R_v \neq \emptyset$ if and only if $uv \in E(G)$; so the edges of G are encoded by intersections of the sets. In an *interval representation* each R_u is a closed interval of the real line. A graph is an *interval graph* if it has an interval representation. We denote the corresponding class of graphs by INT.

We consider two subclasses of interval representations. An interval representation is *proper* if no interval is a proper subset of another interval (meaning $R_u \subseteq R_v$ implies $R_u = R_v$). An interval representation is *unit* if the length of

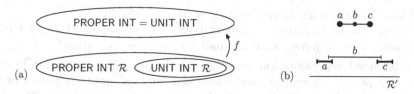

Fig. 1. (a) Relation of the representations and graph classes studied in this paper. The denoted mapping f assigns to a representation the corresponding intersection graph. Roberts' Theorem [20] states that f restricted to unit interval representations is surjective. (b) A partial representation that is extendable as a proper interval representation, but not extendable as a unit interval representation.

each interval is 1. The classes of graphs admitting proper and unit interval representations are called *proper interval graphs* (PROPER INT) and *unit interval graphs* (UNIT INT), respectively. Note that every unit interval representation is also a proper interval representation, and hence UNIT INT $\subseteq$ PROPER INT. It is a well-known fact that indeed equality holds (Roberts' Theorem [20]); see Fig. 1a for an illustration of the relation between the graph classes and their representations studied in this paper.

In an interval representation $\mathcal{R} = \{R_v : v \in V(G)\}$, we denote the left and right endpoint of the interval R_v by ℓ_v and r_v, respectively. For numbered vertices $v_1, \ldots, v_n$, we denote these endpoints by ℓ_i and r_i. Note that several intervals may share an endpoint in a representation. When working with multiple representations, we denote the other one by $\mathcal{R}'$ with intervals $R'_v = [\ell'_v, r'_v]$.

Partial Representation Extension and Bounded Representations. The *recognition* problem of a class $\mathcal{C}$ asks whether an input graph belongs to $\mathcal{C}$, i.e., whether it admits a specific type of representation. We study two generalizations of this problem: The *partial representation extension problem*, introduced in [18], and a new problem called the *bounded representation problem*.

A *partial representation* $\mathcal{R}'$ of G is a representation of an induced subgraph G' of G. A vertex in $V(G')$ is called *pre-drawn*. A representation $\mathcal{R}$ *extends* $\mathcal{R}'$ if $R_u = R'_u$ for each $u \in V(G')$.

> **Problem:** REPEXT($\mathcal{C}$) (Partial Representation Extension of $\mathcal{C}$)
> **Input:** Graph G with partial representation $\mathcal{R}'$.
> **Output:** Does G have a representation $\mathcal{R}$ that extends $\mathcal{R}'$?

Even though PROPER INT = UNIT INT, the problems REPEXT(PROPER INT) and REPEXT(UNIT INT) behave very differently; see Fig. 1b for an example.

Observe that REPEXT(UNIT INT) completely prescribes the representation of the pre-drawn vertices and leaves the representation of the remaining vertices unrestricted. Thus the problem where the position of the interval for each vertex v_i is restricted by upper and lower bounds lbound(v_i) and ubound(v_i) is a strict generalization. A representation $\mathcal{R}$ is called a *bounded representation* if lbound(v_i) $\leq \ell_i \leq$ ubound(v_i) for each vertex v_i.

Problem: BOUNDREP (Bounded Representation of UNIT INT)
   Input: Graph G, rational numbers lbound(v_i), ubound(v_i) for $v_i \in V(G)$.
 Output: Does G have a bounded unit interval representation?

The bounded representation problem can be considered also for interval graphs and proper interval graphs, where the left and right endpoints of the intervals can be restricted individually. A recent paper of Balko et al. [2] proves that this problem is polynomially solvable for these classes. Note that for unit intervals, it suffices to restrict the left endpoint since $r_i = \ell_i + 1$.

Contribution and Outline. In this paper we present five results. The first is a simple linear-time algorithm for REPEXT(PROPER INT), improving over a previous $O(nm)$-time algorithm [18]; it is based on known characterizations.

Theorem 1. REPEXT(PROPER INT) *can be solved in time* $\mathcal{O}(n+m)$.

Second, we give a reduction from the strongly NP-complete problem 3-PARTITION to show that BOUNDREP is NP-complete for disconnected graphs. The main idea is that prescribed intervals partition the real line into gaps of fixed width. Integers are encoded in connected components whose unit interval representations require a certain width. By suitably choosing the lower and upper bounds, we enforce that the connected components have to be placed inside the gaps such that they do not overlap.

Theorem 2. BOUNDREP *is NP-complete*.

Third, in Section 3, we give a relatively simple quadratic-time algorithm for the special case of BOUNDREP where the order of the connected components along the real line is fixed. We formulate this problem as a sequence of linear programs (LPs), and we show that each LP reduces to a shortest-path problem, which we solve with the Bellmann-Ford algorithm [7, Chapter 24.4].

The running time is $O(n^2 r + nD(r))$, where r is the total encoding length of the bounds in the input, and $D(r)$ is the time required for multiplying or dividing two numbers whose binary representation has length r. This is due to the fact that the numbers specifying the upper and lower bounds for the intervals can be quite close to each other, requiring that the corresponding rationals have an encoding that is super-polynomial in n. Clearly, two binary numbers whose representations have length r can be added in $O(r)$ time, explaining the term of $O(n^2 r)$ in the running time. However, using Bellmann-Ford for solving the LP requires also the comparison of rational numbers. To be able to do this efficiently, we convert the rational numbers to a common denominator. Hence, the multiplication cost $D(r)$ enters the running time. The best known algorithm achieves $D(r) = \mathcal{O}(r \log r 2^{\log^* r})$ [11].

Fourth, in Section 4, we show how to reduce the dependency on r to obtain a running time of $O(n^2 + nD(r))$, which may be beneficial for instances with bounds that have a long encoding.

Theorem 3. BOUNDREP *with a prescribed ordering* ◄ *of the connected components can be solved in time* $\mathcal{O}(n^2 + nD(r))$, *where r is the size of the input describing bound constraints.*

Our algorithm performs $\mathcal{O}(n^2)$ combinatorial iterations, each taking time $\mathcal{O}(1)$. The additional time $\mathcal{O}(nD(r))$ is used for arithmetic operations with the bounds.

Finally, we note that every instance of REPEXT(UNIT INT) is an instance of BOUNDREP. In Section 5, we show how to derive for these special instances a suitable ordering ◄ of the connected components, resulting in an efficient algorithm for REPEXT(UNIT INT).

Theorem 4. REPEXT(UNIT INT) *can be solved in time* $\mathcal{O}(n^2 + nD(r))$, *where* r *is the size of the input describing positions of pre-drawn intervals.*

All the algorithms described in this paper are also able to certify the extendibility by constructing the required representations. Many proofs and details are omitted and placed in the full version.

2   Preliminaries and Proper Interval Graphs

As usual, we reserve n for the number of vertices and m for the number of edges of the graph G. We denote the set of vertices by $V(G)$ and the set of edges by $E(G)$. For a vertex v, we define $N[v] = \{x : vx \in E(G)\} \cup \{v\}$. We assume that G contains no two vertices u and v such that $N[u] = N[v]$. In the full version of the paper, we show that our algorithms can be modified to handle also the occurrence of such *indistinguishable vertices*.

Unique Ordering. In each proper interval representation, intervals are uniquely ordered from left to right. This ordering $<$ is the order of the left endpoints and at the same time the order of the right endpoints. Deng et al. [10] proved:

Lemma 5 (Deng et al.). *For a connected proper/unit interval graph, the left-to-right ordering $<$ is uniquely determined up to reordering groups of indistinguishable vertices and complete reversal.*

Such an ordering can be computed in linear time [8]. In particular, if there are no indistinguishable vertices, the ordering $<$ is uniquely determined up to reversal. On the other hand, a partial representation $\mathcal{R}'$ induces a partial order $<^{\mathcal{R}'}$ of the vertices of the input graph. It essentially remains to check whether the ordering $<$ or its reversal extends the ordering $<^{\mathcal{R}'}$. This leads to a characterization of the extendible instances of REPEXT(PROPER INT) and the linear-time algorithm of Theorem 1; for details, see full version.

Representations in ε-Grids. For a value $\varepsilon = \frac{1}{K}$, where K is an integer, the ε-grid is the set of points $\{k\varepsilon : k \in \mathbb{Z}\}$. For a given instance of BOUNDREP, we ask which value of ε ensures that we can construct a representation having all endpoints on the ε-grid.

For the standard unit interval graph representation problem a grid of size $\frac{1}{n}$ is sufficient [8]. In the case of BOUNDREP, consider all values lbound(v_i) and ubound(v_i) distinct from $-\infty, +\infty$, and express them as irreducible fractions $\frac{p_1}{q_1}, \frac{p_2}{q_2}, \dots, \frac{p_b}{q_b}$. Using lcm($\cdot$) to denote the least common multiple, we define:

$$\varepsilon' := \frac{1}{\text{lcm}(q_1, q_2, \dots, q_b)}, \quad \text{and} \quad \varepsilon := \frac{\varepsilon'}{n}. \qquad (1)$$

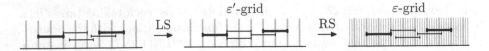

Fig. 2. In the first step, we shift intervals to the left to the ε'-grid. The left shifts of $v_1, \ldots, v_5$ are $(0, 0, \frac{1}{2}\varepsilon', \frac{1}{3}\varepsilon', 0)$. In the second step, we shift to the right in the refined ε-grid. Right shifts have the same relative order as left shifts: $(0, 0, 2\varepsilon, \varepsilon, 0)$.

We show that an ε-grid is sufficient to construct the bounded representation:

Lemma 6. *If there exists a valid representation $\mathcal{R}'$ for an input of the problem* BOUNDREP, *there exists a valid representation $\mathcal{R}$ in which all intervals have endpoints on the ε-grid, where ε is defined by equation* (1).

Proof (Sketch). We construct an ε-grid representation $\mathcal{R}$ from $\mathcal{R}'$ in two steps. First, we shift intervals to the left, and then we shift intervals slightly back to the right. The shifting process is shown in Fig. 2.

The left-shift moves each interval to the left to the closest ε'-grid point. By this, intersections are not removed but new intersections might be introduced (but only in the form of touching pairs of intervals). The right-shift fixes these touching pairs. It is a mapping, RS : $\{v_1, \ldots, v_n\} \rightarrow \{0, \varepsilon, 2\varepsilon, \ldots, (n-1)\varepsilon\}$, having the *right-shift property*: For all pairs (v_i, v_j) with $r_i = \ell_j$, $\mathrm{RS}(v_i) \geq \mathrm{RS}(v_j)$ if and only if $v_i v_j \in E$. This mapping can be constructed from the reversal of the left-shift. It is easy to see that the constructed representation is correct and satisfies the bounds. $\square$

3   LP Algorithm for BoundRep with Prescribed Order

We describe how to solve BOUNDREP in polynomial time for a prescribed ordering $\blacktriangleleft$ of the components using linear programming. According to Lemma 5, the vertices of each component of G can be ordered in at most two different ways. We cannot arbitrarily choose one of the orderings, since neighboring components restrict each other's space. In the algorithm, we process components $C_1 \blacktriangleleft C_2 \blacktriangleleft \cdots \blacktriangleleft C_c$ from left to right. For each component C_t, we calculate the ordering $<$ and its reversal using the algorithm of Corneil et al. [8].

We have two orderings $<$ for C_t, and we solve one linear program for each of them. Let $v_1 < v_2 < \cdots < v_k$ be one of these orderings. We denote the right-most endpoint of a representation of a component C_t by E_t. Additionally, we define $E_0 = -\infty$. Also, we modify all lower bounds by putting $\mathrm{lbound}(v_i) = \max\{\mathrm{lbound}(v_i), E_{t-1} + \varepsilon\}$ for every interval v_i, which forces the representation of C_t to be on the right of the previously constructed representation of C_{t-1}. The linear program has variables $\ell_1, \ldots, \ell_k$, and we minimize the value of E_t. Let ε be defined as in (1). We solve:

Minimize: $E_t := \ell_k + 1,$

subject to: $\ell_i \leq \ell_{i+1},$ $\forall i = 1, \ldots, k-1,$ (2)

$\ell_i \geq \text{lbound}(v_i),$ $\forall i = 1, \ldots, k,$ (3)

$\ell_i \leq \text{ubound}(v_i),$ $\forall i = 1, \ldots, k,$ (4)

$\ell_i \geq \ell_j - 1,$ $\forall v_i v_j \in E, v_i < v_j,$ (5)

$\ell_i + \varepsilon \leq \ell_j - 1,$ $\forall v_i v_j \notin E, v_i < v_j.$ (6)

We solve the same linear program for the other ordering of the vertices of C_t. If none of the two programs is feasible, we report that no bounded representation exists. If at least one of them is feasible, we take the solution minimizing E_t.

Proposition 7. *The* BOUNDREP *problem with prescribed ordering ◄ of connected components can be solved in polynomial time.*

This linear program can easily be transformed into a *system of difference constraints*, which can be solved in time $\mathcal{O}(k^2 r + k D(r))$ by computing minimum-weight shortest paths in a directed graph using the Bellman-Ford algorithm [7, Chapter 24.4]. The result of the next section improves the time complexity for BOUNDREP to $\mathcal{O}(k^2 + k D(r))$.

4   Shifting Algorithm for BoundRep with Fixed Ordering

The goal of this section is to prove Theorem 3. We solve the linear program described in Section 3 by a combinatorial algorithm based on shifting of intervals.

Suppose that we ignore upper bound constraints (4) for a second and we want to find any solution satisfying the remaining constraints of the program. It is easy to construct such a solution, since we can construct any unit interval representation using [8], and then shift this representation enough to the right. The shifting algorithm modifies this initial representation by a series of shifts, and thus constructs an optimal solution of the linear program.

4.1   Structural Properties of Unit Interval Representations

We assume that the unit interval graph is connected. Also, we assume that one left-to-right ordering $<$ of the intervals is prescribed. Let $\mathfrak{Rep}$ denote the set of all ε-grid representations in the ordering $<$ satisfying the lower bounds. As we already discussed, this set is non-empty.

There is a natural partial ordering of these representations: For $\mathcal{R}, \mathcal{R}' \in \mathfrak{Rep}$, we say that $\mathcal{R} \leq \mathcal{R}'$ if and only if $\ell_i \leq \ell_i'$ for every interval $v_i \in V(G)$.

Semilattice Structure. The poset $(\mathfrak{Rep}, \leq)$ is a (meet)-semilattice:

Lemma 8. *Every non-empty $S \subseteq \mathfrak{Rep}$ has an infimum $\inf(S)$.*

The infimum $\mathcal{R}$ has $\ell_i = \min\{\ell_i' : \mathcal{R}' \in S\}$ for every $v_i \in V(G)$. We prove in the full version that $\mathcal{R}$ is the infimum and belongs to $\mathfrak{Rep}$.

We call the infimum $\inf(\mathfrak{Rep})$ the *left-most representation*. Clearly, if this representation satisfies the upper bound constraints, then it is an optimal solution of the linear program. On the other hand, one can easily prove that there exists a representation $\mathcal{R}'$ satisfying both lower and upper bound constraints if and only if the left-most representation satisfies the upper bound constraints. Therefore, we can solve the linear program by constructing the left-most representation.

Left-Shifting of Intervals. Suppose that we construct some initial ε-grid representation that is not the left-most representation. We want to transform this initial representation in $\mathfrak{Rep}$ into the left-most representation by applying the following simple operation called *left-shifting*. The left-shifting operation shifts one interval of the representation by ε to the left such that this shift maintains the correctness of the representation. The main result of this subsection is the following proposition whose proof is in the full version.

Proposition 9. *For $\varepsilon = \frac{1}{K}$ and $K \geq \frac{n}{2}$, an ε-grid representation $\mathcal{R} \in \mathfrak{Rep}$ is the left-most representation if and only if it is not possible to shift any single interval to the left by ε while maintaining correctness of the representation.*

An interval v_i is called *fixed* if it is in the left-most position and cannot be ever shifted more to the left, i.e., $\ell_i = \min\{\ell_i' : \mathcal{R}' \in \mathfrak{Rep}\}$. For example, it is fixed if $\ell_i = \mathrm{lbound}(v_i)$. A representation is the left-most representation if and only if every interval is fixed.

An interval v_i, having $\ell_i \geq \mathrm{lbound}(v_i) + \varepsilon$, can be shifted to the left by ε if it does not make the representation incorrect, and the incorrectness can be obtained in two ways. First, there could be some interval v_j such that $v_j < v_i$, $v_i v_j \notin E(G)$, and $\ell_j + 1 + \varepsilon = \ell_i$; we call v_j a *left obstruction* of v_i. Second, there could be some interval v_j such that $v_i < v_j$, $v_i v_j \in E(G)$, and $\ell_i + 1 = \ell_j$ (so v_i and v_j are touching); then we call v_j a *right obstruction* of v_i. In both cases, we first need to move v_j before moving v_i.

Since $N[u] \neq N[v]$ for each u and v, $\ell_u \neq \ell_v$ in every representation. Therefore each vertex has at most one obstruction of each type, and these obstructions are always the same: If v_i has a left obstruction, it is the first non-neighbor of v_i on the left. If v_i has a right obstruction, it is the right-most neighbor of v_i.

Position Cycle. For each interval in some ε-grid representation with $\varepsilon = \frac{1}{K}$, we can write its position in the form $\ell_i = \alpha_i + \beta_i \varepsilon$, where $\alpha_i \in \mathbb{Z}$, $\beta_i \in \mathbb{Z}_K$. We can depict $\mathbb{Z}_K = \{0, \ldots, K-1\}$ as a cycle with K vertices where the value decreases clockwise. The value β_i assigns to each interval v_i one vertex of the cycle. Together with placed v_i's, we call this the *position cycle*.

The position cycle allows us to visualize and work with left-shifting very intuitively. When an interval v_i is shifted by ε to the left, β_i is cyclically decreased by one, so it is moving clockwise along the cycle. If v_j is the left obstruction of v_i, then $\beta_j = \beta_i - 1$; if v_j is the right obstruction of v_i, then $\beta_i = \beta_j$. So in both cases β_j has to be very close to β_i. For an illustration, see Fig. 3.

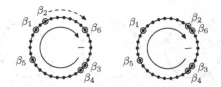

Fig. 3. Examples of position cycles. In the cycle on the left, we can shift β_2 in clockwise direction towards β_6, which gives a new representation whose position cycle is depicted on the right. We note that after left-shifting, v_6 is not necessarily an obstruction of v_2.

4.2   The Shifting Algorithm

The shifting algorithm we describe here solves the linear program of Section 3 in time $\mathcal{O}(k^2 + kD(r))$, where k is the number of vertices of the component and r is the size of the input describing bounds of the component. The left-to-right order $<$ of the vertices is given.

Overview. The algorithm works in three basic steps:

(1) Construct an initial ε-grid representation (in the ordering $<$) having $\ell_i \geq$ lbound(v_i) for all intervals, using the algorithm of Corneil et al. [8].
(2) Shift intervals to the left while maintaining correctness of the representation until the left-most representation is constructed, using Proposition 9.
(3) Check whether the left-most representation satisfies the upper bounds. If so, this representation satisfies all bound constraints and solves the linear program of Section 3. Otherwise, no representation satisfying all bound constraints exists, and thus the linear program has no solution.

Input Size. Since ε can be very small, we do not operate with precise positions on the ε-grid. Instead, we position the intervals on a larger Δ-grid, $\Delta = \frac{1}{n^2}$, and shift them there. Only when some interval becomes fixed, its precise position on the ε-grid is calculated. This allows to reduce the time complexity from $\mathcal{O}(k^2 D(r))$ to $\mathcal{O}(k^2 + kD(r))$. Technical details are in the full version.

Left-Shifting. We deal separately with fixed and unfixed intervals. Unfixed intervals are on the Δ-grid and fixed intervals have precise positions calculated on the ε-grid. We place only unfixed intervals on the position cycle (for the Δ-grid).

We shift unfixed intervals by using gaps in the position cycle: When we shift interval v_i from ℓ_i to ℓ'_i, we decrease β_i to $\beta_\ell + 1$, where β_ℓ is the first β_j we encounter when we move clockwise from β_i. We also check whether this shift is valid with respect to fixed intervals and the lower bound constraint. The interval v_i can become fixed in two ways: Either $\ell'_i \leq$ lbound(v_i) or there is some fixed obstruction v_j to which v_i is shifted (for a left obstruction $\ell'_i \leq \ell_j + 1 + \varepsilon$, for a right obstruction $\ell'_i \leq \ell_j - 1$). All this can be checked in $\mathcal{O}(1)$ time. If v_i becomes fixed, it is removed from the position cycle and its position on the ε-grid is calculated.

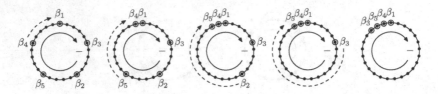

Fig. 4. The position cycle during the first phase, changing from left to right. The first phase clusters the β_i's by moving β_4, β_5, β_2 and β_3 towards β_1. When v_2 is shifted, v_2 becomes fixed and β_2 disappears from the position cycle.

Initial Representation. We start with an initial Δ-grid representation satisfying all lower bounds such that $\ell_i \leq \mathrm{lbound}(v_i) + \Delta$ for at least one interval v_i. Then every other interval can be shifted in total by distance at most $\mathcal{O}(k)$ from the initial position, since the component is connected.

To obtain the initial representation, we use the algorithm in [8], which places the intervals in such a way that β_i's are positioned equidistantly in the position cycle; refer to the left-most position cycle in Fig. 4.

Shifting Phases. The shifting of unfixed intervals proceeds in two phases:

- *The first phase* creates one big gap by clustering all β_i's in one part of the cycle. To do so, we shift intervals in the order given by the position cycle. Of course, some intervals might become fixed and disappear from the position cycle. We obtain one big gap of size at least $n(n-1)$. Again, refer to Fig. 4.
- *In the second phase,* we use this big gap to shift intervals one by one, which also moves the cluster along the position cycle. Again, if some interval becomes fixed, it is removed from the position cycle. The second phase finishes when each interval becomes fixed and the left-most representation is constructed. For an example, see Fig. 5.

Putting Everything Together. We are ready to prove that BOUNDREP with a prescribed ordering $\blacktriangleleft$ can be solved in time $\mathcal{O}(n^2 + nD(r))$:

Proof (Theorem 3, sketch). We process the components $C_1 \blacktriangleleft \cdots \blacktriangleleft C_c$ from left to right, and for each component we solve two LPs using the shifting algorithm described above. To solve the LPs, we construct the left-most representation.

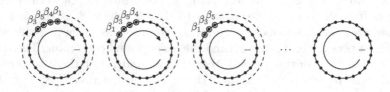

Fig. 5. The position cycle during the second phase, changing from left to right. We shift β_i's across the big gap till all β_i's disappear.

By Proposition 9 the algorithm stops when each interval is fixed, and it indeed constructs the left-most representation. As already argued, for this representation it is sufficient to check the upper bounds.

Concerning complexity, each interval is shifted by distance at most k. The first phase performs $\mathcal{O}(k)$ shifts. In the second phase, each interval is shifted by at least $\frac{n-1}{n}$ unless it becomes fixed. So in total, the second phase performs $\mathcal{O}(k^2)$ shifts. Each shift can be implemented in time $\mathcal{O}(1)$ unless the interval becomes fixed. We need additional time $\mathcal{O}(kD(r))$ for precomputation and to compute exact positions on the ε-grid every time an interval becomes fixed. Thus the total time per component is $\mathcal{O}(k^2 + kD(r))$ and we get $\mathcal{O}(n^2 + nD(r))$ for the entire graph. □

5 Extending Unit Interval Representations

We show that REPEXT(UNIT INT) is a particular instance of BOUNDREP where the ordering ◄ of the components is known, i.e., it can be solved using Theorem 3.

Unlike the recognition problem, REPEXT cannot generally be solved independently for connected components. A connected component C of G is called *located* if it contains at least one pre-drawn interval, and *unlocated* otherwise.

Let $\mathcal{R}$ be any interval representation. For each component C, $\bigcup_{u \in C} R_u$ is a connected segment of the real line and for different components we get disjoint segments. These segments are ordered from left to right, giving a linear ordering ◄ of the components.

Proof (Theorem 4). The graph G contains located and unlocated components. Unlocated components can be placed far to the right and we can deal with them using a standard recognition algorithm.

Concerning located components $C_1, \ldots, C_c$, they have to be ordered from left to right according to the left-to-right ordering of the pre-drawn intervals (otherwise the problem has no solution). This gives the required ordering ◄. We straightforwardly construct the instance of BOUNDREP with this ◄ as follows. For each pre-drawn interval v_i at position ℓ_i, we put lbound(v_i) = ubound(v_i) = ℓ_i. For the rest of intervals, we set no bounds. Clearly, this instance of BOUNDREP is equivalent to the original REPEXT(UNIT INT) problem. □

Open Problem. Our main open question is whether there exists an algorithm for REPEXT(UNIT INT) with running time $o(n^2 + nD(r))$.

Acknowledgments. The first, second and sixth authors are supported by ESF Eurogiga project GraDR as GAČR GIG/11/E023, the first author also by GAČR 14-14179S and the first two authors by Charles University as GAUK 196213. The fourth author is supported by a fellowship within the Postdoc-Program of the German Academic Exchange Service (DAAD), the sixth author by projects NEXLIZ - CZ.1.07/2.3.00/30.0038, which is co-financed by the European Social Fund and the state budget of the Czech Republic, and ESF EuroGIGA project ComPoSe as F.R.S.-FNRS - EUROGIGA NR 13604.

References

1. Angelini, P., Di Battista, G., Frati, F., Jelínek, V., Kratochvíl, J., Patrignani, M., Rutter, I.: Testing planarity of partially embedded graphs. In: SODA 2010: Proc. 21st Annu. ACM-SIAM Sympos. Discr. Alg., pp. 202–221 (2010)
2. Balko, M., Klavík, P., Otachi, Y.: Bounded representations of interval and proper interval graphs. In: Cai, L., Cheng, S.-W., Lam, T.-W. (eds.) Algorithms and Computation. LNCS, vol. 8283, pp. 535–546. Springer, Heidelberg (2013)
3. Bläsius, T., Rutter, I.: Simultaneous PQ-ordering with applications to constrained embedding problems. In: SODA 2013: Proc. 24th Annu. ACM-SIAM Sympos. Discr. Alg., pp. 1030–1043 (2013)
4. Booth, K.S., Lueker, G.S.: Testing for the consecutive ones property, interval graphs, and planarity using PQ-tree algorithms. J. Comput. System Sci. 13, 335–379 (1976)
5. Chaplick, S., Dorbec, P., Kratochvíl, J., Montassier, M., Stacho, J.: Contact representations of planar graph: Rebuilding is hard. In: WG 2014 (to appear, 2014)
6. Chaplick, S., Fulek, R., Klavík, P.: Extending partial representations of circle graphs. In: Wismath, S., Wolff, A. (eds.) GD 2013. LNCS, vol. 8242, pp. 131–142. Springer, Heidelberg (2013)
7. Cormen, T.H., Leiserson, C.E., Rivest, R.L., Stein, C.: Introduction to Algorithms, 3rd edn. The MIT Press (2009)
8. Corneil, D.G., Kim, H., Natarajan, S., Olariu, S., Sprague, A.P.: Simple linear time recognition of unit interval graphs. Inform. Process. Lett. 55(2), 99–104 (1995)
9. Corneil, D.G., Olariu, S., Stewart, L.: The LBFS structure and recognition of interval graphs. SIAM J. Discrete Math. 23(4), 1905–1953 (2009)
10. Deng, X., Hell, P., Huang, J.: Linear-time representation algorithms for proper circular-arc graphs and proper interval graphs. SIAM J. Comput. 25(2), 390–403 (1996)
11. Fürer, M.: Faster integer multiplication. SIAM J. Comput. 39(3), 979–1005 (2009)
12. Gilmore, P.C., Hoffman, A.J.: A characterization of comparability graphs and of interval graphs. Can. J. Math. 16, 539–548 (1964)
13. Hajós, G.: Über eine Art von Graphen. Internationale Mathematische Nachrichten 11, 65 (1957)
14. Jampani, K., Lubiw, A.: The simultaneous representation problem for chordal, comparability and permutation graphs. J. Graph Algorithms Appl. 16(2), 283–315 (2012)
15. Klavík, P., Kratochvíl, J., Krawczyk, T., Walczak, B.: Extending partial representations of function graphs and permutation graphs. In: Epstein, L., Ferragina, P. (eds.) ESA 2012. LNCS, vol. 7501, pp. 671–682. Springer, Heidelberg (2012)
16. Klavík, P., Kratochvíl, J., Otachi, Y., Saitoh, T.: Extending partial representations of subclasses of chordal graphs. In: Chao, K.-M., Hsu, T.-S., Lee, D.-T. (eds.) ISAAC 2012. LNCS, vol. 7676, pp. 444–454. Springer, Heidelberg (2012)
17. Klavík, P., Kratochvíl, J., Otachi, Y., Saitoh, T., Vyskočil, T.: Linear-time algorithm for partial representation extension of interval graphs (2012) (in preparation)
18. Klavík, P., Kratochvíl, J., Vyskočil, T.: Extending partial representations of interval graphs. In: Ogihara, M., Tarui, J. (eds.) TAMC 2011. LNCS, vol. 6648, pp. 276–285. Springer, Heidelberg (2011)
19. Patrignani, M.: On extending a partial straight-line drawing. Int. J. Found. Comput. Sci. 17(5), 1061–1070 (2006)
20. Roberts, F.S.: Indifference graphs. In: Harary, F. (ed.) Proof Techniques in Graph Theory, pp. 139–146. Academic Press (1969)

Minimum Tree Supports for Hypergraphs and Low-Concurrency Euler Diagrams

Boris Klemz, Tamara Mchedlidze, and Martin Nöllenburg

Institute of Theoretical Informatics, Karlsruhe Institute of Technology (KIT), Germany

Abstract. In this paper we present an $O(n^2(m+\log n))$-time algorithm for computing a minimum-weight tree support (if one exists) of a hypergraph $H = (V, S)$ with n vertices and m hyperedges. This improves the previously best known algorithm with running time $O(n^4 m^2)$. A support of H is a graph G on V such that each hyperedge in S induces a connected subgraph in G. If G is a tree, it is called a tree support and it is a minimum tree support if its edge weight is minimum for a given edge weight function. Tree supports of hypergraphs have several applications, from social network analysis and network design problems to the visualization of hypergraphs and Euler diagrams. We show in particular how a minimum-weight tree support can be used to generate an area-proportional Euler diagram that satisfies typical well-formedness conditions and additionally minimizes the number of concurrent curves of the set boundaries in the Euler diagram.

1 Introduction

A hypergraph $H = (V, S)$ is a generalization of a graph that consists of a set of vertices V and a set of hyperedges S, which are arbitrary non-empty subsets of V (in contrast to graph edges, which are defined as pairs of vertices). Thus S is a subset of the power set $\mathcal{P}(V) = 2^V$. A graph $G = (V, E)$ is called a *support* of a hypergraph $H = (V, S)$ on the same vertex set if every hyperedge $s \in S$ induces a connected subgraph in G.

Sparse support graphs are interesting from the perspective of network design as they represent realizations of hypergraphs as graphs, in which the vertices of each hyperedge induce a connected component. Korach and Stern [16] introduced the problem of finding a *minimum(-weight) tree support* (MTS) for a given hypergraph $H = (V, S)$ and a given edge-weight function $w: \binom{V}{2} \to \mathbb{R}$ for the support graph. Here, an MTS is a support $T = (V, E)$ that is a tree with minimum total edge weight $\sum_{e \in E} w(e)$. Not every hypergraph has a tree support, but the decision problem can be solved in linear time by testing whether its dual hypergraph is acyclic [14]. If a hypergraph has a tree support, it is called a *tree-hypergraph*. Korach and Stern [16] gave an algorithm to compute an MTS in $O(|V|^4|S|^2)$ time (if it exists). They later presented another polynomial-time algorithm for a restricted variation, in which they ask for a tree support of minimum weight such that each subtree induced by a hyperedge is a star [17].

Hypergraphs and hypergraph supports are not as frequently used and studied as graphs themselves, but they still have many real-world applications. For example, in social network analysis, minimum tree supports are used to compute maximum-likelihood

R Ravi and I.L. Gørtz (Eds.): SWAT 2014, LNCS 8503, pp. 265–276, 2014.
© Springer International Publishing Switzerland 2014

social networks that serve as models to explain a collection of observed disease outbreaks that are modeled as hyperedges of the infected persons [1, 2]. In topic-based
peer-to-peer publish/subscribe systems [7, 13] the input is a set of users V and a set
of topics S, where each topic $t \in S$ is a subset of users (i.e., a hyperedge) interested
in the topic. The task is to design an overlay network G on V (called *minimum topic-
connected overlay*) with the minimum number of edges so that each topic forms a connected subgraph in G thus enabling private communication within each topic. If the
underlying hypergraph $H = (V, S)$ admits a tree support then this minimum overlay
network will be a tree; if establishing edges in G is linked with a cost, the task is again
to find an MTS. The unweighted problem is also known as *subset interconnection design*, which generally asks for a support graph with the minimum number of edges, i.e.,
not necessarily a tree. It has applications in the design of reconfigurable networks, e.g.,
vacuum systems, in which valves correspond to edges in the support and their number
needs to be minimized [6, 10, 11].

Of particular interest for hypergraph visualizations are *planar supports*. Johnson and
Pollak [14] showed that a hypergraph is *vertex-planar* if and only if it has a planar support. A vertex-planar hypergraph has a representation of the vertices as faces in a planar
subdivision such that for each hyperedge the union of the faces corresponding to the vertices incident to that hyperedge is a connected region. Simple and compact subdivision
drawings [15] are an interesting restriction of vertex-planar hypergraph representations
that puts additional constraint on the geometry of hyperedge representations. Johnson
and Pollak [14] proved, however, that deciding the existence of a planar support is NP-
complete; Buchin et al. [4, 5] extended the NP-completeness to testing the existence
of a 2-outerplanar support. On the other hand, it can be decided in polynomial time,
whether a hypergraph has a path-, cycle-, tree-, or cactus-support [3, 14, 16].

Contributions. In this paper we study minimum tree supports from a perspective that
was initially motivated by generating (area-proportional) Euler diagrams with low concurrency of set contours. Euler diagrams are set visualizations and thus closely related
to hypergraph visualizations. Section 2 describes the background of Euler diagrams, defines our algorithmic problem in that context, and sketches a solution approach based
on minimum tree supports. In Section 3 we introduce some necessary definitions and
notations, before we present our main technical contribution in Section 4. Our result is
initially tailored for the problem to generate low-concurrency Euler diagrams. Hence
we first transform an abstract Euler diagram description D into a so-called *labeled hypergraph $H(D)$* and then present an algorithm that computes an MTS for $H(D)$ in
$O(n^2m)$ time, where n is the number of vertices and m is the number of hyperedges.
The algorithm itself is a simple modification of Kruskal's algorithm for incrementally
constructing minimum spanning trees, but its correctness proof relies on several crucial
properties of tree supports and the special order in which the algorithm adds edges to
the growing tree support. Finally, in Section 5 we generalize our result and show that
every hypergraph can easily be translated into an equivalent labeled hypergraph. Then
we can apply our algorithm for labeled hypergraphs to compute minimum tree supports
for arbitrary hypergraphs that admit a tree support. This improves the result of Korach
and Stern [16], who gave an algorithm with running time $O(n^4m^2)$, to an algorithm
with time complexity $O(n^2(m + \log n))$.

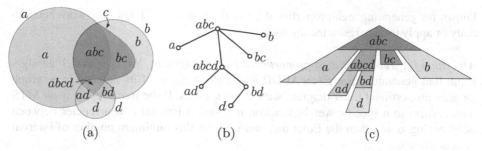

Fig. 1. (a) Euler diagram realizing AEDD $D = (\{a, b, c, d\}, \{a, b, d, ad, bd, bc, abc, abcd\})$, (b) tree support T for $H(D)$, (c) area-proportional Euler diagram based on T

2 Euler Diagrams

An *Euler diagram* $\mathcal{D}$ is a visualization of a set system as a collection of simple closed curves, whose interiors represent the sets, see Fig. 1(a) for an example. The arrangement of curves forms a subdivision of the plane and each face is called a (concrete) *zone* of the Euler diagram. We define an *abstract Euler diagram description* (AEDD) as a pair $D = (L, Z)$, where L is a set of *labels* (each representing one set in a set system) and $Z \subseteq \mathcal{P}(L)$ is a set of label subsets that we call *zones* (each representing a non-empty intersection of a particular set of labels). We say that an Euler diagram $\mathcal{D}$ *realizes* an AEDD D if there is a bijection φ between L and the set of curves in $\mathcal{D}$, as well as between Z and the set of concrete zones of $\mathcal{D}$ such that for each zone $z \in Z$ the concrete zone $\varphi(z)$ is in the interior of curve $\varphi(l)$ for each $l \in z$ and exterior to $\varphi(l')$ for each $l' \in L \setminus z$. AEDDs are closely related to hypergraphs since we can interpret each zone as a vertex and each label l as a hyperedge containing all zones that carry the label l. Similarly, an Euler diagram that realizes a given AEDD is a subdivision drawing of its corresponding hypergraph.

Euler diagrams must adhere to certain *well-formedness conditions* [8,12] that control the visual appearance of the diagrams, e.g., a zone with a certain set of labels L' may exist if and only if there exists an element that is contained in all sets corresponding to L' and that is not contained in any other set. Moreover, every zone must be uniquely labeled, i.e., there cannot be two distinct zones that lie in the interior of exactly the same set of curves. Other common well-formedness criteria require *convex zones*, disallow *triple points*, i.e., points that lie in the intersection of three or more curves, or disallow *concurrent curves* that run partially in parallel, i.e., connected intersections of two distinct curves c and c' that contain more than just one point. There is a number of algorithms and complexity results for generating Euler diagrams with certain well-formedness properties [8,12,18]. Since one can easily come up with AEDDs that require concurrencies (see the example in Fig. 1), it is an interesting problem to generally allow concurrencies, but minimize their total number in the diagram.

Another interesting variation of Euler diagrams are *area-proportional* Euler diagrams. Given an AEDD $D = (L, Z)$ together with a weight function $A: Z \to \mathbb{R}^+$ on the zones, the task is to find an area-proportional Euler diagram that realizes D such that the area of each concrete zone with label set $z \in Z$ is $A(z)$. Some algorithms are known for generating area-proportional Euler diagrams [9, 19], but they work heuristically or apply to very restricted inputs only.

Algorithm for tree-based area-proportional Euler diagrams. We now sketch an algorithm that generates for a given AEDD and a tree support of its induced hypergraph an area-proportional Euler diagram with convex zones. If the tree support is an MTS with respect to a specific weight function measuring internal concurrencies between neighboring zones, then the Euler diagram realizes this minimum number of internal concurrencies.

Let $D = (L, Z)$ be an AEDD. We define the *labeled hypergraph $H(D) = (Z, S(L))$* as the hypergraph that contains a vertex for each zone of D and a hyperedge for each label of D. Each vertex $z \in Z$ is associated with the set of labels of its underlying zone. In the following we use the abbreviated notation $z = abc$ for $z = \{a, b, c\}$ that simply concatenates all labels in the zone. For each label $l \in L$ we create the hyperedge $s(l) = \{z \in Z \mid l \in z\}$, which defines the hyperedge set $S(L) = \{s(l) \mid l \in L\}$. In Section 4 we describe an algorithm that computes a minimum tree support T for a labeled hypergraph $H(D)$ and an arbitrary edge weight function $w \colon \binom{V}{2} \to \mathbb{R}$ (assuming that $H(D)$ admits a tree support). For our purposes we define w as the *concurrency* function of the AEDD D, i.e., we set $w(z, z') = |(z \cup z') \setminus (z \cap z')|$. This function counts the number of concurrent curves that run between zones z and z' if they will be selected as neighboring faces in the Euler diagram.

Now let's assume that we are given an AEDD $D = (L, Z)$ and an MTS T for its labeled hypergraph $H(D)$ provided with the concurrency function. We construct an area-proportional Euler diagram as follows (see Fig. 1). Let r be an arbitrary root of T and create a convex polygon of area $A(r)$, e.g., a triangle. Let $z_1, \ldots, z_t$ be the children of r and choose one edge σ of the root polygon. We create disjoint subsegments of σ and disjoint wedges based on these subsegments, each of which is reserved for the zones in the t subtrees of r. For each z_i we create a trapezoid of area $A(z_i)$ at the base of the i-th wedge. Then we recurse using the respective sides opposite to σ as the new base edges in the construction. It is clear that this produces convex, area-proportional faces. Since T is a support, the union of the zones of each label is connected. Moreover, since we used the concurrency function to minimize the weight of T, we have minimized the total number of concurrencies of curves running between adjacent zones.

3 Preliminaries

Let $D = (L, Z)$ be an abstract Euler diagram description, where $|L| = \lambda$. Recall that the labeled hypergraph for D is denoted by $H(D) = (Z, S(L))$. If $L' \subseteq L$ is a subset of labels, we denote the corresponding hyperedge in $H(D)$ as $S(L') = \{s(\ell) \mid \ell \in L'\}$.

In order to construct a tree support for a hypergraph we define the so-called *skeleton* $G = (Z, E)$ of $H(D) = (Z, S(L))$, which is defined as a complete weighted graph on vertex set Z, where each edge $e = \{u, v\} \in E$ is associated with the *cardinality* $c(e) = |u \cap v|$. Each tree support for $H(D)$ is a spanning subtree of the skeleton G. Since λ is the number of distinct labels in L, the maximum cardinality of an edge of G is $\lambda - 1$. We denote by E_i the set of all edges of G with cardinality i and we

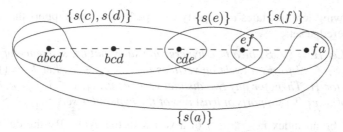

Fig. 2. A path P in G with positive cardinality and a sequence of hyperedge sets $\{s(c), s(d)\}, \{s(e)\}, \{s(f)\}, \{s(a)\}$ defined by P. This sequence forms a cycle on vertices $abcd$, cde, ef, and fa.

set $E_{\geq i} = \bigcup_{j=i}^{\lambda-1} E_j$. For a path P in G we define the *cardinality* of P as the smallest cardinality of its edges. For a tree T we denote by $p(u, v, T)$ the unique path connecting vertices u and v in T.

Recall that we want to compute an MTS with respect to some edge weight function $w\colon \binom{Z}{2} \to \mathbb{R}$, e.g., the concurrency function defined in Section 2. We use w as an edge weight function of the skeleton G.

For a hypergraph $H = (V, S)$ we say that an edge $\{u, v\}$ of its skeleton G *supports* a hyperedge $s \in S$, if both $u, v \in s$. An edge $\{u, v\}$ *supports* a set of hyperedges $S' \subseteq S$ if it supports all the hyperedges of S'. Let T be a spanning tree of G and P be a path in T. Path P is called *supporting* if for each hyperedge $s \in S$, the set $s \cap P$ is either empty or contains a set of consecutive vertices of P. Otherwise, P is called *non-supporting*. A non-supporting path of T is *minimal* if each sub-path of P is supporting. By recalling the definition of a tree support, we observe that a spanning tree T of G is a tree support for $H(D)$ if and only if each path of T is supporting. From this fact and the definition of the cardinality of a path in G we derive the following.

Property 1. Let $e = \{u, v\}$ be an edge in G and let T be a tree support for $H(D)$. Any edge of the path $p(u, v, T)$ supports the set $S(u \cap v)$, i.e., the path $p(u, v, T)$ has cardinality at least $c(e)$.

Let $S_1, \ldots, S_k \subseteq S(L)$ be a sequence of hyperedge sets and let $z_1, \ldots, z_k \in Z$ be a sequence of vertices of $H(D)$ such that $z_i \in s$, $\forall s \in S_i \cup S_{i+1}$, $i = 1, \ldots, k - 1$ and $z_k \in s$, $\forall s \in S_k \cup S_1$. Then we say that the sequence $S_1, \ldots, S_k$ forms a *cycle* on vertices $z_1, \ldots, z_k$. In Fig. 2 the hyperedge sets $\{s(c), s(d)\}, \{s(e)\}, \{s(f)\}, \{s(a)\}$ form a cycle on vertices $abcd, cde, ef, fa$.

A path P in G with non-zero cardinality *defines* a sequence of hyperedge sets as follows (see Fig. 2). Two consecutive vertices of P belong to at least one common hyperedge, since the cardinality of the edge is greater than zero. Include into S_1 those hyperedges that contain the longest initial part of P. Remove all edges of P that are in S_1 and continue recursively. Notice that if the end-vertices of P belong to a common hyperedge, which does not contain at least one internal vertex of P, then there exists a non-trivial sequence of hyperedge sets that forms a cycle on a certain subset of vertices of P.

The following lemma states a property of a path in a tree support that contains a cycle of hyperedge sets.

Lemma 1. *Let $H = (V, S)$ be a tree-hypergraph such that the sequence of hyperedge sets $S_1, \ldots, S_k \subseteq S$ forms a cycle on vertices $v_1, \ldots, v_k \in V$. Let $T = (V, E)$ be a tree support for H. Then, for any two distinct vertices $v_i, v_j, 1 \leq i, j \leq k$, every edge of the path $p(v_i, v_j, T)$ supports at least two of the hyperedge sets $S_1, \ldots, S_k$.*

Proof. Let t be an index $i \leq t < j$ of a vertex in the cycle. By the definition of a cycle formed by the sequence of hyperedge sets, $v_t \in s, \forall s \in S_t \cup S_{t+1}$ and $v_{t+1} \in s$, $\forall s \in S_{t+1} \cup S_{t+2}$. Therefore, both end-vertices of the subpath $p(v_t, v_{t+1}, T)$ belong to each $s \in S_{t+1}$, and therefore each edge of $p(v_t, v_{t+1}, T)$ supports S_{t+1}. (Note that all index computations are performed modulo k.)

Since T is a tree, the removal of any edge of $p(v_t, v_{t+1}, T)$ from T produces two subtrees $T_1 = (V_1, E_1)$ and $T_2 = (V_2, E_2)$, such that $V_1 \cap V_2 = \emptyset$ and the cycle vertices $v_i, \ldots, v_t \in V_1$ and $v_{t+1}, \ldots, v_j \in V_2$. Let $a \geq j$ or $a < i$ be the index such that $v_a \in V_2$ and $v_{a+1} \in V_1$. The cycle among hyperedge sets $S_1, \ldots, S_k$ implies that $v_a, v_{a+1} \in s$ for every hyperedge $s \in S_{a+1}$. Since T is a tree support, every edge of the path $p(v_a, v_{a+1}, T)$ supports S_{a+1}. Thus, every edge of $p(v_t, v_{t+1}, T)$ supports S_{a+1}. We conclude the proof by observing that indices $a + 1$ and $t + 1$ are distinct, since $v_{t+1} \in V_2$ and $v_{a+1} \in V_1$. □

The following lemmata are used as tools in the following section.

Lemma 2. *Let $G = (V, E)$ be a connected graph, let $T = (V, E_0 \cup E_1 \cup \cdots \cup E_t)$ be a spanning-tree of G with $E_i \cap E_j = \emptyset$ for every $0 \leq i \neq j \leq t$ and let the subgraph $T_k = (V_k, E_k), 1 \leq k \leq t$, of T induced by E_k be connected. For any forest consisting of trees $T'_1 = (V_1, E'_1), \ldots, T'_t = (V_t, E'_t)$ the graph $T' = (V, E_0 \cup E'_1 \cup \cdots \cup E'_t)$ is a spanning-tree of G.*

Proof. We show that after substitution of the edges of T_1 by the edges of T'_1, the resulting graph $\tilde{T} = (V, E_0 \cup E'_1 \cup E_2 \cdots \cup E_t)$ is a spanning tree of G. The result then follows by applying this procedure to the remaining $T_2, \ldots, T_t$. It is trivial to see that $\tilde{T}$ is a spanning connected subgraph of G. Assume for the sake of contradiction that $\tilde{T}$ is not a tree, i.e., it contains a cycle. If we substitute the maximal paths of this cycle that belong to T'_1 by paths in T_1, we obtain a (not necessarily simple) cycle in T, which is a contradiction. □

Lemma 3. *Any tree support of $H(D)$ contains the edge set $E_{\lambda-1}$ as a subset.*

Proof. Notice that the statement is trivially true if $E_{\lambda-1}$ is empty. So we assume that $E_{\lambda-1} \neq \emptyset$. An edge belongs to $E_{\lambda-1}$ if it connects a vertex $z \in Z$, containing all labels of L, and a vertex $z' \in Z$, containing $|L| - 1$ labels. Notice that no path between z and z' in G can be supporting, except for the edge $\{z, z'\}$ itself. Therefore, edge $\{z, z'\}$ must be in any tree support. □

4   Minimum Tree Supports for Labeled Hypergraphs

In this section we present the Algorithm MINIMUM TREE SUPPORT (MTS) that takes as an input a labeled hypergraph $H(D) = (Z, S(L))$ for the AEDD $D = (L, Z)$, as

---

Algorithm 1. MINIMUM TREE SUPPORT (MTS)

Input: labeled hypergraph $H(D) = (Z, S(L))$ for AEDD $D = (L, Z)$,
edge weight function $w \colon E \to \mathbb{R}$ for $E = \binom{Z}{2}$
Output: minimum tree support T for $H(D)$ or infeasibility notification

1 **if** $H(D)$ *has no tree support* **then return** *infeasible* partition E into sets E_i,
   $i = 0, \dots, |L| - 1$, of edges with equal cardinality i
2 $F \leftarrow \emptyset$
3 **for** $i \leftarrow |L| - 1$ **to** 0 **do**
4 $\quad$ **foreach** *edge* $e = \{u, v\} \in E_i$ *in non-decreasing order of weights* **do**
5 $\quad\quad$ **if** u *and* v *belong to different connected components of* F **then**
6 $\quad\quad\quad$ $F \leftarrow F \cup \{e\}$

7 **return** F

---

well as the weight function $w \colon E \to \mathbb{R}$ for the skeleton $G = (V, E)$ of $H(D)$, and produces a minimum tree support T for $H(D)$. The algorithm grows an initially empty forest $F = \emptyset$ and implements $|L|$ hierarchy steps. Recall that $E_i \subseteq E$ are all edges of G with cardinality i. During step $i = |L| - 1, \dots, 0$, the algorithm adds to F a subset of the edges of E_i, which we denote by F_i. Recall that $E_{\geq i} = \bigcup_{j=i}^{\lambda-1} E_j$. Analogously to this notation, we set $F_{\geq i} = \bigcup_{j=i}^{\lambda-1} F_j$. Thus $F_{\geq i}$ are the edges added to F in the steps $|L| - 1$ down to i. Observe that the check at line 5 ensures that $F_{\geq i}, i = |L| - 1, \dots, 0$ is a forest. Recalling the definition of the cardinality of a path, we derive the following:

Property 2. Any two vertices $u, v \in Z$ that are connected by a path of cardinality at least k in G, are connected by a path in $F_{\geq k}$.

In the following we first prove that if $H(D)$ is a tree-hypergraph then the output of the Algorithm MTS is a tree support for $H(D)$ (Lemma 4) and then prove that it is actually a minimum tree support (Lemma 5). We conclude the correctness and analyze the running time of Algorithm MTS in Theorem 1.

Lemma 4. *If $H(D)$ is a tree-hypergraph, then Algorithm MTS computes a tree support of $H(D)$.*

Proof. By induction on $i = \lambda - 1, \dots, 1$, we show that there exists a tree support $T_{\geq i}$ for $H(D)$ that extends the forest $F_{\geq i}$. Observe that the base case follows from Lemma 3. As an induction hypothesis we assume that there exists a tree support $T_{\geq i+1}$ for $H(D)$ that extends the forest $F_{\geq i+1}$. Let $T_{\geq i} \equiv (T_{\geq i+1} \setminus E_{\geq i}) \cup F_{\geq i}$. In order to show that $T_{\geq i}$ is a tree support for $H(D)$ we prove that: (a) $T_{\geq i}$ is a spanning tree of G, (b) $T_{\geq i}$ is a support for $H(D)$.

(a) Consider a connected component C of $T_{\geq i+1} \cap E_{\geq i}$. By Property 2, any two vertices of C are also connected in the forest $F_{\geq i}$. Let C' be a connected component of $F_{\geq i}$, and $e = \{u, v\} \in C'$. By Property 1, u and v are connected in $T_{\geq i+1}$ (which is a tree support) by a path of cardinality at least $c(e)$ and therefore by a path in $T_{\geq i+1} \cap E_{\geq i}$. Thus, connected components of $T_{\geq i+1} \cap E_{\geq i}$ and $F_{\geq i}$ have the same vertex sets. Therefore, by Lemma 2, we have that $(T_{\geq i+1} \setminus E_{\geq i}) \cup F_{\geq i}$, and therefore $T_{\geq i}$, is a spanning tree of G.

(b) Recall that $T_{\geq i} \equiv (T_{\geq i+1} \setminus E_{\geq i}) \cup F_{\geq i}$. Let $F_{\geq i}^C$ denote a connected component of $F_{\geq i}$ and let G^C denote the subgraph of G induced by the vertices of $F_{\geq i}^C$. We observe that, in order to show that $T_{\geq i}$ is a tree support for $H(D)$, it is enough to show that $F_{\geq i}^C$ is a tree support for the hypergraph induced by G^C. Assume for the sake of contradiction that $F_{\geq i}^C$ is not a tree support for the hypergraph induced by G^C. Then, there exists a minimal non-supporting path $p(u, v, F_{\geq i}^C)$. Therefore, there exists a hyperedge s that contains u and v, but does not contain any internal vertex of $p(u, v, F_{\geq i}^C)$. Let $S_1, \ldots, S_k$ be a sequence of hyperedge sets defined by path $p(u, v, F_{\geq i}^C)$ such that $S_1, \ldots, S_k$ together with a hyperedge set S' containing s define a cycle. Notice that $|S_j| \geq i, \forall j, 1 \leq j \leq k$, since these sets are formed by the path $p(u, v, F_{\geq i}^C)$. Also observe that there is no index $j, 1 \leq j \leq k$ such that $S' \subseteq S_j$, since $s \in S'$, and $s \notin S_j$.

Recall from the proof of statement (a), that the connected components of $T_{\geq i+1} \cap E_{\geq i}$ and $F_{\geq i}$ have the same vertex sets. Thus, there exists a path $p(u, v, T_{\geq i+1} \cap E_{\geq i})$. Since $T_{\geq i+1}$ is a tree support, and the sequence $S', S_1, \ldots, S_k$ of hyperedge sets in G forms a cycle, we infer by Lemma 1, that each edge e of $p(u, v, T_{\geq i+1} \cap E_{\geq i})$ supports at least two of these hyperedge sets, one of which is S'. Let S_j, $1 \leq j \leq k$ be the second hyperedge set supported by e. Recall that $S' \not\subseteq S_j$, therefore $|S' \cup S_j| > |S_j|$. Thus, $c(e) \geq |S' \cup S_j| > |S_j| \geq i$, i.e. $c(e) \geq i+1$, for each $e \in p(u, v, T_{\geq i+1} \cap E_{\geq i})$, implying that $e \in F_{\geq i+1}$. By recalling that $T_{\geq i+1}$ extends $F_{\geq i+1}$ and that $p(u, v, F_{\geq i}^C)$ is a path in $F_{\geq i}$, which contains $F_{\geq i+1}$, we conclude that $p(u, v, F_{\geq i}^C) = p(u, v, T_{\geq i+1} \cap E_{\geq i})$. Thus $T_{\geq i+1}$ also contains a non-supporting path $p(u, v, T_{\geq i+1} \cap E_{\geq i})$, which is a contradiction to the induction hypothesis that $T_{\geq i+1}$ is a tree support. $\qquad \square$

Lemma 5. *If $H(D)$ is a tree-hypergraph and w an edge-weight function for the skeleton, then the tree support computed by Algorithm MTS is a minimum tree support.*

Proof. The proof is again by induction over the hierarchy steps of Algorithm MTS. We show that after each hierarchy step i there is a minimum tree support that extends the forest $F_{\geq i}$. It is easy to see that this is true after the first hierarchy step $\lambda - 1$ as we know that $F_{\lambda-1} = E_{\lambda-1}$ and that any tree support of $H(D)$ contains $E_{\lambda-1}$ by Lemma 3.

So let $i < \lambda - 1$ and assume by induction that there is a minimum tree support T_{i+1} that extends $F_{\geq i+1}$. Hierarchy step i considers the edge set E_i of edges with cardinality i. If $E_i = \emptyset$ or no edges are added in step i, we have $F_{\geq i} = F_{\geq i+1}$ and the statement holds immediately. So let $E_i \neq \emptyset$ and $F_i \neq \emptyset$. We show that after each edge addition in the current hierarchy step there is a minimum tree support that extends the current forest F assuming that this was true before the edge was added. Let $e = \{u, v\}$ be the next edge to be added by the algorithm and let $\hat{T}$ be a minimum tree support extending the forest F, where $e \notin F$. If $e \in \hat{T}$ there is nothing to show, so assume $e \notin \hat{T}$.

Then $\hat{T} \cup \{e\}$ contains a cycle $\hat{K}$. We further know from Lemma 4 that the final tree T, computed by Algorithm MTS, extends $F \cup \{e\}$ and is a tree support. We show that there is an edge $\hat{e} \in \hat{K} \setminus T$, for which there is a cycle K in $T \cup \{\hat{e}\}$ that contains both e and $\hat{e}$. Firstly, the set $\hat{K} \setminus T$ is not empty since otherwise T would contain a cycle. Assume that no edge in $\hat{K} \setminus T$ has the desired property. Let $(\{u, v\}, \{v_1, v_2\}, \{v_2, v_3\}, \ldots, \{v_{k-1}, v_k\})$ be the edge sequence of $\hat{K}$, where $v = v_1$ and $u = v_k$, and let $1 \leq f_1 < \cdots < f_l \leq k$ be the indices of all edges $e_{f_j} = \{v_{f_j}, v_{f_j+1}\} \in \hat{K} \setminus T$ $(1 \leq j \leq l)$. If we replace each such edge e_{f_j} by the path $p(v_{f_j}, v_{f_j+1}, T)$ we obtain a (not necessarily simple) cycle in T, which is a contradiction to T being a tree.

So let K be a cycle in $T \cup \{\hat{e}\}$ that contains the edges $e = \{u, v\}$ and $\hat{e} = \{\hat{u}, \hat{v}\}$. Since T is a tree support, all edges of the path $p(\hat{u}, \hat{v}, T) = K \setminus \{\hat{e}\}$ must support the hyperedge set $S(\hat{u} \cap \hat{v})$. In particular, edge e supports $S(\hat{u} \cap \hat{v})$. Analogously, $\hat{T}$ is a tree support and thus all edges of the path $p(u, v, \hat{T}) = \hat{K} \setminus \{e\}$, including the edge $\hat{e}$, support the hyperedge set $S(u \cap v)$. It follows that $u \cap v = \hat{u} \cap \hat{v}$ and thus all edges of $\hat{K}$ support $S(\hat{u} \cap \hat{v})$.

We define the tree $\hat{T}_e = (\hat{T} \setminus \{\hat{e}\}) \cup \{e\}$ that replaces $\hat{e}$ by e and claim that it is also a tree support of $H(D)$. For any two vertices $x, y \in Z$ with $\hat{e} \in p(x, y, \hat{T})$ the hyperedge set that has to be supported by every edge of $p(x, y, \hat{T})$ is $S(x \cap y) \subseteq S(\hat{u} \cap \hat{v})$. Since the edges of path $p(x, y, \hat{T}_e)$ are contained in $p(x, y, \hat{T}) \cup \hat{K}$, they also support $S(x \cap y)$. Thus we have showed that there is a tree support, namely $\hat{T}_e$, that extends $F \cup \{e\}$.

It remains to show that $\hat{T}_e$ is a *minimum* tree support. We first observe that both edges e and $\hat{e}$ have the same cardinality $c(e) = c(\hat{e}) = i$ and thus $e, \hat{e} \in E_i$ are both considered in hierarchy step i of the algorithm. Our algorithm, however, considers e before $\hat{e}$, which means that $w(e) \leq w(\hat{e})$. Since $\hat{T}$ is a minimum tree support by the induction hypothesis and $w(\hat{T}_e) \leq w(\hat{T})$, we obtain that $\hat{T}_e$ is also a minimum tree support. This is true for every edge addition in hierarchy step i, so in particular for $F = F_{\geq i}$ after the last edge addition in this hierarchy step. But this already concludes the inductive argument for the whole hierarchy step and shows together with Lemma 4 that the result $T = F_{\geq 0}$ of algorithm MTS is indeed a minimum tree support of $H(D)$. $\quad\square$

Theorem 1. *Given a labeled hypergraph $H(D)$ with n vertices and m hyperedges for an AEDD D Algorithm MTS computes in $O(n^2 m)$ time a minimum tree support T or reports that no tree support exists.*

Proof. Algorithm MTS starts by checking whether $H(D)$ has a tree support using the feasibility check proposed by Johnson and Pollak [14]. If the test fails the algorithm reports this result; otherwise $H(D)$ is a tree-hypergraph and thus we know by Lemma 5 that the resulting tree T is a minimum tree support. This proves the correctness.

The feasibility test of Johnson and Pollak [14] in line 1 of the algorithm is based on testing whether the dual hypergraph $H(D)^*$ of $H(D)$ is acyclic. The dual hypergraph of $H(D)$ can be constructed in $O(nm)$ time and has a vertex for each hyperedge of $H(D)$ and a hyperedge for each vertex of $H(D)$, which contains all hyperedges incident to that vertex. The acyclicity of $H(D)^*$ can be tested in $O(nm)$ time [20].

The next step in line 1 of the algorithm is to partition the edge set E into subsets based on the edge cardinalities. For each edge $\{u, v\} \in E$ computing the cardinality of the intersection $u \cap v$ takes $O(m)$ time, since each vertex consists of at most m labels. Since we have $O(n^2)$ edges this takes $O(n^2 m)$ time in total.

Finally, in lines 2–7 we run a modified version of Kruskal's algorithm to compute a minimum spanning tree. Unlike the original algorithm, we do not sort the whole edge set E by non-decreasing weights, but rather perform $|L|$ hierarchy steps, in which we consider the edges of each subset E_i in the edge partition separately in non-decreasing weight order. This modification, however, does not affect the running time and thus the last part of Algorithm MTS takes $O(|E| \log |Z|)$ time, just as computing a minimum spanning tree by Kruskal's algorithm. The set E is of size $O(n^2)$ and vertices in Z are subsets of the label set L, i.e., $\log |Z|$ is of size $O(m)$. Thus the total running time of Algorithm MTS is $O(n^2 m)$.   □

5 Minimum Tree Supports for Hypergraphs

Labeled hypergraphs, in particular for abstract Euler diagram descriptions as considered in the previous section, seem to be of limited interest at first sight. So it is a natural question to ask for a minimum tree support of a general tree-hypergraph $H = (V, S)$. As discussed in Section 1, Korach and Stern [16] showed that this problem can be solved efficiently in $O(n^4 m^2)$ time, where $n = |V|$ and $m = |S|$.

In this section we generalize Theorem 1 to arbitrary hypergraphs, and thus improve the best known running time from $O(n^4 m^2)$ to $O(n^2(m + \log n))$. The tool to achieve this is to define a mapping that transforms an arbitrary hypergraph to an equivalent labeled hypergraph so that we can apply Algorithm MTS.

Theorem 2. *Given a hypergraph H with n vertices and m hyperedges and an edge weight function $w \colon \binom{V}{2} \to \mathbb{R}$ we can compute in $O(n^2(m + \log n))$ time a minimum tree support T of H or report that no tree support exists.*

Proof. An important difference between an arbitrary hypergraph H and the labeled hypergraph $H(D)$ for an AEDD D is that $H(D)$ contains at most one zone for each possible subset of labels, whereas H may contain any number of vertices that have exactly the same hyperedge incidences. This forces us to slightly modify Algorithm MTS and its analysis.

Let $H = (V, S)$ be a hypergraph. We start by describing the mapping μ, which maps H to an equivalent labeled hypergraph. We define the label set $L_S = \{l_1, \ldots, l_m\}$, which contains one unique label $l_i = l(s_i)$ for each hyperedge $s_i \in S$. Each vertex $v \in V$ is mapped to an indexed label set $(v, \mu(v)) = (v, \{l(s_i) \mid v \in s_i\})$, where $\mu(v)$ contains the labels of all hyperedges containing v. We explicitly allow that two distinct vertices $v \neq v'$ are mapped to the same label set $\mu(v) = \mu(v')$, but their indexed label sets $(v, \mu(v))$ and $(v', \mu(v'))$ are distinguishable. Similarly, we map each hyperedge $s \in S$ to a set of indexed label sets $\mu(s) = \{(v, \mu(v)) \mid v \in s\}$. We use the notation $\mu(V)$ to denote the set $\{(v, \mu(v)) \mid v \in V\}$ and $\mu(S)$ to denote the set $\{\mu(s) \mid s \in S\}$. This defines a labeled hypergraph $\mu(H) = (\mu(V), \mu(S))$, which is isomorphic to the labeled hypergraph of the AEDD $D = (L_S, \mu(V))$ if no two vertices in V are incident to exactly the same hyperedges. We further define a new edge weight function $\mu(w)$ as $\mu(w)((u, \mu(u)), (v, \mu(v))) = w(u, v)$.

Since our construction simply replaces each vertex of H by an indexed label set indicating its incident hyperedges it is obvious that each tree support of H is in one-to-one correspondence to a tree support of $\mu(H)$, in particular an MTS of $\mu(H)$ is also an MTS of H. Thus we can apply Algorithm MTS to the labeled hypergraph $\mu(H)$ and obtain a minimum tree support T for H.

We need to pay attention to one minor issue in the correctness proof of the algorithm. In Section 4 the base case of the inductive proofs started with edges of cardinality $m - 1$. Now we might have edges of cardinality m, namely if multiple vertices are contained in every hyperedge in S. If we run Algorithm MTS with an extra hierarchy step for the cardinality-m edges it computes a minimum spanning tree of the vertex set $\mu(V_m) = \{(v, \mu(v)) \in \mu(V) \mid \mu(v) = S\}$. Using the fact shown by Korach and Stern [16] that every element of the hyperedge intersection closure of H (which contains S and all intersections of subsets of S) induces a connected subtree in every tree support of H, we know that every minimum tree support of $\mu(H)$ must contain a minimum spanning tree of $\mu(V_m)$. This serves as the new base case of the induction; the remainder of the correctness proofs in Section 4 continues to hold.

For the running time analysis we again need to pay attention to a small detail related to vertices with the same label set. Lines 2–7 of Algorithm MTS are a modification of Kruskal's algorithm and thus need $O(n^2 \log n)$ time on a complete graph with n vertices. But since we may have more than one vertex with the same label set it is no longer true in general that $\log n \in O(m)$. Thus the modification of Algorithm MTS takes $O(n^2(m + \log n))$ time to compute the MTS T for $\mu(H)$.

Finally, it remains to argue that $\mu(H)$ can be computed in the same time bound. For creating $\mu(V)$ and $\mu(S)$ we simply scan all hyperedges in S and append their labels to the contained vertices. This can be done in $O(nm)$ time since each hyperedge contains $O(n)$ vertices. Hyperedges in $\mu(S)$ are not explicitly represented as sets of label sets, but rather as sets of pointers to the vertices in $\mu(V)$. □

6 Conclusion

We have studied the problem of computing minimum tree supports for hypergraphs and we have seen that our algorithm for the special case of labeled hypergraphs induced by abstract Euler diagram descriptions easily generalizes to arbitrary hypergraphs. We improved the previously best known running time for computing minimum tree supports [16] from $O(n^4 m^2)$ to $O(n^2(m + \log n))$. Moreover, we described an application of minimum tree supports for generating area-proportional Euler diagrams with convex zones and minimum internal concurrencies for abstract Euler diagram descriptions with a labeled tree-hypergraph.

Other types of sparse supports like outerplanar supports give rise to interesting open questions. For example, the complexity of deciding whether a given hypergraph has an outerplanar support is open [5]. On the practical side, it is interesting to study algorithms for generating well-formed Euler diagrams based on outerplanar supports or other larger classes of supports.

References

1. Angluin, D., Aspnes, J., Reyzin, L.: Inferring social networks from outbreaks. In: Hutter, M., Stephan, F., Vovk, V., Zeugmann, T. (eds.) Algorithmic Learning Theory. LNCS, vol. 6331, pp. 104–118. Springer, Heidelberg (2010)

2. Angluin, D., Aspnes, J., Reyzin, L.: Network construction with subgraph connectivity constraints. J. Comb. Optim. (2013)
3. Brandes, U., Cornelsen, S., Pampel, B., Sallaberry, A.: Blocks of hypergraphs applied to hypergraphs and outerplanarity. In: Iliopoulos, C.S., Smyth, W.F. (eds.) IWOCA 2010. LNCS, vol. 6460, pp. 201–211. Springer, Heidelberg (2011)
4. Buchin, K., van Kreveld, M., Meijer, H., Speckmann, B., Verbeek, K.: On planar supports for hypergraphs. Technical Report UU-CS-2009-035, Utrecht University (2009)
5. Buchin, K., van Kreveld, M., Meijer, H., Speckmann, B., Verbeek, K.: On planar supports for hypergraphs. In: Eppstein, D., Gansner, E.R. (eds.) GD 2009. LNCS, vol. 5849, pp. 345–356. Springer, Heidelberg (2010)
6. Chen, J., Komusiewicz, C., Niedermeier, R., Sorge, M., Suchý, O., Weller, M.: Effective and efficient data reduction for the subset interconnection design problem. In: Cai, L., Cheng, S.-W., Lam, T.-W. (eds.) Algorithms and Computation. LNCS, vol. 8283, pp. 361–371. Springer, Heidelberg (2013)
7. Chockler, G., Melamed, R., Tock, Y., Vitenberg, R.: Constructing scalable overlays for pubsub with many topics. In: Principles of Distributed Computing (PODC 2007), pp. 109–118 (2007)
8. Chow, S.: Generating and Drawing Area-Proportional Euler and Venn Diagrams. PhD thesis, University of Victoria (2007)
9. Chow, S., Ruskey, F.: Drawing area-proportional Venn and Euler diagrams. In: Liotta, G. (ed.) GD 2003. LNCS, vol. 2912, pp. 466–477. Springer, Heidelberg (2004)
10. Du, D.-Z., Kelley, D.F.: On complexity of subset interconnection designs. J. Global Optim. 6, 193–205 (1995)
11. Fan, H., Hundt, C., Wu, Y.-L., Ernst, J.: Algorithms and implementation for interconnection graph problem. In: Yang, B., Du, D.-Z., Wang, C.A. (eds.) COCOA 2008. LNCS, vol. 5165, pp. 201–210. Springer, Heidelberg (2008)
12. Flower, J., Fish, A., Howse, J.: Euler diagram generation. J. Visual Languages and Computing 19(6), 675–694 (2008)
13. Hosoda, J., Hromkovič, J., Izumi, T., Ono, H., Steinová, M., Wada, K.: On the approximability and hardness of minimum topic connected overlay and its special instances. Theoretical Computer Science 429, 144–154 (2012)
14. Johnson, D.S., Pollak, H.O.: Hypergraph planarity and the complexity of drawing Venn diagrams. J. Graph Theory 11(3), 309–325 (1987)
15. Kaufmann, M., van Kreveld, M., Speckmann, B.: Subdivision drawings of hypergraphs. In: Tollis, I.G., Patrignani, M. (eds.) GD 2008. LNCS, vol. 5417, pp. 396–407. Springer, Heidelberg (2009)
16. Korach, E., Stern, M.: The clustering matroid and the optimal clustering tree. Mathematical Programming 98(1-3), 385–414 (2003)
17. Korach, E., Stern, M.: The complete optimal stars-clustering-tree problem. Discrete Applied Mathematics 156, 444–450 (2008)
18. Rodgers, P.J., Zhang, L., Fish, A.: General Euler diagram generation. In: Stapleton, G., Howse, J., Lee, J. (eds.) Diagrams 2008. LNCS (LNAI), vol. 5223, pp. 13–27. Springer, Heidelberg (2008)
19. Stapleton, G., Rodgers, P., Howse, J.: A general method for drawing area-proportional Euler diagrams. J. Visual Languages and Computing 22(6), 426–442 (2011)
20. Tarjan, R.E., Yannakakis, M.: Simple linear-time algorithms to test chordality of graphs, test acyclicity of hypergraphs, and selectively reduce acyclic hypergraphs. SIAM J. Comput. 13(3), 566–579 (1984)

Additive Spanners: A Simple Construction

Mathias Bæk Tejs Knudsen[*]

University of Copenhagen

Abstract. We consider additive spanners of unweighted undirected graphs. Let G be a graph and H a subgraph of G. The most naïve way to construct an additive k-spanner of G is the following: As long as H is not an additive k-spanner repeat: Find a pair $(u, v) \in H$ that violates the spanner-condition and a shortest path from u to v in G. Add the edges of this path to H.

We show that, with a very simple initial graph H, this naïve method gives additive 6- and 2-spanners of sizes matching the best known upper bounds. For additive 2-spanners we start with $H = \varnothing$ and end with $O(n^{3/2})$ edges in the spanner. For additive 6-spanners we start with H containing $\lfloor n^{1/3} \rfloor$ arbitrary edges incident to each node and end with a spanner of size $O(n^{4/3})$.

1 Introduction

Additive spanners are subgraphs that preserve the distances in the graph up to an additive positive constant. Given an unweighted undirected graph G, a subgraph H is an additive k-spanner if for every pair of nodes u, v it is true that

$$d_G(u, v) \leqslant d_H(u, v) \leqslant d_G(u, v) + k$$

In this paper we only consider purely additive spanners, which are k-spanners where $k = O(1)$. Throughout this paper every graph will be unweighted and undirected.

Many people have considered a variant of this problem, namely multiplicative spanners and even mixes between additive and multiplicative spanners [1,2,3]. The problem of finding a k-spanner of smallest size has received a lot of attention. Most notably, given a graph with n nodes Dor et al. [4] prove that it has a 2-spanner of size $O(n^{3/2})$, Baswana et al. [5] prove that it has a 6-spanner of size $O(n^{4/3})$, and Chechik [6] proves that it has a 4-spanner of size $O(n^{7/5} \log^{1/5} n)$. Woodrufff [7] shows that for every constant k there exist graphs with n nodes such that every $(2k - 1)$-spanner must have at least $\Omega(n^{1+1/k})$ edges. This implies that the construction of 2-spanners are optimal. Whether there exists an algorithm for constructing $O(1)$-spanners with $O(n^{1+\varepsilon})$ edges for some $\varepsilon < 1/3$ is unknown and is an important open problem.

[*] Research partly supported by Thorup's Advanced Grant from the Danish Council for Independent Research under the Sapere Aude research carrier programme and by the FNU project AlgoDisc - Discrete Mathematics, Algorithms, and Data Structrues.

R Ravi and I.L. Gørtz (Eds.): SWAT 2014, LNCS 8503, pp. 277–281, 2014.
© Springer International Publishing Switzerland 2014

Let G be a graph and H a subgraph of G. Consider the following algorithm: As long as there exists a pair of nodes u, v such that $d_H(u, v) > d_G(u, v) + k$, find a shortest path from u to v in G and add the edges on the path to H. This process will be referred to as k-**spanner-completion**. After k-spanner-completion, H will be a k-spanner of G. Thus, given a graph G, a general way to construct a k-spanner for G is the following: Firstly, find a simple subgraph of G. Secondly use k-spanner-completion on this subgraph. The main contribution of this paper is:

Theorem 1. *Let G be a graph with n nodes and H the subgraph containing all nodes but no edges of G. For each node add $\lfloor n^{1/3} \rfloor$ edges adjacent to that node to H (or, if the degree is less, add all edges incident to that node). After 6-spanner-completion H will have at most $O(n^{4/3})$ edges.*

It is well-known that a graph with n nodes has a 6-spanner of size $O(n^{4/3})$ [5]. The techniques employed in our proof of correctness are similar to those in [5]. The creation of the initial graph H corresponds to the clustering in [5] and the 6-spanner-completion corresponds to their path-buying algorithm. For completeness we show that the same method gives a 2-spanner of size $O(n^{3/2})$. This fact is already known due to [4] and is matched by a lower bound from [7].

Theorem 2. *Let G be a graph with n nodes and H the subgraph where all edges are removed. Upon 2-spanner-completion H has at most $O(n^{3/2})$ edges.*

2   Creating a 6-Spanner

The algorithm for creating a 6-spanner was described in the abstract and the introduction.

For a given graph G, a 6-spanner of G can be created by strating with some subgraph H of G and applying 6-spanner-completion to H. Theorem 1 states that for a suitable starting choice of H we get a spanner of size $O(n^{4/3})$. The purpose of this section is to show that the 6-spanner created has no more than $O(n^{4/3})$ edges. This will imply that the construction (in terms of the size of the 6-spanner) matches the best known upper bound [5].

Proof (of Theorem 1). Inserting (at most) $\lfloor n^{1/3} \rfloor$ edges per node will only add $n \lfloor n^{1/3} \rfloor = O(n^{4/3})$ edges to H. Therefore it is only necessary to prove that 6-spanner-completion adds no more than $O(n^{4/3})$ edges.

Let $v(H)$ and $c(H)$ be defined by:

$$v(H) = \sum_{u,v \in V(G)} \max\{0, d_G(u,v) - d_H(u,v) + 5\}, \quad c(H) = \#E(H)$$

Say that a shortest path, p, from u to v is added to H, and let H_0 be the subgraph before the edges are added. Let the path consist of the nodes:

$$u = w_0, w_1, \ldots, w_r = v, r \in \mathbb{N}$$

Let $u' = w_i$ be the node w_i with the smallest i such that $\deg_{H_0}(w_i) \geqslant \lfloor n^{1/3} \rfloor$. Likewise let $v' = w_j$ be the node w_j the largest j such that $\deg_{H_0}(w_j) \geqslant \lfloor n^{1/3} \rfloor$. Remember that if $\deg_{H_0}(w_i) < \lfloor n^{1/3} \rfloor$ then all the edges adjacent to w_i are already in H_0. This implies that $d_{H_0}(u', v') > d_G(u', v') + 6$ since $d_{H_0}(u, v) > d_G(u, v) + 6$.

Say that t new edges are added to H. Then there must be at least t nodes on p with degree $> n^{1/3}$. Since every node can be adjacent to no more than 3 nodes on p (since it is a shortest path) there must be $\Omega(n^{1/3}t)$ nodes adjacent to p in H. Let z and w be neighbours to u' and v' in H respectively and let r be any node adjacent to p in H. Let s be a node on p such that r and s are adjacent in H. See Figure 1 for an illustration.

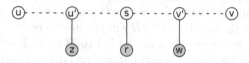

Fig. 1. The dashed line denotes the shortest path from u to v. The solid lines denote edges.

By the triangle inequality we see that:

$$d_H(z, r) + d_H(r, w) \leqslant d_G(u', v') + 4$$

But on the other hand:

$$d_{H_0}(z, r) + d_{H_0}(r, w) \geqslant d_{H_0}(z, w) \geqslant d_{H_0}(u', v') - 2 > d_G(u', v') + 4$$

Combining these two inequalities we obtain $d_{H_0}(z, r) > d_H(z, r)$ or $d_{H_0}(r, w) > d_H(r, w)$. And from the triangle inequality $d_G(z, r) + 5 > d_H(z, r)$ and $d_G(r, w) + 5 > d_H(r, w)$. Since u' and v' have at least $n^{1/3}$ neighbours and there are $\Omega(n^{1/3}t)$ nodes in H adjacent to p, the definition of $v(H)$ implies that:

$$v(H) - v(H_0) \geqslant \Omega(t(n^{1/3})^2)$$

And since $c(H) - c(H_0) = t$:

$$\frac{v(H) - v(H_0)}{c(H) - c(H_0)} \geqslant \Omega(n^{2/3})$$

Since $v(H) \leqslant O(n^2)$ this implies that $c(H)$ increases with no more than $O(n^2/n^{2/3}) = O(n^{4/3})$ in total when all shortest paths are inserted. Hence $c(H) = O(n^{4/3})$ when the 6-spanner-completion is finished which yields the conclusion. □

3 Creating a 2-Spanner

For completeness we show that 2-spanner-completion gives spanners with $O(n^{3/2})$ edges. This matches the upper bound from [4] and the lower bound from [7].

Proof (of Theorem 2). Let G be a graph with n nodes. Whenever H is a spanner of G, define $v(H)$ and $c(H)$ as:

$$v(H) = \sum_{u,v \in V(G)} \max\{0, d_G(u,v) - d_H(u,v) + 3\}, \quad c(H) = \sum_{v \in V(G)} (\deg_H(v))^2$$

It is easy to see that $0 \leqslant v(H) \leqslant 3n^2$ and by Cauchy-Schwartz's inequality $\sqrt{c(H) \cdot n} \geqslant 2\#E(H)$. The goal will be to prove that when the algorithm terminates $c(H) = O(n^2)$, since this implies that $\#E(H) = O(n^{3/2})$. This is done by proving that in each step of the algorithm $c(H) - 12v(H)$ will not increase. Since $v(H) = O(n^2)$ this means that $c(H) = O(n^2)$ which ends the proof. Therefore it is sufficient to check that $c(H) - 12v(H)$ never increases.

Consider a step where new edges are added to H on a shortest path from u to v of length t. Let H_0 be the subgraph before the edges are added. Assume that u, v violates the 2-spanner condition in H_0, i.e. $d_{H_0}(u,v) > d_G(u,v) + 2$. Let the shortest path consist of the nodes:

$$u = w_0, w_1, \ldots, w_{t-1}, w_t = v$$

It is obvious that:

$$c(H) - c(H_0) \leqslant \sum_{i=0}^{t} (\deg_H(w_i))^2 - (\deg_H(w_i) - 2)^2 \leqslant 4 \sum_{i=0}^{t} \deg_H(w_i)$$

Every node cannot be adjacent to more than 3 nodes on the shortest path, since otherwise it would not be a shortest path. Using this insight we can bound the number of nodes which in H are adjacent to or on the shortest path from below by:

$$\frac{1}{3} \sum_{i=0}^{t} \deg_H(w_i)$$

Now let z be a node in H adjacent or on to the shortest path. Obviously:

$$d_H(u,z) + d_H(z,v) \leqslant d_G(u,v) + 2$$

Furthermore $d_{H_0}(u,z) + d_{H_0}(z,v) > d_G(u,v) + 2$ since otherwise there would exist a path from u to v in H_0 of length $\leqslant d_G(u,v) + 2$. Hence:

$$d_H(u,z) + d_H(z,v) < d_{H_0}(u,z) + d_{H_0}(z,v)$$

Now let z be a node on the shortest path which is adjacent to w_i in H (every node on the path will also be adjacent in H to such a node). Then by the triangle inequality:

$$\begin{aligned} d_H(u,z) &\leqslant d_H(u,w_i) + d_H(w_i,z) &= d_G(u,w_i) + 1 \\ &\leqslant d_G(u,z) + d_G(z,w_i) + 1 = d_G(u,z) + 2 \end{aligned}$$

And likewise $d_H(z,v) \leqslant d_G(z,v) + 2$. Combining these two observations yields:

$$\sum_{w \in V} \max\{0, d_G(z,w) - d_H(z,w) + 3\} < \sum_{w \in V} \max\{0, d_G(z,w) - d_{H_0}(z,w) + 3\}$$

Since this holds for every node in H adjacent to or on the shortest path this means that:

$$v(H) - v(H_0) \geqslant \frac{1}{3} \sum_{i=0}^{t} \deg_H(w_i)$$

Combining this with the bound on $c(H) - c(H_0)$ gives:

$$(c(H) - 12v(H)) - (c(H_0) - 12v(H_0)) \leqslant 0$$

which finishes the proof. □

References

1. Pettie, S.: Low distortion spanners. In: Arge, L., Cachin, C., Jurdziński, T., Tarlecki, A. (eds.) ICALP 2007. LNCS, vol. 4596, pp. 78–89. Springer, Heidelberg (2007)
2. Elkin, M., Peleg, D.: $(1 + \varepsilon, \beta)$-spanner constructions for general graphs. SIAM Journal on Computing 33(3), 608–631 (2004); See also STOC 2001
3. Thorup, M., Zwick, U.: Spanners and emulators with sublinear distance errors. In: Proc. 17th ACM/SIAM Symposium on Discrete Algorithms (SODA), pp. 802–809 (2006)
4. Dor, D., Halperin, S., Zwick, U.: All-pairs almost shortest paths. SIAM Journal on Computing 29(5), 1740–1759 (2000); See also FOCS 1996
5. Baswana, S., Kavitha, T., Mehlhorn, K., Pettie, S.: New constructions of (α, β)-spanners and purely additive spanners. In: Proc. 16th ACM/SIAM Symposium on Discrete Algorithms (SODA), pp. 672–681 (2005)
6. Chechik, S.: New additive spanners. In: Proc. 24th ACM/SIAM Symposium on Discrete Algorithms (SODA), pp. 498–512 (2013)
7. Woodruff, D.P.: Lower bounds for additive spanners, emulators, and more. In: Proc. 47th IEEE Symposium on Foundations of Computer Science (FOCS), pp. 389–398 (2006)

Assigning Channels
via the Meet-in-the-Middle Approach[*]

Łukasz Kowalik and Arkadiusz Socała

University of Warsaw, Poland

Abstract. We study the complexity of the CHANNEL ASSIGNMENT problem. By applying the meet-in-the-middle approach we get an algorithm for the ℓ-bounded CHANNEL ASSIGNMENT (when the edge weights are bounded by ℓ) running in time $O^*((2\sqrt{\ell+1})^n)$. This is the first algorithm which breaks the $(O(\ell))^n$ barrier. We extend this algorithm to the counting variant, at the cost of slightly higher polynomial factor.

A major open problem asks whether CHANNEL ASSIGNMENT admits a $O(c^n)$-time algorithm, for a constant c independent of ℓ. We consider a similar question for GENERALIZED T-COLORING, a CSP problem that generalizes CHANNEL ASSIGNMENT. We show that GENERALIZED T-COLORING does not admit a $2^{2^{o(\sqrt{n})}} \text{poly}(r)$-time algorithm, where r is the size of the instance.

1 Introduction

In the CHANNEL ASSIGNMENT problem, we are given a symmetric weight function $w : V^2 \to \mathbb{N}$ (we assume that $0 \in \mathbb{N}$). The elements of V will be called vertices (as w induces a graph on the vertex set V with edges corresponding to positive values of w). We say that w is ℓ-bounded when for every $x, y \in V$ we have $w(x, y) \leq \ell$. An assignment $c : V \to \{1, \ldots, s\}$ is called *proper* when for each pair of vertices x, y we have $|c(x) - c(y)| \geq w(x, y)$. The number s is called the *span* of c. The goal is to find a proper assignment of minimum span. Note that the special case when w is 1-bounded corresponds to the classical graph coloring problem. It is therefore natural to associate the instance of the channel assignment problem with an edge-weighted graph $G = (V, E)$ where $E = \{uv : w(u, v) > 0\}$ with edge weights $w_E : E \to \mathbb{N}$ such that $w_E(xy) = w(x, y)$ for every $xy \in E$ (in what follows we abuse the notation slightly and use the same letter w for both the function defined on V^2 and E). The minimum span is called also the span of (G, w) and denoted by $\text{span}(G, w)$.

It is interesting to realize the place CHANNEL ASSIGNMENT in a kind of hierarchy of constraint satisfaction problems. We have already seen that it is a generalization of the classical graph coloring. It is also a special case of the constraint satisfaction problem (CSP). In CSP, we are given a vertex set V, a constraint set $\mathcal{C}$ and a number of colors d. Each constraint is a set of pairs of the

---

[*] Research supported by National Science Centre of Poland, grant number UMO-2013/09/B/ST6/03136.

R Ravi and I.L. Gørtz (Eds.): SWAT 2014, LNCS 8503, pp. 282–293, 2014.

form (v,t) where $v \in V$ and $t \in \{1,\dots,d\}$. An assignment $c : V \to \{1,\dots,d\}$ is *proper* if every constraint $A \in \mathcal{C}$ is satisfied, i.e. there exists $(v,t) \in A$ such that $c(v) \neq t$. The goal is to determine whether there is a proper assignment. Note that CHANNEL ASSIGNMENT corresponds to CSP where $d = s$ and every edge uv of weight $w(uv)$ in the instance of CHANNEL ASSIGNMENT corresponds to the set of constraints of the form $\{(u,t_1),(v,t_2)\}$ where $|t_1 - t_2| < w(uv)$.

Since graph coloring is solvable in time $O^*(2^n)$ [1] it is natural to ask whether CHANNEL ASSIGNMENT is solvable in time $O^*(c^n)$, for some constant c. Unfortunately, the answer is unknown at the moment and the best algorithm known so far runs in $O^*(n!)$ time (see McDiarmid [10]). However, there has been some progress on the ℓ-bounded variant. McDiarmid [10] came up with an $O^*((2\ell + 1)^n)$-time algorithm which has been next improved by Kral [9] to $O^*((\ell+2)^n)$ and to $O^*((\ell+1)^n)$ by Cygan and Kowalik [2]. These are all dynamic programming (and hence exponential space) algorithms, and the last one applies the fast zeta transform to get a minor speed-up. Interestingly, all these works show also algorithms which *count* all proper assignments of span at most s within the same running time (up to polynomial factors) as the decision algorithm.

It is a major open problem (see [9,2,6]) to find such a $O(c^n)$-time algorithm for c independent of ℓ or prove that it does not exist under a reasonable complexity assumption. A complexity assumption commonly used in such cases is the Exponential Time Hypothesis (ETH), introduced by Impagliazzo and Paturi [7]. It states that 3-CNF-SAT cannot be computed in time $2^{o(n)}$, where n is the number of variables in the input formula. The open problem mentioned above becomes even more interesting when we realize that under ETH, CSP does not have a $O^*(c^n)$-time algorithm for a constant c independent of d, as proved by Traxler [11].

Our Results. Our main result is a new $O^*((2\sqrt{\ell+1})^n)$-time algorithm for the ℓ-bounded CHANNEL ASSIGNMENT problem. Note that this is the first algorithm which breaks the $(O(\ell))^n$ barrier. Our algorithm follows the meet-in-the-middle approach (see e.g. Horowitz and Sahni [5]) and is surprisingly simple, so we hope it can become a yet another clean illustration of this beautiful technique. We show also its (more technical) counting version, which runs within the same time (up to a polynomial factor).

Although we were not able to show that the unrestricted CHANNEL ASSIGNMENT does not admit a $O(c^n)$-time for a constant c under, say ETH, we were able to shed some more light at this issue. Let us consider some more problems in the CSP hierarchy. In the T-COLORING, introduced by Hale [4], we are given a graph $G = (V, E)$, a set $T \subseteq \mathbb{N}$, and a number $s \in \mathbb{N}$. An assignment $c : V \to \{1,\dots,s\}$ is proper when for every edge $uv \in E$ we have $|c(u)-c(v)| \notin T$. As usual, the goal is to determine whether there exists a proper assignment. Like CHANNEL ASSIGNMENT, T-COLORING is a special case of CSP and generalizes graph coloring, but it is incomparable with CHANNEL ASSIGNMENT. However, Fiala, Král' and Škrekovski introduced which is a common generalization of vertex list-coloring (a variant of the classical graph coloring where each vertex has a list, i.e., a set of allowed colors), CHANNEL ASSIGNMENT and T-COLORING.

The instance of the GENERALIZED LIST T-COLORING is a triple (G, Λ, t, s) where $G = (V, E)$ is a graph, $\Lambda : V \to 2^{\mathbb{N}}$, $t : E \to 2^{\mathbb{N}}$ and $s \in \mathbb{N}$, where $\mathbb{N}$ denotes the set of all nonnegative integers. An assignment $c : V \to \{1, \ldots, s\}$ is proper when for every $v \in V$ we have $c(v) \in \Lambda(v)$, and for every edge $uv \in E$ we have $|c(u) - c(v)| \notin t(uv)$. As usual, the goal is to determine whether there exists a proper assignment. Similarly as in the case of CHANNEL ASSIGNMENT, we say that the instance of GENERALIZED LIST T-COLORING is ℓ-bounded if $\max \bigcup_{e \in E} t(e) \leq \ell$. Very recently, the GENERALIZED LIST T-COLORING was considered by Junosza-Szaniawski and Rzążewski [8]. They show GENERALIZED LIST T-COLORING can be solved in $O^*((\ell + 2)^n)$ time, which matches the time complexity of the algorithm of Cygan and Kowalik [2] for CHANNEL ASSIGN-MENT (note that an ℓ-bounded instance of CHANNEL ASSIGNMENT can be seen as an $(\ell - 1)$-bounded instance of GENERALIZED LIST T-COLORING). In this work we show that most likely one cannot hope for am $O^*(c^n)$-time algorithm for GENERALIZED LIST T-COLORING. We even consider a special case of GENER-ALIZED LIST T-COLORING, i.e. the non-list version where every vertex is allowed to have any color, so the instance is just a triple (G, t, s). We call it GENER-ALIZED T-COLORING. We show that, under ETH, GENERALIZED T-COLORING does not admit a $2^{2^{o(\sqrt{n})}} \text{poly}(r)$-time algorithm, where r is the size of the instance (including all the bits needed to represent the sets $t(e)$ for all $e \in E$). Note that this rules out an $O(n!)$ algorithm as well.

Organization of the Paper. In Section 2 we describe an $O^*((\ell + 2)^n)$-time dynamic programming algorithm for ℓ-bounded CHANNEL ASSIGNMENT. It is then used as a subroutine in the $O^*((2\sqrt{\ell + 1})^n)$-time algorithm described in Section 3. Due to space limitations, its extension to counting proper assignments of given span is deferred to a journal version. In section 4 we discuss hardness of GENERALIZED T-COLORING under ETH. Due to space limitations, proofs of claims marked by ★ are deferred to a journal version.

Notation. Throughout the paper n denotes the number of the vertices of the graph under consideration. For an integer k, by $[k]$ we denote the set $\{1, 2, \ldots, k\}$. Finally, $\uplus$ is the disjoint sum of sets i.e. the standard sum of sets $\cup$ but with an additional assumption that the sets are disjoint.

2   Yet Another $O^*((\ell + 2)^n)$-Time Dynamic Programming

In this section we provide a $O^*((\ell + 2)^n)$-time dynamic programming algorithm for CHANNEL ASSIGNMENT. It uses a different approach than e.g. the algorithm of Kral, and will be used as a subroutine in our faster algorithm.

For a subset $X \subseteq V$ and a function $f : X \to [\ell + 1]$ let $\mathcal{A}_{X,f}$ be the set of all proper assignments $c : X \to \mathbb{N}$ of the graph $G[X]$ subject to the condition that for every $x \in X$ we have $c(x) \geq f(x)$.

For every subset $X \subseteq V$ and $f : X \to [\ell + 1]$ we compute the value of $T[X, f]$ which is equal to the minimum span of an assignment from $\mathcal{A}_{X,f}$. Clearly, the minimum span of (G, w) equals to $T[V, f_1]$ where f_1 is the constant function which assigns 1 to every vertex.

The values of $T[X, f]$ are computed by dynamic programming as follows. First we initialize $T[\emptyset, e_{[\ell+1]}] = 0$ (where $e_{[\ell+1]}$ is the only function $f : \emptyset \rightarrow [\ell + 1]$). Next, we iterate over all non-empty subsets of V in the order of nondecreasing cardinality. In order to determine the value of $T[X, f]$ we use the recurrence relation formulated in the following lemma.

Informally, it uses the observation that there is a minimum-span assignment c such that the vertex $v \in X$ with minimum color $c(v)$ is *left-shifted*, i.e. $c(v) = f(v)$. Hence we can check all possibilities for v and then the colors of all the other vertices from X have lower bounds in range $\{f(v), \dots, f(v)+\ell\}$, so we can translate the range back down to $\{1, \dots, \ell+1\}$ and use the previously computed values of $T[X \setminus \{v\}, \cdot]$.

Lemma 2.1. *For a subset $X \subseteq V$, a function $f : X \rightarrow [\ell + 1]$ and a vertex v define the function $f_v : X \setminus \{v\} \rightarrow [\ell+1]$ given by the formula*

$$f_v(x) = 1 + \max\{w(v, x), f(x) - f(v)\} \qquad \text{for every } x \in X \setminus \{v\}.$$

Then,

$$T[X, f] = \min_{v \in X}(f(v) + T[X \setminus \{v\}, f_v] - 1), \tag{1}$$

Proof. Fix $v \in X$. Denote $\mathcal{A}_{X,f,v} = \{c \in \mathcal{A}_{X,f} : c(v) = f(v) = \min_{x \in X} f(x)\}$. Then, for every assignment $c \in \mathcal{A}_{X,f,v}$, for every $x \in X \setminus \{v\}$ we have $c(x) \geq f(v) + \max\{w(v, x), f(x) - f(v)\}$. Hence, the minimum span of an assignment from $\mathcal{A}_{X,f,v}$ is equal to $f(v) + T[X \setminus \{v\}, f_v] - 1$. It suffices to show that there is an assignment $c^* \in \mathcal{A}_{X,f}$ of minimum span such that $c^*(v) \in \mathcal{A}_{X,f,v}$ for some $v \in X$. Consider an arbitrary assignment $c^* \in \mathcal{A}_{X,f}$ of minimum span. Let $x \in X$ be the vertex of minimum color, i.e. $c^*(x)$ is minimum. If $c^*(x) = f(x)$ we are done. Otherwise consider a new assignment c^{**} which is the same as c^* everywhere except for x and $c^{**}(x) = f(x)$; then c^{**} is proper since $c^*(x)$ is minimal and clearly $c^{**} \in \mathcal{A}_{X,f}$. The span of c^{**} is not greater than the span of c^* (actually they are the same since c^* has minimal span), so the claim follows. $\square$

The size of the array T is $\sum_{i=0}^{n} \binom{n}{i}(\ell + 1)^i = (\ell + 2)^n$. Computing a single value based on previously computed values for smaller sets takes $O(n^2)$ time, hence the total computation takes $O((\ell+2)^n n^2)$ time. As described, it gives the minimum span only, but we can retrieve the corresponding assignment within the same running time using standard techniques.

3   The Meet-in-the-Middle Speed-Up

In this section we present our main result, an algorithm for ℓ-bounded CHANNEL ASSIGNMENT that applies the meet-in-the-middle technique. Roughly, the idea is to find partial solutions for all possible *halves* of the vertex set and then merge the partial solutions efficiently to solve the full instance.

For the clarity of the presentation we assume n is even (otherwise we just add a dummy isolated vertex). Before we describe the algorithm let us introduce some notation. For a set $X \subseteq V$, by $\overline{X}$ we denote $V \setminus X$. Moreover, for a function $f : X \to [\ell + 1]$ we define function $\overline{f} : \overline{X} \to [\ell + 1]$ such that for every $v \in \overline{X}$,

$$\overline{f}(v) = 1 + \max(\{1 + w(uv) - f(u) \ : \ uv \in E, \ u \in X\} \cup \{0\}).$$

The values $T[X, f]$ are defined as in Section 2. Our algorithm is based on the following observation.

Lemma 3.1. *The span of* (G, w) *is equal to*

$$\min(T[X, f] + T[\overline{X}, \overline{f}] - 1),$$

where the minimum is over all pairs (X, f) *where* $X \in \binom{V}{n/2}$ *and* $f : X \to [\ell+1]$.

Proof. Let $c^* : V \to \mathbb{N}$ be a proper assignment of minimum span s. Order the vertices of $V = \{v_1, \ldots, v_n\}$ so that for every $i = 1, \ldots, n - 1$ we have $c^*(v_i) \leq c^*(v_{i+1})$. Consider the subset $X = \{v_1, \ldots, v_{n/2}\}$. Let $s_1 = c^*(v_{n/2})$. Define $f : X \to [\ell + 1]$ such that $f(x) = 1 + \min\{s_1 - c^*(x), \ell\}$ for every $x \in X$. From the definition of T we have $T[X, f] \leq s_1$ (because the assignment $x \mapsto 1 + s_1 - c^*(x)$ belongs to $\mathcal{A}_{X,f}$ and has span s_1). Moreover, note that for every $v \in \overline{X}$ it holds that

$$\begin{aligned}
c^*(v) &\geq \max(\{c^*(u) + w(uv) \ : \ uv \in E, \ u \in X\} \cup \{s_1\}) \\
&= \max(\{s_1 + w(uv) - f(u) + 1 \ : \ uv \in E, \ u \in X\} \cup \{s_1\}) \\
&= s_1 - 1 + \overline{f}(v).
\end{aligned}$$

It follows that $s = \max_{v \in \overline{X}} c^*(v) \geq s_1 - 1 + T[\overline{X}, \overline{f}] \geq T[X, f] + T[\overline{X}, \overline{f}] - 1$.

Finally we show that $s > T[X, f] + T[\overline{X}, \overline{f}] - 1$ contradicts the optimality of c^*. Let $c_1 \in \mathcal{A}_{X,f}$ be an assignment of span $T[X, f]$ and let $c_2 \in \mathcal{A}_{\overline{X}, \overline{f}}$ be an assignment of span $T[\overline{X}, \overline{f}]$. Consider the following assignment $c : V \to \mathbb{N}$.

$$c(x) = \begin{cases} 1 + T[X, f] - c_1(x) & \text{for } x \in X \\ T[X, f] + c_2(x) - 1 & \text{for } x \in \overline{X} \end{cases}$$

One can check that from the definition of $\overline{f}$ it follows that c is a proper assignment. Moreover, the span of c is equal to $T[X, f] + T[\overline{X}, \overline{f}] - 1$. Hence, if $s > T[X, f] + T[\overline{X}, \overline{f}] - 1$ then c^* is not optimal, a contradiction. $\square$

From Lemma 3.1 we immediately obtain the following algorithm for computing the span of (G, w):

1. Compute the values of $T[X, f]$ for all $X \in \binom{V}{\leq n/2}$ and $f : X \to [\ell + 1]$ using the algorithm from Section 2.
2. Find the span of (G, w) using the formula from Lemma 3.1.

Note that Step 1 takes time proportional to $\sum_{i=0}^{n/2} \binom{n}{i} (\ell + 1)^i n^2 = O(2^n (\ell + 1)^{n/2} n^2)$. The size of array T is clearly $O(2^n (\ell + 1)^{n/2})$. In Step 2 we compute a minimum of $\binom{n}{n/2}(\ell+1)^{n/2} = O(2^n(\ell+1)^{n/2})$ values. Hence the total time is $O(2^n(\ell+1)^{n/2}n^2)$. As described, the above algorithm gives the minimum span only, but we can retrieve the corresponding assignment within the same running time using standard techniques. We have just proved the following theorem.

Theorem 3.2. *For every ℓ-bounded weight function the channel assignment problem can be solved in $O(2^n(\ell+1)^{n/2}n^2)$ time.*

4   Hardness of Generalized T-Coloring

In this section we give a lower bounds for the time complexity of Generalized T-Coloring, under ETH. To this end we present a reduction from SetCover. The instance of the decision version of SetCover consists of a family of sets $\mathcal{S} = \{S_1, \ldots, S_m\}$ and a number k. The set $U = \bigcup \mathcal{S}$ is called *the universe* and we denote $n = |U|$. The goal is to decide whether there is a subfamily $\mathcal{C} \subseteq \mathcal{S}$ of size at most k such that $\bigcup \mathcal{C} = U$ (then we say the instance is *positive*).

In the following lemma we reduce Set Cover to the decision version of Generalized T-Coloring, where for a given instance (G, w) and a number s we ask whether there is a proper assignment of span at most s (then we say the instance is *positive*). We say that an instance $(\mathcal{S}, k)$ of SetCover is equivalent to an instance (G, w, k) of Generalized T-Coloring when $(\mathcal{S}, k)$ is positive iff (G, w, k) is positive. For every edge e of G, every pair (e, d) for $d \in t(e)$ is called a *constraint*.

Lemma 4.1. *Let $(\mathcal{S}, k)$ be an instance of SetCover with m sets and universe of size n and let $A \in [1, m]$ and $B \in [1, n]$ be two reals. Then we can generate in polynomial time an equivalent instance of Generalized T-Coloring which has $O\left(\frac{n}{B} + \frac{m}{A} \cdot \max\{1, \log A\}\right)$ vertices, $O^*\left(2^A \cdot m^B\right)$ constraints and is $O\left(2^A \cdot m^B\right)$-bounded.*

Proof. For convenience we assume that A and B are natural numbers, since otherwise we round A and B down and the whole construction and its analysis is the same, up to some details.

In the proof we consider coloring of the vertices as placing the vertices on a number line in such a way that every vertex is placed in the coordinate equal to its color.

Let $\mathcal{S} = \{S_1, \ldots, S_m\}$. We are going to construct a complex instance $(G = (V, E), t, s)$ of Generalized T-Coloring. We describe it step-by-step and show some of its properties.

We begin by putting vertices v_L and v_R in V and $t(v_L v_R) = \{0, \ldots, s-2\}$, i.e. in every proper assignment v_L has color 1 and v_R has color s, or the other way around; w.l.o.g. we assume the first possibility. We specify s later.

In what follows, whenever we put a new vertex v in V, we will specify the set $A(v)$ of its *allowed* colors. Formally, this corresponds to putting $t(v_L v) = \{d \in \{0, \ldots, s-1\} : d+1 \notin A(v)\}$.

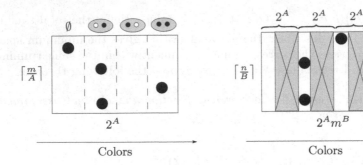

Fig. 1. The set choice module   **Fig. 2.** The witness module. (The grey areas are the gaps between the m^B potentially allowed positions.)

Our instance will consist of three separate modules (the set choice module, the witness module and the parsimonious module). By separate we mean they have disjoint sets of vertices V_S, V_U and V_P and moreover they have disjoint sets of allowed colors, i.e. for $i, j \in \{S, U, P\}$, when $x \in V_i$ and $y \in V_j$ for $i \neq j$ then $A(x) \cap A(y) = \emptyset$. However the modules will interfere with each other by forbidding some distances between pairs of vertices from two different modules.

The Set Choice Module. The first module represents the sets in $\mathcal{S}$. For every $i = 1, \ldots, \lceil \frac{m}{A} \rceil$ the set V_S contains a vertex s_i. Vertex s_i represents the A sets

$$\mathcal{S}_i = \{S_{(i-1) \cdot A+1}, S_{(i-2) \cdot A+2}, \ldots, S_{i \cdot A}\}$$

(and the last vertex $s_{\lceil m/A \rceil}$ represents $\mathcal{S}_{\lceil m/A \rceil} = S_{(\lceil m/A \rceil - 1)A+1}, \ldots, S_m)$. We also put $A(s_i) = \{1, \ldots, 2^A\}$ for every $s_i \in V_S$. The intuition is that the color $c \in [2^A]$ of a vertex s_i corresponds to a subset $\mathcal{S}_i(c) \subseteq \mathcal{S}_i$, i.e. the choice of sets from $\mathcal{S}_i$ to the solution of SETCOVER.

The Witness Module. Let denote the elements of the universe as $e_1, e_2, \ldots, e_n$. For every $i = 1, \ldots, \lceil \frac{n}{B} \rceil$ the set V_U contains a vertex u_i. Vertex u_i represents the B elements

$$U_i = \{e_{(i-1) \cdot B+1}, e_{(i-2) \cdot B+2}, \ldots, e_{i \cdot B}\}$$

(and the last vertex $u_{\lceil n/B \rceil}$ represents $U_{\lceil n/B \rceil} = e_{(\lceil n/B \rceil - 1)B+1}, \ldots, e_n)$.

This time vertices V_U do not need to have the same sets of allowed colors, but for every $u \in V_U$ we have $A(u) \subseteq \{1 + i \cdot 2^A : i = 1, \ldots, m^B\}$. Note that every vertex has at most m^B allowed colors and there are gaps of length $2^A - 1$ where no vertex is going to be assigned.

We say that a sequence $(S_{w_1}, \ldots, S_{w_B}) \in \mathcal{S}^B$ is a *witness* for a vertex $u_i \in V_U$ when

$$U_i \subseteq \bigcup_{j=1}^{B} S_{w_j}.$$

For every $i = 1, \ldots, m^B$ color $1 + i \cdot 2^A$ corresponds to the i-th sequence in the set $\mathcal{S}^B$ (say, in the lexicographic order of indices); we denote this sequence by

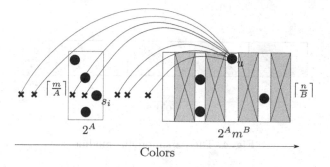

$$2^A \qquad\qquad\qquad 2^A m^B$$

Colors

Fig. 3. The interaction between a vertex s_i in the set choice module and a vertex u in the witness module. All the drawn arcs are forbidden distances between s_i and u. Note that for every possible color $1 + j \cdot 2^A$ of u the subset of $[2^A]$ excluded by the forbidden distances in $t(us_i)$ is exactly $F_{i,j}$.

$\mathcal{W}_i$. Then, for every $u \in V_U$,

$$A(u) = \{1 + i \cdot 2^A \ : \ \mathcal{W}_i \text{ is a witness for u, } i = 1, \ldots, m^B\}.$$

The intuition should be clear: color of a vertex $u_i \in V_U$ in a proper assignment represents the choice of at most B sets in the solution of SETCOVER which cover U_i.

The Interaction between the Set Choice Module and the Witness Module. As we have seen, every assignment c of colors to the vertices determines a choice of a subfamily $\mathcal{S}(c) \subseteq \mathcal{S}$, where $\mathcal{S}(c) = \bigcup_{i=1}^{\lceil m/A \rceil} \mathcal{S}_i(c(i))$. Similarly, c determines a choice of a subfamily $\mathcal{S}'(c) \subseteq \mathcal{S}$, where $\mathcal{S}'(c) = \bigcup_{u \in V_U} \mathcal{W}_{c(u)}$. It should be clear that we want to force that in every proper assignment $\mathcal{S}'(c) \subseteq \mathcal{S}(c)$. To this end we introduce edges between the two modules.

For $i = 1, \ldots, \lceil \frac{m}{A} \rceil$ and $j = 1, \ldots, m^B$ define the following set of forbidden colors

$$F_{i,j} = \{c \in [2^A] \ : \ \mathcal{W}_j \cap \mathcal{S}_i \nsubseteq \mathcal{S}_i(c)\}.$$

The intuition is the following: If a proper assignment colors a vertex $u_i \in V_U$ with color $1 + j \cdot 2^A$ (i.e. it assigns the witness $\mathcal{W}_j$ to the set U_i) then it cannot color the vertex s_i with colors from $F_{i,j}$ (i.e. choose this subsets of $\mathcal{S}_i$ corresponding to these colors), for otherwise $\mathcal{S}'(c) \nsubseteq \mathcal{S}(c)$.

Claim 1. Consider any proper assignment $c : V \to [s]$. If for every $i = 1, \ldots, \lceil \frac{m}{A} \rceil$ we have $c(s_i) \notin \bigcup_{u \in V_U} F_{i,c(u)}$, then $\mathcal{S}'(c) \subseteq \mathcal{S}(c)$.
Proof of the Claim: Consider a set $S_t \in \mathcal{W}_{c(u)}$ for an arbitrary $u \in V_U$. Then $S_t \in \mathcal{S}_i$ for some i. From the assumption, $c(s_i) \notin F_{i,c(u)}$, so $\mathcal{W}_{c(u)} \cap \mathcal{S}_i \subseteq \mathcal{S}_i(c)$. Hence, $S_t \in \mathcal{S}(c)$, as required.

Hence we would like to add some forbidden distances to our instance to make the assumption of Clam 1 hold. To this end, for every $u \in V_U$ and every $s_i \in V_S$ we put

$$t(us_i) = \bigcup_{j=1}^{m^B} \{1 + j \cdot 2^A - f \ : \ f \in F_{i,j}\}.$$

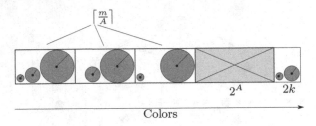

Fig. 4. The parsimonious module

In other words, for every possible color $1 + j \cdot 2^A$ of u we forbid all distances between u and s_i that would result in coloring s_i with $F_{i,j}$. Then indeed the assumption from Claim 1 holds.

Claim 2. For any proper assignment $c : V \to [s]$ we have $\mathcal{S}'(c) \subseteq \mathcal{S}(c)$.

Proof of the Claim: We need to verify the assumption in Claim 1. Assume for the contradiction that for some i and some $u \in V_U$ we have $c(s_i) \in F_{i,c(u)}$. Recall that in a proper assignment $c(u) = 1 + j \cdot 2^A$ for some $j = 1, \ldots, m^B$. Then $|c(u) - c(s_i)| = 1 + j \cdot 2^A - c(s_i) \in t(us_i)$, a contradiction.

Claim 3. For any proper assignment $c : V \to [s]$ we have $\mathcal{S}(c)$ covers the universe.

Proof of the Claim: This is an immediate corollary from Claim 2 and the fact that every vertex $u \in V_U$ is colored with a color from $A(u)$.

Claim 4. For every cover $\mathcal{C} \subseteq \mathcal{S}$ of the universe, there is a proper assignment $c : V \to [s]$ such that $\mathcal{S}(c) = \mathcal{C}$.

Proof of the Claim: We color v_L and v_R with 1 and s, and every vertex s_i with the color from $[2^A]$ corresponding to the subset $\mathcal{S}_i \cap \mathcal{C}$ of $\mathcal{S}_i$. For every set U_i for every $e \in U_i$ we pick a set $S_e \in \mathcal{C}$ that contains e and we build a witness $\mathcal{W}$ from the sets S_e. We color u_i with the color $1 + j \cdot 2^A$, where j is the number of $\mathcal{W}$ in the lexicographic order of all witnesses. It remains to check that the resulting assignment c is proper. The only nontrivial issue is whether for every $u \in V_U$ and $s_i \in V_S$ we have $|c(u) - c(s_i)| \notin t(us_i)$. It is clear that $|c(u) - c(s_i)| \notin \{c(u) - f \ : \ f \in F_{i,j}\}$, where j is such that $c(u) = 1 + j \cdot 2^A$. However, for every $j' \neq j$ the set $\{1 + j' \cdot 2^A - f \ : \ f \in F_{i,j'}\}$ is disjoint from 2^A (this is where we make use of the 'gaps' of length $2^A - 1$).

Bounding the Number of Sets Chosen to the Solution. The last thing we need in a proper assignment c is to keep the number of the sets in $\mathcal{S}(c)$ bounded by k. To this end we use the parsimonious module with the vertex set V_P.

The third parsimonious module consists of $\lceil \frac{m}{A} \rceil$ consecutive submodules and an additional free space of length $2k$ (meaning that for every $v \in V_P$ the set of allowed colors $A(v)$ contains this free space. Between those submodules and the additional free space we put a gap of length 2^A, where no vertex can be assigned. The intuition is that in a proper assignment c the i-th submodule represents the number of sets from $\mathcal{S}_i$ chosen to the solution, i.e. $|\mathcal{S}_i(c_i)|$.

More precisely, $V_P = \biguplus_{i=1}^{\lceil m/A \rceil} V_i$, where V_i is a set of $1 + \lfloor \log A \rfloor$ vertices representing numbers $2^0, 2^1, \ldots, 2^{\lfloor \log A \rfloor}$. Let For a vertex $x \in V_P$ let $r(x)$ denote the number represented by x. For every two vertices $x, y \in V_P$ we define

$$t(xy) = \{0, \ldots, r(x) + r(y) - 1\}.$$

It follows that we can interpret those vertices as disjoint disks with radii equal to the represented numbers (see Fig. 4). Let $q = (1 + m^B)2^A$, i.e. q is the number of colors used by the first two modules. For every i, we define i-th slot as the set of colors $\{q + 1 + (i-1) \cdot 4A, \ldots, q + i \cdot 4A\}$. Note that the length of each slot is $4A$. Define also the free space as $Q = \{q + \lceil m/A \rceil \cdot 4A + 2^A + 1, \ldots, q + \lceil m/A \rceil \cdot 4A + 2^A + 2k\}$. Each vertex $x \in V_i$ is either in i-th slot or in the free space Q. However, x has exactly one allowed color in the i-th slot chosen so that we can put all the disks in the i-th slot and they will be disjoint. Let j be such that $r(x) = 2^j$. Then we denote the allowed color by $a_x = q + (i-1) \cdot 4A + \sum_{r < j} 2 \cdot 2^r + 2^j$. In precise terms, $A(x) = \{a_x\} \cup Q$.

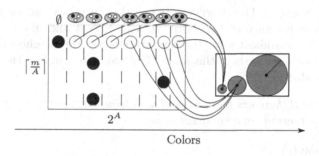

Fig. 5. The interaction between the set choice module and one of the submodules of the parsimonious module. Note that colors in $[2^A]$ are ordered according to the cardinality of the chosen collection of sets $(0, 1, 1, 1, 2, 2, 2, 3)$.

Vertices of the i-th submodule have some edges to the vertex s_i of the set choice module. As we mentioned, for a proper assignment c the i-th submodule is going to be a counter representing the number of sets in $S_i(c)$; in fact the vertex representing 2^j corresponds to the j-th bit of the counter. So if $r(x) = 2^j$ for $x \in V_i$, then $t(s_i x)$ contains all distances d such that $a_x - d$ is a color b from 2^A such that the j-th bit of $|S_i(b)|$ is 1. Hence, in a proper assignment c, if the j-th bit of the number of sets in $S_i(c)$ is 1 then x is thrown away from the i-th slot and it is colored by a color from the free space Q. However, the $|Q| = 2k$ so the sum of the radii of the disks thrown out from its slots is at most k. It follows that the total number of the chosen sets is also at most k. Also, if there is a cover $C \subseteq S$ of the universe such that $|C| \leq k$, then for every i, if $|C \cap S_i|$ has 1 on the j-th bit we put the vertex of V_i representing 2^j in Q. It is clear that since $|C| \leq k$ we have enough space for them in Q. Moreover, we do not violate any edge between these vertices and V_S because of the gap 2^A inside the parsimonious module. Together with Claim 4 it implies that (S, k) is a YES-instance of SETCOVER iff (G, t, s) is a YES-instance of GENERALIZED T-COLORING, provided that s is sufficiently large to provide disjoint intervals of colors for all the modules. From the construction we infer that it is sufficient to put $s = 2^A + 2^A \cdot m^B + 4A \cdot \lceil \frac{m}{A} \rceil + 2^A + 2k$.

Calculating the Parameters. Note that $s = O(2^A m^B)$ and in particular our instance is $O(2^A m^B)$-bounded. Moreover, $|V| = \lceil \frac{n}{B} \rceil + \lceil \frac{m}{A} \rceil + \lceil \frac{m}{A} \rceil \cdot (1 + \lfloor \log A \rfloor) + 2 = O\left(\frac{n}{B} + \frac{m}{A} \cdot (1 + \log A)\right) = O\left(\frac{n}{B} + \frac{m}{A} \cdot \max\{1, \log A\}\right)$. Finally, the total number of constraints is bounded by $O^*((\frac{n}{B} + \frac{m}{A} \cdot \max\{1, \log A\})^2 \cdot (2^A \cdot m^B)) = O^*\left(2^A \cdot m^B\right)$, i.e., the number of pairs of the vertices times the maximum forbidden distance $s - 1$. It ends the proof. □

Corollary 4.2. *Let (G, k) be an instance of* DOMINATING SET *where G is a graph on n vertices and $k \in \mathbb{N}$. Then, for any real number $A \in [1, n]$ we can generate in polynomial time an instance of* GENERALIZED T-COLORING *with $O\left(\frac{n}{A} \cdot \max\{1, \log A\}\right)$ vertices and with $O^*\left((2n)^A\right)$ constraints and such that all the numbers in the instance have $O\left(A \cdot \max\{1, \log n\}\right)$ bits.*

Proof. The instance of DOMINATING SET with n vertices can be transformed to an equivalent instance of SET COVER with n sets and also n elements of the universe in a standard way (the sets are exactly the neighborhoods of the vertices). The number k stays the same. Therefore we can use the Lemma 4.1 with $A = B$ and $m = n$. □

Theorem 4.3. *If there exists an algorithm solving* GENERALIZED T-COLORING *in one of the following time complexities:*

(i) $2^{2^{o(\sqrt{n})}} \operatorname{poly}(r)$,

(ii) (★) $2^{n \cdot o\left(\frac{\log l}{(\log \log l)^2}\right)} \operatorname{poly}(r)$,

where n is the number of vertices in the input graph and r is the bit size of the input, then there exists an algorithm solving DOMINATING SET *in time $2^{o(n)}$.*

Proof. (i). Let us assume that we have an algorithm solving GENERALIZED T-COLORING in time $2^{2^{f(n)}} \operatorname{poly}(r)$ where f is some function such that $f(n) = o(\sqrt{n})$. We can assume without loss of generality that f is positive and nondecreasing. Let C be a constant such that Corollary 4.2 will give us always at most $C \cdot \frac{n}{A} \cdot \max\{1, \log A\}$ vertices. Let α be a positive nondecreasing function such that $\alpha(n) \leq \frac{\sqrt{n}}{f(Cn)}$ and $\alpha(n) = \omega(1)$. Such a function always exists because $\frac{\sqrt{n}}{f(Cn)} = \frac{1}{\sqrt{C}} \cdot \frac{\sqrt{Cn}}{f(Cn)} = \omega(1)$. For every instance of DOMINATING SET with n vertices we can take $A = \frac{n}{\alpha(\log^2 n) \log n}$ and use Corollary 4.2 to obtain an instance of GENERALIZED T-COLORING with $O\left(\frac{n}{A} \log A\right) = O\left(\alpha\left(\log^2 n\right) \log^2 n\right)$ vertices and

$$O^*\left((2n)^A\right) = O^*\left(2^{A + A \log n}\right) = O^*\left(2^{O\left(\frac{n}{\alpha(\log^2 n)}\right)}\right) = O^*\left(2^{o(n)}\right) = 2^{o(n)} \text{ con-}$$

straints. Moreover the numbers in the instance have polynomial size, so the size of the whole instance is $2^{o(n)}$. Thus this instance can be built in $\operatorname{poly}\left(n, 2^{o(n)}\right) = 2^{o(n)}$ time. Then we can solve this instance in $2^{2^{f\left(C \cdot \frac{n}{A} \log A\right)}} \operatorname{poly}\left(2^{o(n)}\right)$ time. But

$$f\left(C \cdot \tfrac{n}{A} \log A\right) \quad \leq \quad f\left(C \cdot \alpha \left(\log^2 n\right) \log^2 n\right) \quad \leq \quad \frac{\sqrt{\alpha(\log^2 n)} \log^2 n}{\alpha(\alpha(\log^2 n) \log^2 n)} \quad =$$

$$\log n \cdot \frac{\sqrt{\alpha(\log^2 n)}}{\alpha(\alpha(\log^2 n) \log^2 n)} = \log n \cdot \frac{\sqrt{\alpha(\log^2 n)}}{\alpha(\log^2 n)} \cdot \frac{\alpha(\log^2 n)}{\alpha(\alpha(\log^2 n) \log^2 n)} \leq \frac{\log n}{\sqrt{\alpha(\log^2 n)}} = o\left(\log n\right).$$

So the time of the whole procedure is $2^{o(n)} + 2^{2^{o(\log n)}} \mathrm{poly}\left(2^{o(n)}\right) = 2^{o(n)}$. $\square$

Corollary 4.4. *There is no algorithm solving an n-vertex instance of* GENER-ALIZED *T*-COLORING *with bit size r in any of the listed time complexities*

$$- \; 2^{2^{o(\sqrt{n})}} \mathrm{poly}(r),$$

$$- \; 2^{n \cdot o\left(\frac{\log l}{\log^2 \log l}\right)} \mathrm{poly}(r),$$

unless the Exponential Time Hypothesis fails.

Proof. Under the ETH assumption there is no algorithm solving DOMINATING SET in time $2^{o(n)}$ where n is a number of the vertices (See [3]). Therefore the claim follows immediately from Theorem 4.3. $\square$

Regarding the first claim the theorem above, recall that there is a $2^{O(n \log l)} \mathrm{poly}(r)$-time algorithm for GENERALIZED *T*-COLORING, see [8].

References

1. Björklund, A., Husfeldt, T., Koivisto, M.: Set partitioning via inclusion-exclusion. SIAM J. Comput. 39(2), 546–563 (2009)
2. Cygan, M., Kowalik, L.: Channel assignment via fast zeta transform. Inf. Process. Lett. 111(15), 727–730 (2011)
3. Fomin, F.V., Kratsch, D., Woeginger, G.J.: Exact (exponential) algorithms for the dominating set problem. In: Hromkovič, J., Nagl, M., Westfechtel, B. (eds.) WG 2004. LNCS, vol. 3353, pp. 245–256. Springer, Heidelberg (2004)
4. Hale, W.: Frequency assignment: Theory and applications. Proceedings of the IEEE 68(12), 1497–1514 (1980)
5. Horowitz, E., Sahni, S.: Computing partitions with applications to the knapsack problem. J. ACM 21(2), 277–292 (1974)
6. Husfeldt, T., Paturi, R., Sorkin, G.B., Williams, R.: Exponential Algorithms: Algorithms and Complexity Beyond Polynomial Time (Dagstuhl Seminar 13331). Dagstuhl Reports 3(8), 40–72 (2013)
7. Impagliazzo, R., Paturi, R.: On the complexity of k-sat. J. Comput. Syst. Sci. 62(2), 367–375 (2001)
8. Junosza-Szaniawski, K., Rzążewski, P.: An exact algorithm for the generalized list T-coloring problem. CoRR, abs/1311.0603 (2013)
9. Král, D.: An exact algorithm for the channel assignment problem. Discrete Applied Mathematics 145(2), 326–331 (2005)
10. McDiarmid, C.J.H.: On the span in channel assignment problems: bounds, computing and counting. Discrete Mathematics 266(1-3), 387–397 (2003)
11. Traxler, P.: The time complexity of constraint satisfaction. In: Grohe, M., Niedermeier, R. (eds.) IWPEC 2008. LNCS, vol. 5018, pp. 190–201. Springer, Heidelberg (2008)

Consistent Subset Sampling*

Konstantin Kutzkov and Rasmus Pagh

IT University of Copenhagen, Denmark

Abstract. Consistent sampling is a technique for specifying, in small space, a subset S of a potentially large universe U such that the elements in S satisfy a suitably chosen sampling condition. Given a subset $\mathcal{I} \subseteq U$ it should be possible to quickly compute $\mathcal{I} \cap S$, i.e., the elements in $\mathcal{I}$ satisfying the sampling condition. Consistent sampling has important applications in similarity estimation, and estimation of the number of distinct items in a data stream.

In this paper we generalize consistent sampling to the setting where we are interested in sampling size-k subsets occurring in some set in a collection of sets of bounded size b, where k is a small integer. This can be done by applying standard consistent sampling to the k-subsets of each set, but that approach requires time $\Theta(b^k)$. Using a carefully designed hash function, for a given sampling probability $p \in (0, 1]$, we show how to improve the time complexity to $\Theta(b^{\lceil k/2 \rceil} \log \log b + pb^k)$ in expectation, while maintaining strong concentration bounds for the sample. The space usage of our method is $\Theta(b^{\lceil k/4 \rceil})$.

We demonstrate the utility of our technique by applying it to several well-studied data mining problems. We show how to efficiently estimate the number of frequent k-itemsets in a stream of transactions and the number of bipartite cliques in a graph given as incidence stream. Further, building upon a recent work by Campagna et al., we show that our approach can be applied to frequent itemset mining in a parallel or distributed setting. We also present applications in graph stream mining.

1   Introduction

Consistent sampling is an important technique for constructing randomized sketches (or "summaries") of large data sets. The basic idea is to decide whether to sample an element x depending on whether a certain sampling condition is satisfied. Usually, consistent sampling is implemented using suitably defined hash functions and x is sampled if its hash value $h(x)$ is below some threshold. If x is encountered several times, it is therefore either *never* sampled or *always* sampled. The set of items to sample is described by the definition of the hash function, which is typically small.

Consistent sampling comes in two basic variations: In one variation (sometimes referred to as *subsampling*) there is a fixed sampling probability $p \in (0, 1)$,

---

* This work is supported by the Danish National Research Foundation under the Sapere Aude program.

R Ravi and I.L. Gørtz (Eds.): SWAT 2014, LNCS 8503, pp. 294–305, 2014.

and elements in a set must be sampled with this probability. In the alternative model the sample size is fixed, and the sampling probability must be scaled to achieve the desired sample size.

Depending on the strength of the hash function used, the sample will exhibit many of the properties of a random sample (see e.g. [5,18]). One of the most famous applications of consistent sampling [6] is estimating the *Jaccard similarity* of two sets by the similarity of consistent samples, using the same hash function. Another well-known application is reducing the number of distinct items considered to $\Theta(1/\varepsilon^2)$ in order to make an $(1 \pm \varepsilon)$-approximation of the total number of distinct items (see [21] for the state-of-the-art result).

In this paper we consider consistent sampling of certain *implicitly defined* sets. That is, we sample from a set much larger than the size of the explicitly given database. Our main focus is on streams of sets, where we want to sample subsets of the sets in the stream.

We demonstrate the usability of our technique by designing new algorithms for several well-studied counting problems in the streaming setting. We present the first nontrivial algorithm for the problem of estimating the number of frequent k-itemsets with rigorously understood complexity and error guarantee and also give a new algorithm for counting bipartite cliques in a graph given as an incidence stream. Also, using a technique presented in [9], we show that our approach can be easily parallelized and applied to frequent itemset mining algorithms based on hashing [10,11].

2 Preliminaries

Notation. Let $\mathcal{C} = T_1, .., T_m$ be a collection of m subsets of a ground set $\mathcal{I}$, $T_j \subseteq \mathcal{I}$, where $\mathcal{I} = \{1, \ldots, n\}$ is a set of *elements*. The sets T_j each contain at most b elements, i.e., $|T_j| \leq b$, and in the following are called *b-sets*. Let further $S \subseteq \mathcal{I}$ be a given subset. If $|S| = k$ we call S a *k-subset*. We assume that the b-sets are explicitly given as input while a k-subset can be any subset of $\mathcal{I}$ of cardinality k. In particular, a b-set with b elements contains $\binom{b}{k}$ distinct k-subsets for $k \leq b$. The *frequency* of a given k-subset is the number of b-sets containing it.

In order to simplify the presentation, we assume a lexicographic order on the elements in $\mathcal{I}$ and a unique representation of subsets as ordered vectors of elements. However, we will continue to use standard set operators to express computations on these vectors. In our algorithm we will consider only lexicographically ordered k-subsets. For two subsets I_1, I_2 we write $I_1 < I_2$ iff $i_1 < i_2$ $\forall i_1 \in I_1, i_2 \in I_2$.

The set of k-subsets of $\mathcal{I}$ is written as $\mathcal{I}^k$ and similarly, for a given b-set T_j, we write T_j^k for the family of k-subsets occurring in T_j. A family of k-subsets $\mathcal{S} \subset \mathcal{I}^k$ is called a *consistent sample* for a given sampling condition P if for each b-set T_i the set $\mathcal{S} \cap T_i^k$ is sampled, i.e., all elements satisfying the sampling condition P that occur in T_i are sampled. The sampling condition P will be defined later.

Let $[q]$ denote the set $\{0, \ldots, q-1\}$ for $q \in \mathbb{N}$. A hash function $h : \mathcal{E} \to [q]$ is t-wise independent iff $\mathbf{Pr}[h(e_1) = c_1 \wedge h(e_2) = c_2 \wedge \cdots \wedge h(e_t) = c_t] = q^{-t}$ for distinct elements $e_i \in \mathcal{E}$, $1 \le i \le t$, and $c_i \in [q]$. We denote by $p = 1/q$ the sampling probability we use in our algorithm. Throughout the paper we will often exchange p and $1/q$.

We assume the standard computation model and further we assume that one element of $\mathcal{I}$ can be written in one machine word.

Example. In order to simplify further reading let us consider a concrete data mining problem. Let $\mathcal{T}$ be a stream of m transactions $T_1, T_2, \ldots, T_m$ each of size b. Each such transaction is a subset of the ground set of items $\mathcal{I}$. We consider the problem of finding the set of frequent k-itemsets, i.e., subsets of k items occurring in at least t transactions for a user-defined $t \le m$. As a concrete example consider a supermarket. The set of items are all offered goods and transactions are customers baskets. Frequent 2-itemsets will provide knowledge about goods that are frequently bought together.

The problem can be phrased in terms of the above described abstraction by associating transactions with b-sets and k-itemsets with k-subsets. Assume we want to sample 2-itemsets. A consistent sample can be described as follows: for a hash function $h : \mathcal{I} \to [q]$ we define S to be the set of all 2-subsets (i, j) such that $h(i) + h(j) = 0 \bmod q$. In each b-set we can then generate all $\binom{b}{2}$ 2-subsets and check which of them satisfy the so defined sampling condition. For a suitably defined hash function, one can show that resulting sample is "random enough" and can provide important information about the data, for example, we can use it to estimate the number of 2-itemsets occurring above a certain number of times.

3   Our Contribution

3.1   Time-Space Trade-Offs Revisited

Streaming algorithms have traditionally been mainly concerned with space usage. An algorithm with a superior space usage, for example polylogarithmic, has been considered superior to an algorithm using more space but less computation time. We would like to challenge this view, especially for time complexities that are in the polynomial (rather than polylogarithmic) range. The purpose of a scalable algorithm is to allow the largest possible problem sizes to be handled (in terms of relevant problem parameters). A streaming algorithm may fail either because the processing time is too high, or because it uses more space than what is available. Typically, streaming algorithms should work in space that is small enough to fit in fast cache memory, but there is no real advantage to using only 10% of the cache. Looking at high-end processors over the last 20 years, see for example http://en.wikipedia.org/wiki/Comparison_of_Intel_Processors, reveals that the largest system cache capacity and the number of instructions per second have developed rather similarly (with the doubling time for space being about 25% larger than the doubling time for number of instructions). Assuming that this trend continues, a future processor with x times more processing power will have about $x^{0.8}$ times larger cache. So informally, whenever we have $S = o(T^{0.8})$ for an algorithm using time T and space S the space will not be the asymptotic bottleneck.

3.2 Main Result

In this paper we consider consistent sampling of certain implicitly defined sets, focusing on size-k subsets in a collection of b-sets. The sampling is consistent in the sense that each occurrence of a k-subset satisfying the sampling condition is recorded in the sample.

Theorem 1. *For each integer $k \geq 2$ there is an algorithm computing a consistent, pairwise independent sample of k-subsets from a given b-set in expected time $O(b^{\lceil k/2 \rceil} \log \log b + pb^k)$ and space $O(b^{\lceil k/4 \rceil})$ for a given sampling probability p, such that $1/p = O(b^k)$ and p can be described in one word. An element of the sample is specified in $O(k)$ words.*

Note that for the space complexity we do not consider the size of the computed sample. We will do this when presenting concrete applications of our approach.

For low sampling rates our method, which is based on hash collisions among $k/2$-subsets, is a quadratic improvement in running time compared to the naïve method that iterates through all k-subsets in a given b-set. In addition, we obtain a quadratic improvement in space usage compared to the direct application of the hashing idea. Storing a single $2k$-wise independent hash function suffices to specify a sample, where every pair of k-subsets are sampled independently.

An important consequence of our consistent sampling algorithm is that it can be applied to b-sets revealed one at a time, thus it is well-suited for streaming problems.

4 Our Approach

4.1 Intuition

A naïve consistent sampling approach works as follows: Define a pairwise independent hash function $h : \mathcal{I}^k \to [q]$, for a given b-set T generate all $\binom{b}{k}$ k-subsets $I_k \in T^k$ and sample a subset I_k iff $h(I_k) = 0$. Clearly, to decide which I_k are sampled the running time is $O(b^k)$ and the space is $O(b)$ since the space needed for the description of the hash function for reasonably small sampling probability p is negligible. A natural question is whether a better time complexity is possible.

Our idea is instead of explicitly considering all k-subsets occurring in a given b-set, to hash all elements to a value in $[q]$, $q = \lceil 1/p \rceil$ for a given sampling probability p. We show that the sampling of k-subsets is *pairwise independent* and for many concrete applications this is sufficient to consider the resulting sample "random enough".

The construction of the hash function is at the heart of our algorithm and allows us to exploit several tricks in order to improve the running time. Let us for simplicity assume k is even. Then we sample a given k-subset if the sum (mod q) of the hash values of its first $k/2$ elements equals the sum of the hash values of its last $k/2$ elements modulo q. The simple idea is to sort all $k/2$-subsets according to hash value and then look for collisions. Using a technique similar to the one of Schroeppel and Shamir for the knapsack problem [23], we show how by a clever use of priority queues one can design an algorithm with much better time complexity than the naïve method and quadratic improvement in the space complexity of the sorting approach.

4.2   The Hash Function

Now we explain how we sample a given k-subset. Let $h : \mathcal{I} \to [q]$ be a $2k$-wise independent hash function, $k \geq 2$. It is well-known, see for example [12], that such a function can be described in $O(k)$ words for a reasonable sampling probability, i.e., a sampling probability that can be described in one machine word.

We take a k-subset $(a_1, \ldots, a_{\lfloor k/2 \rfloor}, a_{\lfloor k/2 \rfloor + 1}, \ldots, a_k)$ in the sample iff $(h(a_1) + \cdots + h(a_{\lfloor k/2 \rfloor})) \bmod q = (h(a_{\lfloor k/2 \rfloor + 1}) + \cdots + h(a_k)) \bmod q$. Note that we have assumed a unique representation of subsets as sorted vectors and thus the sampling condition is uniquely defined.

For a given k-subset $I = (a_i, a_{i+1} \ldots, \ldots, a_{i+k-1})$, $i \in \mathcal{I}$, we denote by $h(a_i, a_{i+1} \ldots, \ldots, a_{i+k-1}))$ the value $(h(a_1) + h(a_{i+1}) + \cdots + h(a_{i+k-1})) \bmod q$. We define the random variable X_I to indicate whether a given k-subset $I = (a_1, \ldots, a_k)$ will be considered for sampling:

$$X_I = \begin{cases} 1, & \text{if } h(a_1 \ldots a_{\lfloor k/2 \rfloor}) = h(a_{\lfloor k/2 \rfloor + 1} \ldots a_k), \\ 0, & \text{otherwise} \end{cases}$$

The following lemmas allow us to assume that from our sample we can obtain a reliable estimate with high probability:

Lemma 1. *Let I be a t-subset with $t \leq k$. Then for a given $r \in [q]$, $\mathbf{Pr}[h(I) = r] = 1/q$.*

Proof. Since h is $2k$-wise independent and uniform each of the $t \leq k$ distinct elements is hashed to a value between 0 and $q-1$ uniformly and independently from the remaining $t-1$ elements. Thus, the sum (mod q) of the hash values of I's t elements is equal with probability $1/q$ to r.   □

Lemma 2. *For a given k-subset I, $\mathbf{Pr}[X_I = 1] = 1/q$.*

Proof. Let $I = I_l \cup I_r$ with $|I_l| = \lfloor k/2 \rfloor$ and $|I_r| = \lceil k/2 \rceil$. The hash value of each subset is uniquely defined, h is $2k$-wise independent, and together with the result of Lemma 1 we have $\mathbf{Pr}[h(I_l) = h(I_r) = r] = 1/q^2$ for a particular $r \in [q]$. Thus, we have $\mathbf{Pr}[h(I_l) = h(I_r) = 0 \vee \cdots \vee h(I_l) = h(I_r) = q - 1] = \sum_{i=0}^{q-1} \mathbf{Pr}[h(I_l) = h(I_r) = i] = 1/q$.   □

Lemma 3. *Let I_1 and I_2 be two distinct k-subsets. Then the random variables X_{I_1} and X_{I_2} are independent.*

Proof. We show that $\mathbf{Pr}[X_{I_1} = 1 \wedge X_{I_2} = 1] = \mathbf{Pr}[X_{I_1} = 1]\mathbf{Pr}[X_{I_2} = 1] = 1/q^2$ for arbitrarily chosen k-subsets I_1, I_2. This will imply pairwise independence on the events that two given k-subsets are sampled since for a given k-subset I, $\mathbf{Pr}[X_I = 1] = 1/q$ as shown in Lemma 1.

Let $I_1 = I_{1l} \cup I_{1r}$ and $I_2 = I_{2l} \cup I_{2r}$ with $|I_{il}| = \lfloor k/2 \rfloor$ and $|I_{ir}| = \lceil k/2 \rceil$. Let us assume without loss of generality that $h(I_{1l}) = r_1$ and $h(I_{2l}) = r_2$ for some $r_i \in [q]$. As shown in the previous lemmas for fixed r_1 and r_2, $\mathbf{Pr}[h(I_{1r}) = r_1] = \mathbf{Pr}[h(I_{2r}) = r_2] = 1/q$. Since h is $2k$-wise independent, all elements in $I_{1l} \cup I_{1r} \cup I_{2l} \cup I_{2r}$ are hashed independently of each other. Thus, it is easy to see that the event we hash I_{2r} to r_2 is independent from the event that we have hashed I_{1r} to r_1, thus the statement follows. $\square$

The above lemmas imply that our sampling will be uniform and pairwise independent.

4.3   The Algorithm

A pseudocode description of our algorithm is given in Figure 1. We explain how the algorithm works with a simple example. Assume we want to sample 8-subsets from a b-set $(a_1, \ldots, a_b)$ with $b > 8$. We want to find all 8-subsets $(a_1, \ldots, a_8)$ for which it holds $h(a_1, \ldots, a_4) = h(a_5, \ldots, a_8)$. As discussed, we assume a lexicographic order on the elements in $\mathcal{I}$ and we further assume b-sets are sorted according to this total order. The assumption can be removed by preprocessing and sorting the input. Since $\mathcal{I}$ is discrete, one can assume that each b-set can be sorted by the Han-Thorup algorithm in $O(b\sqrt{\log \log b})$ expected time and space $O(b)$ [16] (for the general case of sampling k-subsets even for $k = 2$ this will not dominate the complexity claimed in Theorem 1). In the following we assume the elements in each b-set are sorted. Recall we have assumed a total order on subsets, and all subsets we consider are sorted according to this total order. We will also consider only sorted subsets for sampling.

We simulate a sorting algorithm in order to find all 4-subsets with equal hash values. Let the set of 2-subsets be H. First, in CONSISTENTSUBSETSAMPLING we generate all $\binom{b}{2}$ 2-subsets and sort them according to their hash value in a circular list L guaranteeing access in expected constant time. We also build a priority queue P containing $\binom{b}{2}$ 4-subsets as follows: For each 2-subset $(a_i, a_j) \in H$ we find the 2-subset $(a_k, a_\ell) \in L$ such that $h(a_i, a_j, a_k, a_\ell)$ is minimized and keep track of the position of (a_k, a_ℓ) in L. Then we successively output all 4-subsets sorted according to their hash value from the priority queue by calling OUTPUT-NEXT. For a given collection of 4-subsets with the same hash value we generate all valid 8-subsets, i.e., we find all combinations yielding lexicographically ordered 8-subsets. Note that the "head" 2-subsets from H are never changed while we only update the "tail" 2-subsets with new 2-subsets from L. During the process we also check whether all elements in the newly created 4-subsets are different.

CONSISTENTSUBSETSAMPLING

Input: b-set $T \subset \mathcal{I}$, a $2k$-wise independent $h : \mathcal{I} \to [q]$
  Let $H = T^{k/4}$ be the $k/4$-subsets occurring in T.
  Sort all $k/4$-subsets from H in a circular list L according to hash value.
  Build a priority queue P with $k/2$-subsets $I = I_H \cup I_L$ according to hash value, for
  $I_H \in H$, $I_L \in L$.
  for $i \in [q]$ **do**
    $T_i^{k/2} = \text{OutputNext}(P, L, i)$
    Generate all k-subsets from $T_i^{k/2}$ satisfying the sampling condition (and consisting of k different elements).

OUTPUTNEXT

Input: a circular list L, a priority queue P of $k/2$-subsets $I = (I_H \cup I_L)$ compared by
  hash value $h(I)$, $i \in \mathbb{N}$
  while there is $k/2$-subset with a hash value i **do**
    Output the next $k/2$-subset $I = (I_H \cup I_L)$ from P.
    if $\text{cnt}(I_H) < L.\text{length}$ **then**
      Replace I by $I_H \cup I'_L$ in P where I'_L is the $k/4$-subset following I_L in L.
      Update the hash value of $I_H \cup I'_L$ and restore the PQ invariant.
      $\text{cnt}(I_H)$++.
    else
      Remove $I = (I_H \cup I_L)$ from P and restore the PQ invariant.

Fig. 1. A high-level pseudocode description of the algorithm. For simplicity we assume that k is a multiple of 4. The letter H stands for "head", these are the $k/4$-subsets that will constitute the first half of $k/2$-subsets in P. We will always update the second half with a $k/4$-subset from L.

In OUTPUTNEXT we simulate a heapsort-like algorithm for 4-subsets. We do not keep explicitly all 4-subsets in P but at most $\binom{b}{2}$ 4-subsets at a time. Once we output a given 4-subset (a_i, a_j, a_k, a_ℓ) from P, we replace it with $(a_i, a_j, a'_k, a'_\ell)$ where (a'_k, a'_ℓ) is the 2-subset in L following (a_k, a_ℓ). We also keep track whether we have not already traversed L for each 2-subset in H. If this is the case, we remove the 4-subset (a_i, a_j, a_k, a_ℓ) from P and the number of recorder entries in P is decreased by 1. At the end we update P and maintain the priority queue invariant.

In the following lemmas we will prove the correctness of the algorithm for general k and will analyze its running time. This will yield our main Theorem 1.

Lemma 4. *For $k \geq 4$ with $k \bmod 4 = 0$ CONSISTENT SUBSET SAMPLING outputs the $k/2$-subsets from a given b-set in sorted order according to their hash value in expected time $O(b^{k/2} \log \log b)$ and space $O(b^{k/4})$.*

Proof. Let T be the given b-subset and $\mathcal{S} = I_1, \ldots, I_{\binom{b}{k/2}}$ be the $k/2$-subsets occurring in T sorted according to hash value. For correctness we first show that the following invariant holds: After the j smallest $k/2$-subsets have been output from P, P contains the $(j+1)$th smallest $k/2$-subset in S, ties resolved arbitrarily. For $j = 1$ the statement holds by construction. Assume now that it holds for some $j \geq 1$ and we output the jth smallest $k/2$-subset $I = I_H \cup I_L$ for $k/4$-subsets I_H and I_L. We then replace it by $I' = I_H \cup I'_L$ where I'_L is the $k/4$-subset in L following I_L, or, if L has been already traversed, remove I from P. If P contains the $(j + 1)$th smallest $k/2$-subset, then the invariant holds. Otherwise, we show that it must be that the $(j + 1)$th smallest $k/2$-subset is I'. Since L is sorted and we traverse it in increasing order, the $k/2$-subsets $I_H \cup I_L$ with a fixed head I_H that remain to be considered have all a bigger hash value than I. The same reasoning applies to all other $k/2$-subsets in P, and since no of them is the $(j + 1)$-th smallest $k/2$-subset, the only possibility is that indeed I' is the $(j + 1)$-th smallest $k/2$-subset.

In L we need to explicitly store $O(b^{k/4})$ subsets. Clearly, we can assume that we access the elements in L in constant time. The time and space complexity depend on how we implement the priority queue P. We observe that for a hash function range in $b^{O(1)}$ the keys on which we compare the 2-subsets are from a universe of size $b^{O(1)}$. Thus, we can implement P as a y-fast trie [24] in $O(b^{k/4})$ space supporting updates in $O(\log \log b)$ time. This yields the claimed bounds.   □

Note however, that the number of $k/2$-subsets with the same hash value might be $\omega(b^{k/4})$. We next guarantee that the worst case space usage is $O(b^{k/4})$.

Lemma 5. *For a given $r \in [q]$, $k \bmod 4 = 0$, and sampling probability $p \in (0, 1]$, we generate all k-subsets from a set of $k/2$-subsets with a hash value r that satisfy the sampling condition in expected time $O(p^2 b^k)$ and space $O(b^{k/4})$.*

Proof. We use the following implicit representation of $k/2$-subsets with the same hash value. For a given $k/4$-subset I_P the $k/4$-subsets I_L in L occurring in $k/2$-subsets $I_P \cup I_L$ with the same hash value are contained in a subsequence of L. Therefore, instead of explicitly storing all $k/2$-subsets, for each $k/4$-subset I_P we store two indices i and j indicating that $h(I_P \cup L[k]) = r$ for $i \leq k \leq j$. Clearly, this guarantees a space usage of $O(b^{k/4})$.

We expect $pb^{k/2}$ $k/2$-subsets to have hash value r, thus the number of k-subsets that will satisfy the sampling condition is $O(p^2 b^k)$.   □

The above lemmas prove Theorem 1 for the case $k \bmod 4 = 0$. One generalizes to arbitrary $k \geq 4$ in the following way:

For even k with $k \bmod 4 = 2$, meaning that $k/2$ is odd, it is easy to see that we need a circular list with all $\lfloor k/4 \rfloor$-subsets but the priority queue will contain $\binom{b}{\lceil k/4 \rceil}$ pairs of $k/2$-subsets (which are concatenations of $\lceil k/4 \rceil$-subsets and $\lfloor k/4 \rfloor$-subsets). For odd k we want to sample all k-subsets for which the sum of the hash values of the first $\lfloor k/2 \rfloor$ elements equals the sum of the hash values of the last $\lceil k/2 \rceil$ elements. We can run two copies of OUTPUTNEXT in parallel, one will output the $\lceil k/2 \rceil$-subsets with hash value r and the other one the $\lfloor k/2 \rfloor$-subsets with hash value r for all $r \in [q]$. Then we can generate all k-subsets satisfying the sampling condition as outlined in Lemma 5 with the only difference that we will combine $\lceil k/2 \rceil$-subsets with $\lfloor k/2 \rfloor$-subsets output by each copy of OUTPUTNEXT. Clearly, the space complexity is bounded by $O(b^{\lceil k/4 \rceil})$

and the expected running time is $O(b^{\lceil k/2 \rceil} + pb^k)$. This completes the proof of Theorem 1.

A Time-Space Trade-Off. A better space complexity can be achieved by increasing the running time. The following theorem generalizes our result.

Theorem 2. *For any $k \geq 2$ and $\ell \leq k/2$ we can compute a consistent, pairwise independent sample of k-subsets from a given b-set in expected time $O(b^{\lceil k/2+\ell \rceil} \log \log b) + pb^k)$ and space $O(b^{\lceil (k-2\ell)/4 \rceil} + b)$ for a given sampling probability p, such that $1/p = O(b^k)$ and p can be described in one word.*

Proof. We need space $O(b)$ to store the b-set. Assume that we iterate over 2ℓ-subsets $(a_1, \ldots, a_{2\ell})$, without storing them and their hash values. We assume that we have fixed ℓ elements among the first $\lfloor k/2 \rfloor$ elements, and ℓ elements among the last $\lceil k/2 \rceil$ ones. We compute the value $h^{\ell} = (h(a_1) + \cdots + h(a_{\ell}) - h(a_{\lfloor k/2 \rfloor+1}) - \cdots - h(a_{\lfloor k/2 \rfloor+\ell}))$ mod q. We now want to determine all $(k - 2\ell)$-subsets for which the sum of the hash values of the first $\lfloor k/2 \rfloor - \ell$ elements equals the sum of the hash values of the last $\lceil k/2 \rceil - \ell$ elements minus the value h^{ℓ}. Essentially, we can sort $(k-2\ell)$-subsets according to their hash value in the same way as before and the only difference is that we subtract h^{ℓ} from the hash value of the last $\lceil k/2 \rceil - \ell$ elements. Thus, we can use two priority queues, where in the second one we have subtracted h^{ℓ} from the hash value of each $(\lceil k/2 \rceil - \ell)$-subset, output the minima and look up for collisions. Disregarding the space for storing the b-set, the outlined modification requires time $O(b^{\lceil k/2+\ell \rceil} \log \log b)$ and space $O(b^{\lceil (k-2\ell)/4 \rceil})$ to process a given b-set. □

Discussion. Let us consider the scalability of our approach to larger values of b, assuming that the time is not dominated by iterating through the sample. If we are given a processor that is x times more powerful, this will allow us to increase the value of b by a factor $x^{1/\lceil k/2 \rceil}$. This will work because the space usage of our approach will only rise by a factor $\sqrt{x}$, and, as already discussed, we expect a factor $x^{0.8}$ more space to be available. An algorithm using space $b^{\lceil k/2 \rceil}$ would likely be space-bounded rather than time-bounded, and thus only be able to increase b by a factor of $x^{0.8/\lceil k/2 \rceil}$. At the other end of the spectrum an algorithm using time x^k and constant space would only be able to increase b by a factor $x^{1/k}$.

5   Applications of Consistent Subset Sampling

In this section we discuss several algorithmic applications of Consistent Subset Sampling for well-studied data mining problems. The reader is referred to the full version of the paper[1] for more details.

A fundamental problem in data mining is the problem of frequent itemsets mining in transactional data streams. For example, transactions correspond to

---

[1] http://arxiv.org/pdf/1404.4693.pdf

market baskets and we want to detects sets of goods that are frequently bought together, see [15] for an overview. A low frequency threshold may lead to an explosion of the number of frequent itemsets, therefore a good prediction of their number is needed [14]. Known approaches for the problem are all based on some heuristics [2,19] and the worst case running time is exponential. By associating transactions with b-sets, Consistent Subset Sampling naturally applies to the problem. As a result, we obtain the first algorithm with rigorously understood complexity and approximation guarantees. The next theorem is our main result:

Theorem 3. *Let $\mathcal{T}$ be a stream of m transactions of size at most b over a set of n items and f and z be the number of frequent and different k-itemsets, $k \geq 2$, in $\mathcal{T}$, respectively. For any $\alpha, \varepsilon, \delta > 0$ there exists a randomized algorithm running in expected time $O(mb^{\lceil k/2 \rceil} \log \log b + \frac{\log m \log \delta^{-1}}{\alpha \varepsilon^2})$ and space $O(b^{\lceil k/4 \rceil} + \frac{\log m \log \delta^{-1}}{\alpha \varepsilon^2})$ in one pass over $\mathcal{T}$ returning a value $\tilde{f}$ such that*

- *if $f/z \geq \alpha$, then $\tilde{f}$ is an (ε, δ)-approximation of f.*
- *otherwise, if $f/z < \alpha$, then $\tilde{f} \leq (1+\varepsilon)f$ with probability at least $1 - \delta$.*

Extending a recent technique by Campagna and the authors [9], we show how to use Consistent Subset Sampling to parallelize frequent items mining algorithm like [10,11] when applied to transactional data streams. Here, instead of estimating the number of frequent itemsets, we show how to distribute the computation such that we achieve good load balancing among different processors.

A second application is in the area of graph mining where a graph is provided as a stream of edges. The problem of estimating the number of fixed-size subgraphs in *incidence list streams*, i.e., a stream of edges where all edges incident to a vertex are provided one after another, has become very popular in the last decade [3,4,7,8,20,20]. By associating b-sets with the set of a vertex neighbors, we design new algorithms for the estimation of the number of k-cliques in incidence list streams for bounded degree graphs. Also, we present the first algorithm for the estimation of the number of (i^+, j)-bipartite cliques, i.e., bipartite cliques with j vertices on the right-hand side and *at least* i vertices on the left hand side. We argue that this is a problem with important real-life applications and show that a straightforward applications of Consistent Subset Sampling yields the following result:

Theorem 4. *Let $G = (V, E)$ be a graph with n vertices, m edges and bounded degree Δ revealed as a stream of incidence lists. Let further $K_{i^+,j}$ be the number of (i^+, j)-bicliques in G and A_j the number of j-adjacencies in G for $i \geq 1, j \geq 2$. For any $\gamma, \varepsilon, \delta \in (0, 1]$ there exits a randomized algorithm running in expected time $O(n\Delta^{\lceil j/2 \rceil} \log \log \Delta + \frac{\log n \log \delta^{-1}}{\gamma \varepsilon^2})$ and space $O(\Delta^{\lceil j/4 \rceil} + \frac{\log n \log \delta^{-1}}{\gamma \varepsilon^2})$ in one pass over the graph returning a value $\tilde{K}_{i^+,j}$ such that*

- *if $K_{i^+,j}/A_j \geq \gamma$, $\tilde{K}_{i^+,j}$ is an (ε, δ)-approximation of $K_{i^+,j}$.*
- *otherwise, if $K_{i^+,j}/A_j < \gamma$, $\tilde{K}_{i^+,j} \leq (1 + \varepsilon)K_{i^+,j}$ with probability at least $1 - \delta$.*

6   Conclusions

Finally, we make a few remarks about possible improvements in the running time of our consistent sampling technique. As one can see, the algorithmic core of our approach is closely related to the d-SUM problem where one is given an array of n integers and the question is to find d integers that sum up to 0. The best known randomized algorithm for 3-SUM runs in time $O(n^2 (\log \log n)^2 / \log^2 n)[1]$, thus it is difficult to hope to design an algorithm enumerating all 3-subsets satisfying the sampling condition much faster than in $O(b^2)$ steps. Moreover, Pătraşcu and Williams [22] showed that solving d-SUM in time $n^{o(d)}$ would imply an algorithm for the 3-SAT problem running in time $O(2^{o(n)})$ contradicting the *exponential time hypothesis* [17]. It is even an open problem whether one can solve d-SUM in time $O(n^{\lceil d/2 \rceil - \alpha})$ for $d \geq 3$ and some constant $\alpha > 0$ [25].

In a recent work Dinur et al. [13] presented a new "dissection" technique for achieving a better time-space trade-off for the computational complexity of various problems. Using the approach from [13], we can slightly improve the results from Theorem 2. However, the details are beyond the scope of the present paper.

References

1. Baran, I., Demaine, E.D., Pătraşcu, M.: Subquadratic Algorithms for 3SUM. Algorithmica 50(4), 584–596 (2008)
2. Boley, M., Grosskreutz, H.: A Randomized Approach for Approximating the Number of Frequent Sets. In: ICDM 2008, pp. 43–52 (2008)
3. Becchetti, L., Boldi, P., Castillo, C., Gionis, A.: Efficient semi-streaming algorithms for local triangle counting in massive graphs. In: KDD 2008, pp. 16–24 (2008)
4. Bordino, I., Donato, D., Gionis, A., Leonardi, S.: Mining Large Networks with Subgraph Counting. In: ICDM 2008, pp. 737–742 (2008)
5. Broder, A.Z., Charikar, M., Frieze, A.M., Mitzenmacher, M.: Min-Wise Independent Permutations. J. Comput. Syst. Sci. 60(3), 630–659 (2000)
6. Broder, A.Z., Glassman, S.C., Manasse, M.S., Zweig, G.: Syntactic Clustering of the Web. Computer Networks 29(8-13), 1157–1166 (1997)
7. Buriol, L.S., Frahling, G., Leonardi, S., Marchetti-Spaccamela, A., Sohler, C.: Counting triangles in data streams. In: PODS 2006, pp. 253–262 (2006)
8. Buriol, L.S., Frahling, G., Leonardi, S., Sohler, C.: Estimating Clustering Indexes in Data Streams. In: Arge, L., Hoffmann, M., Welzl, E. (eds.) ESA 2007. LNCS, vol. 4698, pp. 618–632. Springer, Heidelberg (2007)
9. Campagna, A., Kutzkov, K., Pagh, R.: On Parallelizing Matrix Multiplication by the Column-Row Method. In: ALENEX 2013, pp. 122–132 (2013)
10. Charikar, M., Chen, K., Farach-Colton, M.: Finding frequent items in data streams. Theor. Comput. Sci. 312(1), 3–15 (2004)
11. Cormode, G., Muthukrishnan, S.: An improved data stream summary: The count-min sketch and its applications. J. Algorithms 55(1), 58–75 (2005)
12. Dietzfelbinger, M., Gil, J., Matias, Y., Pippenger, N.: Polynomial Hash Functions Are Reliable (Extended Abstract). In: Kuich, W. (ed.) ICALP 1992. LNCS, vol. 623, pp. 235–246. Springer, Heidelberg (1992)

13. Dinur, I., Dunkelman, O., Keller, N., Shamir, A.: Efficient Dissection of Composite Problems, with Applications to Cryptanalysis, Knapsacks, and Combinatorial Search Problems. In: Safavi-Naini, R., Canetti, R. (eds.) CRYPTO 2012. LNCS, vol. 7417, pp. 719–740. Springer, Heidelberg (2012)
14. Geerts, F., Goethals, B., Van den Bussche, J.: Tight upper bounds on the number of candidate patterns. ACM Trans. Database Syst. 30(2), 333–363 (2005)
15. Han, J., Kamber, M.: Data Mining: Concepts and Techniques. Morgan Kaufmann (2000)
16. Han, Y., Thorup, M.: Integer Sorting in $O(n\sqrt{\log\log n})$ Expected Time and Linear Space. In: FOCS 2002, pp. 135–144 (2002)
17. Impagliazzo, R., Paturi, R., Zane, F.: Which Problems Have Strongly Exponential Complexity? J. Comput. Syst. Sci. 63(4), 512–530 (2001)
18. Indyk, P.: A Small Approximately Min-Wise Independent Family of Hash Functions. J. Algorithms 38(1), 84–90 (2001)
19. Jin, R., McCallen, S., Breitbart, Y., Fuhry, D., Wang, D.: Estimating the number of frequent itemsets in a large database. In: EDBT, pp. 505–516 (2009)
20. Kane, D.M., Mehlhorn, K., Sauerwald, T., Sun, H.: Counting Arbitrary Subgraphs in Data Streams. In: Czumaj, A., Mehlhorn, K., Pitts, A., Wattenhofer, R. (eds.) ICALP 2012, Part II. LNCS, vol. 7392, pp. 598–609. Springer, Heidelberg (2012)
21. Kane, D.M., Nelson, J., Woodruff, D.P.: An optimal algorithm for the distinct elements problem. In: PODS 2010, pp. 41–52 (2010)
22. Pătraşcu, M., Williams, R.: On the Possibility of Faster SAT Algorithms. In: SODA 2010, pp. 1065–1075 (2010)
23. Schroeppel, R., Shamir, A.: A $T = O(2^{n/2}), S = O(2^{n/4})$ Algorithm for Certain NP-Complete Problems. SIAM J. Comput. 10(3), 456–464 (1981)
24. Willard, D.E.: Log-Logarithmic Worst-Case Range Queries are Possible in Space $\Theta(N)$. Inf. Process. Lett. 17(2), 81–84 (1983)
25. Woeginger, G.J.: Space and Time Complexity of Exact Algorithms: Some Open Problems (Invited Talk). In: Downey, R.G., Fellows, M.R., Dehne, F. (eds.) IWPEC 2004. LNCS, vol. 3162, pp. 281–290. Springer, Heidelberg (2004)

Triangle Counting in Dynamic Graph Streams*

Konstantin Kutzkov and Rasmus Pagh

IT University of Copenhagen, Denmark

Abstract. Estimating the number of triangles in graph streams using a limited amount of memory has become a popular topic in the last decade. Different variations of the problem have been studied, depending on whether the graph edges are provided in an arbitrary order or as incidence lists. However, with a few exceptions, the algorithms have considered *insert-only* streams. We present a new algorithm estimating the number of triangles in *dynamic* graph streams where edges can be both inserted and deleted. We show that our algorithm achieves better time and space complexity than previous solutions for various graph classes, for example sparse graphs with a relatively small number of triangles. Also, for graphs with constant transitivity coefficient, a common situation in real graphs, this is the first algorithm achieving constant processing time per edge. The result is achieved by a novel approach combining sampling of vertex triples and sparsification of the input graph.

1 Introduction

Many relationships between real life objects can be abstractly represented as graphs. The discovery of certain structural properties in a graph, which abstractly describes a given real-life problem, can often provide important insights into the nature of the original problem. The number of triangles, and the closely related clustering and transitivity coefficients, have proved to be an important measure used in applications ranging from social network analysis and spam detection to motif detection in protein interaction networks. We refer to [23] for a detailed discussion on the applications of triangle counting.

The best known algorithm for triangle counting in the RAM model runs in time $O(m^{\frac{2\omega}{\omega+1}})$ [4] where ω is the matrix multiplication exponent, the best known bound is $\omega = 2.3727$ [24]. However, this algorithm is mainly of theoretical importance since exact fast matrix multiplication algorithms do not admit an efficient implementation for input matrices of reasonable size.

The last decade has witnessed a rapid growth of available data. This has led to a shift in attitudes in algorithmic research and solutions storing the whole input in main memory are not any more considered a feasible choice for many real-life problems. Classical algorithms have been adjusted in order to cope with the new requirements and many new techniques have been developed [17].

---

* This work is supported by the Danish National Research Foundation under the Sapere Aude program.

Approximate Triangle Counting in Streamed Graphs. For many applications one is satisfied with a good approximation of the number of triangles instead of their exact number, thus researchers have designed randomized approximation algorithms returning with high probability a precise estimate using only small amount of main memory. Two models of streamed graphs have been considered. In the *incidence list stream* model the edges incident to each vertex arrive consecutively and in the *adjacency stream* model edges arrive in arbitrary order. Also, it has been distinguished between algorithms using only a single pass over the input, and algorithms assuming that the input graph can be persistently stored on a secondary device and multiple passes are allowed. The one-pass algorithms with the best known space complexity and constant processing time per edge, both in the incidence list stream and adjacency stream model, are due to Buriol et al. [8], and when several passes are allowed – by Kolountzakis et al. [14]. For an overview of results and developed techniques we refer to [23].

Dynamic graph streams have a wider range of applications. Consider for example a social network like Facebook where one is allowed to befriend and "unfriend" other members, or join and leave groups of interest. Estimating the number of triangles in a network is a main building block in algorithms for the detection of emerging communities [7], and thus it is required that triangle counting algorithms can also handle edge deletions. The problem of designing triangle counting algorithms for dynamic streams matching the space and time complexity of algorithms for insert-only streams has been presented as an open question in the 2006 IITK Workshop on Algorithms for Data Streams [15]. The best known algorithms for insert-only streams work by sampling a non-empty subgraph on three vertices from the stream (e.g. an edge (u, v) and a vertex w). Then one checks whether the arriving edges will complete the sampled subgraph to a triangle (we look for (u, w) and (v, w)). The approach does not work for dynamic streams because an edge in the sampled subgraph might be deleted later. Proposed solutions [1,16] have explored different ideas. These approaches, however, only partially resolve the open problem from [15] because of high processing time per edge update, see Section 3 for more details.

Our Contribution. In this work we propose a method to adjust sampling to work in dynamic streams and show that for graphs with constant transitivity coefficient, a ubiquitous assumption for real-life graphs, we can achieve constant processing time per edge. At a very high level, the main technical contribution of the present work can be summarized as follows.

For dynamic graph streams sampling-based approaches fail because we don't know how many of the sampled subgraphs will survive after edges have been deleted. On the other hand, graph sparsification approaches [19,22,23] can handle edge deletions but the theoretical guarantees on the complexity of the algorithms depend on specific properties of the underlying graph, e.g., the maximum number of triangles an edge is part of. The main contribution in the present work is a novel technique for sampling 2-paths *after* the stream has been processed. It is based on the combination of standard 2-path sampling with graph sparsification. The main technical challenge is to show that sampling at random a 2-path in a sparsified graph is (almost) equivalent to sampling at random a 2-path in the original graph. In the course of the analysis, we also obtain combinatorial results about general graphs that might be of independent interest.

2   Preliminaries

Notation. A simple undirected graph without loops is denoted as $G = (V, E)$ with $V = \{1, 2, \ldots, n\}$ being a set of vertices and E a set of edges. The edges are provided as a stream of insertions and deletions in arbitrary order. We assume the strict turnstile model where each edge can be deleted only after being inserted. We assume that n is known in advance[1] and that the number of edges cannot exceed m. For an edge connecting the vertices u and v we write (u, v) and u and v are the *endpoints* of the edge (u, v). Vertex u is *neighbor* of v and vice versa and $N(u)$ is the set of u's neighbors. We say that edge (u, v) is *isolated* if $|N(u)| = |N(v)| = 1$. We consider only edges (u, v) with $u < v$. A *2-path centered at* v, (u, v, w), consists of the edges (u, v) and (v, w). A k-clique in G is a subgraph of G on k vertices $v_1, \ldots, v_k$ such that $(v_i, v_j) \in E$ for all $1 \le i < j \le k$. A 3-clique on u, v, w is called a *triangle* on u, v, w, and is denoted as $\langle u, v, w \rangle$. We denote by $P_2(v)$ the number of 2-paths centered at a vertex v, and $P_2(G) = \sum_{v \in V} P_2(v)$ and $T_3(G)$ the number of 2-paths and number of triangles in G, respectively. We will omit G when clear from the context.

We say that two 2-paths are *independent* if they have at most one common vertex. The *transitivity coefficient* of G is

$$\alpha(G) = \frac{3T_3}{\sum_{v \in V} \binom{d_v}{2}} = \frac{3T_3}{P_2},$$

i.e., the ratio of 2-paths in G contained in a triangle to all 2-paths in G. When clear from the context, we will omit G.

Hashing. A family $\mathcal{F}$ of functions from U to a finite set S is k-*wise independent* if for a function $f : U \to S$ chosen uniformly at random from $\mathcal{F}$ it holds

$$\mathbf{Pr}[f(u_1) = c_1 \wedge f(u_2) = c_2 \wedge \cdots \wedge f(u_k) = c_k] = 1/s^k$$

for $s = |S|$, distinct $u_i \in U$ and any $c_i \in S$ and $k \in \mathbb{N}$. We will call a function chosen uniformly at random from a k-wise independent family k-*wise independent function* and a function $f : U \to S$ *fully random* if f is $|U|$-wise independent. We will say that a function $f : U \to S$ *behaves like a fully random function* if for any set of input from U, with high probability f has the same probability distribution as a fully random function.

We will say that an algorithm returns an (ε, δ)-*approximation* of some quantity q if it returns a value $\tilde{q}$ such that $(1 - \varepsilon)q \le \tilde{q} \le (1 + \varepsilon)q$ with probability at least $1 - \delta$ for every $0 < \varepsilon, \delta < 1$.

---

[1] Our results hold when the n vertices come from some arbitrary universe U and are known in advance. We omit this generalization due to lack of space.

3 The New Approach

The following theorem is our main result.

Theorem 1. *Let $G = (V, E)$ be a graph given as a stream of edge insertions and deletions with no isolated edges and vertices, $V = \{1, 2, \ldots, n\}$ and $|E| \leq m$. Let P_2, T_3 and α be the number of 2-paths, number of triangles and the transitivity coefficient of G, respectively. Let $\varepsilon, \delta \in (0, 1)$ be user defined and $b = \max(n, P_2/n)$. Assuming fully random hash functions, there exists a one-pass algorithm running in expected space $O(\frac{m}{\sqrt{b}\varepsilon^3\alpha} \log \frac{1}{\delta})$ and $O(\frac{1}{\varepsilon^2\alpha} \log \frac{1}{\delta})$ processing time per edge. After processing the stream, an (ε, δ)-approximation of T_3 can be computed in expected time $O(\frac{\log n}{\varepsilon^2\alpha} \log \frac{1}{\delta})$ and worst case time $O(\frac{\log^2 n}{\varepsilon^2\alpha} \log \frac{1}{\delta})$ with high probability.*

(For simplicity, we assume that there are no isolated edges in G. More generally, the result holds by replacing n with n_C, where n_C is the number of vertices in connected components with at least two edges. We recall again that we assume m and n can be described in $O(1)$ words.)

Table 1. Overview of time and space bounds. It holds $b = \max(n, P_2/n)$.

	Space	Update time
Ahn et al.[1]	$O(\frac{mn}{\varepsilon^2 T_3} 1 \log \frac{1}{\delta})$	$O(n \log n)$
Manjunath et al. [16]	$O(\frac{m^3}{\varepsilon^2 T_3^2} \log \frac{1}{\delta})$	$O(\frac{m^3}{\varepsilon^2 T_3^2} \log \frac{1}{\delta})$
This work	$O(\frac{m}{\sqrt{b}\varepsilon^3\alpha} \log \frac{1}{\delta})$	$O(\frac{1}{\varepsilon^2\alpha} \log \frac{1}{\delta})$

Table 2. Comparison of the theoretical guarantees for the per edge processing time for varying z

	$n \log n$	m^3/T_3^2
$z < 1/2$	$T_3 = \omega(C^2/(n^{2z} \log n))$	$T_3 = o(Cn^{2-z})$
$1/2 < z < 1$	$T_3 = \omega(C^2/(n \log n))$	$T_3 = o(Cn^{3-3z})$
$z > 1$	$T_3 = \omega(C^2/(n \log n))$	$T_3 = o(C)$

Before presenting the algorithm, let us compare the above to the bounds in [1,16]. The algorithm in [1] estimates T_3 by applying ℓ_0 sampling [11] to non-empty subgraphs on 3 vertices. There are $O(mn)$ such subgraphs, thus $O(\frac{mn}{\varepsilon^2 T_3} \log \frac{1}{\delta})$ samples are needed for an (ε, δ)-approximation. However, each edge insertion or deletion results in the update of $n - 2$ non-empty subgraphs on 3 vertices. Using the ℓ_0 sampling algorithm from [12], this results in processing time of $O(n \log n)$ per edge. The algorithm by Manjunath et al. [16] estimates the number of triangles (and more generally of cycles of fixed length) in streamed graphs by computing complex valued sketches of the stream. Each of them yields an unbiased estimator of T_3. The average of $O(\frac{m^3}{\varepsilon^2 T_3^2} \log \frac{1}{\delta})$ estimators is an (ε, δ)-approximation of T_3. However, each new edge insertion or deletion has to update all estimators, resulting in update time of $O(\frac{m^3}{\varepsilon^2 T_3^2} \log \frac{1}{\delta})$. The algorithm was generalized to counting arbitrary subgraphs of fixed size in [13].

The time and space bounds are summarized in Table 1. Comparing our space complexity to the bounds in [1,16], we see that for several graph classes our algorithm is more time and space efficient. (We ignore ε and δ and logarithmic factors in n for the space complexity.) For d-regular graphs the processing time per edge is better than $O(n \log n)$ for $T_3 = \omega(d^2/\log n)$, and better than $O(m^3/T_2^3)$ for $T_3 = o(n^2 d)$. Our space bound is better than $O(mn/T_3)$ when $d = o(n^{1/4})$, and better than $O(m^3/T_3^2)$ for $T_3 = o(\max(n^{3/2}, nd))$. Most real-life graphs exhibit a skewed degree distribution adhering to some form of power law, see for example [2]. Assume vertices are sorted according to their degree in decreasing order such that the ith vertex has degree C/i^z for some $C \leq n$, and constant $z > 0$, i.e., we have Zipfian distribution with parameter z. It holds $\sum_{i=1}^{n} i^{-z} = O(n^{1-z})$ for $z < 1$ and $\sum_{i=1}^{n} i^{-z} = O(1)$ for $z > 1$. Table 2 summarizes for which values of T_3 our algorithm achieves faster processing time than [1,16], and Table 3 – for which values of C our algorithm is more space-efficient than [1], and for which values of T_3 – more space-efficient than [16].

However, the above values are for arbitrary graphs adhering to a certain degree distribution. We consider the main advantage of the new algorithm to be that it achieves constant processing time per edge for graphs with constant transitivity coefficient. This is a common assumption for real-life networks, see for instance [3,8]. Note that fast update is essential for real life applications. Consider for example the Facebook graph. In May 2011, for less than eight years existence, there were about 69 billion friendship links [6]. This means an average of above 300 new links per second, without counting deletions and peak hours.

In the full version of the paper [2] we compare the theoretical guarantees for several real life graphs. While such a comparison is far from being a rigorous experimental evaluation, it clearly indicates that the processing time per edge in [1,16] is prohibitively large and the assumption that the transitivity coefficient is constant is justified. Also, for graphs with a relatively small number of triangles our algorithm is much more space-efficient.

Table 3. Comparison of the theoretical guarantees for the space usage for varying z

	mn/T_3	m^3/T_3^2
$z < 1/2$	$C = o(n^{1/4+z})$	$T_3 = o(\max(n^{3/2}, Cn^{1-z}))$
$1/2 < z < 1$	$C = o(n^{3/4})$	$T_3 = o(n^{5/2-2z})$
$z > 1$	$C = o(n^{3/4})$	$T_3 = o(n^{1/2})$

3.1   The Main Idea

The main idea behind our algorithm is to design of a new sampling technique for dynamic graph streams. It exploits a combination of the algorithms by Buriol et al. [8] for the incidence stream model, and the Doulion algorithm [22] and its improvement [19]. Let us briefly describe the approaches.

---

[2] http://arxiv.org/pdf/1404.4696.pdf

The Buriol et al. Algorithm for Incidence List Streams. Assume we know the total number of 2-paths in G. One chooses at random one of them, say (u, v, w), and checks whether the edge (u, w) appears later in the stream. For a triangle $\langle u, v, w \rangle$ the three 2-paths (u, v, w), (w, u, v), (v, w, u) appear in the incidence list stream, thus the probability that we sample a triangle is exactly α. One chooses independently at random K 2-paths and using standard techniques shows that for $K = O(\frac{1}{\varepsilon^2 \alpha} \log \frac{1}{\delta})$ we compute an (ε, δ)-approximation of $\alpha(G)$. One can get rid of the assumption that the number of 2-paths is known in advance by running $O(\log n)$ copies of the algorithm in parallel, each guessing the right value. The reader is referred to the original work for more details. For incidence streams, the number of 2-paths in G can be computed exactly by updating a single counter, thus $\tilde{T}_3 = \tilde{\alpha} P_2$ is an (ε, δ)-approximation of T_3.

Doulion and Monochromatic Sampling. The Doulion algorithm [22] is a simple and intuitive sparsification approach. Each edge is sampled independently with probability p and added to a sparsified graph G_S. We expect pm edges to be sampled and a triangle survives in G_S with probability p^3, thus multiplying the number of triangles in G_S by $1/p^3$ we obtain an estimate of T_3. The algorithm was improved in [19] by using *monochromatic sampling*. Instead of throwing a biased coin for each edge, we uniformly at random color each vertex with one of $1/p$ colors. Then we keep an edge in the sparsified graph iff its endpoints have the same color. A triangle survives in G_S with probability p^2. It is shown that for a fully random coloring the variance of the estimator is better than in Doulion. However, in both algorithms it depends on the maximum number of triangles an edge is part of, and one might need constant sampling probability in order to obtain an (ε, δ)-approximation on T_3. The algorithm can be applied to dynamic streams because one counts the number of triangles in the sparsified graph after all edges have been processed. However, it can be expensive to obtain an estimate since the exact number of triangles in G_S is required.

Combining the above Approaches. The basic idea behind the new algorithm is to use the estimator of Buriol et al. for the incidence stream model: (i) estimate the transitivity coefficient $\alpha(G)$ by choosing a sufficiently large number of 2-paths at random and check which of them are part of a triangle, and (ii) estimate the number of 2-paths P_2 in the graph. We first observe that estimating P_2 in dynamic graph streams can be reduced to second moment estimation of streams of items in the turnstile model, see e.g. [21]. For (i), we will estimate $\alpha(G)$ by adjusting the monochromatic sampling approach. Its main advantage compared to the sampling of edges separately is that if we have sampled the 2-path (u, v, w), then we must also have sampled the edge (u, w), if existent. So, the idea is to use monochromatic sampling and then in the sparsified graph to pick up at random a 2-path and check whether it is part of a triangle. Instead of random coloring of the vertices, we will use a suitably defined hash function and we will choose a sampling probability guaranteeing that for a graph with no isolated edges (or rather a small number of isolated edges) the sparsified graph will contain a sufficiently big number of 2-paths. A 2-path in the sparsified graph picked up at random, will then be used to estimate $\alpha(G)$. Thus, unlike in [8], we sample *after* the stream has been processed and this allows to handle edge deletions. The main technical obstacles are to analyze the required sampling probability p and to show that this sampling approach indeed provides an unbiased estimator of $\alpha(G)$. We will obtain bounds on p and show that even if the estimator might be biased, the bias can be made arbitrarily small and one can still achieve an (ε, δ)-approximation of $\alpha(G)$. Also, we present an implementation for storing a sparsified graph G_S such that each edge is added or deleted in constant time and a random 2-path in G_S, if existent, can be picked up without explicitly considering all 2-paths in G_S.

3.2   The Algorithm

Pseudocode description of the algorithm is given in Figure 1. We assume that the graph is given as a stream $\mathcal{S}$ of pairs $((u,v), \$)$, where $(u,v) \in E$ and $\$ \in \{+, -\}$ with the obvious meaning that the edge (u,v) is inserted or deleted from G. In ESTIMATENUMBEROFTWOPATHS each incoming pair $((u,v), \$)$ is treated as the insertion, respectively deletion, of two items u and v, and these update a second moment estimator SME, working as a blackbox algorithm. We refer to the proof of Lemma 1 for more details. In SPARSIFYGRAPH we assume access to a fully random coloring hash function $f : V \to C$. Each edge (u,v) is inserted/deleted to/from a sparsified graph G_S iff $f(u) = f(v)$. At the end G_S consists of all monochromatic edges that have not been deleted. In ESTIMATENUMBEROFTRIANGLES we run in parallel the algorithm estimating P_2 and K copies of SAMPLERANDOM2PATH. For each G_S^i, $1 \leq i \leq K$, with at least s pairwise independent 2-paths we choose at random a 2-path and check whether it is a triangle. (Note that we require the existence of s *pairwise independent* 2-paths but we choose a 2-path at random from *all* 2-paths in G_S.) The ratio of triangles to all sampled 2-paths and the estimate of P_2 are then used to estimate T_3. In the next section we obtain bounds on the user defined parameters C, K and s. In Lemma 6 we present en efficient implementation of G_S that guarantees constant time updates and allows the sampling of a random 2-path in expected time $O(\log n)$ and worst case time $O(\log^2 n)$ with high probability.

3.3   Theoretical Analysis

We will prove the main result in several lemmas. Due to lack of space, proofs which are not essential for the understanding of the main ideas can be found in the full version of the paper. The next lemma provides an estimate of P_2 using an estimator for the second frequency moment of data streams [21].

Lemma 1. *Let G be a graph with no isolated edges given as a stream of edge insertions and deletions. There exists an algorithm returning an (ε, δ)-approximation of the number of 2-paths in G in one pass over the stream of edges which needs $O(\frac{1}{\varepsilon^2} \log \frac{1}{\delta})$ space and $O(\log \frac{1}{\delta})$ processing time per edge.*

ESTIMATENUMBEROFTWOPATHS

Input: stream of edge deletions and insertions $\mathcal{S}$, algorithm SME estimating the second moment items streams

1. $m = 0$
2. **for** each $((u, v), \$)$ in $\mathcal{S}$ **do**
3.   **if** $\$= +$ **then**
4. $m = m + 1$
5. $SME.update(u, 1)$, $SME.update(v, 1)$
6.   **else**
7. $m = m - 1$
8. $SME.update(u, -1)$, $SME.update(v, -1)$
9. **return** $SME.estimate/2 - m$

SPARSIFYGRAPH

Input: stream of edge deletions and insertions $\mathcal{S}$, coloring function $f : V \to C$

1. $G_S = \emptyset$
2. **for** each $((u, v), \$) \in \mathcal{S}$ **do**
3.   **if** $f(u) = f(v)$ **then**
4.   **if** $\$= +$ **then**
5. $G_S = G_S \cup (u, v)$.
6.   **else**
7. $G_S = G_S \backslash (u, v)$.
8. Return G_S.

SAMPLERANDOM2PATH

Input: sparsified graph G_S

1. choose at random a 2-path (u, v, w) in G_S
2. **if** the vertices $\{u, v, w\}$ form a triangle **then**
3.   **return** 1
4. **else**
5.   **return** 0

ESTIMATENUMBEROFTRIANGLES

Input: streamed graph $\mathcal{S}$, set of K independent fully random coloring functions $\mathcal{F}$, algorithm SME estimating the second moment of streams of items, threshold s

1. run in parallel ESTIMATENUMBEROFTWOPATHS($\mathcal{S}, SME$) and let $\tilde{P}_2$ be the returned estimate
2. run in parallel K copies of SPARSIFYGRAPH($\mathcal{S}, f_i$), $f_i \in \mathcal{F}$
3. $\ell = 0$
4. **for** each G_S^i with at least s pairwise independent 2-paths **do**
5. $X+ = $ SAMPLERANDOM2PATH(G_S^i)
6. $\ell+ = 1$
7. $\tilde{\alpha} = X/\ell$
8. **return** $\frac{\tilde{\alpha}\tilde{P}_2}{3}$

Fig. 1. Estimating the number of 2-paths in G, the transitivity coefficient and the number of triangles

The next two lemmas will show a lower bound on the number of pairwise independent 2-paths in a graph without isolated edges. The results are needed in order to obtain bounds on the required sampling probability. First we show that a graph without isolated edges contains a linear number of pairwise independent 2-paths.

Lemma 2. *Let $G = (V, E)$ be a graph over n vertices without isolated edges. Then there exist at least $\Omega(n)$ pairwise independent 2-paths.*

The next result gives a lower bound on the number of pairwise independent 2-paths in terms of the total number of 2-paths. For denser graphs it implies the existence of $\omega(n)$ pairwise independent 2-paths.

Lemma 3. *Let the number of 2-paths in a graph $G = (V, E)$ be P_2. There exist $\Omega(P_2/n)$ pairwise independent 2-paths.*

Next we obtain bounds on the sampling probability such that there are sufficiently many pairwise independent 2-paths in G_S. As we show later, this is needed to guarantee that SAMPLERANDOM2PATH will return an almost unbiased estimator of the transitivity coefficient. The events for two 2-paths being monochromatic are independent, thus the next lemma follows from Lemma 2 and Chebyshev's inequality. Note that we still don't need the coloring function f to be fully random.

Lemma 4. *Let f be 6-wise independent and $p \geq \frac{5\sqrt{3}}{\varepsilon\sqrt{b}}$ for $b = \max(n, P_2/n)$ and $\varepsilon \in (0, 1]$. Then with probability at least 3/4 SPARSIFYGRAPH returns G_S such that there are at least $18/\varepsilon^2$ pairwise independent 2-paths in G_S.*

Lemma 5. *Assume we run ESTIMATENUMBEROFTRIANGLES with $s = 18/\varepsilon^2$ and let X be the value returned by SAMPLERANDOM2PATH. Then $(1 - \varepsilon)\alpha \leq \mathbb{E}[X] \leq (1 + \varepsilon)\alpha$.*

Proof. We analyze how much differs the probability between 2-paths to be selected by SAMPLERANDOM2PATH. Consider a given 2-path (u, v, w). It will be sampled if the following three events occur:

1. (u, v, w) is monochromatic, i.e., it is in the sparsified graph G_S.
2. There are $i \geq 18/\varepsilon^2$ pairwise independent 2-paths in G_S.
3. (u, v, w) is selected by SAMPLERANDOM2PATH.

The first event occurs with probability p^2. Since f is fully random, the condition that (u, v, w) is monochromatic does not alter the probability for any 2-path independent from (u, v, w) to be also monochromatic. The probability to be in G_S changes only for 2-paths containing two vertices from $\{u, v, w\}$, which in turn changes the number of 2-paths in G_S and thus probability for (u, v, w) to be picked up by SAMPLERANDOM2PATH. In the following we denote by p_{G_S} the probability that a given 2-path is monochromatic and there are at least $18/\varepsilon^2$ pairwise independent 2-paths in G_S, note that p_{G_S} is equal for all 2-paths.

Consider a fixed coloring to $V\backslash\{u,v,w\}$. We analyze the difference in the number of monochromatic 2-paths depending whether $f(u) = f(v) = f(w)$ or not. There are two types of 2-paths that can become monochromatic conditioning on $f(u) = f(v) = f(w)$: either (i) 2-paths with two endpoints in $\{u,v,w\}$ centered at some $\{u,v,w\}$, or (ii) 2-paths with two vertices in $\{u,v,w\}$ centerer at a vertex $x \notin \{u,v,w\}$. For the first case assume w.l.o.g. there is a 2-path $(u,v,w) \in G_S$ centered at v and let $d_v^{f(v)} = |\{z \in N(v)\backslash\{u,w\} : f(z) = f(v)\}|$, i.e., $d_v^{f(v)}$ is the number of v's neighbors different from u and w, having the same color as v. Thus, the number of monochromatic 2-paths centered at v varies by $2d_v^{f(v)}$ conditioning on the assumption that $f(u) = f(v) = f(w)$. The same reasoning applies also to the 2-paths centered at u and w. For the second case consider the vertices u and v. Conditioning on $f(u) = f(w)$, we additionally add to G_S 2-paths (u,x_i,w) for which $f(x_i) = f(u) = f(w)$ and $x_i \in N(u) \cap N(w)$. The number of such 2-paths is at most $\min(d_u^{f(u)}, d_w^{f(w)})$. The same reasoning applies to any pair of vertices from $\{u,v,w\}$. Therefore, depending on whether $f(u) = f(v) = f(w)$ or not, the number of monochromatic 2-paths centered at a vertex from $\{u,v,w\}$ varies between

$$\sum_{y\in\{u,v,w\}} \binom{d_y^{f(y)}}{2} \quad \text{and} \quad \sum_{y\in\{u,v,w\}} \binom{d_y^{f(y)}}{2} + 3d_y^{f(y)}.$$

Set $k = 18/\varepsilon^2$. Consider now two different, but not necessarily independent, 2-paths $(u_1,v_1,w_1), (u_2,v_2,w_2) \in G$. We analyze the probability for each of them to be selected by SAMPLERANDOM2PATH. Let $\mathcal{C}$ be a partial coloring to $V\backslash\{u_j,v_j,w_j\}$, $j = 1,2$. If $\mathcal{C}$ is completed to a coloring of all vertices such that both (u_1,v_1,w_1) and (u_2,v_2,w_2) are monochromatic, then clearly they are picked up with the same probability. Assume that with probability p_i, $i-1$ 2-paths are colored monochromatic by $\mathcal{C}$ and consider extensions of $\mathcal{C}$ that make exactly one of (u_1,v_1,w_1) and (u_2,v_2,w_2) monochromatic. Under the assumption there are $i \geq k$ 2-paths in G_S and following the above discussion about the number of 2-paths with at least two vertices from $\{u_j,v_j,w_j\}$, we see that the number of monochromatic 2-paths can vary between i and $i + 3\sqrt{2i}$. Thus, the probability for (u_1,v_1,w_1) and (u_2,v_2,w_2) to be sampled varies between

$$p_{G_S} \sum_{i\geq k} \frac{p_i}{i} \quad \text{and} \quad p_{G_S} \sum_{i\geq k} \frac{p_i}{i + 3\sqrt{2i}}.$$

We assume G_S contains at least k 2-paths, thus $\sum_{i\geq k} p_i = 1$ and there exists $r \geq k, r \in \mathbb{R}$ such that $\sum_{i\geq k} p_i i^{-1} = 1/r$. Thus we bound

$$\sum_{i\geq k} \frac{p_i}{i + 3\sqrt{2i}} = \sum_{i\geq k} \frac{p_i}{i(1 + 3\sqrt{2/i})} \geq \frac{1}{1 + 3\sqrt{2/k}} \sum_{i\geq k} \frac{p_i}{i} = \frac{1}{r(1 + 3\sqrt{2/k})}.$$

Since the function f is fully random, each coloring is equally probable. The above reasoning applies to any pair of 2-paths in G, thus for any 2-path the probability to be sampled varies between

$$\frac{p_{G_S}}{r} \quad \text{and} \quad \frac{p_{G_S}}{(1 + \sqrt{18/k})r} = \frac{p_{G_S}}{(1 + \varepsilon)r}.$$

Assume first the extreme case that 2-paths which are not part of a triangle are sampled with probability $\frac{1}{r}$ and 2-paths part of a triangle with probability $\frac{1}{(1+\varepsilon)r}$. We have $X = \sum_{(u,v,w) \in P_2} I_{(u,v,w)}$, where $I_{(u,v,w)}$ is an indicator random variable denoting whether (u, v, w) is part of a triangle. Thus

$$\mathbb{E}[X] \geq \frac{p_{G_S} 3 T_3}{(1 + \varepsilon)r} \frac{r}{p_{G_S} P_2} = \frac{\alpha}{1 + \varepsilon} \geq (1 - \varepsilon)\alpha.$$

On the other extreme, assuming that we select 2-paths part of triangles with probability $\frac{1}{r}$ and 2-paths not part of a triangle with probability $\frac{1}{r(1+\varepsilon)}$, using similar reasoning we obtain $\mathbb{E}[X] \leq (1 + \varepsilon)\alpha$. $\square$

Applying a variation of rejection sampling, in the next lemma we show how to store a sparsified graph G_S such that we efficiently sample a 2-path uniformly at random and G_S is updated in constant time.

Lemma 6. *Let $G_S = (V, E_S)$ be a sparsified graph over m' monochromatic edges. There exists an implementation of G_S in space $O(m')$ such that an edge can be inserted to or deleted from G_S in constant time with high probability. A random 2-path, if existent, can be selected from G_S in expected time $O(\log n)$ and $O(\log^2 n)$ time with high probability.*

Now we have all components in order to prove the main result.

Proof. (of Theorem 1).

Assume EstimateNumberOfTriangles runs K copies in parallel of Spar-sifyGraph with $p = \frac{5\sqrt{3}}{\varepsilon\sqrt{b}}$ for $b = \max(n, P_2/n)$. By Lemma 4 with probability $3/4$ we have a sparsified graph with at least $s = 18/\varepsilon^2$ pairwise independent 2-paths. Thus, we expect to obtain from $3K/4$ of them an indicator random variable. A standard application of Chernoff's inequality yields that with probability $O(2^{-K/36})$ we will have $\ell \geq K/2$ indicator random variables X_i denoting whether the sampled 2-path is part of a triangle. By Lemma 5 we have $(1 - \varepsilon)\alpha \leq \mathbb{E}[X_i] \leq (1 + \varepsilon)\alpha$ and as an estimate of α we return $\sum_{i=1}^{\ell} X_i/\ell$. Observe that $(1 + \varepsilon/3)^2 \leq 1 + \varepsilon$, respectively $(1 - \varepsilon/3)^2 \geq 1 - \varepsilon$. From the above discussion and applying Chernoff's inequality and the union bound, we see that for $K = \frac{36}{\varepsilon^2 \alpha} \log \frac{2}{\delta}$, we obtain an $(\varepsilon, \delta/2)$-approximation of α.

By Lemma 1 we can compute an $(\varepsilon, \delta/2)$-approximation of the number of 2-paths in space $O(\frac{1}{\varepsilon^2} \log \frac{1}{\delta})$ and $O(\log \frac{1}{\delta})$ per edge processing time. It is trivial to show that this implies an $(3\varepsilon, \delta)$-approximation of the number of triangles for $\varepsilon < 1/3$. Clearly, one can rescale ε in the above, i.e. $\varepsilon = \varepsilon/3$, such that EstimateNumberOfTriangles returns an (ε, δ)-approximation.

By Lemma 6, each sparsified graph with m' edges uses space $O(m')$ and each update takes constant time with high probability, thus we obtain that each edge is processed with high probability in time $O(K)$. Each monochromatic edge and its color can be represented in $O(\log n)$ bits.

By Lemma 6, in expected time $O(\log n)$ and worst case time $O(\log^2 n)$ with high probability we sample uniformly at random a 2-path from each G_S with at least $18/\varepsilon^2$ pairwise independent 2-paths.   □

References

1. Ahn, K.J., Guha, S., McGregor, A.: Graph sketches: sparsification, spanners, and subgraphs. In: PODS 2012, pp. 5–14 (2012)
2. Aiello, W., Chung, F.R.K., Lu, L.: A random graph model for massive graphs. In: STOC 2000, pp. 171–180 (2000)
3. Albert, R., Barabási, A.-L.: Statistical mechanics of complex networks. Rev. Mod. Phys. 74, 47–97 (2002)
4. Alon, N., Yuster, R., Zwick, U.: Finding and Counting Given Length Cycles. Algorithmica 17(3), 209–223 (1997)
5. Arbitman, Y., Naor, M., Segev, G.: Backyard Cuckoo Hashing: Constant Worst-Case Operations with a Succinct Representation. In: FOCS 2010, pp. 787–796 (2010)
6. Backstrom, L., Boldi, P., Rosa, M., Ugander, J., Vigna, S.: Four degrees of separation. In: WebSci 2012, pp. 33–42 (2012)
7. Berry, J.W., Hendrickson, B., LaViolette, R., Phillips, C.A.: Tolerating the Community Detection Resolution Limit with Edge Weighting. Phys. Rev. E 83(5)
8. Buriol, L.S., Frahling, G., Leonardi, S., Marchetti-Spaccamela, A., Sohler, C.: Counting triangles in data streams. In: PODS 2006, pp. 253–262 (2006)
9. Carter, L., Wegman, M.N.: Universal Classes of Hash Functions. J. Comput. Syst. Sci. 18(2), 143–154 (1979)
10. Dietzfelbinger, M.: Design Strategies for Minimal Perfect Hash Functions. In: Hromkovič, J., Královič, R., Nunkesser, M., Widmayer, P. (eds.) SAGA 2007. LNCS, vol. 4665, pp. 2–17. Springer, Heidelberg (2007)
11. Frahling, G., Indyk, P., Sohler, C.: Sampling in dynamic data streams and applications. In: Symposium on Computational Geometry 2005, pp. 142–149 (2005)
12. Jowhari, H., Saglam, M., Tardos, G.: Tight bounds for Lp samplers, finding duplicates in streams, and related problems. In: PODS 2011, pp. 49–58 (2011)
13. Kane, D.M., Mehlhorn, K., Sauerwald, T., Sun, H.: Counting Arbitrary Subgraphs in Data Streams. In: Czumaj, A., Mehlhorn, K., Pitts, A., Wattenhofer, R. (eds.) ICALP 2012, Part II. LNCS, vol. 7392, pp. 598–609. Springer, Heidelberg (2012)
14. Kolountzakis, M.N., Miller, G.L., Peng, R., Tsourakakis, C.E.: Efficient Triangle Counting in Large Graphs via Degree-based Vertex Partitioning. Internet Mathematics 8(1-2), 161–185 (2012)
15. Leonardi, S.: List of Open Problems in Sublinear Algorithms: Problem 11, http://sublinear.info/11
16. Manjunath, M., Mehlhorn, K., Panagiotou, K., Sun, H.: Approximate Counting of Cycles in Streams. In: Demetrescu, C., Halldórsson, M.M. (eds.) ESA 2011. LNCS, vol. 6942, pp. 677–688. Springer, Heidelberg (2011)
17. Muthukrishnan, S.: Data Streams: Algorithms and Applications. Foundations and Trends in Theoretical Computer Science 1(2) (2005)
18. Pagh, A., Pagh, R.: Uniform Hashing in Constant Time and Optimal Space. SIAM J. Comput. 38(1), 85–96 (2008)

19. Pagh, R., Tsourakakis, C.E.: Colorful triangle counting and a MapReduce implementation. Inf. Process. Lett. 112(7), 277–281 (2012)
20. Pătraşcu, M., Thorup, M.: The Power of Simple Tabulation Hashing. J. ACM 59(3), 14 (2012)
21. Thorup, M., Zhang, Y.: Tabulation based 4-universal hashing with applications to second moment estimation. In: SODA 2004, pp. 615–624 (2004)
22. Tsourakakis, C.E., Kang, U., Miller, G.L., Faloutsos, C.: DOULION: Counting triangles in massive graphs with a coin. In: KDD 2009, pp. 837–846 (2009)
23. Tsourakakis, C.E., Kolountzakis, M.N., Miller, G.L.: Triangle Sparsifiers. J. of Graph Algorithms and Appl. 15(6), 703–726 (2011)
24. Vassilevska Williams, V.: Multiplying matrices faster than Coppersmith-Winograd. In: STOC 2012, pp. 887–898 (2012)

Linear Time LexDFS on Cocomparability Graphs*

Ekkehard Köhler[1] and Lalla Mouatadid[2]

[1] Brandenburg University of Technology, 03044 Cottbus, Germany
ekoehler@math.tu-cottbus.de
[2] University of Toronto, Toronto ON M5S 2J7, Canada
lalla@cs.toronto.edu

Abstract. Lexicographic depth first search (LexDFS) is a graph search protocol which has already proved to be a powerful tool on cocomparability graphs. Cocomparability graphs have been well studied by investigating their complements (comparability graphs) and their corresponding posets. Recently however LexDFS has led to a number of elegant polynomial and near linear time algorithms on cocomparability graphs when used as a preprocessing step [2,3,11]. The nonlinear runtime of some of these results is a consequence of complexity of this preprocessing step. We present the first linear time algorithm to compute a LexDFS cocomparability ordering, therefore answering a problem raised in [2] and helping achieve the first linear time algorithms for the minimum path cover problem, and thus the Hamilton path problem, the maximum independent set problem and the minimum clique cover for this graph family.

Keywords: lexicographic depth first search, cocomparability graphs, graph searching, posets, hamiltonian path.

1 Introduction

Graph searching is a very useful and widely used tool that gave rise to a number of efficient and easily implementable algorithms. Lexicographic breadth first search (LexBFS) for instance, is a well known graph search protocol which has led to elegant algorithms on various graph families, as illustrated in [1,7]. Recently another graph search protocol, lexicographic depth first search (LexDFS), was introduced and has already proved to be a powerful tool on cocomparability graphs [2,3,11]. Indeed, since LexDFS was introduced [4], many problems, such as computing a maximum cardinality independent set, a minimum clique cover or a minimum path cover, now have near-linear time solutions for cocomparability graphs. These successful approaches share a strategy: They start with creating a so-called cocomparability ordering of the graph and preprocess it with a LexDFS sweep and then basically extend, or slightly modify, the linear time algorithms that work for interval graphs (a subfamily of cocomparability graphs).

---

* This is an extended abstract. The full version is available in [13].

R Ravi and I.L. Gørtz (Eds.): SWAT 2014, LNCS 8503, pp. 319–330, 2014.
© Springer International Publishing Switzerland 2014

The nonlinear runtime of the algorithms on cocomparability graphs is is forced by the nonlinearity of LexDFS.

In this paper, we present the first linear time algorithm to compute a LexDFS cocomparability ordering. Therefore, as immediate corollaries, we now have the first linear time algorithm to compute a minimum path cover, and thus a Hamilton path if one exists, and the first linear time algorithm to compute a maximum independent set and a minimum clique cover for cocomparability graphs. We will also show how to specifically compute a LexDFS$^+$ ordering in linear time. LexDFS$^+$ is a variant of LexDFS that is needed in [2,3,11].

Cocomparability graphs are a family of perfect graphs whose complements, comparability graphs, admit a transitive orientation of the edges. That is, for every three vertices x, y, z, if the edges xy, yz are oriented $x \to y \to z$ then $xz \in E$ and $x \to z$. Cocomparability graphs and partially ordered sets, or *posets*, are closely related. In Section 2, we explain this relationship and how algorithms on cocomparability graphs immediately translate into algorithms on posets. Cocomparability graphs have been well studied by investigating their complements (comparability graphs) and their corresponding posets [6]. Recently however, there has been a growing motivation to exploit the structure of cocomparability graphs in order to design algorithms that do not require the computation of the complement or the poset. For this approach the use of LexDFS has proven to be quite sucessful [2,3,11].

A key point in our algorithm for computing a LexDFS ordering is a partition refinement approach of the layers of a corresponding poset of the cocomparability graph, *without* computing the poset itself. This refinement is *in situ* and performed *backwards*, as opposed to the well known forward partition refinement used to compute a LexBFS ordering [7]. We discuss the technique of partition refinement in more details in Section 2. The paper is organized as follows: Section 2 gives the necessary background and relevant definitions. In Section 3, we present the LexDFS algorithm, and in 4 we prove its correctness and show how to compute in linear time a LexDFS$^+$ ordering. In Section 5 we present our concluding remarks. Due to a lack of space, we omit some proofs and the implementation details, and refer the reader to [13] for the full version of the paper.

2   Background

We assume the reader to be familiar with basic graph notation. All the graphs considered in this paper are finite, simple, and undirected, unless explicitly stated otherwise. For a vertex v, $N(v) = \{u | uv \in E\}$; and we say v is *simplicial* if $N(v)$ is a clique. An *ordering* σ of V is a bijection $\sigma: [1...n] \to V$. Given an ordering $\sigma = v_1, v_2, ..., v_n$, we write $v_i \prec_\sigma v_j$ if v_i appears before v_j in σ, i.e., $i < j$; and $N^+(v_i) = \{v_j | v_i v_j \in E \text{ and } i < j\}$; we denote by $\sigma^- = v_n, v_{n-1}, ..., v_2, v_1$ the reverse ordering of σ. Given a cocomparability graph $G(V, E)$, an ordering σ is a *cocomparability ordering* (or an umbrella free ordering) of G if for any triple $a \prec_\sigma b \prec_\sigma c$ where $ac \in E$, we either have $ab \in E$ or $bc \in E$, or both

[9]. If neither ab and bc are edges, we say that the umbrella ac *flies* over b. Note that the cocomparability ordering is just the equivalent to transitivity in the complement. In [10], McConnell and Spinrad presented an algorithm that computes a cocomparability ordering in $\mathcal{O}(m+n)$ time, where $m = |E|$ and $n = |V|$.

A poset $P(V, \prec)$ is an irreflexive, antisymmetric and transitive relation on the set V. We say that two elements $a, b \in V$ are comparable if $a \prec b$ or $b \prec a$, otherwise they are incomparable. A linear extension $\mathcal{L}$ of P is a total ordering of V which respects the order imposed by $\prec$. As already mentioned, posets and cocomparability graphs are closely related. In fact, if $G(V, E)$ is a comparability graph, then G together with a transitive orientation of E can equivalently be represented by a poset $P(V, \prec)$ where $uv \in E$ if and only if u and v are comparable in P. This implies that σ, a cocomparability ordering of $\overline{G}$ is a linear extension of P. Notice that for every poset $P(V, \prec)$, there exists a unique comparability graph $G(V, E)$, and thus a unique cocomparability graph. Conversely, every transitive orientation of a comparability graph, and thus every cocomparability ordering in the complement, is a linear extension of a poset P.

A graph search is a mechanism for visiting vertices of a given graph in a certain manner. We say that two or more vertices are *tied* if at a given step of the graph search, these vertices are all eligible to be visited next. In 2008, Corneil and Krueger [4] introduced LexDFS, a graph search that extends depth first search by assigning lexicographic labels to the vertices in order to break ties. Algorithm 1 is the generic LexDFS algorithm, as presented in [4] and the ordering $\sigma = a, b, c, d, e$ is a LexDFS ordering starting at vertex a for the cocomparability graph in Fig. 1. LexDFS admits the following vertex ordering characterization, known as the *4 Point Condition*:

Theorem 1. *[4] σ is a LexDFS ordering of a graph $G(V, E)$ if and only if for every triple $a \prec_\sigma b \prec_\sigma c$ where $ac \in E, ab \notin E$, there must exists a vertex d such that $a \prec_\sigma d \prec_\sigma b$ and $db \in E, dc \notin E$.*

LexDFS$^+$ is the LexDFS variant with the additional '*rightmost*' tie breaking rule. That is, given a vertex ordering σ of G, the ordering $\tau = $ LexDFS$^+(\sigma)$ is a LexDFS of G where ties between eligible vertices are broken by choosing the rightmost vertex in σ. Therefore, by definition, LexDFS$^+$ always starts by the rightmost vertex in σ. For an example, look at the graph in Fig. 1. If we compute $\tau =$LexDFS$^+(\sigma)$, for this example, we have to start with vertex e, i.e. $\tau(1) = e$; obviously, $\tau(2) = c$. For $\tau(3)$ in a regular LexDFS a, b and d are tied. However, for LexDFS$^+(\sigma)$, $\tau(3) = d$, since d is rightmost in σ among a, b, and d. It was shown in [2] that if $G(V, E)$ is a cocomparability graph, and σ a cocomparability ordering of G, then the LexDFS ordering $\tau = LexDFS^+(\sigma)$ is also a cocomparability ordering of G. It is easy to see that if σ is a LexDFS cocomparability ordering, then for any triple $a \prec_\sigma b \prec_\sigma c$ where c is a nonsimplicial vertex and $ab \notin E, ac, bc \in E$, there exists a vertex d such that $a \prec_\sigma d \prec_\sigma b$ and ad, $db \in E$ and $dc \notin E$ [2]. Indeed the edge ad destroys the umbrella ac over d.

As was already pointed out in the introduction, the key idea of our algorithm to determine a LexDFS of a cocomparability graph is the backward in situ

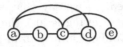

Fig. 1. $G(V, E)$ a cocomparability graph

Algorithm 1. LexDFS

Input: a graph $G(V, E)$ and a start vertex s
Output: a LexDFS ordering σ of V
1: assign the label ϵ to all vertices
2: $label(s) \leftarrow \{0\}$
3: **for** $i \leftarrow 1$ to n **do**
4:     pick an unnumbered vertex v with lexicographically largest label
5:     $\sigma(i) \leftarrow v$ ▷ v is assigned the number i
6:     **foreach** unnumbered vertex w adjacent to v **do**
7:         prepend i to $label(w)$
8:     **end for**
9: **end for**

---

partition refinement of the layers of a corresponding poset, without computing the poset itself. Given a set S, we call $\mathcal{P} = (P_1, P_2, ..., P_k)$ a partition of S if for all P_i, P_j, $i \neq j$, $P_i \cap P_j = \emptyset$ and $\bigcup_{i=1}^{k} P_i = S$. Given a set $T \subseteq S$, we say that T refines $\mathcal{P}$ when every partition class $P_i \in \mathcal{P}$ is replaced with subpartition classes $A_i = P_i \cap T$ and $B_i = P_i \backslash A_i$. This technique, known as *partition refinement* has led to a simple and elegant implementation of LexBFS in linear time [7]. The LexBFS partition refinement algorithm is as follows: Initially $\mathcal{P} = (V)$; select a start vertex s where $N(s)$ refines $\mathcal{P}$ by placing $A = V \cap N(s)$ before $B = V \backslash A$. The vertex whose neighbourhood is used to refine the partition classes is called a *pivot*. Pick the next pivot v amongst vertices in A; and use $N(v)$ for refining A then B and maintaining the order of the partition classes created so far: $(A \cap N(v), A \backslash N(v), B \cap N(v), B \backslash N(v))$ in this order. This process is repeated until all partition classes have been refined.

This refinement can be seen as a *forward* refinement in the sense that pivots are selected left to right, i.e., from the A's sets then the B's set, and the refinement is *in situ*, meaning the A_i's always precede the B_i's. In other words, pivots do not reorder the already created subpartitions of P. That is , if $N(u)$ was used to refine P to $A = P \cap N(u)$ and $B = P \backslash A$, and v is the next pivot then $N(v)$ is used to refine A first to $A \cap N(v)$ followed by $A \backslash N(v)$, next $N(v)$ refines B to $B \cap N(v)$ followed by $B \backslash N(v)$, and the subpartitions of A always precede the subpartitions of B. This in situ refinement results in a linear time implementation for LexBFS. Also for LexDFS, one can define a partition refinement scheme, but this partition refinement is not in situ and can be seen as a *backward* refinement in the sense that pivots are selected right to left. Consider a pivot v; due to the depth first search character of LexDFS, v has to pull *all* its neighbours to the front, i.e., $A \cap N(v)$ followed by $B \cap N(v)$ both precede $A \backslash N(v)$ followed by $B \backslash N(v)$. This sorting of the partition classes is an obstacle to a linear time implementation for LexDFS. In fact, to the best of our knowledge, there is no linear time implementation of LexDFS. The best known algorithm takes $\mathcal{O}(\min(n^2, n + m \log \log n))$ time and uses the above explained non in situ partition refinement together with van Emde Boas trees [12].

3   The Algorithm

Before presenting our algorithm in detail we first give an overview. Let $G(V, E)$ be a cocomparability graph. We first compute a cocomparability ordering σ using the algorithm in [10]. Then, we assign a label, denoted $\#(v)$, to each vertex v in σ. We use these labels to compute, for every vertex v, the length of a largest chain succeeding v in the corresponding poset of the complement of G. Roughly speaking, we then partition V by iteratively placing vertices with smallest label into the same partition set. In a comparability graph one could finds these sets by iteratively removing the set of maximal elements of the poset. Since we work on the complement, this has to be done using only edges of the cocomparability graph, i.e. non-edges of the comparability graph. Once all the vertices have their initial labeling, we iteratively create a partition $\mathcal{P}$ of V wherein each step i, the partition class P_i consists of the vertices of minimum label value. When a vertex v is added to a partition class P_i, we say that v has been *visited*.

Since σ is a cocomparability ordering, and thus a linear extension of a poset, the P_1 vertices are exactly the elements in the linear extension with no upper cover. Therefore they are just the maximal elements of the partial order defined by σ in $\overline{G}$. Similarly, when all the P_1 have been visited, i.e., 'removed', P_2 is just the set of maximal elements in the partial order of $\overline{G} \backslash P_1$, and so on. Creating the partition classes is indeed equivalent to removing the maximal elements of a poset corresponding to $\overline{G}$ one layer at a time.

The final step of the algorithm is the partition refinement where we refine each partition class one at a time in a specific manner. In particular, each partition class P_i is assigned a set S_i of pivots that will be used to refine P_i only. The set S_i is implemented as a stack, and the order in which the pivots are pushed onto S_i is crucial. When v is taken from S_i to be the next pivot, $N(v)$ performs an in situ refinement on P_i. We use τ_i to denote the final (refined) ordering of P_i. When all partition classes have been refined, we concatenate all the τ_i's in order, i.e., $\tau = \tau_1 \cdot \tau_2 \cdot \ldots \cdot \tau_p$ where $\cdot$ denotes concatenation, and use τ to denote the final ordering. Our main theorem is the following:

Theorem 2. *Let $G(V, E)$ be a cocomparability graph, τ is a LexDFS cocomparability ordering that can be computed in $\mathcal{O}(m + n)$.*

3.1   Vertex Labelling

Let $G(V, E)$ be an undirected cocomparability graph, and let $\sigma = v_1 \prec_\sigma v_2 \prec_\sigma \ldots \prec_\sigma v_n$ be a cocomparability ordering of G returned by the algorithm in [10]; we use $\sigma = ccorder(G)$ to denote such an algorithm. For every vertex $v \in V$, we assign a label $\#(v)$ initialized to the number of nonneighbours of v to its right in σ: $\#(v) = |\{u|uv \notin E$ and $v \prec_\sigma u\}|$. Given such a labelling of the vertices, we create a partition of V denoted by $\mathcal{P} = \bigcup_{i=1}^p P_i$ in the following manner: Initially all vertices are marked unvisited, P_1 is the set of vertices with the smallest $\#$ label value. Now all vertices in P_1 are marked to be visited. For all unvisited vertices u and for all $v \in P_1$, such that $uv \in E$, $\#(u)$ is incremented by 1. To create P_2, again select the set of unvisited vertices of smallest $\#$ value. These vertices in P_2 are marked to be visited and for each such $v \in P_2$ and unvisited u adjacent to v, $\#(u)$ is incremented by 1. We increment i and repeat this operation of creating a partition class of the vertices with the smallest label until all vertices belong to a partition class.

Algorithm 2. PartitionClasses

Input: a cocomparability graph $G(V, E)$
Output: partition $\mathcal{P}$ of V with p partition classes, and p
1: $\sigma \leftarrow ccorder(G(V, E))$ ▷ As computed in [10]
2: $S \leftarrow \emptyset$
3: **for** $i \leftarrow n$ downto 1 **do**
4:     $\#(v_i) \leftarrow (n - i) - |S \cap N(v_i)|$ ▷ Initial labelling $\#(v)$
5:     $S \leftarrow S \cup \{v_i\}$
6: **end for**
7: $U \leftarrow V$ ▷ U the set of unvisited vertices
8: $i \leftarrow 1$
9: **while** U not empty **do**
10:     $P_i \leftarrow \{v|\#(v) = min(\#(U))\}$ ▷ Creating Partition Classes
11:     $U \leftarrow U \backslash P_i$
12:     **for** $v \in P_i$ **do**
13:         **for** $u \in U$ and $uv \in E$ **do**
14:             $\#(u) \leftarrow \#(u) + 1$
15:         **end for**
16:     **end for**
17:     $i \leftarrow i + 1$
18: **end while**
19: $p \leftarrow i - 1$
20: **return** $\mathcal{P} \leftarrow (P_1, P_2, ..., P_p)$ and p

---

Algorithm 2 is a formal description of the algorithm which takes a cocomparability graph $G(V, E)$ as input and returns the partition $\mathcal{P} = (P_1, P_2, ..., P_p)$. Let $\pi = P_1 \cdot P_2 \cdot ... \cdot P_p$ be the order of V resulting from Algorithm 2 such that $\forall x \in P_i, y \in P_{j>i}$, we have $x \prec_\pi y$. The order inside each P_i is arbitrary. Consider the graph in Fig. 2 with a valid cocomparability vertex ordering. The numbers below the vertices are their labels as computed by Algorithm 2. $PartitionClasses(G(V, E))$ would produce the following partition: $\mathcal{P} = \{P_1 = \{h, k\}, P_2 = \{j\}, P_3 = \{g, i\}, P_4 = \{d, e, f\}, P_5 = \{a, b, c\}\}$.

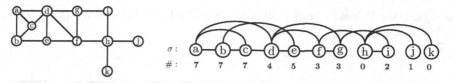

Fig. 2. $G(V, E)$, a cocomparability order σ of V, and the initial labelling $\#$ of V

3.2 Partition Refinement

Once all the partition classes are computed, we reorder the adjacency list of each v according to π in $\mathcal{O}(m + n)$ time [13], then construct a new ordering of V by refining $\mathcal{P}$. Our refinement is slightly different than the generic partition refinement algorithm presented in [8] and briefly explained in Section 2.

Algorithm 3. Refine

Input: a partition class P ordered by π and its corresponding ordered list of pivots S
Output: refinement τ of P
1:   $Q_1 \leftarrow P, k \leftarrow 1$
2: **while** S not empty **do**
3:     $j \leftarrow 1$
4:     $v \leftarrow S.pop$ ▷ S is implemented as a stack
5:     **for** $i \leftarrow 1$ to k **do**
6:        **if** $|Q_i \cap N(v)| = 0$ or $|Q_i \cap N(v)| = |Q_i|$ **then**
7:           $Q'_j \leftarrow Q_i$
8:           $j \leftarrow j + 1$
9:        **else**
10:          $Q'_j \leftarrow Q_i \cap N(v)$
11:          $Q'_{j+1} \leftarrow Q_i \backslash N(v)$
12:          $j \leftarrow j + 2$
13:        **end if**
14:     **end for**
15:     $k \leftarrow j - 1$
16:     **for** $i \leftarrow 1$ to k **do**   ▷ Rename the new partitions for the next pivot
17:        $Q_i \leftarrow Q'_i$
18:     **end for**
19: **end while**
20: **return** $\tau \leftarrow Q_1 \cdot Q_2 \cdot ... \cdot Q_k$ ▷
      $x \in Q_i, y \in Q_{j>i} \implies x \prec_\tau y$ and $x, y \in Q_i \implies x \prec_\tau y$ iff $x \prec_\pi y$

In the remainder of the paper, we will use $\#^*(v_i)$ to refer to the initial value of v_i's label, i.e., the number of nonneighbours of v to its right in σ; and $\#_k(v_i)$ to denote the label value of v_i when P_k is being created, i.e., at iteration k. Given the partition $\mathcal{P} = (P_1, P_2, ..., P_p)$ returned by Algorithm 2, we associate a set S_i to each P_i, where S_i is a set of pivots that will be used to refine P_i. We say that a partition class P_i is *processed* when it has been refined. We use τ_i to denote the final ordering of P_i after it has been refined, i.e. $\tau_i = Refine(P_i, S_i)$. If a partition class P_i has an empty pivot set S_i, then for τ_i the (arbitrary) ordering of P_i in π is used.

Algorithm 4. UpdatePivots

Input: a newly refined partition class P_j and its index j
Output: updated pivot lists for the upcoming partition classes, i.e. for P_i, $i > j$
  1: **for** $v \in P_j$ **do**   ▷ in the τ_j order
  2:     **if** v has neighbours in $P_{i>j}$ **then**
  3:         $S_i.push(v)$ ▷ Update the pivot list of P_i
  4:     **end if**
  5: **end for**

The sets S_i are implemented as stacks and are created as follows: $S_1 = \emptyset$ and P_1 is considered processed. For all $P_{i>1}$, we scan τ_1 from *left to right* and for each $v \in \tau_1$ and every $u \in P_{i>1}$ where $uv \in E$, we push v in S_i. In general, every time a partition class $P_{j<i}$ is refined, i.e., τ_j has been produced, we scan τ_j from left to right, and for every $v \in \tau_j$ with neighbours in $P_{i>j}$, we push v into S_i. To refine P_i, we pop elements of S_i one at a time, and for each $v \in S_i$, v is the pivot that refines P_i by reordering P_i into the subpartitions $P_i \cap N(v)$ followed by $P_i \backslash N(v)$. The next pivot out of S_i performs an *in situ* refinement of the current subpartitions of P_i.

Let $u_j^1, u_j^2, ..., u_j^k$ be the left to right ordering of the vertices inside τ_j. We mentioned in Section 1 that not only this refinement is in situ, but also *backwards*. Backwards in two ways: First, the pivots of τ_j have a higher priority, i.e. a stronger pull, than the pivots of $\tau_{k<j}$, and second the pivots are pushed down the stack S_i in $\tau_1 \cdot \tau_2 \cdot ... \cdot \tau_{i-1}$ order (left to right) and thus are popped in reverse order. Therefore we maintain the priority of the pivots in the backward order: $(\tau_1 \cdot \tau_2 \cdot ... \cdot \tau_{i-1})^-$, but also the priority of the pivots inside each $\tau_{j<i}$ in the backward order τ_j. That is for any two vertices $u_j^a, u_j^b \in \tau_j$ where $a < b$, if u_j^a and u_j^b are both pivots for P_i, then u_j^b refines P_i first before $u_j^{a<b}$. Note that this is very similar to standard partition refinement with the difference that in standard partition refinement, P_i is first refined by τ_1 then τ_2, τ_3 and so on. Here we start refining with the last vertex in τ_{i-1}, then τ_{i-2}, and so on up to τ_1. This opposite refinement shows the key difference between LexDFS and LexBFS. Whereas in a LexBFS order the earliest neighbours have the strongest pull and the latest neighbours the weakest, in a LexDFS the last vertices are more influencial then the earlier visited ones. Algorithm 3, *Refine*, takes P_i and S_i as input, and returns the new ordering τ_i of P_i. Algorithm 4, *UpdatePivots*, takes τ_j as input, the refined ordering of P_j, and updates the stacks S_i for all unprocessed partition classes $P_{i>j}$. Let τ denote the final ordering of all the refined partition classes, i.e., $\tau = \tau_1 \cdot \tau_2 \cdot ... \cdot \tau_p$. Refining the partition $\mathcal{P} = \{P_1 = \{h, k\}, P_2 = \{j\}, P_3 = \{g, i\}, P_4 = \{d, e, f\}, P_5 = \{a, b, c\}\}$ of the graph in Fig. 2, we ge the ordering $\tau = h, k, j, i, g, f, d, e, b, c, a$.

3.3   The Complete Algorithm

We are now ready to present the complete algorithm *CCLexDFS*.

Algorithm 5. CCLexDFS

Input: a cocomparability graph $G(V, E)$
Output: a LexDFS order τ of G that is also a cocomparability order of G
 1: $\tau \leftarrow \emptyset$
 2: $(\mathcal{P}, p) \leftarrow PartitionClasses(G)$ ▷ Compute the partition classes
 3: $S_1, ..., S_p \leftarrow \emptyset$
 4: **for** $i \leftarrow 1$ to p **do**
 5:   $\tau_i \leftarrow Refine(P_i, S_i)$ ▷ Refine the partition classes
 6:   $UpdatePivots(\tau_i, i)$ ▷ Update the pivot sets
 7:   $\tau \leftarrow \tau \cdot \tau_i$
 8: **end for**
 9: **return** τ

---

4   Correctness of the Algorithm

We denote the set of partition classes P_1 to P_{i-1} by $\mathcal{P}_i = (P_1, P_2, ..., P_{i-1})$.

Lemma 1. *For any u, v such that $u \prec_\sigma v$ and $uv \notin E$: $\#^*(u) > \#^*(v)$ and at any step i, $\#_i(u) > \#_i(v)$.*

Proof. Since $u \prec_\sigma v$ and $uv \notin E$, any vertex w with $v \prec_\sigma w, wv \notin E$ implies $wu \notin E$ otherwise uw flies over v contradicting σ being a cocomparability ordering. Thus w contributes equally to $\#^*(u)$ and $\#^*(v)$. Moreover, since $u \prec_\sigma v, uv \notin E$, v also contributes to $\#^*(u)$. Therefore $\#^*(u) > \#^*(v)$.

Suppose at a step $i, \#_i(u) < \#_i(v)$ then a vertex $z \in P_{j<i} \in \mathcal{P}_i$ must have closed the gap in $\#^*(u) > \#^*(v)$ by contributing to $\#_j(v)$ but not to $\#_j(u)$. Let $z \in P_{j<i}$ be such a vertex, we are only interested in the case when $zv \in E$ and $zu \notin E$, since adding z to $P_j \in \mathcal{P}_i$ would have incremented $\#_j(v)$, making it closer to $\#_j(u)$. Notice that $u \prec_\sigma z$, otherwise zv flies over u which contradicts σ being a cocomparability ordering . Therefore z contributes one to $\#^*(u)$, thus not reducing the gap between u and v's labels and $\#_i(u) > \#_i(v)$.   □

Lemma 2. *For $1 \le i \le p$, P_i is the set of maximal elements in the poset $P \backslash \bigcup_{j=1}^{i-1} P_j$.*

Lemma 3. *Every partition class $P_i \in \mathcal{P}$ returned by Algorithm 1 is a clique.*

Lemma 4. *(The Flipping Lemma): Let σ be a cocomparability order, and τ the corresponding ordering created from σ and returned by Algorithm 5. For every $uv \notin E, u \prec_\sigma v \iff v \prec_\tau u$.*

Proof. As we are assigning vertices to their partition classes, let u and v be the left most pair of vertices in τ to satisfy $uv \notin E$ and $u \prec_\sigma v$ and $u \prec_\tau v$. By Lemma 1, $\#^*(u) > \#^*(v)$, and by Lemma 3, u and v belong to two different partition classes; P_i and $P_{j>i}$ respectively since $u \prec_\tau v$. Therefore when P_i was created $\#_i(u) < \#_i(v)$, which contradicts Lemma 1. Therefore $v \prec_\tau u$.

For sufficiency, using the contraposition we know that $(u \prec_\tau v \Rightarrow v \prec_\sigma u)$ if and only if $(u \prec_\sigma v \Rightarrow v \prec_\tau u)$. Thereby completing the proof. $\square$

Corollary 1. τ *is a cocomparability order of* G.

Proof. As in [2], if τ is not a cocomparability order as witnessed by $x \prec_\tau y \prec_\tau z$ and $xz \in E, xy, yz \notin E$, then by the Flipping Lemma we have $y \prec_\sigma x$ and $z \prec_\sigma y$, which implies that the umbrella zx flies over y in σ, contradicting the fact that σ is a cocomparability order. $\square$

We refer the reader to [13] for the proofs of Lemma 2 and 3, as well as the implementation details. We are now ready to prove Theorem 2. Namely, that the ordering τ produced by Algorithm 5 is a LexDFS cocomparability order of G.

Proof (of Theorem 2). By Corollary 1, we know that τ is a cocomparability order of G. Suppose it is not a LexDFS order. Therefore for some triple a, b, c with $a \prec_\tau b \prec_\tau c$, $ac \in E$ and $ab \notin E$, there doesn't exist a vertex d as required by the 4 Point Condition (Theorem 1). Since τ is a cocomparability order, $bc \in E$ to destroy the umbrella ac over b.

Suppose b and c belong to the same partition class P_i. Since $ab \notin E$ and $a \prec_\tau b$, $a \in P_{j<i}$. Since $ac \in E$, a is a pivot with respect to P_i and thus b and c could not have been in the same subclass since a would have pulled c before b. Since $b \prec_\tau c$, a pivot u that pulled b in front of c must exist, i.e. $ub \in E, uc \notin E$. Pick u to be the rightmost pivot to b to satisfy this configuration. Notice that $a \prec_\tau u$. Otherwise, since the refinement is backwards, a would have refined P_i before u thus pulling c in front of b. Therefore $a \prec_\tau u$. This means that u plays the role of d with respect to LexDFS, a contradiction to our assumption; therefore b and c must be in different partition classes.

Since $b \prec_\tau c$, $b \in P_i$ and $c \in P_{j>i}$, when P_i was created $\#_i(b) < \#_i(c)$. We investigate how this gap could have occurred given $ac, bc \in E$ and $ab \notin E$. Without loss of generality, let a, b, c be the left most triple in τ that does not satisfy the 4 Point Condition, and consider the vertices that have contributed to $\#_i(b)$ and $\#_i(c)$. Let u be one of these vertices. Thus u increased $\#_i(b), \#_i(c)$ either by being a non adjacent right neighbour of b or c in the initial ordering σ (Algorithm 2, line 6) or u changed $\#(b), \#(c)$ when u was assigned to a partition class (Algorithm 2, lines 12-16).

If u is contained in a set of $\mathcal{P}_i$, then $u \prec_\tau b \prec_\tau c$. Consider all the possible adjacencies between u, b and c. If $ub \in E$ and $uc \notin E$, then either $a \prec_\tau u$ in which case u plays the role of d as required by LexDFS; or $u \prec_\tau a$ in which case u increments $\#_j(b)$ at iteration j when u was assigned to $P_{j<i}$, a set in $\mathcal{P}_i$, and u also contributes to $\#^*(c)$ by the Flipping Lemma. Thus u contributes equally to the labels of b and c. If $ub, uc \in E$ then u increments both b's and c's labels when it is assigned to a partition class; and if $ub, uc \notin E$, then by the Flipping Lemma, u contributes to both $\#^*(b)$ and $\#^*(c)$. But in all three cases u does

not reduce the gap between b and c's labels. Therefore $ub \notin E, uc \in E$. However by the Flipping Lemma $b \prec_\sigma u$, and thus u contributes to $\#^*(b)$ since $ub \notin E$, but also u increments c's label since $u \in \mathcal{P}_i$ and $uc \in E$; again not reducing the gap. Therefore u must be in $V \setminus \mathcal{P}_i$.

If u is not in a set of $\mathcal{P}_i$, then u has not been assigned to a partition class yet. Since u is responsible for the gap $\#_i(b) < \#_i(c)$, u created this gap when b and c were assigned their initial labels $\#^*(b)$ and $\#^*(c)$. For u to create such a gap, u must contribute to c's initial label $\#^*(c)$ and not contribute to $\#^*(b)$. In other words, $uc \notin E$ and $c \prec_\sigma u$. Therefore by the Flipping Lemma, $u \prec_\tau c$. Moreover, for u to not contribute to $\#^*(b)$, we either have $ub \in E$ or $ub \notin E$ but $u \prec_\sigma b$. Notice that this latter case is impossible, since $u \prec_\sigma b \Rightarrow b \prec_\tau u$ (by the Flipping Lemma); but also $u \prec_\tau c$ causing bc to fly over u in τ and contradicting Corollary 1. Thus $ub \in E, uc \notin E$ and $c \prec_\sigma u$. Since u is not in a set of $\mathcal{P}_i$, $b \prec_\tau u$, otherwise u plays the role of d with respect to LexDFS. Moreover, $au \in E$ since τ is a cocomparability order; and the triple a, b, u must satisfy the LexDFS ordering otherwise we contradict the choice of a, b, c as $u \prec_\tau c$. Therefore there must exist a vertex w such that $a \prec_\tau w \prec_\tau b, wb \in E, wu \notin E$; this forces the edge aw in order to avoid the umbrella au over w. If $wc \notin E$, then w plays the role of d as required by LexDFS for the triple a, b, c, and if $wc \in E$, then the umbrella wc flies over u, contradicting τ being a cocomparability order. Therefore there must always exists a vertex that satisfies the LexDFS ordering for $\#_i(b) < \#_i(c)$ to hold; and thus τ is a LexDFS cocomparability order of G.   □

Corollary 2. *Prior to the partition refinement step, if the vertices inside each partition class were ordered according to σ^-, then the resulting τ is a $LexDFS^+(\sigma)$.*

See [13] for the proof.

5   Conclusion and Open Problems

We have presented the first linear time algorithm to determine a LexDFS cocomparability order, therefore answering a question raised in [2], and also overcoming the bottleneck in the near linear time algorithms in [2,3]. It is still an open question whether there exists a linear time implementation for LexDFS on arbitrary graphs. Our implementation exploits the poset structure of the cocomparability graph. In fact, computing the partition classes is equivalent to computing the layers of the corresponding poset. It is fairly straightforward to see that if $G(V, E)$ is a *comparability* graph, a LexDFS($\overline{G}$) cocomparability ordering can also be computed in time linear in the size of G. The details to extend the algorithm to compute such an ordering will be provided in the journal paper. Clearly this leads to the obvious question of whether this algorithm can also be modified to compute a LexDFS ordering of a comparability graph.

Looking at the power of LexDFS on cocomparability, and how it has led to simple and elegant algorithms on this graph family when LexBFS has failed, simply by extending the existing algorithms on interval graphs; it is natural to ask whether there are other problems that can be solved using a similar approach: First a LexDFS preprocessing, then extending the algorithm for interval graphs.

Moreover with this algorithm in hand now, preprocessing is 'easy', which raises the question of possible multisweep LexDFS algorithms. Multisweeps algorithms perform a constant number of sweeps (i.e., graph searches) where each sweep generally reveals more structural properties about the graph. LexDFS has not been used in a multisweep manner yet, we raise the question of whether a second LexDFS sweep reveals more structure that was not seen through the previous sweep. If so, are there problems that can benefit from this structure?

Stepping away from cocomparability graphs but still looking at structured graph families, it is natural to ask whether LexDFS can be implemented in linear time for other restricted graph families, such as asteroidal triple free graphs, a graph family that contains cocomparability graphs. But also whether there are other applications to LexDFS in other graph classes. Graph searches have been exploited on various graph families, it is necessary to explore the possible insights LexDFS has to offer, in contrast with these other graph searches.

Acknowledgments. The authors thank Derek Corneil for his helpful suggestions and valuable comments.

References

1. Brandstädt, A., Dragan, F.F., Nicolai, F.: LexBFS-orderings and powers of chordal graphs. Discrete Mathematics 171(1), 27–42 (1997)
2. Corneil, D.G., Dalton, B., Habib, M.: LDFS-based certifying algorithm for the minimum path cover problem on cocomparability graphs. SIAM Journal on Computing 42(3), 792–807 (2013)
3. Corneil, D.G., Dusart, J., Habib, M., Köhler, E.: On the power of graph searching for cocomparability graphs (in preparation)
4. Corneil, D.G., Krueger, R.M.: A unified view of graph searching. SIAM Journal on Discrete Mathematics 22(4), 1259–1276 (2008)
5. Corneil, D.G., Olariu, S., Stewart, L.: The LBFS structure and recognition of interval graphs. SIAM Journal on Discrete Mathematics 23(4), 1905–1953 (2009)
6. Golumbic, M.C.: Algorithmic graph theory and perfect graphs, vol. 57. Elsevier (2004)
7. Habib, M., McConnell, R.M., Paul, C., Viennot, L.: LexBFS and partition refinement, with applications to transitive orientation and consecutive ones testing. Theoretical Computer Science 234 (2000)
8. Habib, M., Paul, C., Viennot, L.: Partition refinement techniques: An interesting algorithmic tool kit. International Journal of Foundations of Computer Science 10(02), 147–170 (1999)
9. Kratsch, D., Stewart, L.: Domination on cocomparability graphs. SIAM Journal on Discrete Mathematics 6(3), 400–417 (1993)
10. McConnell, R.M., Spinrad, J.P.: Modular decomposition and transitive orientation. Discrete Mathematics 201(1), 189–241 (1999)
11. Mertzios, G.B., Corneil, D.G.: A simple polynomial algorithm for the longest path problem on cocomparability graphs. SIAM Journal on Discrete Mathematics 26(3), 940–963 (2012)
12. Spinrad, J.P.: Efficient implementation of lexicographic depth first search (submitted)
13. Köhler, E., Mouatadid, L.: Linear time lexdfs on cocomparability graphs. available on arXiv at http://arxiv.org/pdf/1404.5996v1.pdf

Quantum Algorithms for Matrix Products over Semirings

François Le Gall[1] and Harumichi Nishimura[2]

[1] Graduate School of Information Science and Technology,
The University of Tokyo, Japan
[2] Graduate School of Information Science, Nagoya University, Japan

Abstract. In this paper we construct quantum algorithms for matrix products over several algebraic structures called semirings, including the $(\max, \min)$-matrix product, the distance matrix product and the Boolean matrix product. In particular, we obtain the following results.

- We construct a quantum algorithm computing the product of two $n \times n$ matrices over the $(\max, \min)$ semiring with time complexity $O(n^{2.473})$. In comparison, the best known classical algorithm for the same problem has complexity $O(n^{2.687})$. As an application, we obtain a $O(n^{2.473})$-time quantum algorithm for computing the all-pairs bottleneck paths of a graph with n vertices, while classically the best upper bound for this task is $O(n^{2.687})$.
- We construct a quantum algorithm computing the ℓ most significant bits of each entry of the distance product of two $n \times n$ matrices in time $O(2^{0.64\ell} n^{2.46})$. In comparison, prior to the present work, the best known classical algorithm for the same problem had complexity $O(2^{\ell} n^{2.69})$. Our techniques lead to further improvements for classical algorithms as well, reducing the classical complexity to $O(2^{0.96\ell} n^{2.69})$, which gives a sublinear dependency on 2^{ℓ}.

The above two algorithms are the first quantum algorithms that perform better than the $\tilde{O}(n^{5/2})$-time straightforward quantum algorithm based on quantum search for matrix multiplication over these semirings. We also consider the Boolean semiring, and construct a quantum algorithm computing the product of two $n \times n$ Boolean matrices that outperforms the best known classical algorithms for sparse matrices.

1 Introduction

Background. Matrix multiplication over semirings has a multitude of applications in computer science, and in particular in the area of graph algorithms (e.g., [5,18,19,20,21,23]). One example is Boolean matrix multiplication, related for instance to the computation of the transitive closure of a graph, where the product of two $n \times n$ Boolean matrices A and B is defined as the $n \times n$ Boolean matrix $C = A \cdot B$ such that $C[i, j] = 1$ if and only if there exists a $k \in \{1, \ldots, n\}$ such that $A[i, k] = B[k, j] = 1$.

More generally, given a set $R \subseteq \mathbb{Z} \cup \{-\infty, \infty\}$ and two binary operations $\oplus \colon R \times R \to R$ and $\odot \colon R \times R \to R$, the structure $(R, \oplus, \odot)$ is a semiring if it

R Ravi and I.L. Gørtz (Eds.): SWAT 2014, LNCS 8503, pp. 331–343, 2014.

behaves like a ring except that there is no requirement on the existence of an inverse with respect to the operation $\oplus$. Given two $n \times n$ matrices A and B over R, the matrix product over $(R, \oplus, \odot)$ is the $n \times n$ matrix C defined as $C[i,j] = \bigoplus_{k=1}^{n} (A[i,k] \odot B[k,j])$ for any $(i,j) \in \{1, \ldots, n\} \times \{1, \ldots, n\}$. The Boolean matrix product is simply the matrix product over the semiring $(\{0,1\}, \vee, \wedge)$. The $(\max, \min)$-product and the distance product, which both have applications to a multitude of tasks in graph theory such as constructing fast algorithms for all-pairs paths problems (see, e.g., [19]), are the matrix products over the semiring $(\mathbb{Z} \cup \{-\infty, \infty\}, \max, \min)$ and the semiring $(\mathbb{Z} \cup \{\infty\}, \min, +)$, respectively.

Whenever the operation $\oplus$ is such that a term as $\bigoplus_{k=1}^{n} x_k$ can be computed in $\tilde{O}(\sqrt{n})$ time using quantum techniques (e.g., for $\oplus = \vee$ using Grover's algorithm [8] or for $\oplus = \min$ and $\oplus = \max$ using quantum algorithms for minimum finding [7]) and each operation $\odot$ can be implemented in $\text{polylog}(n)$ time, the product of two $n \times n$ matrices over the semiring $(R, \oplus, \odot)$ can be computed in time $\tilde{O}(n^{5/2})$ on a quantum computer.[1] This is true for instance for the Boolean matrix product, and for both the $(\max, \min)$ and distance matrix products.

A fundamental question is whether we can do better than those $\tilde{O}(n^{5/2})$-time straightforward quantum algorithms. For the Boolean matrix product, the answer is affirmative since it can be computed classically in time $\tilde{O}(n^\omega)$, where $\omega < 2.373$ is the exponent of square matrix multiplication over a field. However, Boolean matrix product appears to be an exception, and for most semirings it is not known if matrix multiplication can be done in $\tilde{O}(n^\omega)$-time. For instance, the best known classical algorithm for the $(\max, \min)$-product, by Duan and Pettie [5], has time complexity $\tilde{O}(n^{(3+\omega)/2}) = O(n^{2.687})$ while, for the distance product, no truly subcubic classical algorithm is even known.

Our Results. We construct in this paper the first quantum algorithms with exponent strictly smaller than $5/2$ for matrix multiplication over several semirings.

We first obtain (in Section 4.1) the following result for multiplication over the $(\max, \min)$ semiring.

Theorem 1. *There exists a quantum algorithm that computes, with high probability, the* $(\max, \min)$-*product of two* $n \times n$ *matrices in time* $O(n^{2.473})$.

In comparison, the best known classical algorithm for the $(\max, \min)$-product, by Duan and Pettie [5], has time complexity $\tilde{O}(n^{(3+\omega)/2}) = O(n^{2.687})$, as mentioned above. The $(\max, \min)$-product has mainly been studied in the field in fuzzy logic [6] under the name *composition of relations* and in the context of computing the all-pairs bottleneck paths of a graph (i.e., computing, for all pairs (s,t) of vertices in a graph, the maximum flow that can be routed between s and t). More precisely, it is well known (see, e.g., [5,18,21]) that if the $(\max, \min)$-product of two $n \times n$ matrices can be computed in time $T(n)$, then the all-pairs bottleneck paths of a graph with n vertices can be computed in time $\tilde{O}(T(n))$. As an application of Theorem 1, we thus obtain a $O(n^{2.473})$-time quantum algorithm

---

[1] In this paper the notation $\tilde{O}(\cdot)$ suppresses the $n^{o(1)}$ factors.

computing the all-pairs bottleneck paths of a graph of n vertices, while classically the best upper bound for this task is $O(n^{2.687})$, again from [5].

In order to prove Theorem 1, we construct a quantum algorithm that computes the product of two $n \times n$ matrices over the existence dominance semiring (defined in Section 2) in time $\tilde{O}(n^{(5+\omega)/3}) \leq O(n^{2.458})$. The dominance product has applications in computational geometry [17] and graph algorithms [20] and, in comparison, the best known classical algorithm for this product [23] has complexity $O(n^{2.684})$. Computing efficiently the existence dominance product is, nevertheless, not enough for our purpose. We introduce (in Section 3) a new generalization of it that we call the *generalized existence dominance product*, and construct both quantum and classical algorithms that compute this product.

We also show (in Section 4.2) how these results for the generalized existence dominance product can be used to construct classical and quantum algorithms computing the ℓ most significant bits of each entry of the distance product of two $n \times n$ matrices. In the quantum setting, we obtain time complexity $\tilde{O}\left(2^{0.640\ell}n^{(5+\omega)/3}\right) \leq O(2^{0.640\ell}n^{2.458})$. In comparison, prior to the present work, the best known classical algorithm for the same problem by Vassilevska and Williams [20] had time complexity $\tilde{O}\left(2^{\ell}n^{(3+\omega)/2}\right) \leq O(2^{\ell}n^{2.687})$, with a slight improvement on the exponent of n obtained later by Yuster [23]. We obtain an improvement for this classical time complexity as well, reducing it to $\tilde{O}\left(2^{0.960\ell}n^{(3+\omega)/2}\right)$, which gives a sublinear dependency on 2^{ℓ}.

These results are, to the best of our knowledge, the first quantum algorithms for matrix multiplication over semirings other than the Boolean semiring improving over the straightforward $\tilde{O}(n^{5/2})$-time quantum algorithm, and the first nontrivial quantum algorithms offering a speedup with respect to the best classical algorithms for matrix multiplication when no assumptions are made on the sparsity of the matrices involved (sparse matrix multiplication is discussed below). This shows that, while quantum algorithms may not be able to outperform the classical $\tilde{O}(n^{\omega})$-time algorithm for matrix multiplication of (dense) matrices over a ring, they can offer a speedup for matrix multiplication over other algebraic structures.

We finally investigate under which conditions quantum algorithms faster than the best known classical algorithms can be constructed for Boolean matrix multiplication. This question has been recently studied extensively in the output-sensitive scenario [3,10,12,13], for which quantum algorithms multiplying two $n \times n$ Boolean matrices with query complexity $\tilde{O}(n\sqrt{\lambda})$ and time complexity $\tilde{O}(n\sqrt{\lambda} + \lambda\sqrt{n})$ were constructed, where λ denotes the number of non-zero entries in the output matrix. In this work, we focus on the case where the input matrices are sparse (but not necessarily the output matrix), and obtain the following result.

Theorem 2 (simplified version). *Let A and B be two $n \times n$ Boolean matrices each containing at most m non-zero entries. There exists a quantum algorithm that computes, with high probability, the Boolean matrix product $A \cdot B$ and has time complexity*

$$\begin{cases} \tilde{O}(n^2) & \text{if } m \le n^{1.151}, \\ \tilde{O}\left(m^{0.517}n^{1.406}\right) & \text{if } n^{1.151} \le m \le n^{\omega-1/2}, \\ \tilde{O}(n^\omega) & \text{if } n^{\omega-1/2} \le m \le n^2. \end{cases}$$

In comparison, the best known classical algorithm, by Yuster and Zwick [24], has complexity $\tilde{O}(n^2)$ if $m \le n^{1.151}$, $\tilde{O}(m^{0.697}n^{1.199})$ if $n^{1.151} \le m \le n^{(1+\omega)/2}$, and $\tilde{O}(n^\omega)$ if $n^{(1+\omega)/2} \le m \le n^2$. Our algorithm performs better when $n^{1.151} < m < n^{\omega-1/2}$. For instance, if $m = O(n^{(1+\omega)/2}) = O(n^{1.686\cdots})$, then our algorithm has complexity $O(n^{2.277})$, while the algorithm in [24] has complexity $\tilde{O}(n^\omega)$. The complete statement of Theorem 2, and its proof, are given in the full version of the present paper [15].

Our main quantum tool is rather standard: quantum enumeration, a variant of Grover's search algorithm. We use this technique in various ways to improve the combinatorial steps in several classical approaches [1,5,21,24] that are based on a combination of algebraic steps (computing some matrix products over a field) and combinatorial steps. Moreover, the speedup obtained by quantum enumeration enables us to depart from these original approaches and optimize the combinatorial and algebraic steps in different ways, for instance relying on rectangular matrix multiplication instead of square matrix multiplication. On the other hand, several subtle but crucial issues appear when trying to apply quantum enumeration, such as how to store and access information computed during the preprocessing steps, which induces complications and requires the introduction of new algorithmic ideas. We end up with algorithms fairly remote from these original approaches, where most steps are tailored for the use of quantum enumeration.

2 Preliminaries

Rectangular Matrix Multiplication over Fields. For any $k_1, k_2, k_3 > 0$, let $\omega(k_1, k_2, k_3)$ represent the minimal value τ such that, over a field, the product of an $n^{k_1} \times n^{k_2}$ matrix by an $n^{k_2} \times n^{k_3}$ matrix can be computed with $\tilde{O}(n^\tau)$ arithmetic operations. The value $\omega(1, 1, 1)$ is denoted by ω, and the current best upper bound on ω is $\omega < 2.373$, see [14,22]. Other important quantities are the value $\alpha = \sup\{k \mid \omega(1, k, 1) = 2\}$ and the value $\beta = (\omega - 2)/(1 - \alpha)$. The current best lower bound on α is $\alpha > 0.302$, see [11]. The following facts are known, and will be used in this paper. We refer to [4,9] for details.

Fact 1. $\omega(1, k, 1) = 2$ for $k \le \alpha$ and $\omega(1, k, 1) \le 2 + \beta(k - \alpha)$ for $\alpha \le k \le 1$.

Fact 2. *The following relations hold for any values $k_1, k_2, k_3 > 0$: (i) for any $k > 0$, $\omega(kk_1, kk_2, kk_3) = k\omega(k_1, k_2, k_3)$; (ii) $\omega(k_{\pi(1)}, k_{\pi(2)}, k_{\pi(3)}) = \omega(k_1, k_2, k_3)$ for any permutation π over $\{1, 2, 3\}$; (iii) $\omega(k_1, k_2, 1 + k_3) \le \omega(k_1, k_2, 1) + k_3$; (iv) $\omega(k_1, k_2, k_3) \ge \max\{k_1 + k_2, k_1 + k_3, k_2 + k_3\}$.*

Matrix Products over Semirings. We define below two matrix products over semirings considered in Sections 3 and 4, respectively, additionally to the

Boolean product, the (max, min)-product and the distance product defined in the introduction. These products were also used in [5,20,21].

Definition 1. *Let A be an $n \times n$ matrix with entries in $\mathbb{Z} \cup \{\infty\}$ and B be an $n \times n$ matrix with entries in $\mathbb{Z} \cup \{-\infty\}$. The existence dominance product of A and B, denoted $A * B$, is the $n \times n$ Boolean matrix C such that $C[i, j] = 1$ if and only if there exists some $k \in \{1, \ldots, n\}$ such that $A[i, k] \leq B[k, j]$. The product $A \lhd B$ is the $n \times n$ matrix C such that $C[i, j] = -\infty$ if $A[i, k] > B[k, j]$ for all $k \in \{1, \ldots, n\}$, and $C[i, j] = \max_k\{A[i, k] \mid A[i, k] \leq B[k, j]\}$ otherwise.*

It is easy to check, as mentioned for instance in [5,21], that computing the (max, min)-product reduces to computing the product $\lhd$. Indeed if C denotes the (max, min)-product of two matrices A and B, then for any $(i, j) \in \{1, \ldots, n\} \times \{1, \ldots, n\}$ we can write $C[i, j] = \max\left\{(A \lhd B)[i, j], (B^T \lhd A^T)[j, i]\right\}$, where A^T and B^T denote the transposes of A and B, respectively. Matrix products over the semirings $(\min, \max)$, $(\min, \leq)$ and $(\max, \geq)$ studied, for instance, in [19], similarly reduce to computing the product $\lhd$.

Quantum Algorithms for Matrix Multiplication. We assume that a quantum algorithm can access any entry of the input matrix in a random access way, similarly to the standard model used in [3,10,12,13] for Boolean matrix multiplication.

We will use variants of Grover's search algorithm, as described for instance in [2], to find elements satisfying some conditions inside a search space of size N. Concretely, suppose that a Boolean function $f \colon \{1, \ldots, N\} \to \{0, 1\}$ is given and that we want to find a solution, i.e., an element $x \in \{1, \ldots, n\}$ such that $f(x) = 1$. Consider the quantum search procedure (called safe Grover search in [16]) obtained by repeating Grover's standard search a logarithmic number of times, and checking if a solution has been found. This quantum procedure outputs one solution with probability at least $1 - 1/\text{poly}(N)$ if a solution exists, and always rejects if no solution exists. Its time complexity is $\tilde{O}(\sqrt{N/\max(1, t)})$, where t denotes the number of solutions, if the function f can be evaluated in $\tilde{O}(1)$ time. By repeating this procedure and striking out solutions as soon as they are found, one can find all the solutions with probability at least $1 - 1/\text{poly}(N)$ using $\tilde{O}(\sqrt{N/t} + \sqrt{N/(t-1)} + \cdots + \sqrt{N/1}) = \tilde{O}(\sqrt{N(t+1)})$ computational steps. We call this procedure *quantum enumeration*.

3 Existence Dominance Matrix Multiplication

In this section we present a quantum algorithm that computes the existence dominance product of two matrices A and B. The underlying idea of our algorithm is similar to the idea in the best classical algorithm for the same problem by Duan and Pettie [5]: use a search step to find some of the entries of $A * B$, and rely on classical algebraic algorithms to find the other entries. We naturally use quantum search to implement the first part, and perform careful modifications of their approach to improve the complexity in the quantum setting, taking advantage of the features of quantum enumeration. There are two notable differences: The first one is that the algebraic part of our quantum algorithms uses rectangular matrix multiplication, while [5] uses square matrix multiplication. The second and crucial difference is that, for applications in later sections, we give a quantum algorithm that can handle a new (and more general) version of the existence dominance product, defined on set of matrices, which we call the *generalized existence dominance product* and define below.

Definition 2. *Let u, v be two positive integers, and S be the set $S = \{1, \ldots, u\} \times \{1, \ldots, v\}$. Let $\prec$ be the lexicographic order over $S \cup \{(0,0)\}$ (i.e., $(i,j) \prec (i',j')$ if and only if $i < i'$ or $(i = i'$ and $j < j'))$. Consider u matrices $A^{(1)}, \ldots, A^{(u)}$, each of size $n \times n$ with entries in $\mathbb{Z} \cup \{\infty\}$, and v matrices $B^{(1)}, \ldots, B^{(v)}$, each of size $n \times n$ with entries in $\mathbb{Z} \cup \{-\infty\}$. For each $(i,j) \in \{1, \ldots, n\} \times \{1, \ldots, n\}$ define the set $S_{ij} \subseteq S \cup \{(0,0)\}$ as follows:*

$$S_{ij} = \{(x,y) \in S \mid A^{(x)} * B^{(y)}[i,j] = 1\} \cup \{(0,0)\}.$$

The generalized existence dominance product of these matrices is the $n \times n$ matrix C with entries in $S \cup \{(0,0)\}$ defined as follows: for all $(i,j) \in \{1, \ldots, n\} \times \{1, \ldots, n\}$ the entry $C[i,j]$ is the maximum element in S_{ij}, where the maximum refers to the lexicographic order.

Note that the case $u = v = 1$ corresponds to the standard existence dominance product, since $C[i,j] = (1,1)$ if $A^{(1)} * B^{(1)}[i,j] = 1$ and $C[i,j] = (0,0)$ if $A^{(1)} * B^{(1)}[i,j] = 0$.

Proposition 1. *Let $A^{(1)}, \ldots, A^{(u)}$ be u matrices of size $n \times n$ with entries in $\mathbb{Z} \cup \{\infty\}$, and $B^{(1)}, \ldots, B^{(v)}$ be v matrices of size $n \times n$ with entries in $\mathbb{Z} \cup \{-\infty\}$. Let $m_1 \in \{1, \ldots, n^2 u\}$ denote the total number of finite entries in the matrices $A^{(1)}, \ldots, A^{(u)}$, and $m_2 \in \{1, \ldots, n^2 v\}$ denote the total number of finite entries in the matrices $B^{(1)}, \ldots, B^{(v)}$. For any parameter $t \in \{1, \ldots, m_1\}$, there exists a quantum algorithm that computes, with high probability, their generalized existence dominance product in time*

$$\tilde{O}\left(\sqrt{\frac{m_1 m_2 n}{t}} + \sqrt{\frac{m_1 m_2 uv}{tn}} + n^{\omega(1+\log_n u, 1+\log_n t, 1+\log_n v)}\right).$$

Proof. Let $t \in \{1, \ldots, m_1\}$ be a parameter to be chosen later. Let L be the list of all finite entries in $A^{(1)}, \ldots, A^{(u)}$ sorted in increasing order. Decompose L into t successive parts $L_1, \ldots, L_t$, each containing at most $\lceil m_1/t \rceil$ entries. For each $x \in \{1, \ldots, u\}$ and each $r \in \{1, \ldots, t\}$ we construct two $n \times n$ matrices $A_r^{(x)}, \bar{A}_r^{(x)}$ as follows: for all $(i,j) \in \{1, \ldots, n\} \times \{1, \ldots, n\}$,

$$A_r^{(x)}[i,j] = \begin{cases} A^{(x)}[i,j] \text{ if } A^{(x)}[i,j] \in L_r, \\ \infty \text{ otherwise,} \end{cases} \qquad \bar{A}_r^{(x)}[i,j] = \begin{cases} 1 \text{ if } A^{(x)}[i,j] \in L_r, \\ 0 \text{ otherwise.} \end{cases}$$

Similarly, for each $y \in \{1, \ldots, v\}$ and each $r \in \{1, \ldots, t\}$ we construct two $n \times n$ matrices $B_r^{(y)}, \bar{B}_r^{(y)}$ as follows: for all $(i,j) \in \{1, \ldots, n\} \times \{1, \ldots, n\}$,

$$B_r^{(y)}[i,j] = \begin{cases} B^{(y)}[i,j] \text{ if } \min L_r \leq B^{(y)}[i,j] < \max L_r, \\ -\infty \text{ otherwise,} \end{cases}$$

$$\bar{B}_r^{(y)}[i,j] = \begin{cases} 1 \text{ if } B^{(y)}[i,j] \geq \max L_r, \\ 0 \text{ otherwise.} \end{cases}$$

The cost of this (classical) preprocessing step is $O(n^2 t(u + v))$ time.

It is easy to see that, for each $x \in \{1, \ldots, u\}$ and $y \in \{1, \ldots, v\}$, the following equality holds (where the operators $+$ and $\sum$ refer to the entry-wise OR):

$$A^{(x)} * B^{(y)} = \sum_{r=1}^{l} \left(\bar{A}_r^{(x)} \cdot \bar{B}_r^{(y)} \right) + \sum_{r=1}^{t} \left(A_r^{(x)} * B_r^{(y)} \right). \tag{1}$$

Indeed, the second term compares entries that are in a same part L_r, while the first term takes into consideration entries in distinct parts. Define two $n \times n$ matrices C_1 and C_2 with entries in $S \cup \{(0,0)\}$ as follows: for all $(i,j) \in \{1, \ldots, n\} \times \{1, \ldots, n\}$,

$$C_1[i,j] = \max \left\{ \{(0,0)\} \cup \{(x,y) \in S \mid \sum_{r=1}^{t} \bar{A}_r^{(x)} \cdot \bar{B}_r^{(y)}[i,j] = 1\} \right\}, \tag{2}$$

$$C_2[i,j] = \max \left\{ \{(0,0)\} \cup \{(x,y) \in S \mid \sum_{r=1}^{t} A_r^{(x)} * B_r^{(y)}[i,j] = 1\} \right\}. \tag{3}$$

From Equation (1), the generalized existence dominance product C satisfies $C[i,j] = \max\{C_1[i,j], C_2[i,j]\}$ for all $(i,j) \in \{1, \ldots, n\} \times \{1, \ldots, n\}$. The matrix C can then be computed in time $O(n^2)$ from C_1 and C_2.

The matrix C_1 can clearly be computed in time $O(n^2 uv)$ if all the terms $\sum_r \bar{A}_r^{(x)} \cdot \bar{B}_r^{(y)}$ are known. We can obtain all these uv terms by computing the following Boolean product of an $nu \times nt$ matrix by an $nt \times nv$ matrix (both matrices can be constructed in time $\tilde{O}(n^2 t(u + v))$).

$$\begin{bmatrix} \bar{A}_1^{(1)} & \cdots & \bar{A}_t^{(1)} \\ \vdots & & \vdots \\ \bar{A}_1^{(u)} & \cdots & \bar{A}_t^{(u)} \end{bmatrix} \cdot \begin{bmatrix} \bar{B}_1^{(1)} & \cdots\cdots & \bar{B}_1^{(v)} \\ \vdots & & \vdots \\ \bar{B}_t^{(1)} & \cdots\cdots & \bar{B}_t^{(v)} \end{bmatrix}$$

The cost of this matrix multiplication is $\tilde{O}\left(n^{\omega(1+\log_n u, 1+\log_n t, 1+\log_n v)}\right)$. From item (iv) of Fact 2, we conclude that the matrix C_1 can be computed in time

$$\tilde{O}\left(n^2 uv + n^2 t(u + v) + n^{\omega(1+\log_n u, 1+\log_n t, 1+\log_n v)}\right)$$

$$= \tilde{O}\left(n^{\omega(1+\log_n u, 1+\log_n t, 1+\log_n v)}\right).$$

We now explain how to compute the matrix C_2. Intuitively, the main difficulty is that Equation (3) cannot be used directly since we do not know how to compute the dominance product $*$ efficiently. Lemma 1 below shows that it is possible to replace this dominance product by a Boolean product if we replace the matrices $A_r^{(x)}$ and $B_r^{(y)}$ by some Boolean matrices $\hat{A}_r^{(x)}$ and $\hat{B}_r^{(y)}$ (compare Equation (3) with Equation (4) below). This lemma further shows that the latter matrices can be computed efficiently by a quantum algorithm (based on quantum search). Actually, for technical reasons we additionally need to replace the term $\{(0,0)\}$ in Equation (3) by the term $\{D[i,j]\}$ in Equation (4), where D is a matrix that can also be computed efficiently using a quantum algorithm. While this lemma is the main technical part of the proof of this proposition, due to space constraints its proof is omitted (we refer to [15] for all details).

Lemma 1. *There exists a quantum algorithm that, with high probability, outputs*

- *tu Boolean matrices $\hat{A}_r^{(x)}$, each of size $n \times 2n$, for all $x \in \{1,\ldots,u\}$ and $r \in \{1,\ldots,t\}$,*
- *tv Boolean matrices $\hat{B}_r^{(y)}$, each of size $2n \times n$, for all $y \in \{1,\ldots,v\}$ and $r \in \{1,\ldots,t\}$,*
- *a matrix D of size $n \times n$ with entries in $S \cup \{(0,0)\} = (\{1,\ldots,u\} \times \{1,\ldots,v\}) \cup \{(0,0)\}$,*

such that

$$C_2[i,j] = \max \left\{ \{D[i,j]\} \cup \{(x,y) \in S \mid \sum_{r=1}^{t} \hat{A}_r^{(x)} \cdot \hat{B}_r^{(y)}[i,j] = 1\} \right\} \quad (4)$$

for all $(i,j) \in \{1,\ldots,n\} \times \{1,\ldots,n\}$. The time complexity of this quantum algorithm is

$$\tilde{O}\left(n^2 t(u+v) + \sqrt{\frac{m_1 m_2 n}{t}} + \sqrt{\frac{m_1 m_2 uv}{tn}} \right).$$

After applying the quantum algorithm of Lemma 1, we can obtain the matrix C_2, similarly to the computation of C_1, if we know all the terms $\sum_r \hat{A}_r^{(x)} \cdot \hat{B}_r^{(y)}$. we obtain all these uv terms by computing the following Boolean product of an $nu \times nt$ matrix by an $nt \times nv$ matrix.

$$\begin{bmatrix} \hat{A}_1^{(1)} & \cdots & \hat{A}_t^{(1)} \\ \vdots & & \vdots \\ \hat{A}_1^{(u)} & \cdots & \hat{A}_t^{(u)} \end{bmatrix} \cdot \begin{bmatrix} \hat{B}_1^{(1)} & \cdots\cdots & \hat{B}_1^{(v)} \\ \vdots & & \vdots \\ \hat{B}_t^{(1)} & \cdots\cdots & \hat{B}_t^{(v)} \end{bmatrix}$$

The cost of this matrix multiplication is $\tilde{O}\left(n^{\omega(1+\log_n u, 1+\log_n t, 1+\log_n v)} \right)$. The total cost of computing the matrix C_2 is thus

$$\tilde{O}\left(n^2 t(u+v) + \sqrt{\frac{m_1 m_2 n}{t}} + \sqrt{\frac{m_1 m_2 uv}{tn}} + n^{\omega(1+\log_n u, 1+\log_n t, 1+\log_n v)} \right),$$

which is the desired bound since the term $n^2 t(u+v)$ is negligible here by item (iv) of Fact 2.   □

We can also give a classical version of the algorithm of Proposition 1, as stated in the following proposition (see [15] for a proof).

Proposition 2. *There exists a classical algorithm that computes the generalized existence dominance product in time* $\tilde{O}\left(\frac{m_1 m_2}{tn} + n^{\omega(1+\log_n u, 1+\log_n t, 1+\log_n v)}\right)$, *for any parameter* $t \in \{1, \ldots, m_1\}$.

We now consider the case $u = v = 1$ corresponding to the standard existence dominance product. By optimizing the choice of the parameter t in Proposition 1, we obtain the following theorem.

Theorem 3. *Let A be an $n \times n$ matrix with entries in $\mathbb{Z} \cup \{\infty\}$ containing at most m_1 non-(∞) entries, and B be an $n \times n$ matrix with entries in $\mathbb{Z} \cup \{-\infty\}$ containing at most m_2 non-$(-\infty)$ entries. There exists a quantum algorithm that computes, with high probability, the existence dominance product of A and B in time $\tilde{O}(\sqrt{m_1 m_2 n^{1-\mu}})$, where μ is the solution of the equation $\mu + 2\omega(1, 1 + \mu, 1) = 1 + \log_n(m_1 m_2)$. In particular, this time complexity is upper bounded by $\tilde{O}\left((m_1 m_2)^{1/3} n^{(\omega+1)/3}\right)$.*

Proof. The complexity of the algorithm of Proposition 1 is minimized for $t = n^\mu$, where μ is the solution of the equation $\mu + 2\omega(1, 1 + \mu, 1) = 1 + \log_n(m_1 m_2)$. We can use items (ii) and (iii) of Fact 2 to obtain the upper bound $\omega(1, 1 + \mu, 1) \leq \omega + \mu$, and optimize the complexity of the algorithm by taking $t = \lceil (m_1 m_2)^{1/3} n^{(1-2\omega)/3} \rceil$, which gives the upper bound claimed in the second part of the theorem. □

In the case of completely dense input matrices (i.e., $m_1 \approx n^2$ and $m_2 \approx n^2$), the second part of Theorem 3 shows that the complexity of the algorithm is $\tilde{O}(n^{(5+\omega)/3}) \leq O(n^{2.458})$.

4   Applications: (max, min)-Product, Distance Product

4.1   Quantum Algorithm for the (max, min)-Product

In this subsection we present a quantum algorithm for the matrix product $\lhd$, which immediately gives a quantum algorithm with the same complexity for the (max, min)-product as explained in Section 2, and then gives Theorem 1. Our algorithm first exploits the methodology by Vassilevska et al. [21] to reduce the computation of the product $\lhd$ to the computation of several sparse dominance products. The main technical difficulty to overcome is that, unlike in the classical case, computing all the sparse dominance products successively becomes too costly (i.e., the cost exceeds the complexity of all the other parts of the quantum algorithm). Instead, we show that it is sufficient to obtain a small fraction of the entries in each dominance product and that this task reduces to the computation of a generalized existence dominance product, and then use the quantum techniques of Proposition 1 to obtain precisely only those entries.

Theorem 4. *There exists a quantum algorithm that computes, for any two $n \times n$ matrices A and B with entries respectively in $\mathbb{Z} \cup \{\infty\}$ and $\mathbb{Z} \cup \{-\infty\}$, the product $A \lhd B$ with high probability in time $\tilde{O}(n^{(5-\gamma)/2})$, where γ is the solution of the equation $\gamma + 2\omega(1+\gamma, 1+\gamma, 1) = 5$. In particular, this complexity is upper bounded by $O(n^{2.473})$.*

Proof. Let $g \in \{1, \ldots, n\}$ be a parameter to be chosen later. For each $i \in \{1, \ldots, n\}$, we sort the entries in the i-th row of A in increasing order and divide the list into $s = \lceil n/g \rceil$ successive parts $R_1^i, \ldots, R_s^i$ with at most g entries in each part. For each $r \in \{1, \ldots, s\}$, define the $n \times n$ matrix A_r as follows: $A_r[i,j] = A[i,j]$ if $A[i,j] \in R_r^i$ and $A_r[i,j] = \infty$ otherwise. The cost of this (classical) preprocessing is $O(n^2 s)$ time.

We describe below the quantum algorithm that computes $C = A \lhd B$.

Step 1. For each $(i,j) \in \{1, \ldots, n\} \times \{1, \ldots, n\}$, we compute the largest $r \in \{1, \ldots, s\}$ such that $(A_r * B)[i,j] = 1$, if such an r exists. This is done by using the quantum algorithm of Proposition 1 with $u = s$, $v = 1$, $A^{(r)} = A_r$ for each $r \in \{1, \ldots, s\}$ and $B^{(1)} = B$. Note that $m_1 \leq s \times (ng) = O(n^2)$ and $m_2 \leq n^2$. The complexity of this step is thus

$$\tilde{O}\left(\frac{n^{5/2}}{\sqrt{t}} + n^{\omega(1 + \log_n s, 1 + \log_n t, 1)}\right)$$

for any parameter $t \in \{1, \ldots, n^2\}$. We want to minimize this expression. Let us write $t = n^\gamma$ and $g = n^\delta$. For a fixed δ, the first term is a decreasing function of γ, while the second term is an increasing function of γ. The expression is thus minimized for the value of γ solution of the equation

$$\omega(2 - \delta, 1 + \gamma, 1) = (5 - \gamma)/2, \tag{5}$$

in which case the expression becomes $\tilde{O}(n^{(5-\gamma)/2})$.

Step 2. Note that at Step 1 we also obtain all $(i,j) \in \{1, \ldots, n\} \times \{1, \ldots, n\}$ such that no r satisfying $(A_r * B)[i,j] = 1$ exists. For all those (i,j), we set $C[i,j] = -\infty$. For all other (i,j), we will denote by r_{ij} the value found at Step 1. We now know that

$$C[i,j] = \max_{k: A[i,k] \in R_{r_{ij}}^i} \{A_{r_{ij}}[i,k] \mid A_{r_{ij}}[i,k] \leq B[k,j]\},$$

and $C[i,j]$ can be computed in time $\tilde{O}(\sqrt{g})$ using the quantum algorithm for maximum finding [7], since $|R_{r_{ij}}^i| \leq g$. The complexity of Step 2 is thus $\tilde{O}(n^2 \sqrt{g})$.

This algorithm computes, with high probability, all the entries of $C = A \lhd B$. Its complexity is

$$\tilde{O}\left(n^2 s + n^{(5-\gamma)/2} + n^2 \sqrt{g}\right) = \tilde{O}\left(n^{(5-\gamma)/2} + n^{2+\delta/2}\right),$$

since the term $n^2 s = n^{3-\delta}$ is negligible with respect to $n^{(5-\gamma)/2} = n^{\omega(2-\delta,1+\gamma,1)}$ by item (iv) of Fact 2. This expression is minimized for δ and γ satisfying $\delta + \gamma = 1$. Injecting this constraint into Equation (5), we find that the optimal value of γ is the solution of the equation $\gamma + 2\omega(1+\gamma, 1+\gamma, 1) = 5$, as claimed. Using items (i) and (ii) of Fact 2 and Fact 1, we obtain

$$5 = \gamma + 2(1+\gamma)\omega\left(1, 1, \frac{1}{1+\gamma}\right) \le \gamma + 2(1+\gamma)\left(2 + \beta\left(\frac{1}{1+\gamma} - \alpha\right)\right)$$
$$= (4 + 2\beta - 2\alpha\beta) + (5 - 2\alpha\beta)\gamma$$

and then $\gamma \ge \frac{1+2\beta-2\beta}{5-2\alpha\beta}$. The complexity is thus $\tilde{O}\left(n^{(12-6\alpha\beta+\beta)/(5-2\alpha\beta)}\right) \le O(n^{2.473})$. $\qquad\square$

4.2   Quantum Algorithm for the Distance Product

In this subsection we present a quantum algorithm that computes the most significant bits of the distance product of two matrices, as defined below.

Let A and B be two $n \times n$ matrices with entries in $\mathbb{Z} \cup \{\infty\}$. Let W be a power of two such that the value of each finite entry of their distance product C is upper bounded by W. For instance, one can take the smallest power of two larger than $\max_{i,j}\{A[i,j]\} + \max_{i,j}\{B[i,j]\}$, where the maxima are over the finite entries of the matrices. Each non-negative finite entry of C can then be expressed using $\log_2(W)$ bits: the entry $C[i,j]$ can be expressed as $C[i,j] = \sum_{k=1}^{\log_2(W)} C[i,j]_k \frac{W}{2^k}$ for bits $C[i,j]_1, \ldots, C[i,j]_{\log_2(W)}$. For any $\ell \in \{1, \ldots, \log_2(W)\}$, we say that an algorithm computes the ℓ most significant bits of each entry if, for all $(i,j) \in \{1, \ldots, n\} \times \{1, \ldots, n\}$ such that $C[i,j]$ is finite and non-negative, the algorithm outputs all the bits $C[i,j]_1, C[i,j]_2, \cdots, C[i,j]_\ell$. Vassilevska and Williams [20] have studied this problem, and shown how to reduce the computation of the ℓ most significant bits to the computation of $O(2^\ell)$ existence dominance matrix products of $n \times n$ matrices. By combining this with the $\tilde{O}(n^{(3+\omega)/2})$-time algorithm for dominance product from [17], they obtained a classical algorithm that computes the ℓ most significant bits of each entry of the distance product of A and B in time $\tilde{O}\left(2^\ell n^{(3+\omega)/2}\right) \le \tilde{O}\left(2^\ell n^{2.687}\right)$.

Here is the main result of this subsection, whose proof is given in [15], obtained by reducing the computation of the ℓ most significant bits to computing a generalized existence dominance product.

Theorem 5. *There exists a quantum algorithm that computes, for any two $n \times n$ matrices A and B with entries in $\mathbb{Z} \cup \{\infty\}$, the ℓ most significant bits of each entry of the distance product of A and B in time $\tilde{O}\left(2^{0.640\ell} n^{(5+\omega)/3}\right) \le O(2^{0.640\ell} n^{2.458})$ with high probability.*

Similarly, we can obtain a better classical algorithm as shown in the following theorem. We refer to [15] for details.

Theorem 6. *There exists a classical algorithm that computes, for any two $n \times n$ matrices A and B with entries in $\mathbb{Z} \cup \{\infty\}$, the ℓ most significant bits of each entry of the distance product of A and B in time $\tilde{O}\left(2^{0.960\ell} n^{(3+\omega)/2}\right) \leq O(2^{0.960\ell} n^{2.687})$.*

References

1. Amossen, R.R., Pagh, R.: Faster join-projects and sparse matrix multiplications. In: Proceedings of ICDT, pp. 121–126 (2009)
2. Boyer, M., Brassard, G., Høyer, P., Tapp, A.: Tight bounds on quantum searching. Fortschritte der Physik 46(4-5), 493–505 (1998)
3. Buhrman, H., Špalek, R.: Quantum verification of matrix products. In: Proceedings of SODA, pp. 880–889 (2006)
4. Bürgisser, P., Clausen, M., Shokrollahi, M.A.: Algebraic complexity theory. Springer (1997)
5. Duan, R., Pettie, S.: Fast algorithms for (max, min)-matrix multiplication and bottleneck shortest paths. In: Proceedings of SODA, pp. 384–391 (2009)
6. Dubois, D., Prade, H.: Fuzzy sets and systems: Theory and applications. Academic Press (1980)
7. Dürr, C., Høyer, P.: A quantum algorithm for finding the minimum. arXiv:quant-ph/9607014 (1996)
8. Grover, L.K.: A fast quantum mechanical algorithm for database search. In: Proceedings of STOC, pp. 212–219 (1996)
9. Huang, X., Pan, V.Y.: Fast rectangular matrix multiplication and applications. Journal of Complexity 14(2), 257–299 (1998)
10. Jeffery, S., Kothari, R., Magniez, F.: Improving quantum query complexity of Boolean matrix multiplication using graph collision. In: Czumaj, A., Mehlhorn, K., Pitts, A., Wattenhofer, R. (eds.) ICALP 2012, Part I. LNCS, vol. 7391, pp. 522–532. Springer, Heidelberg (2012)
11. Le Gall, F.: Faster algorithms for rectangular matrix multiplication. In: Proceedings of FOCS, pp. 514–523 (2012)
12. Le Gall, F.: Improved output-sensitive quantum algorithms for Boolean matrix multiplication. In: Proceedings of SODA, pp. 1464–1476 (2012)
13. Le Gall, F.: A time-efficient output-sensitive quantum algorithm for Boolean matrix multiplication. In: Chao, K.-M., Hsu, T.-S., Lee, D.-T. (eds.) ISAAC 2012. LNCS, vol. 7676, pp. 639–648. Springer, Heidelberg (2012)
14. Le Gall, F.: Powers of tensors and fast matrix multiplication. In: Proceedings of ISSAC (to appear, 2014)
15. Le Gall, F., Nishimura, H.: Quantum algorithms for matrix products over semirings. Full version of the present paper, available as arXiv:1310.3898
16. Magniez, F., Santha, M., Szegedy, M.: Quantum algorithms for the triangle problem. SIAM Journal on Computing 37(2), 413–424 (2007)
17. Matoušek, J.: Computing dominances in E^n. Information Processing Letters 38(5), 277–278 (1991)
18. Shapira, A., Yuster, R., Zwick, U.: All-pairs bottleneck paths in vertex weighted graphs. In: Proceedings of SODA, pp. 978–985 (2007)
19. Vassilevska, V.: Efficient Algorithms for Path Problems in Weighted Graphs. PhD thesis, Carnegie Mellon University (2008)
20. Vassilevska, V., Williams, R.: Finding a maximum weight triangle in $n^{3-\delta}$ time, with applications. In: Proceedings of STOC, pp. 225–231 (2006)

21. Vassilevska, V., Williams, R., Yuster, R.: All pairs bottleneck paths and max-min matrix products in truly subcubic time. Theory of Computing 5(1), 173–189 (2009)
22. Vassilevska Williams, V.: Multiplying matrices faster than Coppersmith-Winograd. In: Proceedings of STOC, pp. 887–898 (2012)
23. Yuster, R.: Efficient algorithms on sets of permutations, dominance, and real-weighted APSP. In: Proceedings of SODA, pp. 950–957 (2009)
24. Yuster, R., Zwick, U.: Fast sparse matrix multiplication. ACM Transactions on Algorithms 1(1), 2–13 (2005)

Ranked Document Selection*

J. Ian Munro[1], Gonzalo Navarro[2], Rahul Shah[3], and Sharma V. Thankachan[1]

[1] Cheriton School of CS, Univ. Waterloo, Canada
{imunro,thanks}@uwaterloo.ca
[2] Dept. of CS, Univ. Chile, Chile
gnavarro@dcc.uchile.cl
[3] School of EECS, Louisiana State Univ., USA
rahul@csc.lsu.edu

Abstract. Let $\mathcal{D}$ be a collection of string documents of n characters in total. The *top-k document retrieval problem* is to preprocess $\mathcal{D}$ into a data structure that, given a query (P, k), can return the k documents of $\mathcal{D}$ most relevant to pattern P. The relevance of a document d for a pattern P is given by a predefined ranking function $w(P, d)$. Linear space and optimal query time solutions already exist for this problem.

In this paper we consider a novel problem, *document selection* queries, which aim to report the kth document most relevant to P (instead of reporting all top-k documents). We present a data structure using $O(n \log^\epsilon n)$ space, for any constant $\epsilon > 0$, answering selection queries in time $O(\log k / \log \log n)$, and a linear-space data structure answering queries in time $O(\log k)$, given the locus node of P in a (generalized) suffix tree of $\mathcal{D}$. We also prove that it is unlikely that a succinct-space solution for this problem exists with poly-logarithmic query time.

1 Introduction and Related Work

Document retrieval is a special branch of pattern matching related to information retrieval and web searching. In this problem, the data consists of a collection of text *documents*, and the queries refer to documents rather than text positions [12]. In this paper we focus on arguably the most important of those problems, called *top-k document retrieval*: Given $\mathcal{D} = \{d_1, d_2, d_3, ..., d_D\}$, of total length $n = \sum_{i=1}^{D} |d_i|$, preprocess it into a data structure that, given a pattern P and a threshold k, retrieves the k documents from $\mathcal{D}$ that are more most *relevant* to P, in decreasing order of relevance. The relevance of a document d with respect to P is captured using any function $w(P, d)$ of the starting positions of the occurrences of P in d. A popular example of relevance is the *term frequency* metric, that is, the number of occurrences of P in d. This a well studied problem, and the best known linear space data structure can answer queries in optimal time $O(k)$ [17], once the locus node of P in a generalized suffix tree of $\mathcal{D}$ is found.

* Funded in part by NSERC of Canada and the Canada Research Chairs program, Fondecyt Grant 1-140796, Chile, and NSF Grants CCF–1017623, CCF–1218904.

R Ravi and I.L. Gørtz (Eds.): SWAT 2014, LNCS 8503, pp. 344–356, 2014.
© Springer International Publishing Switzerland 2014

In this paper we study a new related problem called *document selection*, where we must return the kth document of $\mathcal{D}$ most relevant to P, that is, the kth element returned by a top-k query (breaking ties arbitrarily).

We present three results, depending on the amount of space used: (1) We give a data structure that uses $O(n \log^{\epsilon} n)$ space, for any constant $\epsilon > 0$, and answers queries in time $O(\log k / \log \log n)$. (2) We give a linear-space data structure that answers queries in $O(\log k)$ time. (3) We prove that it is highly unlikely that the problem can be solved in less than linear space within poly-logarithmic time, via a reduction from the *position restricted substring searching* problem [9,5].

Document selection is useful for various advanced queries. When a user browses ranked results of a query and asks for the next set of results, we need to report the top-k_2 documents that are not top-k_1. Instead of computing a top-k_2 query in time $O(k_2)$, which is nonoptimal if $k_2 - k_1 = o(k_2)$, our results allow solving this query in $O((k_2 - k_1) \log k_2)$ time and linear space. Another possible query is to count the number K of documents d with $w(P, d) \geq \tau$, given P and τ. This can be answered via doubling search using document selection queries, in time $O(\log^2 K)$, assuming $w(P, d)$ can be computed in constant time given the locus of P. Similarly, we can count or list the documents d with $w(P, d) \in [\tau_1, \tau_2]$. Such queries are important in bioinformatics, for example for motif mining or for avoiding sequences where P is "over-expressed", and for data mining in general, for example to estimate the distribution of relevance scores of certain patterns.

Related Work. The notion of relevance-based string retrieval was introduced by Muthukrishnan [11], who proposed and solved various problem but not top-k document retrieval. The first data structure for this problem, under the term frequency measure and using $O(n \log n)$ words of space, was given by Hon et al. [4]. Later, Hon et al. [7] introduced a linear space structure ($O(n)$ words), that works for general weight functions as described earlier, with query time $O(p + k \log k)$. This was improved to $O(p + k)$ [13], and finally to the optimal $O(k)$ [17], all using linear space. Those times are in addition to the time for finding the locus node of P, $\mathtt{locus}(P)$, in the generalized suffix tree of $\mathcal{D}$, GST.

The problem has also been studied in scenarios where less than linear space (i.e., $o(n \log n)$ *bits*) can be used. For example, it is possible to solve the problem efficiently using $n \log \sigma + o(n \log \sigma)$ bits [14,18], where σ is the alphabet size of the text (thus $n \log \sigma$ bits are used to represent the text itself). The results are mostly tailored to the term frequency measure of relevance, and achieve times of the form $O(k \operatorname{polylog} n)$. See [12,3,6] for more details.

2   The Top-k Framework

This section briefly describes the linear-space framework of Hon et al. [7] for *top-k* queries. The generalized suffix tree (GST) of a document collection $\mathcal{D} = \{d_1, d_2, d_3, \ldots, d_D\}$ is the combined compact trie of all the non-empty suffixes of all the documents [19]. The total number of leaves in GST is same as the total length n of all the documents. For each node j in GST, $prefix(j)$ is the string obtained by concatenating the edge labels on the path from the root to node j. The highest node v satisfying that P is a prefix of $prefix(v)$ is called the *locus* of P and denoted $\texttt{locus}(P) = v$.

Let ℓ_i represent the ith leftmost leaf node in GST. We say that a node is *marked* with a document d if it is either a leaf node whose corresponding suffix belongs to d, or it is the lowest common ancestor (LCA) of two such leaves. This implies that the number of nodes marked with document d is exactly equal to the number of nodes in the suffix tree of d (at most $2|d|$). A node can be marked with multiple documents. For each node j and each of its marking documents d, define a *link* to be a quadruple $(origin = j, target, doc = d, weight = w(prefix(j), d))$, where *target* is the lowest proper ancestor of node j marked with d (a dummy parent of the root node is added, marked with all the documents). Since the number of links with document $doc = d$ is at most $2|d|$, the total number of links is $\leq \sum_{i=1}^{D} 2|d_i| \leq 2n$. The following is a crucial observation by Hon et al. [7].

Lemma 1. *For each document d that contains a pattern P, there is a unique link with* origin *in the subtree of* $\texttt{locus}(P)$, *a proper ancestor of* $\texttt{locus}(P)$ *as its* target, *and* weight $w(P, d)$.

We say that a link is *stabbed* by a node j if its origin is in the subtree of j (j itself included) and its target is a proper ancestor of j. Therefore, the problem of finding the kth most relevant document for P can be reduced to finding the kth highest weighted link stabbed by $\texttt{locus}(P)$.

3   Super-Linear Space Structure

In this section we start by introducing a basic data structure that uses $O(n \log n)$ words and answers queries in $O(\log n)$ time. Then we enhance it to a structure that uses $O(n \log^{1+\epsilon} n)$ words, for any constant $\epsilon > 0$, and $O(\log n / \log \log n)$ time. The basic structure will be used in Section 4 to achieve linear space within the same time, whereas the enhanced one will be reduced to $O(n \log^\epsilon n)$ words. In Section 5 we show how how the linear-space structure can be improved to answer queries in time $O(\log k)$ and the enhanced structure in time $O(\log k / \log \log n)$, thus reaching our final results.

3.1   The Basic Structure

We prove the following result.

Lemma 2. *Given the* GST *of a text collection of total length n, we can build an $O(n \log n)$-word structure that, given $\texttt{locus}(P)$ and k, answers the document selection query in time $O(\log n)$.*

Let N represent the set of nodes in GST and S represent the set of links $(origin, target, doc, weight)$ in GST, as described in Section 2. Next we construct a balanced binary tree $\mathcal{T}$ of $|S|$ leaves, so that the ith highest weighted link (ties broken arbitrarily) is associated with the ith leftmost leaf of $\mathcal{T}$. Notice that $n \leq |S| \leq 2n$. We use $S(x)$ to denote the set of links associated with the leaves in the subtree of node $x \in \mathcal{T}$. Further, let $N(x)$ denote the set of nodes in GST that are (i) either the origin or the target of a link in $S(x)$, or (ii) the LCA of two such nodes. Clearly $|N(x)| = \Theta(|S(x)|) = \Theta(n/2^{depth(x)})$, where $depth(x)$ is the number of ancestors of x (depth of root is 0).

With every node $x \in \mathcal{T}$, we associate a tree structure $\mathsf{GST}(x)$. $\mathsf{GST}(x)$ is the subtree of GST obtained by retaining only the nodes in $N(x)$, so that node v is the parent of node w in $\mathsf{GST}(x)$ iff v is the lowest proper ancestor of w in GST that also belongs to $N(x)$. The number of nodes and edges in $\mathsf{GST}(x)$ is $\Theta(n/2^{depth(x)})$.

Notice that the same node $w \in$ GST may appear in several $\mathsf{GST}(\cdot)$'s. With each node $w \in \mathsf{GST}(x)$ we associate the following information:

- $stab.count_x(w)$: The number of links in $S(x)$ that are stabbed by w.
- $left.ptr_x(w)$: Let x_L be the left child of x (in $\mathcal{T}$). Let w_L be the highest node in the subtree of w (in $\mathsf{GST}(x)$) that appears also in $\mathsf{GST}(x_L)$ (w_L can be w itself). Then $left.ptr_x(w)$ is a pointer from $w \in \mathsf{GST}(x)$ to $w_L \in \mathsf{GST}(x_L)$. If there exists no such node w_L, then $left.ptr_x(w)$ is null.
- $right.ptr_x(w)$: Analogous to $left.ptr_x(w)$, now considering x_R, the right child of $x \in \mathcal{T}$, and w_R being the highest node in the subtree of $w \in \mathsf{GST}(x)$ that appears also in $\mathsf{GST}(x_R)$.

Note that the space needed for maintaining $\mathsf{GST}(x)$ and the associated information is $O(n/2^{depth(x)})$ words. Added over all the nodes $x \in \mathcal{T}$, the total space occupancy of all $\mathsf{GST}(\cdot)$'s is $O(n \log n)$ words. Finally, the following result is crucial for our data structure (the case of w_R and x_R is analogous).

Lemma 3. *Both w and w_L stab the same subset of links of $S(x_L)$.*

Proof. Otherwise, the target of a link in $S(x_L)$ stabbing w_L but not w would be higher than w_L, below w, and belong to $\mathsf{GST}(x_L)$, contradicting the definition of w_L. The same happens with the source of a link stabbing w but not w_L. □

3.2   Query Algorithm for Document Selection

Assume $\mathtt{locus}(P)$ is given. Notice that the tree $\mathsf{GST}(root)$ associated with the *root* of $\mathcal{T}$ is the same GST of the collection. Therefore, $stab.count_{root}(\mathtt{locus}(P))$ gives the number of documents containing P. If the count is less than k, there is no kth document to select. Otherwise, let L^* be the kth highest weighted link stabbed by $\mathtt{locus}(P)$. Our query algorithm traverses $\mathcal{T}$ top-down, starting from *root* and ending at the leaf node associated with link L^*. Then it reports the document d^* corresponding to L^*.

In our query algorithm, we use x to denote a node in $\mathcal{T}$, w to denote a node in $\mathsf{GST}(x)$ and K to denote an integer $\leq k$. First we initialize x to the root of $\mathcal{T}$, w to $\mathtt{locus}(P)$ and K to k. This establishes the invariant that we have to return the Kth highest weighted link in $S(x)$ stabbed by w. Let x_L and x_R

be the left and right children of x. Then we obtain the nodes $w_L \in \mathsf{GST}(x_L)$ and $w_R \in \mathsf{GST}(x_R)$ pointed by $left.ptr_x(w)$ and $right.ptr_x(w)$, respectively. The following values are then computed in constant time.

- $c = stab.count_x(w)$, the number of links in $S(x)$ stabbed by w.
- $c_L = stab.count_{x_L}(w_L)$, the number of links in $S(x_L)$ stabbed by w (or w_L).
- $c_R = stab.count_{x_R}(w_R)$, the number of links in $S(x_R)$ stabbed by w (or w_R).

Notice that $c = c_L + c_R$. If $c_L \geq K$ then, by Lemma 3, the Kth link below $S(x)$ (or $S(x_L)$) stabbed by $w \in \mathsf{GST}(x)$ is the same as the Kth link below $S(x_L)$ stabbed by $w_L \in \mathsf{GST}(x_L)$. Therefore, we maintain the invariant if we continue the traversal in the subtree of $x \leftarrow x_L$ with $\mathsf{GST}(x_L)$ node $w \leftarrow w_L$. On the other hand, if $c_L < K$, then by Lemma 3 the Kth link stabbed by w below $S(x)$ is same as the $(K - c_L)$th link below $S(x_R)$ stabbed by $w_R \in \mathsf{GST}(x_R)$. In this case, we maintain the invariant if we continue the traversal in the subtree of $x \leftarrow x_R$ with $\mathsf{GST}(x_R)$ node $w \leftarrow w_R$ and with $K \leftarrow K - c_L$. We terminate the algorithm when x is a leaf, thus $K = 1$ and x represents L^*. As the height of $\mathcal{T}$ is $O(\log n)$ and the time spent at each node is constant, the total query time is $O(\log n)$ and Lemma 2 is proved.

3.3   An Enhanced Structure

We now prove the following result, which will hold in the RAM model of computation, with a computer word of $w = \Omega(\log n)$ bits.

Lemma 4. *Given the* GST *of a text collection of total length n and any constant $0 < \epsilon \leq 1$, we can build an $O(n \log^{1+\epsilon} n)$-word structure that, given* $\texttt{locus}(P)$ *and k, answers the document selection query in time $O(\log n / \log \log n)$.*

In order to speed up the structure of Lemma 2, we will choose a step $s = \epsilon \log \log n$ and build the $\mathsf{GST}(x)$ structures only for nodes $x \in \mathcal{T}$ whose depth is a multiple of s. Each node $w \in \mathsf{GST}(x)$ for the selected nodes x will store sufficient information for the query algorithm to jump directly to the corresponding node x' at depth $depth(x') = depth(x) + s$, instead of just to x_L or x_R.

Given $x, x' \in \mathcal{T}$ as above (x' in the subtree of x) and $w \in \mathsf{GST}(x)$, we define $w_{x'}$ as the highest node in the subtree of w that appears also in $\mathsf{GST}(x')$. Let us call $x_1, x_2, \ldots, x_{2^s}$ the nodes at depth $depth(x) + s$ that descend from x (or the leaves below x, if they have depth less than $depth(x) + s$), ordered left to right in $\mathcal{T}$ (i.e., from highest to lowest weights in $S(x_i)$).

Associated to each node $w \in \mathsf{GST}(x)$, we store 2^s pointers $ptr_x(w)[i] = w_{x_i}$. We also store the 2^s cumulative values $acc_x(w)[i] = \sum_{j=1}^{i} stab.count_{x_j}(w_{x_j})$; note that $acc_x(w)[2^s] = stab.count_x(w)$. We will store those $acc_x(w)$ values in a fusion tree [1], which takes $O(2^s) = O(\log^\epsilon n)$ words of space and solves predecessor queries in $acc_x(w)$ in constant time. The space is the same used by array $ptr_x(w)$, which added over all the $\mathsf{GST}(\cdot)$'s is $O(n \log^{1+\epsilon} n)$ words (even if only one level out of s in $\mathcal{T}$ stores $\mathsf{GST}(\cdot)$ structures).

Queries now proceed as in Section 3.2, but now we use the fusion tree to determine, given $w \in \mathsf{GST}(x)$, which is the node $x_i \in \mathcal{T}$ that contains the Kth link below $S(x)$ stabbed by w. Therefore we can move directly from x to x_i and from $w \in \mathsf{GST}(x)$ to $w_i \in \mathsf{GST}(x_i)$, where $w_i = ptr_x(w)[i]$. We also update $K \leftarrow K - acc_x(w)[i-1]$ (assume $acc_x(w)[0] = 0$). Thus we complete the query in $O((\log n)/s) = O(\log n/(\epsilon \log \log n))$ constant-time steps and Lemma 4 is proved.

4   Linear Space Structure

In this section we build on the basic structure of Lemma 2 in order to achieve linear space and logarithmic query time. At the end, we reduce the space of the enhanced structure to $O(n \log^\epsilon n)$. The results hold under the RAM model.

Lemma 5. *Given the* GST *of a text collection of total length n, we can build an $O(n)$-word structure that, given* locus(P) *and k, answers the document selection query in time $O(\log n)$.*

To achieve linear space, we replace some of our data structures by succinct ones. We will measure the space in bits, aiming at using $O(n \log n)$ bits overall. The binary tree $\mathcal{T}$ can be maintained in $O(n \log n)$ bits, where each internal node x stores an $O(\log n)$-bit pointer to the corresponding tree $\mathsf{GST}(x)$ and each leaf stores the document identifier corresponding to the associated link. The global GST can also be maintained in $O(n \log n)$ bits. Therefore, the space-consuming component are the $\mathsf{GST}(\cdot)$'s and their associated information.

Using well-known succinct data structures [16], the $\mathsf{GST}(x)$ tree topologies can be represented in $O(1)$ bits per node (i.e., $O(n \log n)$ bits overall) with constant-time support of all the basic navigational operations required in our algorithm. We refer to any node $w \in \mathsf{GST}(x)$ by its pre-order rank, that is, node j means the node with pre-order rank j. The pre-order rank of the root node of any $\mathsf{GST}(x)$ is 1. Next we show how to encode the remaining information associated with each node in $\mathsf{GST}(x)$ using $O(1)$ bits per node.

4.1   Encoding $stab.count_x(j)$

We note that $stab.count_x(j)$ is exactly equal to the number of links of $S(x)$ associated with $\mathsf{GST}(x)$ that originate in the subtree of j minus the number of links in $S(x)$ that target any node in the subtree of j (j belongs to its subtree). We encode this information in two bit vectors: $B_x = 1 0^{\alpha_1} 1 0^{\alpha_2} 1 0^{\alpha_3} \ldots$ and $B'_x = 1 0^{\beta_1} 1 0^{\beta_2} 1 0^{\beta_3} \ldots$, where α_j (resp., β_j) is the number of links of $S(x)$ originating from (resp., targeting at) node j in $\mathsf{GST}(x)$. We augment B_x and B'_x with structures supporting constant-time rank/select queries [10]. Notice that $\sum \alpha_j = \sum \beta_j = O(|S(x)|) = O(|\mathsf{GST}(x)|)$. Therefore, both B_x and B'_x can be represented in $O(1)$ bits per node.

Now we can compute $stab.count_x(j)$ for any j in $O(1)$ time as follows: find the rightmost leaf node j' in the subtree of j in $O(1)$ time using the succinct tree representation of $\mathsf{GST}(x)$ [16]. Then the number n_o of links originating from the subtree of j is equal to the number of 0-bits between the jth and $(j'+1)$th 1-bit in B_x (because j and j' are preorder numbers). Similarly, the number n_t of links targeted at any node in the subtree of j is equal to the number of 0-bits between the jth and $(j'+1)$th 1-bits in B'_x. Using rank/select operations on B_x and B'_x, n_o and n_t are computed in $O(1)$ time and $stab.count_x(j)$ is given by $n_o - n_t$.

4.2   Encoding $left.ptr_x(j)$ and $right.ptr_x(j)$

We show how to encode $left.ptr_x(\cdot)$ for all nodes in $\mathsf{GST}(x)$; $right.ptr_x(j)$ is symmetric. The idea is to maintain a bit vector LP such that $LP[j] = 1$ iff there exists a node $j_L \in \mathsf{GST}(x_L)$ such that both $j \in \mathsf{GST}(x)$ and $j_L \in \mathsf{GST}(x_L)$ represent the same node in GST. We add constant-time rank/select data structures [10] on LP. Since the length of LP is equal to the number of nodes in $\mathsf{GST}(x)$, its space occupancy is $O(1)$ bits per node.

Now, for any given node $j \in \mathsf{GST}(x)$, the node $j_L \in \mathsf{GST}(x_L)$ to which $left.ptr_x(j)$ points is the (unique) highest descendant of j that is marked in LP, thus it can be identified by (1) finding the position j^* of the leftmost 1-bit in $LP[j\ldots]$; (2) checking if node j^* is in the subtree of node j in $\mathsf{GST}(x)$; (3) if so, then $j_L \in \mathsf{GST}(x_L)$ is equal to the number of 1's in $LP[1\ldots j^*]$, otherwise, j_L is null. All these operations require constant time, either using the succinct tree operations or the rank/select data structures. This works because all the nodes in $\mathsf{GST}(x_L)$ appear in $\mathsf{GST}(x)$, in the same order (pre-order).

In summary, the space requirement of our encoding scheme is $O(1)$ bits per node in any $\mathsf{GST}(x)$, thus adding to $O(n \log n)$ bits. The query algorithm, as well as its time complexity, remain the same. This completes the proof of Lemma 5.

4.3   Reducing Space of the Enhanced Structure

The space of the enhanced structure of Section 3.3 can be similarly reduced to $O(n \log^\epsilon n)$ words, obtaining the following result.

Lemma 6. *Given the* GST *of a text collection of total length n and a constant $\epsilon > 0$, we can build an $O(n \log^\epsilon n)$-word structure that, given* locus(P) *and k, answers the document selection query in time $O(\log n / \log\log n)$.*

For this sake, recalling the definition of $x_1, \ldots, x_{2^s}$ of Section 3.3, we will maintain bit vectors LP_i for $i = 1$ to 2^s, so that $LP_i[j] = 1$ iff there exists a node $j_i \in \mathsf{GST}(x_i)$ such that both $j \in \mathsf{GST}(x)$ and $j_i \in \mathsf{GST}(x_i)$ represent the same node in GST. Then each array entry $ptr_x(j)[i]$ is computed using LP_i as in Section 4.2. The total space used by all the LP_i bit vectors is $O(2^s) = O(\log^\epsilon n)$ bits per node, adding up to $O(n \log^{1+\epsilon} n)$ bits in total.

To compute $acc_x(j)[i]$, we store bitmaps $B_{x,1}, \ldots, B_{x,2^s}$ and $B'_{x,1}, \ldots, B'_{x,2^s}$, analogous to B and B' of Section 4.1. In this case, $B_{x,i} = 10^{\alpha_1^i} 10^{\alpha_2^i} 10^{\alpha_3^i} \ldots$, so that $\alpha_j^i = \sum_{r=1}^i s(r)$, where $s(r)$ is the number of links of $S(x_r)$ originating from node $ptr_x(j)[i] \in \mathsf{GST}(x_r)$, and $B'_{x,i} = 10^{\beta_1^i} 10^{\beta_2^i} 10^{\beta_3^i} \ldots$, so that $\beta_j^i = \sum_{r=1}^i t(r)$, where $t(r)$ is the number of links of $S(x_r)$ targeting at node $ptr_x(j)[i] \in \mathsf{GST}(x_r)$.

Then, it holds $acc_x(j)[i] = \alpha^i_j - \beta^i_j$, which is computed in constant time using rank/select operations. Since it holds $\alpha^i_j \leq \alpha_j$ and $\beta^i_j \leq \beta_j$ for all i values, the total space of these $2^s = \log^\epsilon n$ bitmaps adds up to $O(n \log^{1+\epsilon} n)$ bits.

To carry out predecessor searches on the virtual vector $acc_x(j)$, we use succinct SB-trees [2, Lemma 3.3]. Given constant-time access to any $acc_x(j)[i]$, this structure provides predecessor searches in $O(1 + \log(2^s)/\log\log n) = O(1)$ time and use $O(2^s \log\log n) = O(\log^\epsilon n)$ bits per node (by adjusting ϵ). Thus the total space is $O(n \log^{1+\epsilon} n)$ bits as well. This concludes the proof of Lemma 6.

5   Achieving $O(\log k)$ Query Time and Better

In this section we first build on the linear-space data structure of Lemma 5 in order to improve its query time to $O(\log k)$. At the end, we show that the result extends to our superlinear-space data structure of Lemma 6, improving its query time to $O(\log k / \log\log n)$. Thus we start by proving the following theorem.

Theorem 1. *A collection $\mathcal{D}$ of documents can be preprocessed into a linear-space data structure that can answer any document selection query (P, k) in time $O(\log k)$, given the locus of pattern P in the generalized suffix tree of $\mathcal{D}$.*

Notice that the query time $O(\log n)$ in Lemma 5 can be written as $O(\log k)$ for $k > \sqrt{n}$. Therefore, we turn our attention to the case where $k \leq \sqrt{n}$. First, we derive a space-efficient structure $DS(\delta)$, which can answer document selection queries faster, but only for values of k below a predefined parameter $\delta \leq \sqrt{n}$. More precisely, structure $DS(\delta)$ will satisfy the following properties:

Lemma 7. *The structure $DS(\delta)$ uses $O(n(\log \delta + \log\log n))$ bits of space and can answer document selection queries in time $O(\log \delta + \log\log n)$, for $k \leq \delta \leq \sqrt{n}$.*

To obtain the result in Theorem 1, we maintain structures $DS(\delta_i)$ with $\delta_i = \lceil n^{1/2^i} \rceil$ for $i = 1, 2, 3, \ldots, r$, where $\delta_{r+1} \leq \sqrt{\log n} < \delta_r$ (therefore $r < \log\log n$). The total space needed is $O(n \sum_{i=1}^{r} (\log \delta_i + \log\log n)) = O(n \log n)$ bits ($O(n)$ words). When k comes as a query, if $k > \delta_{r+1}$, we first find h, where $\delta_{h+1} < k \leq \delta_h$ and obtain the answer using $DS(\delta_h)$. The resulting time is $O(\log \delta_h + \log\log n) = O(\log k)$. The case where $k < \delta_{r+1}$ is handled separately using other structures in $O(1)$ time (Section 5.2). We now describe the details of $DS(\delta)$.

5.1   Structure $DS(\delta)$

The first step is to identify certain nodes in GST as *marked* nodes and *prime* nodes, based on a parameter $g = \lceil \delta \log n \rceil$ called the *grouping factor*. Every gth leftmost leaf is marked, and the LCA of every two consecutive marked leaves is also marked. Therefore, the number of marked nodes is $\Theta(n/g)$. Nodes with their parent marked are prime. A prime node with at least one marked node in its subtree is a type-1 prime node, otherwise it is a type-2 prime node. Notice that the highest marked node in the subtree of any node is unique, if it exists. Therefore, except the root node, every marked node j^* can be associated with a unique type-1 prime node j', which is the first prime node on the path from j^* to the root. Notice that a node can be both prime and marked.

Let j' be a prime node and j^* be the highest marked node in its subtree (j^* exists only if j' is of type-1, and it can be that $j' = j^*$). We use $G(j'\backslash j^*)$ to represent the subtree of GST rooted at j' after removing the subtree of j^* (j^* is not removed). With a slight abuse of notation, we use $G(j'\backslash j^*)$ to represent the set of nodes within $G(j'\backslash j^*)$ as well. A crucial result [17] is that, for any prime node j', the number of nodes in $G(j'\backslash j^*)$ is $O(g)$.

We define $prime.parent(j)$ of any node j in GST as the first prime node j' on the path from j to the root. Note that $j \in G(j'\backslash j^*)$, otherwise j would be a (strict) descendant of j^* and its corresponding j' would be below j^*.

It is not hard to determine $j' = prime.parent(j)$ in constant time and $O(n)$ bits, by sampling the prime nodes in a succinct tree representation and looking for the lowest sampled ancestor of j [15, Lemma 4.4].

The structure $DS(\delta)$ is a collection of substructures $STR(j')$ associated with every prime node j' in GST. If the input node $\texttt{locus}(P) \in G(j'\backslash j^*)$ and $k \leq \delta$, we obtain the answer using $STR(j')$ in $O(\log g) = O(\log \delta + \log \log n)$ time. Based on the type of j', we have two cases; we describe the simpler one first.

$STR(j')$ Associated with a Type-2 Prime Node j': The structure can be constructed as follows: take $G(j')$, the subtree rooted at node j', and replace the pre-order rank of each node j by $(j - j' + 1)$. Also associate a dummy parent node to the root. Then, among the links defined over GST (Section 2), choose those that originate from the subtree of j' and: (1) Assign a new value to its origin and target, which is its original value minus j' plus 1. The target of some links can be negative; replace those by 0. (2) Replace the weight by a rank-space reduced value in $[1, O|G(j')|]$. Notice that the number of links chosen is $O(|G(j')|)$. (3) Let d be its document identifier. Instead of writing d explicitly in $\lceil \log D \rceil$ bits, use a pointer to one leaf node in $G(j')$, using $\lceil \log |G(j')| \rceil$ bits, where the suffix corresponding to that leaf belongs to document d.

In summary, we have a tree of $(|G(j')| + 1)$ nodes and $O(|G(j')|)$ links associated with it. The information $(origin, target, document, weight)$ associated with each link is encoded in $O(\log |G(j')|)$ bits. Then $STR(j')$ is the structure described in Lemma 5 over these nodes and links. The space required is $O(|G(j')| \log |G(j')|) = O(|G(j')| \log g)$ bits. We maintain structures $STR(j')$ for all type-2 prime nodes j' in total $O(n \log g)$ bits, since a node can be in the subtree of at most one type-2 prime node.

$STR(j')$ Associated with a Type-1 Prime Node j': We first identify the *candidate set* $\mathcal{C}(j')$ of $O(g)$ links, such that for any $k \leq \delta$, the kth link stabbed by any node $j \in G(j'\backslash j^*)$ belongs to $\mathcal{C}(j')$. Clearly we can ignore the links that do not originate from the subtree of j'. The links that do can be categorized into the following types [17]: *near-links* are stabbed by j^*, but not by j'; *far-links* are stabbed by both j^* and j'; *small-links* are targeted at a node in the subtree of j^*; and *fringe-links* are the others.

We include all near-links and fringe-links into $\mathcal{C}(j')$, which are $O(g)$ in number [17, Lemma 8]. All small-links can be ignored as none of them is stabbed by any node in $G(j' \backslash j^*)$. Notice that if any node in $G(j' \backslash j^*)$ stabs a far-link, it indeed stabs all far-links. Therefore, it is sufficient to insert the top-δ far-links into $\mathcal{C}(j')$. Thus, we have $O(g)$ links in $\mathcal{C}(j')$ overall.

Now we perform a rank-space reduction of pre-order rank of nodes in $G(j' \backslash j^*)$ as well as of the information associated with the links in $\mathcal{C}(j')$, as follows:

- The target of those links targeting at any proper ancestor of j' is changed to a dummy parent node of j'. Similarly, the origin of all those links originating in the subtree of j^* is changed to node j^*.
- The pre-order rank of all those nodes in $G(j' \backslash j^*)$, and the corresponding origin and target values of links in $\mathcal{C}(j')$, are changed to a rank-space reduced value in $[0, |G(j' \backslash j^*)|]$. Notice that the new pre-order rank of j' is 1 and that of its dummy parent node is 0. We remark that this mapping (and remapping) can be stored separately in $O(|G(j' \backslash j^*)| \log |G(j' \backslash j^*)|)$ bits.
- The weights of the links are also replaced by rank-space reduced values.
- Let L be a near- or fringe-link in $\mathcal{C}(j')$ with d its corresponding document. Then there must be at least one leaf ℓ in $G(j' \backslash j^*)$ where the suffix corresponding to ℓ belongs to d. Therefore, instead of representing d, we maintain a pointer to ℓ, which takes only $O(\log g)$ bits. This trick will not work for far-links, as the existence of such a leaf node is not guaranteed. Therefore, we spend $\log D$ bits for each far-link, which is still affordable because there are only $O(\delta) = O(g/\log n)$ far-links.

In summary, we have a tree of $(|G(j' \backslash j^*)| + 1) = O(g)$ nodes with $O(g)$ links associated with it. Then $STR(j')$ is the structure described in Lemma 5 over these nodes and links. The space required is $O(g \log g)$ bits. As the number of type-1 prime nodes is $O(n/g)$, the total space to maintain $STR(j')$ for all type-2 primes nodes j' is $O(n \log g)$ bits.

Query Answering: Given node $j = \text{locus}(P)$, we find $j' = prime.parent(j)$. Then we map node j to the corresponding node in $STR(j')$ and obtain the answer by querying $STR(j')$, in $O(\log g) = O(\log \delta + \log \log n)$ time. The answer may come in the form of a node in $STR(j')$, which is mapped back to GST in order to obtain the associated document. This completes the proof of Lemma 7.

5.2   Structure for $k \leq \delta_{r+1}$

First, identify the marked and prime nodes in GST with $g = \delta_{r+1} \log n$. At every prime node j', we explicitly maintain the candidate set $\mathcal{C}(j')$. This takes $O(n)$-word space. Then for any $k \leq \delta_{r+1}$, the kth link stabbed by node j can be encoded as a pointer to the corresponding entry in $\mathcal{C}(prime.parent(j'))$ using $\lceil \log |\mathcal{C}(prime.parent(j'))| \rceil = O(\log g) = O(\log \log n)$ bits. Therefore, the answers for all $k \in [1, \delta_{r+1}]$ for all nodes in GST can be maintained in additional $O(n \cdot \delta_{r+1} \log \log n) = o(n \log n)$ bits of space. Now the kth link (and its document) stabbed by any query node $\texttt{locus}(P)$ can be obtained from $C(prime.parent(\texttt{locus}(P)))$ in $O(1)$ time.

5.3 Speeding Up the Enhanced Structure

The same construction used above can be used to speed up our superlinear-space structure of Lemma 6, simply by using it instead of the linear-space one of Lemma 5 to implement the structures $STR(j')$. The space of the form $O(n \log^\epsilon n)$ words, or $O(n \log^{1+\epsilon} n)$ bits, will become $O(g \log g \log^\epsilon n)$ inside the structures $STR(j')$, because we will maintain the sampling step $s = \epsilon \log \log n$ depending on n, not on g, and use the succinct SB-trees with parameter n, not g. As a result, the total space per value of δ will be $O(n \log g \log^\epsilon n)$ bits, and added over all the values of δ we will have $O(n \log^\epsilon n \sum_{i=1}^{r}(\log \delta_i + \log \log n)) = O(n \log^{1+\epsilon} n)$ bits, or $O(n \log^\epsilon n)$ words. The time, on the other hand, will be $O(1 + \log \delta/(\epsilon \log \log n))$ on $DS(\delta)$, which becomes $O(1 + \log k/(\epsilon \log \log n))$ in terms of k. We have proved our final result for the superlinear structure.

Theorem 2. *A collection $\mathcal{D}$ of documents of total length n can be preprocessed into a data structure using $O(n \log^\epsilon n)$ words of space, for any constant $\epsilon > 0$, which can answer* document selection *queries (P, k) in time $O(1 + \log k/\log \log n)$, given the locus of pattern P in the generalized suffix tree of $\mathcal{D}$.*

6 Hardness of an Efficient Succinct Solution

One could expect to obtain an index using $O(n \log \sigma)$ bits of space, proportional to the $n \log \sigma$ bits needed to store $\mathcal{D}$, as achieved for the top-k document retrieval problem. We show, however, that this is very unlikely unless a significant breakthrough in the current state of the art of computational geometry is obtained.

Theorem 3. *If there exists a data structure using $O(n \log \sigma + D \operatorname{polylog} n)$ bits with query time $O(|P| \operatorname{polylog} n)$ for document selection (σ being the alphabet size), then there exists a linear-space data structure that can answer three-dimensional range reporting queries in poly-logarithmic time per reported point.*

Proof. We reduce from the position restricted substring searching (PRSS) problem, which is defined as follows: Index a given a text $T[1, n]$ over an alphabet set $[1, \sigma]$, such that whenever a pattern P (of length p) and a range $[x, y]$ comes as a query, all those $occ_{x,y}$ occurrences of P in $T[x \ldots y]$ can be reported efficiently. Many indexes offering different space and query time trade-offs exist [9,8].

Hon et al. [5] proved that answering PRSS queries in polylog time and succinct space is at least as hard as performing 3-dimensional orthogonal range reporting in polylog time and linear space. They also showed that if the query pattern is longer than $\alpha = \lceil \log^{2+\epsilon} n \rceil$ for some predefined constant $\epsilon > 0$, an efficient succinct space index can be designed. Therefore, the harder case arises when

$p < \alpha$. We now show how to answer PRSS queries with $p < \alpha$ via document selection queries on the following set: $\mathcal{D} = \{d_1, d_2, d_3, ..., d_{\lceil n/\alpha \rceil}\}$, where $d_i = T[1 + (i-1)\alpha ... (i+1)\alpha]$ and $|d_i| = 2\alpha$, except possibly for $d_{\lceil n/\alpha \rceil - 1}$ and $d_{\lceil n/\alpha \rceil}$. The score function $w(P, d_i)$ is i if P appears at least once in d_i and 0 otherwise. Notice that an occurrence of any pattern of length at most α overlaps with at least one and at most two documents in $\mathcal{D}$. Therefore, the previously defined PRSS query on T can be answered via multiple document selection queries on $\mathcal{D}$ as follows: first report all those documents d_i with $w(P, d_i) \in [\lceil x/\alpha \rceil, \lfloor y/\alpha + 2 \rfloor]$. Then, within all those reported documents, look for other occurrences of P via an exhaustive scanning. If the time for document selection queries is polylog in the total length of all documents in $\mathcal{D}$ (which is at most $2n$), then the time for PRSS query is also bounded by $O((p + occ_{x,y})\text{polylog} \, n)$. Therefore, answering document selection queries in polylog time and succinct space is at least as hard as answering PRSS queries in polylog time and succinct space.   □

References

1. Fredman, M., Willard, D.: Surpassing the information theoretic barrier with fusion trees. J. Comp. Sys. Sci. 47, 424–436 (1993)
2. Grossi, R., Orlandi, A., Raman, R., Rao, S.S.: More haste, less waste: Lowering the redundancy in fully indexable dictionaries. In: STACS, pp. 517–528 (2009)
3. Hon, W.-K., Patil, M., Shah, R., Thankachan, S.V., Vitter, J.S.: Indexes for document retrieval with relevance. In: Brodnik, A., López-Ortiz, A., Raman, V., Viola, A. (eds.) Ianfest-66. LNCS, vol. 8066, pp. 351–362. Springer, Heidelberg (2013)
4. Hon, W.-K., Patil, M., Shah, R., Wu, S.-B.: Efficient index for retrieving top-k most frequent documents. J. Discr. Alg. 8(4), 402–417 (2010)
5. Hon, W.-K., Shah, R., Thankachan, S.V., Vitter, J.S.: On position restricted substring searching in succinct space. J. Discr. Alg. 17, 109–114 (2012)
6. Hon, W.-K., Shah, R., Thankachan, S.V., Vitter, J.S.: Space-efficient framework for top-k string retrieval. J. of the ACM (to appear, 2014)
7. Hon, W.-K., Shah, R., Vitter, J.S.: Space-efficient framework for top-k string retrieval problems. In: FOCS, pp. 713–722 (2009)
8. Lewenstein, M.: Orthogonal range searching for text indexing. In: Brodnik, A., López-Ortiz, A., Raman, V., Viola, A. (eds.) Ianfest-66. LNCS, vol. 8066, pp. 267–302. Springer, Heidelberg (2013)
9. Mäkinen, V., Navarro, G.: Position-restricted substring searching. In: Correa, J.R., Hevia, A., Kiwi, M. (eds.) LATIN 2006. LNCS, vol. 3887, pp. 703–714. Springer, Heidelberg (2006)
10. Munro, I.: Tables. In: FSTTCS, pp. 37–42 (1996)
11. Muthukrishnan, S.: Efficient algorithms for document retrieval problems. In: SODA, pp. 657–666 (2002)
12. Navarro, G.: Spaces, trees and colors: The algorithmic landscape of document retrieval on sequences. ACM Computing Surveys 46(4), article 52 (2014)
13. Navarro, G., Nekrich, Y.: Top-k document retrieval in optimal time and linear space. In: SODA, pp. 1066–1077 (2012)
14. Navarro, G., Thankachan, S.V.: Faster top-k document retrieval in optimal space. In: Kurland, O., Lewenstein, M., Porat, E. (eds.) SPIRE 2013. LNCS, vol. 8214, pp. 255–262. Springer, Heidelberg (2013)

15. Russo, L., Navarro, G., Oliveira, A.: Fully-compressed suffix trees. ACM Trans. Alg. 7(4), art. 53 (2011)
16. Sadakane, K., Navarro, G.: Fully-functional succinct trees. In: SODA, pp. 134–149 (2010)
17. Shah, R., Sheng, C., Thankachan, S.V., Vitter, J.S.: Top-k document retrieval in external memory. In: Bodlaender, H.L., Italiano, G.F. (eds.) ESA 2013. LNCS, vol. 8125, pp. 803–814. Springer, Heidelberg (2013)
18. Tsur, D.: Top-k document retrieval in optimal space. Inf. Process. Lett. 113(12), 440–443 (2013)
19. Weiner, P.: Linear pattern matching algorithms. In: SWAT (FOCS), pp. 1–11 (1973)

Approximation Algorithms for Hitting Triangle-Free Sets of Line Segments[*]

Anup Joshi and N.S. Narayanaswamy

Department of Computer Science and Engineering,
Indian Institute of Technology Madras, India
{anup,swamy}@cse.iitm.ac.in

Abstract. We present polynomial time constant factor approximations on NP-Complete special instances of the Guarding a Set of Segments(GSS) problem. The input to the GSS problem consists of a set of line segments, and the goal is to find a minimum size hitting set of the given set of line segments. We consider the underlying planar graph on the set of intersection points as vertices and the edge set as pairs of vertices which are adjacent on a line segment. Our results are for the subclass of instances of GSS for which the underlying planar graph has girth at least 4. On this class of instances, we show that an optimum solution to the natural hitting set LP can be rounded to yield a 3-factor approximation to the optimum hitting set. The GSS problem remains NP-Complete on the sub-class of such instances. The main technique, that we believe could be quite general, is to round the hitting set LP optimum for special hypergraphs that we identify.

Keywords: Art gallery problem, line segments, hitting sets, approximation algorithm.

1 Introduction

In this paper we consider the Guarding a Set of Segments problem which has been investigated by Brimkov *et al* in [5,6,7].

Guarding a Set of Segments (GSS)

Input   A set of line segments $\mathcal{L} = \{l_1, l_2, \ldots, l_m\}$ in the 2-dimensional plane. Further, each line segment is maximal in the given set of line segments, in the sense that the union of two distinct line segments is not a line segment.

Output A smallest set of points $\mathcal{P}$ referred to as the *cover*, such that for every line segment $l_j \in \mathcal{L}$, $l_j \cap \mathcal{P}$ is non-empty.

GSS is a geometric instance of the corner-stone Set Cover and Hitting Set problems. In Set Cover we are given a set system $(U; \mathcal{S} = \{S_i | 1 \leq i \leq n, S_i \subseteq U\})$,

[*] This work was supported by the Indo-Max Planck Centre for Computer Science Programme in the area of Algebraic and Parameterized Complexity and by the Half-Time Research Assistantship provided by The Ministry of Human Resource Development, Government of India.

R Ravi and I.L. Gørtz (Eds.): SWAT 2014, LNCS 8503, pp. 357–367, 2014.

and we have to find a sub-collection $\mathcal{S}' \subseteq \mathcal{S}$ of minimum size such that $\cup_{S \in \mathcal{S}'} S = U$. It is known to be NP-Complete [14], and Feige [10] showed a threshold of $(1 - \epsilon) \ln n$ below which Set Cover cannot be approximated unless NP has quasi-polynomial time algorithms. Indeed GSS is a geometric instance of a disarmingly attractive version of Set Cover with Intersection 1(SC1) – SC instances in which any pair of sets have an intersection of at most one element. SC1 was shown to be inapproximable within a $o(\log n)$-factor unless $NP \subseteq ZTIME(n^{O(\log \log n)})$ by Kumar *et al.* [15]. To see that a GSS instance $\mathcal{L}$ can be cast as an instance of SC1: Let $\mathcal{I}$ be the set in which each point is an end point of a line segment in $\mathcal{L}$ or an intersection point of two line segments in $\mathcal{L}$. Now, for every line $l_i \in \mathcal{L}$ define a set $S_i = \{p \in I | p \in l_i\}$, and let $\mathcal{S} = \{S_1, S_2, ..., S_n\}$ and the universe $U = \cup_{i=1}^n S_i$. $(U, \mathcal{S})$ is a SC1 formulation as any two sets will intersect on at most one element and there is a bijective correspondence between the solutions of the GSS and the SC1 instance. Line Cover is yet another geometric instance of SC1. In Line Cover the input is a set of lines in the plane, and the question is to find a minimum number of points so that every line passes through atleast one of the points. Megiddo and Tamir showed that Line Cover is NP-hard [17] and it was observed by Kumar *et al.* that Line Cover is Max-SNP Hard [15]. GSS can also be seen as a generalization of planar vertex cover owing to the fact that each planar graph has a planar embedding in which each edge is a line segment and no two adjacent edges form a longer line segment. Planar vertex cover has a Polynomial Time Approximation Scheme (PTAS) by the famous Baker's technique [2] which uses the k-outerplanarity of the given planar graph, and its connection to the treewidth. This proximity in many ways to planar vertex cover can be seen as an important motivation to understand the approximability of GSS.

GSS can also be viewed as an Art Gallery Problem with visibility only between points that are on the same given line segment. One can think of GSS as the problem of guarding the corridors (which are the line segments) of an Art Gallery using as few guards/cameras as possible. GSS was shown to be NP-Complete by Brimkov *et al.* in [6], and they also give a deterministic algorithm to find an optimum solution when the segments together form a tree-like structure. In [7] Brimkov *et al.* have studied the approximability of the greedy heuristic for GSS, and construct a family of instances in which the solution returned by the greedy approach is at least a $\log_2(n)$ factor away from the optimum. This is a tight study of the greedy heuristic. In [5] Brimkov presents a constant-factor approximation algorithm for special instances of GSS (not known to be hard). Art Gallery Problems are very well studied and we point to the papers and books referred to by Brimkov *et al.* [6].

GSS is also a special instance of the extensively studied geometric hitting set problems. In a hitting set problem the input is a range space $\mathcal{R} = (\mathcal{P}, \mathcal{D})$ consisting of a set $\mathcal{P}$ and a set $\mathcal{D}$ of subsets of $\mathcal{P}$ called the ranges. A hitting set is a subset of $\mathcal{P}$ that has a non-empty intersection with every set in $\mathcal{D}$. Clearly, as discussed above SC, SC1, Line Cover, and GSS are all special instances of this problem. Another geometric instance is when $\mathcal{P}$ is a set of points in the plane, and $\mathcal{D}$ is a set of convex polygons containing points of $\mathcal{P}$. For this and many other geometric range spaces the hitting set problem is NP-hard. Recently [18] show PTASes for geometric hitting set on special geometric range spaces using a local search method. The approximation of geometric hitting sets is closely connected to the ϵ-net problem as discovered by Brönnimann and Goodrich [8]. Given a range space $(\mathcal{P}, \mathcal{D})$, $|\mathcal{P}| = n$, an $S \subseteq \mathcal{P}$ is an ϵ-net if for all $D \in \mathcal{D}$ such that $|D| \geq \epsilon n$, $D \cap S \neq \phi$. In the weighted version, sometimes, each point in $\mathcal{P}$ is considered to have a positive weight so that the total weight of points in $\mathcal{P}$ is 1. The weight of each range is the sum of the weights of the elements in it. A weighted ϵ-net is a subset of $\mathcal{P}$ that intersects all ranges of weight more than ϵ. Brönnimann and Goodrich [8] proved that a c-factor approximation to the minimum hitting set can be computed in polynomial time if an ϵ-net of size $\frac{c}{\epsilon}$ for the weighted ϵ-net problem can be obtained. Closely related to this is result by Long [16] who showed that the natural hitting set LP relaxation for geometric instances has an integrality gap of K if and only if there is an ϵ-net of size $\frac{K}{\epsilon}$. This connection between ϵ-nets and hitting sets yield constant factor approximations for some geometric range spaces like the case in which the ranges are half-spaces in $\mathcal{R}^3$ [19]. Haussler and Welzl [13] show that for range spaces of VC-dimension d, there exists an ϵ-net of size $O(\frac{d}{\epsilon} log(\frac{d}{\epsilon}))$. Note that GSS instances have VC dimension 2, apart from other geometric hitting spaces [18]. Indeed for many years [1] it was conjectured that in all natural geometric scenarios of VC-dimension d there always exists an ϵ-net of size $O(\frac{d}{\epsilon})$. Alon [1] disproved this conjecture by constructing for each $C >> 0$, a range space of VC-dimension 2 for which the smallest possible size of an ϵ-net is larger than $\frac{C}{\epsilon}$. The range space $(\mathcal{P}, \mathcal{D})$ is such that $\mathcal{P}$ is a set of points on the plane, and $\mathcal{D}$ is a set of line segments containing points from $\mathcal{P}$. The range spaces deceptively look like instances of the GSS problem, and this forms the second motivation in this paper which is to understand the possibility of ϵ-nets for GSS that are linear in $\frac{1}{\epsilon}$.

Our Work. We first observe that the range spaces constructed by Alon in [1] are not GSS instances. Therefore the possibility of constant factor approximations for GSS by rounding the natural hitting set LP are not ruled out by the super linear lower bound on the size of ϵ-nets proved by Alon. The way we prove it is by considering the natural underlying graph given an instance consisting of points and line segments as the ranges. The natural graph is obtained by viewing the instance as a planar graph, and we observe that in the range space constructed by Alon [1], all the points have high degree (more than 5) in the natural graph. On the other hand, the vertices in the natural graph associated with a GSS instance are all the points of intersection. Since we know that any planar graph has vertices of degree at most 5, it follows that the natural graph associated with a GSS instance must have vertices of degree at most 5. It then follows that Alon's instance of points and lines are not GSS instances. This is presented in Section 3.2.

We attempt to exploit the presence of low degree vertices in the natural graph associated with GSS instances. For this, we consider the hitting set problem in a subclass of hypergraphs for which there is an integer $d \geq 0$, and a vertex ordering $\sigma = v_1, v_2, ..., v_n$, such that in the hypergraph induced on the vertices $v_i, v_{i+1}, ..., v_n$ (for $i = 1...n-1$) the degree of v_i is at most d. On such hypergraphs, we present a way of rounding an optimum solution to the natural hitting set LP to obtain a d-approximation for Hitting Set. These results are in Section 2. We find this result of rounding the hitting set LP based on the above mentioned ordering interesting, and is in the spirit of getting good approximations when the hypergraphs satisfy some special structure (like bounded VC-dimension, see Brönnimann and Goodrich [8]). In particular, the degree based vertex ordering that we have hit upon seems to be new and useful at least in some geometric settings.

We then use the structure of the underlying graph of a GSS instance to obtain an upper bound on the minimum degree of the underlying graph. In particular, we use the well known result that planar graphs of girth at least 4 have a vertex of degree at most 3. These observations give us an ordering of the vertices associated with a GSS instance such that for each $1 \leq i \leq n - 1$, the instance induced on vertices $v_i, v_{i+1}, ..., v_n$ is a GSS instance and the degree of v_i in the GSS instance (viewed as a graph) is at most 3 or 2. This yields a 3-approximation to the hitting set on GSS instances for which the underlying graph has girth at least 4. This is nice because it gives a hope that further structure of girth 3 instances can be handled differently to obtain a constant factor approximation algorithm for the GSS itself. The other way of viewing this result is that if GSS does not have constant factor approximations, then the underlying graph of the hard instances must have girth 3. It must be noted that GSS on the subclass of instances for which the underlying graph has girth at least 4 is NP-complete, as the problem is at least as hard as vertex cover in planar triangle free graphs [12]. While our approach gives a constant factor approximation algorithm for GSS instances which have girth at least 4, we believe that on such instances there is a PTAS. Finally, it is known that outerplanar graphs have a vertex of degree at most 2. Consequently, our approach yields a 2-factor approximation algorithm for GSS instances whose underlying graph is outerplanar, and we do not know the complexity of GSS on the class of such instances. On such instances, we believe that it must be polynomial time solvable by setting up an appropriate dynamic program. These results are in Section 3.

2   Rounding the Hitting Set LP Using a Degree Based Vertex Ordering

Let $H = (V, E)$ be a hypergraph. For a subset $V' \subseteq V$, let $H(V')$ denote the hypergraph (V', E') where $E' = \{e \cap V' | e \in E\}$. $H(V')$ is the induced hypergraph of H on V'. The number of hyperedges containing a vertex is referred to as the *degree of the vertex*. Let $\sigma = v_1, v_2, ..., v_n$ be an ordering of V such that for each $1 \leq i \leq n$, in the hypergraph induced on the set $\{v_i, v_{i+1}, ..., v_n\}$, v_i has degree at most d. Through out this section we assume that the algorithm processes $H = (V, E)$ along with the vertex order σ for the integer value $d \geq 0$. With the hypergraph $H = (V, E)$ we use the standard Hitting Set LP relaxation in our algorithm as given in LP1.

$$\text{minimize} \quad \sum_{v \in V(H)} x_v$$

(LP1)
$$\text{subject to} \quad \sum_{v \in e} x_v \geq 1, \forall e \in E(H)$$

$$x_v \geq 0 \quad , \forall v \in V(H)$$

The dual of LP1 is given below, the variable y_e is the dual variable corresponding to the hyperedge e.

$$\text{maximize} \quad \sum_{e \in E(H)} y_e$$

(LP2)
$$\text{subject to} \quad \sum_{v \in e} y_e \leq 1, \forall v \in V(H)$$

$$y_e \geq 0 \quad , \forall e \in E(H)$$

Algorithm 1 takes as input a hypergraph $H = (V, E)$. It starts by ordering of the set of vertices $V(H)$ given by $\sigma = v_1, v_2, ..., v_n$ such that for each $1 < i \leq n$, the degree of v_i is at most d in the hypergraph induced on $\{v_i, \ldots, v_n\}$. It outputs a hitting set $V' \subseteq V$ of the set of hyperedges $E(H)$. The algorithm iterates over the set of vertices according to the ordering σ. We use the values to the variables x_v given by an optimum solution to LP1. We then use a primal-dual like approach to generate an assignment to the dual variables y_e in LP2. We say primal-dual like approach because we do not have to maintain dual feasibility.

Theorem 1. *Let $H = (V, E)$ be an instance of hitting set and let $\sigma = v_1, \ldots, v_n$ be an ordering of vertices such that for each $1 \leq i \leq n$, degree of v_i in the hypergraph induced on the set $\{v_i, v_{i+1}, \ldots, v_n\}$ is at most d. The set V' is returned by Algorithm 1 in polynomial time and is a d-approximation to the minimum hitting set of H.*

Proof. Algorithm 1 terminates in polynomial time as, each vertex is considered exactly once according to the ordering σ, and the steps performed for each vertex involve only updating the color of some edge and increasing the value of at most d-dual variables incident on the vertex. When the algorithm terminates, all the edges are blue, that is they are hit by V' and therefore V' is a hitting set of the edges in $H = (V, E)$. Further, for each $e \in E(H)$, $\sum_{v \in e} x_v^* \geq 1$, and for each vertex v the value of the dual variable of every red edge incident on it is increased by x_v^*. Due to these two observations, it follows that each $e \in E(H)$ will be colored blue in some iteration. This guarantees that V' is a hitting set. Now, we argue that $|V'| \leq d \sum_{v \in V} x_v^*$. When a vertex v is considered according to the order σ, x_v^* is added to at most d dual variables, therefore, $\sum_{e \in E(H)} y_e$ is increased by at

Algorithm 1. d-Approximation Algorithm for Hitting Set on Hypergraph $H = (V, E)$

Let $\sigma = v_1, \ldots, v_n$, $i = 1$.
Let $x^* = (x_v^* \mid v \in V)$ denote the LP optimum for LP1
For all $e \in E$, let $y_e = 0$ and let $V' = \phi$
Each e is colored red.
{/* red colored edges are not hit by V' */}
while some edge is red in $E(H)$ **do**
    Let $v = \sigma(i)$
    {/*The i-vertex in the ordering σ */}
    if v is the only vertex in some **red** edge e **then**
        Let $y_e = y_e + x_v$ {/* Invariant: $y_e \geq 1$ */}
        $V'' = \{v\}$
    else
        For each **red** edge $e \in E(H)$ and $v \in e$, Let $y_e = y_e + x_v$
        $V'' = \{v \in V(H) \mid \sum\limits_{v \in e, e \text{ red}} y_e \geq 1\}$
    end if
    while $V'' \neq \phi$ **do**
        Let $v' \in V''$ {/* v' is in some red edge */}
        $V' = V' \cup \{v'\}$
        For each **red** edge e and $v' \in e$, change color of e to **blue**.
        {/* blue colored edges are hit by V' */}
        Remove each $v \in V''$ such that $\sum\limits_{v \in e, e \text{ red}} y_e < 1$
    end while
    $i = i + 1$
end while
return V'

---

most dx_v^*. At the end of the algorithm, $\sum\limits_{e \in E(H)} y_e \leq d \sum\limits_{v \in V} x_v^*$. Further, whenever a vertex v' is added to V', it satisfies that property that $\sum\limits_{v' \in e, e \text{ is red}} y_e \geq 1$. Note here that summation is over edges for which v' is the first vertex (according to σ) to hit them. Therefore, the addition of v to the hitting set V' can be charged to these edges that are hit for the first time by v'. Formally, the edges of H can be partitioned into $|V'|$ classes, one corresponding to each $v' \in V$, referred to as $E_{v'}$. Further, for each $v' \in V'$, $\sum\limits_{e \in E_{v'}} y_e \geq 1$. This shows that $|V'| \leq \sum\limits_{e \in E(H)} y_e$. Therefore, $|V'| \leq d \sum\limits_{v \in V} x_v^*$, and this shows that $|V'|$ is a d-approximation to the minimum hitting set of H. $\qquad\square$

It is also important to note that we can get a dual feasible solution with the same bound on $|V'|$, and we do not do it as it is unnecessary to maintain this additional property. Finally, as we end this section, we believe that this rounding technique is of independent interest towards rounding fractional solutions to LP1. The upcoming section definitely supports this belief, as we use the rounding approach in this section to find a constant factor approximation to the optimum hitting set for special classes of line segments.

3 Guarding Special Sets of Segments

We now use a variant of Algorithm 1 to hit a given set of line segments using points of intersection. Given a GSS instance, that is a collection of line segments $\mathcal{L} = \{l_i | 1 \leq i \leq n\}$, the set of intersection points among the segments can be found in polynomial time, see for example [4], [9], and [3]. The range space corresponding to a $\mathcal{L}$ is as follows: U is the set of all intersection points of pairs of line segments in $\mathcal{L}$, and $\mathcal{S} = \{S_l \mid l \in \mathcal{L}, S_l \subseteq U\}$, where for each line segment $l \in \mathcal{L}, S_l = \{v \mid v \in l\}$. Clearly $H = (U, \mathcal{S})$ is a hypergraph. The *underlying graph* for a GSS instance is $G = (V, E)$ such that $V = U$ and an edge $(u, v) \in E(G)$ iff u and v are points adjacent on a line segment. We use standard graph theoretic concepts such as vertex degree, open neighborhood of a vertex, girth, planar and outerplanar graphs as they occur in the book by West [20].

3.1 Exploiting Girth 4 in the Underlying Graph

When the underlying graph of a GSS instance has girth at least 4, we show, that there is a vertex ordering $\sigma = v_1, v_2, ..., v_n$ such that for each $1 \leq i \leq n$, the degree of v_i is at most 3 in the graph induced on $\{v_i, v_{i+1}, \ldots, v_n\}$. We call such instances of GSS as Triangle-Free GSS. As discussed in the introduction the hitting set problem on Triangle-Free GSS instances is NP-hard. We observe that the underlying graph of girth 4, corresponding to a GSS instance, has at most $2n - 4$ edges, where n is the number of points of intersection. This follows from the well-known result in graph theory [20]:

Lemma 1. *Let G be a simple planar graph with $n \geq 3$ vertices, and girth $g \geq 4$. Then the number of edges in G is at most $2n - 4$.*

Lemma 2. *Let G be the underlying graph with n vertices and $m \leq 2n - 4$ edges. There exists an ordering of vertices $\sigma = v_1, \ldots, v_n$ such that for each $1 \leq i \leq n$, the degree of v_i in the graph induced on the set $\{v_i, v_{i+1}, \ldots, v_n\}$ is at most 3.*

Proof. Since the number of edges is at most $2n - 4$, we know that there is a vertex of degree at most 3. Let this vertex be v_1. We now show how to construct a GSS instance after removing v_1 such that the number of edges in the associated underlying graph is at most $2(n - 1) - 4$. If degree of v_1 is exactly 3, in $\mathcal{L}$ there are two possibilities: v_1 is the end point of all the 3 line segments containing v_1; v_1 is the end point of one line segment and is not an end point on exactly one another line segments (forming T-like patterns). If degree of v_1 is 2, then v_1 is the end point of 2 line segments. In the cases when v_1 is the end point of all the line segments in which it is present, consider the GSS instance obtained by removing v_1 and *shortening* all the line segments containing v_1. Clearly, the result is a GSS instance, and the number of edges in the underlying graph is at most $2(n-1) - 4$. In the case when v_1 is the end point of one line segment l_1 and is not the end point of a line segment l_2, consider the GSS instance obtained by shortening l_1 by removing the part of the line segments between v_1 and the next intersection point on l_1. In this GSS instance, v_1 is not a vertex as it is not a point of intersection. Therefore, the underlying graph has $n-1$ vertices, and at most $2(n-1) - 4$ edges. Therefore, the lemma follows by induction on the number of points of intersection.   □

It is important to note that Lemma 2 holds for all underlying graphs that have at most $2n - 4$ edges, and this is guaranteed for the case when the underlying graph has girth 4. We now present a 3-approximation for hitting set of on GSS instances for which the underlying graph has at most $2n-4$ edges. This algorithm uses Algorithm 1 along with one additional step.

Algorithm 2. Hitting set for a GSS instance $\mathcal{L}$ with at most $2n - 4$ edges

---

Solve LP1 for $H = (U, \mathcal{S})$
Let $x^* = (x_v^* | v \in U)$ be the LP optimum.
For each $x_v \geq \frac{1}{2}$, Let $V' = V' \cup \{v\}$.
Remove each $e \in \mathcal{S}$ such that $V' \cap e \neq \phi$.
Invoke Algorithm 1 on $H = (U, \mathcal{S})$. Let U be returned set.
{/*Algorithm 1 will set-up σ as in Lemma 2*/}
return $U \cup V'$

Theorem 2. *Let $\mathcal{L}$ be a GSS instance with n points of intersection and for which the underlying graph has at most $2n - 4$ edges. Algorithm 2 on $\mathcal{L}$ gives a hitting set whose size is at most 3 times the optimum hitting set for $\mathcal{L}$.*

Proof. The set output by the algorithm is indeed a hitting set, as each hyperedge is either hit by V' hit or by U. We now observe that $|V' \cup U| \leq \sum_{v \in V'} 2x_v^* + \sum_{u \in U} 3x_u^*$.

Clearly, $|V'| \leq \sum_{v \in V'} 2x_v^*$, since each $x_v^* \geq \frac{1}{2}$. To see that $|U| \leq \sum_{u \in U} 3x_u^*$, we first observe that Algorithm 1 uses a vertex ordering $\sigma = v_1, \ldots, v_n$ such that for each v_i, its degree in the graph induced on $\{v_i, \ldots, v_n\}$ is at most 3. The degree of a vertex v in the underlying graph, when it is non-zero, also accounts for the number of red hyperedges *with more than one element* incident on the v in Algorithm 1. The number of red hyperedges which contain v as the only element is at most 1, because $x_v^* < \frac{1}{2}$. To see this, if there were two red hyperedges e and e' containing exactly v, then the values to the two dual variables y_e and $y_{e'}$

would have been more than $\frac{1}{2}$. Consequently, the dual inequality corresponding to v would have evaluated at least 1, and v would have already been chosen into the hitting set. This contradicts the premise that e and e' are red. Therefore, the number of red hyperedges which contain v as the only element is at most 1. The bound on $|U|$ now follows from Lemma 1. Hence the theorem.   □

Corollary 1. *Let $\mathcal{L}$ be a triangle free GSS instance with n points of intersection. Algorithm 2 on $\mathcal{L}$ gives a hitting set whose size is at most 3 times the optimum hitting set for $\mathcal{L}$.*

Proof. By Lemma 1, it follows that the GSS instance has at most $2n - 4$ edges. From Theorem 2 it follows that the hitting set output by Algorithm 2 is a 3 approximation to the optimum hitting set.   □

3.2 Why GSS Escapes Alon's ϵ-Net Lower Bound?

It is well known that LP1 has a factor K-integrality gap if and only if the range space has an ϵ-net of size $\frac{K}{\epsilon}$. Alon [1] showed the construction of a set of points and a set of line segments such that any ϵ-net is of size $\Omega(\frac{1}{\epsilon})$. We show here that the instances output by Alon's construction [1] are not instances of GSS. To show this we review the salient features of Alon's construction here.

Theorem 3 ([1]). *For every (large) positive constant C there exist n and $\epsilon > 0$ and a set X of n points in the plane, so that the smallest possible size of an ϵ-net for lines for X is larger than $\frac{C}{\epsilon}$.*

One of the main concepts used in the proof of the above theorem is a combinatorial line which is defined as follows:

Definition 1 (Combinatorial Line). *Let $k \geq 2$ be an integer and $[k]^d$ denote the set of d-dimensional vectors with coordinates in $[k]$, where $[k]$ denotes the set $\{1, 2, \ldots, k\}$. A combinatorial line is a subset $L \subset [k]^d$ such that there is:*

1. *a set of coordinates $I \subset [d] = \{1, \ldots, d\}$*
2. *values $k_i \in [k]$ for all $i \in I$*

and $L = \{l_1, l_2, \ldots, l_k\}$ where $l_j = \{(x_1, \ldots, x_d) : x_i = k_i$ for all $i \in I$ and $x_i = j$ for all $i \in [d] \setminus I\}$

Thus a combinatorial line is a set of k d-dimensional vectors all having some fixed values in the coordinates in I, where the j-th vector has the value j in all other coordinates. This picture helps us in a counting argument that follows. Another result used by Alon's construction relies on the existence of combinatorial lines based on parameters k and d which is guaranteed by the Furstenberg-Katznelson Theorem as follows:

Theorem 4 ([11]). *For any fixed integer k and any fixed $\delta > 0$ there exists an integer $d_0 = d_0(k, \delta)$ so that for any $d \geq d_0$, any set Y of at least δk^d members of $[k]^d$ contains a combinatorial line.*

The lower bound in [1] is based on the following argument: For a large positive constant C, fix an integer $k > 2C$, and select $d_0 = d_0(k, 1/2)$ as given by Theorem 4. Let $n = k^d$ and $\epsilon = \frac{k}{k^d}$. Let $v_1, \ldots, v_d$ be vectors in the plane satisfying the condition of Lemma 2.2 in [1]. Let $X = \{m_1v_1 + m_2v_2 + \ldots + m_dv_d : 1 \leq m_i \leq k, \text{ for all } i\}$. The ranges are the line segments obtained for each combinatorial line L in $[k]^d$. The choice of the $v_1, \ldots, v_d$ guarantees that from each combinatorial line L, the set of k points $\{m_1v_1 + m_2v_2 + \ldots + m_dv_d : (m_1, m_2, \ldots, m_d) \in L\}$ lies on a geometric line segment containing exactly k-points of X.

Observation 1. *For each $k \geq 2$, $d \geq 3$, each point in X is present in at least d lines and at most $2^d - 1$ lines among the lines output by the above construction. Consequently, the points and lines output by Alon's construction do not form an instance of GSS.*

Proof. We present a lower bound on the number of combinatorial lines containing a vector. It is easy to see that the lower bound is d itself, since for $k = 2$, every coordinate of a vector from X can be inverted once giving us d distinct combinatorial lines. The upper bound is $2^d - 1$. For the chosen value of $d \geq 3$ and $k \geq 2$, the upper bound is 7 which is also tight. We demonstrate this observation by choosing values $k = 2$ and $d = 3$. In this case we have a set of 3-dimensional vectors with coordinates in $\{1, 2, 3\}$. Fixing the vector $(1, 1, 1)$ we see that the number of distinct combinatorial lines containing the vector are 7. Similarly, every other vector will be contained in 7 different combinatorial lines, implying that the points on the plane corresponding to these vectors have 7 (geometric) line segments passing through them. Since the underlying graph of these line segments is a planar graph, we know that there are points of intersection of degree at most 5 in the underlying graph. Such points are in at most 5 geometric line segments. However, all the given points in Alon's construction are present in 7 geometric line segments. Therefore, the instances output by Alon's construction [1] are not instances of GSS. Hence the claim.   □

This observation shows that the ϵ-net lower bound for points and lines constructed by Alon does not directly apply to the GSS instances. So, either a different technique is necessary to prove the ϵ-net lower bound for GSS instances or there is a constant factor approximation for the hitting set on GSS instances. If the former is the case, then the lower bound can only be shown by GSS instances for which the underlying graph has girth 3.

Acknowledgements. We would like to thank G. Ramakrishna for many helpful discussions and comments during the many versions of the algorithms in this paper. We would also like thank the referees for their comments that have greatly helped in improving the presentation of the paper.

References

1. Alon, N.: A non-linear lower bound for planar epsilon-nets. In: 2010 51st Annual IEEE Symposium on Foundations of Computer Science (FOCS), pp. 341–346 (2010)
2. Baker, B.S.: Approximation algorithms for np-complete problems on planar graphs. J. ACM 41(1), 153–180 (1994)
3. Balaban, I.J.: An optimal algorithm for finding segments intersections. In: Proceedings of the Eleventh Annual Symposium on Computational Geometry, SCG 1995, pp. 211–219. ACM, New York (1995)
4. Bentley, J., Ottmann, T.: Algorithms for reporting and counting geometric intersections. IEEE Transactions on Computers C-28(9), 643–647 (1979)
5. Brimkov, V.E.: Approximability issues of guarding a set of segments. Int. J. Comput. Math. 90(8), 1653–1667 (2013)
6. Brimkov, V.E., Leach, A., Mastroianni, M., Wu, J.: Guarding a set of line segments in the plane. Theoretical Computer Science 412(15), 1313–1324 (2011)
7. Brimkov, V.E., Leach, A., Wu, J., Mastroianni, M.: Approximation algorithms for a geometric set cover problem. Discrete Applied Mathematics 160, 1039–1052 (2011)
8. Brönnimann, H., Goodrich, M.: Almost optimal set covers in finite vc-dimension. Discrete and Computational Geometry 14(1), 463–479 (1995)
9. Chazelle, B.: Reporting and counting segment intersections. Journal of Computer and System Sciences 32, 156–182 (1986)
10. Feige, U.: A threshold of ln n for approximating set cover. J. ACM 45(4), 634–652 (1998)
11. Furstenberg, H., Katznelson, Y.: A density version of the hales-jewett theorem. Journal d'Analyse Mathématique 57(1), 64–119 (1991)
12. Garey, M.R., Johnson, D.S., Stockmeyer, L.: Some simplified np-complete problems. In: Proceedings of the Sixth Annual ACM Symposium on Theory of Computing, STOC 1974, pp. 47–63. ACM, New York (1974)
13. Haussler, D., Welzl, E.: Epsilon-nets and simplex range queries. In: Proceedings of the Second Annual Symposium on Computational Geometry, SCG 1986, pp. 61–71. ACM, New York (1986)
14. Karp, R.M.: Reducibility Among Combinatorial Problems. In: Miller, R.E., Thatcher, J.W. (eds.) Complexity of Computer Computations, pp. 85–103. Plenum Press (1972)
15. Kumar, V.S.A., Arya, S., Ramesh, H.: Hardness of set cover with intersection 1. In: Welzl, E., Montanari, U., Rolim, J.D.P. (eds.) ICALP 2000. LNCS, vol. 1853, pp. 624–635. Springer, Heidelberg (2000)
16. Long, P.M.: Using the pseudo-dimension to analyze approximation algorithms for integer programming. In: Proceedings of the 7th International Workshop on Algorithms and Data Structures, WADS 2001, pp. 26–37. Springer, London (2001)
17. Megiddo, N., Tamir, A.: On the complexity of locating linear facilities in the plane. Operations Research Letters 1(5), 194–197 (1982)
18. Mustafa, N.H., Ray, S.: Improved results on geometric hitting set problems. Discrete & Computational Geometry 44(4), 883–895 (2010)
19. Pyrga, E., Ray, S.: New existence proofs for ε-nets. In: Proceedings of the Twenty-fourth Annual Symposium on Computational Geometry, SCG 2008, pp. 199–207. ACM, New York (2008)
20. West, D.B.: Introduction to Graph Theory, 2nd edn. Prentice Hall (September 2000)

Reduction Techniques for Graph Isomorphism in the Context of Width Parameters

Yota Otachi[1] and Pascal Schweitzer[2]

[1] Japan Advanced Institute of Science and Technology
School of Information Science, Japan
otachi@jaist.ac.jp
[2] RWTH Aachen University, Aachen, Germany
schweitzer@informatik.rwth-aachen.de

Abstract We study the parameterized complexity of the graph isomorphism problem when parameterized by width parameters related to
tree decompositions. We apply the following technique to obtain fixed-
parameter tractability for such parameters. We first compute an isomorphism invariant set of potential bags for a decomposition and then
apply a restricted version of the Weisfeiler-Lehman algorithm to solve
isomorphism. With this we show fixed-parameter tractability for several
parameters and provide a unified explanation for various isomorphism
results concerned with parameters related to tree decompositions.

As a possibly first step towards intractability results for parameterized
graph isomorphism we develop an fpt Turing-reduction from strong tree
width to the a priori unrelated parameter maximum degree.

1 Introduction

The graph isomorphism problem is the algorithmic task to decide whether two
given graphs are isomorphic, i.e., whether there exists a bijection from the vertices of one graph to the vertices of the other graph preserving adjacency and
non-adjacency. The problem is situated in the complexity class NP. However,
despite extensive research on this problem, the complexity remains unknown. It
is neither known whether the problem is polynomial-time solvable nor whether
it is NP-hard.

In this paper, we are interested in the parameterized complexity of the isomorphism problem. For other aspects related to the isomorphism problem we
refer the reader to other sources (e.g., [18], [23], [26]).

In the parameterized context, for a graph parameter k, such as the maximum
degree of the input graphs, we ask for an algorithm that solves isomorphism
of graphs with parameter at most k. In this context, we are interested in the
existence of algorithms with a running time of $O(f(k)n^c)$ for some constant $c \in \mathbb{N}$
in contrast to algorithms with a running time of $O(n^{f(k)})$. Running times of the
former type are called fpt time and the algorithms are said to be fixed-parameter
tractable algorithms.

R Ravi and I.L. Gørtz (Eds.): SWAT 2014, LNCS 8503, pp. 368–379, 2014.
© Springer International Publishing Switzerland 2014

Related Work. There are various results that show that isomorphism is fixed-parameter tractable with respect to some parameter. Such results exist for the parameters color multiplicity [14] (also known for hypergraphs [1]), eigenvalue multiplicity [10], rooted distance width [28], feedback vertex set number [19], bounded permutation distance [24], tree-depth [6] and connected path distance width [21]. For chordal graphs, tractability results are known for the parameters clique number [17], [20] and the size of simplicial components [27]. Yet, for many parameters, such as maximum degree, tree width[1] and genus, it is not known whether there exist fixed-parameter tractable algorithms solving isomorphism (see [19]). However, no non-tractability results are known. One of the obstacles to understanding the parameterized complexity of graph isomorphism is the uncertainty whether the standard reduction techniques, like showing W[1]-hardness, can be applied (see the discussions in [19] and [28]).

Our Results. We study the parameterized complexity of isomorphism with respect to various parameters related to strong tree decompositions. We first develop a method to obtain fixed-parameter tractable algorithms for parameterized graph isomorphism problems. The underlying technique of many approaches showing such results is to first find a restricted isomorphism invariant family of sets, potential bags, which capture a tree decompositions and to then use these to perform an isomorphism test that uses some form of dynamic programming. It turns out that it is possible to prove that this technique is applicable in general. To prove this general statement, we develop a restricted version of the Weisfeiler-Lehman color refinement algorithm and prove that it successfully decides isomorphism whenever an invariant family of potential bags capturing tree decompositions is available for the input graphs. The algorithm neither computes a decomposition nor does it require a decomposition to be given.

Using the technique, we show tractability of graph isomorphism for the parameters root-connected tree distance width and connected strong tree width. We also provide families of examples showing that neither of the two graph parameters mentioned can be bounded by a function of the other. The two tractability results extend results in [13], [21], and [28] also concerned with restricted forms of strong tree decompositions, and answers a question from [28]. Furthermore, with the technique, it is for example also possible to show that graph isomorphism parameterized by the maximum of the length of a longest geodesic cycle and strong tree width or by the maximum of the chordality and degree is fixed-parameter tractable.

In general, our technique provides a unified explanation for the various results [6,13,17,19,20,21,27,28] all showing that certain restrictions on tree decompositions lead to efficient algorithms for the isomorphism problem. Indeed, all of these approaches can be interpreted as determining some restricted family of potential bags capturing a tree decomposition and then performing some form of dynamic programming to check for isomorphism, that can also be performed

---

[1] We remark that after the submission of this paper, Daniel Lokshtanov, Marcin Pilipczuk, Michał Pilipczuk, and Saket Saurabh have published a preprint showing that graph isomorphism is fixed-parameter tractable with respect to tree width.

by the restricted Weisfeiler-Lehman algorithm. In each of the references above, the dynamic programming is a substantial part of the argumentation, which can now be replaced by the general theorem.

Finally, we show how the technique can be applied to obtain parameterized isomorphism algorithms by exploiting knowledge on the set of potential maximal cliques, of which we already know that it can always be computed in polynomial time in the number of potential maximal cliques.

Our technique also provides a proof of the fact that for graphs of bounded tree width a sufficiently high-dimensional Weisfeiler-Lehman algorithm can be used to determine isomorphism. This fact was first proven by Grohe and Mariño using logic [16] (see also [15]) and provides to date the fastest running time for isomorphism of bounded tree width graphs. Our proof provides a direct argument for this fact, which does not involve logic. We remark that in his book, Toda [26] also gives a dynamic programming algorithm matching the running time of the algorithm of Grohe and Mariño.

In this paper, we also take a first step towards developing means for some form of intractability result. Specifically, for the isomorphism problem we construct an fpt Turing reduction from strong tree width to the a priori unrelated parameter maximum degree. The existence of this reduction in particular implies that if graph isomorphism is fixed-parameter tractable when parameterized by degree then it is also fixed-parameter tractable when parameterized by strong tree width. However, a possibly better interpretation of this result is that isomorphism parameterized by degree is hard, being at least as intractable as isomorphism parameterized by strong tree width.

To obtain the reduction, we reduce the problem to biconnected components, a technique frequently used for isomorphism algorithms concerned with planar graphs (see [8]). However, we require an extended form of such a reduction allowing us to work with graphs equipped with an equivalence relation and equipped with a coloring of the linear orders of the equivalence classes.

Throughout this proceedings version, most proofs have been omitted. For these the reader is referred to [22].

2   Preliminaries

In this paper all graphs are finite, simple, undirected graphs. A *biconnected component* (also called a block) is a maximal connected subgraph not containing a cut-vertex. In particular, the connected graph on 2 vertices is biconnected.

A *strong tree decomposition* of a graph $G = (V, E)$ is a pair $(\{X_i \mid i \in I\}, T = (I, F))$ where $\{X_i \mid i \in I\}$ is a partition of the vertex set V into so-called bags X_i and $T = (I, F)$ is a tree such that the following holds: for all edges $\{u, v\} \in E$, either there is $i \in I$ with $u, v \in X_i$, or there are two adjacent tree vertices $i, i' \in I$ such that $u \in X_i$ and $v \in X_{i'}$. A *connected strong tree decomposition* is a strong tree decomposition for which each bag X_i induces a connected subgraph. The *width* of a strong tree decomposition is the maximum size of a bag of the decomposition.

A strong tree decomposition $(\{X_i \mid i \in I\}, T = (I, F))$ with a distinguished root $r \in I$ is a *tree distance decomposition* if each $v \in X_i$ with $i \neq r$ has a neighbor $u \in X_j$ where j is the parent of i in T rooted at r. A tree distance decomposition with root r is a *root-connected tree distance decomposition* if X_r induces a connected subgraph.

Here, we slightly diverge from the terminology used in [21] to highlight the fact that only the root set must induce a connected graph, and thereby avoid confusion with the term connected strong tree decomposition.

For a class of decompositions $\mathcal{C}$, the $\mathcal{C}$ width of a graph G is the minimal width over all $\mathcal{C}$ decompositions of G. We thus obtain the graph parameters strong tree width, denoted $\mathrm{stw}(G)$, connected strong tree width, denoted $\mathrm{cstw}(G)$, tree distance width, denoted $\mathrm{tdw}(G)$ and root-connected tree distance width, denoted $\mathrm{rctdw}(G)$. The notion of strong tree width was introduced by Seese [25] and is also known as *tree-partition width* [9]. In the context of graph isomorphism, tree distance decompositions were first considered in [28].

For a graph G, there may be several tree distance decompositions with the same root set S. However, there is a unique minimal decomposition (i.e., the partition into bags is at least as fine as any other partition into bags obtained from a tree distance decomposition) with root set S. Given S, this minimal decomposition can be computed in linear time.

Theorem 1 ([28, Theorem 2.1]). *Given a graph G and a set S, one can compute in $O(m)$ time the unique tree distance decomposition with root set S.*

We denote the width of this decomposition by $\mathrm{tdw}_S(G)$. Note that if G is not connected, it may be the case that there is no tree distance decomposition with root set S. To facilitate our proofs and simplify algorithms, we define $\mathrm{tdw}_S(G)$ to be infinite in this case.

For a graph G with distinct non-adjacent vertices s and t an *s-t-separator* is a set of vertices S such that s and t are in different components of $G - S$. An s-t separator is minimal if no proper subset of S is an s-t-separator.

An *fpt Turing reduction* (see [11]) of a parameterized problem P_1 with parameter k_1 to a parameterized problem P_2 with parameter k_2 is a Turing reduction from P_1 to P_2 with fpt running time for which the parameter k_2 of all oracle calls to the problem P_2 is bounded by a computable function in terms of k_1. In other words, a Turing reduction is an fpt-algorithm solving the parameterized problem P_1 with the help of an oracle that solves problem P_2 such that there exists a computable function g such that for all oracle queries $y \in P_2$ posed on an input x with parameter k_1 it holds that the parameter k_2 of y is at most $g(k_1)$.

Suppose we assign to every graph G a subset of the vertices $\mathcal{V}(G) \subseteq V(G)$. We say this assignment is *isomorphism invariant* if for every isomorphism $\pi\colon G_1 \to G_2$ we have $\mathcal{V}(G_2) = \pi(\mathcal{V}(G_1))$. This definition extends to assignments of tuples or sets of vertex sets and also to colored graphs.

3   Tree Decompositions and the Weisfeiler-Lehman Algorithm

In the graph isomorphism literature, for various graph classes, results are known showing that the Weisfeiler-Lehman algorithm yields polynomial time isomorphism algorithms (see [15]). In this section we describe a restricted version of the Weisfeiler-Lehman algorithm and show that it can be used to obtain fixed-parameter tractability results. Intuitively, the k-dimensional Weisfeiler-Lehman algorithm repeatedly recolors k-tuples of vertices by assigning them a color that depends on the multiset of previous colors of adjacent k-tuples, where tuples are adjacent if they differ by at most one entry. Our restricted version of the algorithm performs this recoloring operation only on a restricted set of k-tuples. For more information on the standard Weisfeiler-Lehman algorithm we refer the reader to existing literature (see [2] and [23] for more pointers).

For $k \geq 2$ we now define the *restricted k-dimensional Weisfeiler-Lehman color refinement*. We say a family of sets $\mathcal{V}$ has *width* k if the largest set in $\mathcal{V}$ has size k. Let G be a graph and $\mathcal{V}$ be a family of sets of vertices of G of width at most k'. Let $\mathcal{V}^+$ be the set of k-tuples $(v_1, \ldots, v_k)$ (with entries not necessarily distinct) for which $\{v_1, \ldots, v_{k'}\}$ is in $\mathcal{V}$. For every k-tuple $(v_1, \ldots, v_k)$ in $\mathcal{V}^+$ we define $\mathrm{wl}_0^k[\mathcal{V}, G](v_1, \ldots, v_k)$ as the isomorphism type of the subgraph induced by the ordered tuple $(v_1, \ldots, v_k)$. If the graph is colored then the isomorphism type has to take the coloring into account. More precisely, the coloring wl_0^k is a coloring that satisfies $\mathrm{wl}_0^k[\mathcal{V}, G](v_1, \ldots, v_k) = \mathrm{wl}_0^k[\mathcal{V}, G](v_1', \ldots, v_k')$ if and only if we can map v_i to v_i' and obtain an isomorphism of the colored graphs induced by $\{v_1, \ldots, v_k\}$ and $\{v_1', \ldots, v_k'\}$. If $(v_1, \ldots, v_k) \notin \mathcal{V}^+$ then we define $\mathrm{wl}_0^k[\mathcal{V}, G](v_1, \ldots, v_k)$ to be the empty set $\emptyset$.

Iteratively for $i \geq 0$, we define $\mathrm{wl}_{i+1}^k[\mathcal{V}, G](v_1, \ldots, v_k)$ to be the empty set $\emptyset$ if $(v_1, \ldots, v_k) \notin \mathcal{V}^+$ and to be $\left(\mathrm{wl}_i^k[\mathcal{V}, G](v_1, \ldots, v_k), \mathcal{M}_i^k \right)$ otherwise, where $\mathcal{M}_i^k$ is the multiset given by

$$\mathcal{M}_i^k := \{\!\{ (\mathrm{wl}_i^k[\mathcal{V}, G](x, v_2, \ldots, v_k), \mathrm{wl}_i^k[\mathcal{V}, G](v_1, x, v_3, \ldots, v_k), \ldots,$$

$$\mathrm{wl}_i^k[\mathcal{V}, G](v_1, \ldots, v_{k-1}, x)) \mid x \in V(G) \}\!\}.$$

The process partitions the ordered k-tuples into classes according to their color. Since in each iteration the color of the previous iteration is encoded in the new color, k-tuples which are assigned different colors will continue to have different colors in all subsequent iterations. Therefore the refinement process stabilizes. We define $\mathrm{wl}_\infty^k[\mathcal{V}, G](v_1, v_2, \ldots, v_k)$ as $\mathrm{wl}_i^k[\mathcal{V}, G](v_1, v_2, \ldots, v_k)$ where i is the least positive integer such that the induced partition in step i is equivalent to the induced partition in step $i + 1$. Abusing notation, we may drop the specifications $[\mathcal{V}, G]$ whenever they are apparent from the context.

Lemma 1. *For a graph G and a family $\mathcal{V}$ of sets of vertices of G of width k', the stable partition of the restricted $(k'+c)$-dimensional Weisfeiler-Lehman color refinement can be computed in time $\mathcal{O}\big((k' + c)^2 \cdot |\mathcal{V}^+| n \cdot \log(|\mathcal{V}^+|)\big)$.*

Being a restriction implies that the known examples that cannot be solved by the Weisfeiler-Lehman algorithm [7] can also not be solved by the restricted version. However, we will now prove that the restricted Weisfeiler-Lehman algorithm decides isomorphism of graphs whenever the set $\mathcal{V}(G)$ captures a tree decomposition. To facilitate the proof and to make it more easily applicable in the future, we prove the theorem for tree decompositions, instead of proving it just for strong tree decompositions.

Recall that a *tree decomposition* is a pair $(\{X_i \mid i \in I\}, T = (I, F))$ for which $\bigcup_{i \in I} X_i = V(G)$ and $T = (I, F)$ is a tree such that every vertex is contained in some bag, for adjacent vertices there is a bag containing both of them, and for every vertex v the set of bags containing v induces a connected subtree of T.

Given a graph G we say that a family of sets $\mathcal{V}(G)$ *captures a tree decomposition* T of G if every bag is in $\mathcal{V}(G)$. If G is equipped with an equivalence relation and possibly a tuple-coloring, we additionally require that every equivalence class is contained in a bag of T. We say that a tree decomposition is *semi-smooth* if the intersection of adjacent bags has size at most one smaller that the size of the larger bag (for the decomposition to be smooth one also requires that all bags have the same size, see [3]).

Theorem 2. *Suppose we are given an algorithm computing for every graph G in a graph class C an isomorphism invariant family of vertex sets $\mathcal{V}(G)$ of width at most k' such that $\mathcal{V}(G)$ captures a semi-smooth tree decomposition of G. Then we can decide isomorphism of graphs in C with the $(k' + 3)$-dimensional Weisfeiler-Lehman algorithm restricted to $\mathcal{V}(G)$.*

The previous theorem requires $\mathcal{V}(G)$ to capture a semi-smooth tree decomposition. However, we can extend the theorem to tree decompositions and strong tree decompositions by using the alternative set $\mathcal{V}' = \{B_1 \cup B_2 \mid B_1, B_2 \in \mathcal{V}(G)\}$. This can be seen by the following two observations.

If $\mathcal{V}$ is isomorphism invariant and captures a tree decomposition of G, then $\mathcal{V}'$ is isomorphism invariant and captures a semi-smooth tree decomposition. This follows from the construction that produces a smooth tree decomposition from a tree decomposition given in [3].

Suppose $\mathcal{B}$ is the set of bags of a strong tree decomposition. Then there is a semi-smooth tree decomposition $\mathcal{B}'$ such that for every bag B of $\mathcal{B}'$ there are bags B_1 and B_2 in $\mathcal{B}$ such that $B \subseteq B_1 \cup B_2$. This can be seen with the standard way of constructing a tree decomposition from a strong tree decomposition by inserting for each edge between two bags B_1 and B_2 a path of bags transforming B_1 to B_2 by replacing successively one vertex after the other. We conclude, if $\mathcal{V}$ captures a strong tree decomposition then $\mathcal{V}'$ captures a semi-smooth tree decomposition. This also shows that by setting $\mathcal{V}$ to be the set of all k-tuples of vertices, every graph of tree width at most k has a smooth tree decomposition captured by $\mathcal{V}'$ and shows that the a sufficiently high-dimensional Weisfeiler-Lehman algorithm solves graph isomorphism of graphs of bounded tree width, as mentioned in the introduction.

Corollary 1. *For a parameter k', given an fpt-algorithm that computes for every graph G in a graph class C an isomorphism invariant family of vertex sets $\mathcal{V}(G)$ of width at most k' such that $\mathcal{V}(G)$ captures a tree decomposition (or a strong tree decomposition), isomorphism of graphs in C is fixed parameter tractable in k'.*

We remark that if we were interested in actual running times, in the tree width case it is possible to avoid the increase in width from $\mathcal{V}$ to $\mathcal{V}'$ yielding better running time bounds.

4   Tree Distance Decompositions with Connected Root Bags

As a first application we show that graph isomorphism parameterized by root connected tree distance width is fixed parameter tractable. We say two vertices v_1 and v_2 in a graph are *k-connected*, if there are k internally vertex disjoint paths from v_1 to v_2. We denote this by $v_1 \equiv_k v_2$. The task of checking whether two vertices are k-connected is also known as the Menger Problem and can be solved in polynomial time via a reduction to the maximum flow problem.

Lemma 2. *Let G be a graph containing vertices v_1 and v_2. If $v_1 \equiv_{2k} v_2$, then in every strong tree decomposition of width at most k the vertices v_1 and v_2 are in the same bag.*

A similar result for tree decompositions, stipulating the existence of a bag that contains both v_1 and v_2 in tree decompositions of width at most k whenever $v_1 \equiv_{k+1} v_2$, can be found in [4]. The previous lemma restricts the possible bags in a decomposition and leads to an efficient algorithm for isomorphism parameterized by root-connected tree distance width.

Theorem 3. *The isomorphism problem parameterized by root-connected tree distance width can be solved in fpt time.*

We remark that our algorithm for the theorem does not necessarily compute a connected root set. In fact the number of connected root sets that yield distance decompositions of smallest width cannot be bounded by an fpt function, and they in particular cannot be enumerated in fpt time.

It is also possible to show that the root-connected tree distance width of a graph cannot be bounded in terms of the rooted tree distance width (i.e., only one vertex in the root).

Note that, in contrast to this, when we consider only path distance decompositions, the root-connected path distance width of a graph is bounded by a function of the rooted path distance width as shown in [21].

5   A Reduction from Strong Tree Width to Maximum Degree

To define a reduction from isomorphism parameterized by strong tree width to isomorphism parameterized by maximum degree, we first reduce the problem to biconnected graphs relative to an equivalence relation and a suitable type of coloring compatible with the equivalence relation.

Let R be an equivalence relation on the vertices of a graph G. We define the *quotient graph* of G with respect to R as the graph whose vertex set consists of the equivalence classes and in which two equivalence classes E_1 and E_2 are adjacent if there exists vertices $v_1 \in E_1$ and $v_2 \in E_2$ such that v_1 and v_2 are adjacent. We define *the biconnected components of G relative to R* as the sets of vertices that comprise biconnected components of the quotient graph, i.e., pre-images of biconnected components under the projection to the quotient graph. A *tuple-coloring* is a map that assigns a color to every linear ordering of vertices in an equivalence class. When concerned with isomorphism of tuple-colored graphs equipped with an equivalence relation, we demand that isomorphisms preserve equivalence classes and the tuple-coloring, that is, an isomorphism must map an equivalence class to an equivalence class and colored ordered tuples to ordered tuples of the same color.

We define the *biconnected component tree* (also called block-cut tree) of a graph G with respect to R as the following bipartite graph: the vertices of the one partition class are those equivalence classes that form cut-vertices in the quotient graph. The vertices of the other partition class are the biconnected components of G relative to R. In the tree, there is an edge between an equivalence class and a biconnected component if the corresponding cut-vertex is contained in the corresponding biconnected component in the quotient graph.

The quotient graph can be constructed in time linear in the number of edges of a graph. Since the biconnected components of the quotient graph can then also be computed in polynomial time, the biconnected components and the component tree with respect to R can be computed in polynomial time.

Lemma 3. *The isomorphism problem of graphs with an equivalence relation on the vertices Turing-reduces to isomorphism of tuple-colored biconnected components of the input graphs relative to the equivalence relation. The running time is bounded by a polynomial in n and $k!$, where n is the size of the input graphs and k is the size of the largest equivalence class.*

From the theorem we obtain as corollary that to decide isomorphism of graphs in a hereditary graph class, i.e., a class closed under taking induced subgraphs, it suffices to be able to decide isomorphism of biconnected vertex-colored graphs.

Corollary 2. *The graph isomorphism problem of vertex-colored graphs in a hereditary graph class $\mathcal{C}$ polynomial-time Turing-reduces to the isomorphism problem of biconnected vertex-colored graphs in $\mathcal{C}$.*

For general graph isomorphism, for every integer k, it is possible to reduce the isomorphism problem to isomorphism of k-connected graphs by simply adding universal vertices adjacent to all other vertices. However, for hereditary graph classes, under application of this technique or similar gadget constructions, the graphs may not necessarily remain within the class. In fact, if there is reduction to 3-connected graphs analogous to Corollary 2, then graph isomorphism would be polynomial-time solvable in general. This can be seen by considering the isomorphism-complete class of bipartite graphs in which in one bipartition class every vertex has degree at most 2. This class does not contain any 3-connected graphs that have components with more than 2 vertices.

We will now employ the relation $\equiv_{2k}$ defined in Section 4. However, this relation is not necessarily an equivalence relation. Let $\equiv_{2k}^{+}$ be the transitive closure of the relation $\equiv_{2k}$. It turns out that graphs that are biconnected relative to $\equiv_{2k}^{+}$ have bounded degree.

Lemma 4. *For $k \geq 2$, if a graph G with $\mathrm{stw}(G) \leq k$ is biconnected relative to $\equiv_{2k}^{+}$ then G has a maximum degree of at most $2k^2(k-1) + k - 1$.*

Lemma 5. *The isomorphism problem of tuple-colored graphs of degree at most d with an equivalence relation on the vertices with no equivalence class having more than k elements reduces to isomorphism of uncolored graphs of degree at most $\mathcal{O}(d + k!)$. The running time is polynomial in n and $k!$, where n is the size of the input graphs.*

For the lemma, it is essential that equivalence classes do not intersect. By coloring partially overlapping sets, it is possible to encode hypergraphs, and thus to encode graphs, even with sets of size at most 2.

Together, the lemmas of this section can be used to assemble a reduction from strong tree width to maximum degree.

Theorem 4. *There is an fpt Turing-reduction from isomorphism parameterized by strong tree width to isomorphism parameterized by maximum degree.*

Proof. We first compute $\equiv_{2k}$ and the transitive closure $\equiv_{2k}^{+}$ which can be done in polynomial time. If the largest equivalence class has size greater than k we reject the input as infeasible containing a graph of strong tree width larger than k. We then reduce via Lemma 3 to biconnected graphs relative to $\equiv_{2k}^{+}$. By Lemma 4 the biconnected components have bounded degree. By Lemma 5 we can then reduce the isomorphism problem of the tuple-colored biconnected components to isomorphism of graphs of bounded degree.   □

6   Applications to fpt Isomorphism Results

In this section we combine the two results from the previous sections to obtain further fpt isomorphism algorithms.

Theorem 5. *Graph isomorphism parameterized by connected strong tree width can be solved in fpt time.*

Proof. By Theorem 4 the problem reduces to isomorphism of the tuple-colored biconnected components. Let G be a graph of connected strong tree width at most k of bounded degree. To apply Corollary 1 we first describe a set of potential bags capturing a strong tree decomposition of G computable in fpt time and having an fpt size bound. For this consider the family $\mathcal{V}(G)$ of sets of size at most k that project to a connected subgraph in the quotient graph relative to $\equiv_{2k}^{+}$. Since the degree of G is bounded, the number of such sets is bounded by an fpt number. The theorem now follows from Corollary 1.   □

We have shown fixed-parameter tractability for isomorphism with respect to the parameters connected strong tree width and root-connected tree distance width. In turns out that these parameters are unrelated, i.e., that neither of these parameters can be bounded by a function of the other.

As further examples of applications of our technique we can obtain fixed-parameter tractability results of other parameters as follows. A geodesic cycle in a graph G is a cycle C such that the distance between every two vertices in C is the same as the distance in G. The chordality of a graph is the length of the longest induced cycles.

Theorem 6. *Graph isomorphism when parameterized by the maximum of the length of a geodesic cycle and strong tree width can be solved in fpt time.*

Corollary 3. *Graph isomorphism when parameterized by the maximum of the chordality and degree can be solved in fpt time.*

Further applications of the theorems can be obtained by considering the set of potential maximal cliques. A *potential maximal clique* of a graph G is a set of vertices that is a bag in some minimal tree-decomposition of G. The set of potential maximal cliques can be computed in polynomial time in the size of the set itself [5]. Moreover, this set is isomorphism invariant and the subset of potential maximal cliques of size k captures a tree decomposition of G of minimal width. We can thus apply Corollary 1 to the potential maximal cliques.

Theorem 7. *If for a parameterized graph class there is an fpt bound on the number of potential maximal bags then isomorphism is fixed parameter tractable in the maximum of the parameter and the tree width.*

Equivalently, in the theorem it suffices to have a bound on the number of minimal s-t-separators, since by a theorem of Bouchitté and Todinca [5] the number of potential maximal cliques is polynomially bounded the number of minimal s-t-separators.

Using results from [12] we can apply the theorem to various graph classes. Indeed, it is known that the number of minimal s-t-separators is polynomially bounded for weakly chordal, polygonal circle, circular-arc and d-trapezoid graphs (see [12, Section 5]). So for all these classes, isomorphism is fixed parameter tractable when parameterized by tree width. Moreover, for weakly chordal graphs and circular-arc graphs the tree width of H-minor free graphs is bounded by a function of the number of vertices in H (see also [12, Section 5]), so isomorphism of H-minor free weakly chordal graphs and H-minor free circular-arc graphs is fixed parameter tractable when parameterized by the size of H.

7   Conclusion

In this paper we show that, in order to perform isomorphism tests, it suffices to compute an invariant set of potential bags that is comprehensive enough to express a tree decomposition. Indeed, by applying the restricted Weisfeiler-Lehman, we do not need to worry about how to perform the isomorphism test, nor how to compute a decomposition.

For various other results this means that their isomorphism testing part can be replaced by the general theorem and only the part analyzing the graph class remains. For example in [20] and [17] it is shown that for chordal graphs of bounded clique number a tree model can be computed in cubic time. This tree model is unique up to the ordering of children and gives rise to an invariant family of sets of vertices capturing a tree-decomposition.

In [6] it is shown that for graphs of tree-depth at most k the number of vertices that can be chosen as the root in a tree-depth decomposition is bounded by a function of k, recursively applying this gives rise to invariant family of sets of vertices capturing a tree-decomposition.

Furthermore we demonstrated how the theorems can be used in conjunction with the set of potential maximal cliques. The advantage here is that it is known in general that this set can be efficiently computed.

However, our technique can also be applied to tree decompositions, where the long question whether graph isomorphism is fixed parameter tractable when parameterized by tree width remains.

On the other hand, our parameterized reduction to bounded degree is only valid for strong tree width, and whether such a reduction exists for tree width remains open. Finally, as mentioned in the introduction, no non-tractability results are known in this context, and parameterized reduction could be a method to establish hardness criteria.

References

1. Arvind, V., Das, B., Köbler, J., Toda, S.: Colored hypergraph isomorphism is fixed parameter tractable. In: FSTTCS, pp. 327–337 (2010)
2. Berkholz, C., Bonsma, P., Grohe, M.: Tight lower and upper bounds for the complexity of canonical colour refinement. In: Bodlaender, H.L., Italiano, G.F. (eds.) ESA 2013. LNCS, vol. 8125, pp. 145–156. Springer, Heidelberg (2013)
3. Bodlaender, H.L.: A linear-time algorithm for finding tree-decompositions of small treewidth. SIAM J. Comput. 25(6), 1305–1317 (1996)
4. Bodlaender, H.L.: Necessary edges in k-chordalisations of graphs. J. Comb. Optim. 7(3), 283–290 (2003)
5. Bouchitté, V., Todinca, I.: Listing all potential maximal cliques of a graph. Theor. Comput. Sci. 276(1-2), 17–32 (2002)
6. Bouland, A., Dawar, A., Kopczyński, E.: On tractable parameterizations of graph isomorphism. In: Thilikos, D.M., Woeginger, G.J. (eds.) IPEC 2012. LNCS, vol. 7535, pp. 218–230. Springer, Heidelberg (2012)

7. Cai, J.-Y., Fürer, M., Immerman, N.: An optimal lower bound on the number of variables for graph identification. Combinatorica 12(4), 389–410 (1992)
8. Datta, S., Limaye, N., Nimbhorkar, P., Thierauf, T., Wagner, F.: Planar graph isomorphism is in log-space. In: IEEE Conference on Computational Complexity, pp. 203–214 (2009)
9. Ding, G., Oporowski, B.: On tree-partitions of graphs. Discrete Math. 149(1-3), 45–58 (1996)
10. Evdokimov, S., Ponomarenko, I.N.: Isomorphism of coloured graphs with slowly increasing multiplicity of jordan blocks. Combinatorica 19(3), 321–333 (1999)
11. Flum, J., Grohe, M.: Parameterized Complexity Theory (Texts in Theoretical Computer Science). An EATCS Series. Springer, London (2006)
12. Fomin, F.V., Todinca, I., Villanger, Y.: Large induced subgraphs via triangulations and cmso. In: SODA, pp. 582–583 (2014)
13. Fuhlbrück, F.: Fixed-parameter tractability of the graph isomorphism and canonization problems. Diploma thesis, Humboldt-Universität zu Berlin (2013)
14. Furst, M.L., Hopcroft, J.E., Luks, E.M.: Polynomial-time algorithms for permutation groups. In: FOCS, pp. 36–41 (1980)
15. Grohe, M.: Fixed-point definability and polynomial time on graphs with excluded minors. In: LICS, pp. 179–188 (2010)
16. Grohe, M.: Definability and descriptive complexity on databases of bounded treewidth. In: Beeri, C., Bruneman, P. (eds.) ICDT 1999. LNCS, vol. 1540, pp. 70–82. Springer, Heidelberg (1998)
17. Klawe, M., Corneil, D., Proskurowski, A.: Isomorphism testing in hookup classes. SIAM Journal on Algebraic Discrete Methods 3(2), 260–274 (1982)
18. Köbler, J., Schöning, U., Torán, J.: The graph isomorphism problem: Its structural complexity. Birkhäuser Verlag, Basel (1993)
19. Kratsch, S., Schweitzer, P.: Isomorphism for graphs of bounded feedback vertex set number. In: Kaplan, H. (ed.) SWAT 2010. LNCS, vol. 6139, pp. 81–92. Springer, Heidelberg (2010)
20. Nagoya, T.: Counting graph isomorphisms among chordal graphs with restricted clique number. In: Eades, P., Takaoka, T. (eds.) ISAAC 2001. LNCS, vol. 2223, pp. 136–147. Springer, Heidelberg (2001)
21. Otachi, Y.: Isomorphism for graphs of bounded connected-path-distance-width. In: Chao, K.-M., Hsu, T.-S., Lee, D.-T. (eds.) ISAAC 2012. LNCS, vol. 7676, pp. 455–464. Springer, Heidelberg (2012)
22. Otachi, Y., Schweitzer, P.: full version of the paper. arXiv:1403.7238 [cs.DM] (2014)
23. Schweitzer, P.: Problems of unknown complexity: graph isomorphism and Ramsey theoretic numbers. Phd thesis, Universität des Saarlandes, Saarbrücken, Germany (July 2009)
24. Schweitzer, P.: Isomorphism of (mis)labeled graphs. In: Demetrescu, C., Halldórsson, M.M. (eds.) ESA 2011. LNCS, vol. 6942, pp. 370–381. Springer, Heidelberg (2011)
25. Seese, D.: Tree-partite graphs and the complexity of algorithms. In: Budach, L. (ed.) Fundamentals of Computation Theory, FCT 1985. LNCS, vol. 199, pp. 412–421. Springer, Heidelberg (1985)
26. Toda, S.: Gurafu Doukeisei Hantei Mondai (The Graph Isomorphism Decision Problem). Nihon University, Tokyo, Japan (2001) (in Japanese)
27. Toda, S.: Computing automorphism groups of chordal graphs whose simplicial components are of small size. IEICE Transactions 89-D(8), 2388–2401 (2006)
28. Yamazaki, K., Bodlaender, H.L., de Fluiter, B., Thilikos, D.M.: Isomorphism for graphs of bounded distance width. Algorithmica 24(2), 105–127 (1999)

Approximate Counting of Matchings
in $(3,3)$-Hypergraphs*

Andrzej Dudek[1,**], Marek Karpinski[2,***], Andrzej Ruciński[3,†],
and Edyta Szymańska[3,‡]

[1] Western Michigan University, Kalamazoo, MI, USA
andrzej.dudek@wmich.edu
[2] Department of Computer Science, University of Bonn, Germany
marek@cs.uni-bonn.de
[3] Faculty of Mathematics and Computer Science, Adam Mickiewicz University,
Poznań, Poland
{rucinski,edka}@amu.edu.pl

Abstract. We design a fully polynomial time approximation scheme (FPTAS) for counting the number of matchings (packings) in arbitrary 3-uniform hypergraphs of maximum degree three, referred to as $(3,3)$-hypergraphs. It is the first polynomial time approximation scheme for that problem, which includes also, as a special case, the 3D Matching counting problem for 3-partite $(3,3)$-hypergraphs. The proof technique of this paper uses the general correlation decay technique and a new combinatorial analysis of the underlying structures of the intersection graphs. The proof method could be also of independent interest.

1 Introduction

The computational status of approximate counting of matchings in hypergraphs has been open for some time now, contrary to the existence of polynomial time approximation schemes for graphs. The matching (packing) counting problems in hypergraphs occur naturally in the higher dimensional free energy problems, like in the monomer-trimer systems discussed, e.g, by Heilmann [9]. The corresponding optimization versions of hypergraph matching problem relate also to various allocations problems.

This paper aims at shedding some light on the approximation complexity of that problem in 3-uniform hypergraphs of maximum vertex degree three (called

---

* Part of research of the 3rd and 4th authors done at Emory University, Atlanta and another part during their visits to the Institut Mittag-Leffler (Djursholm, Sweden).
** Research supported by Simons Foundation Grant #244712 and by a grant from the Faculty Research and Creative Activities Award (FRACAA), Western Michigan University.
*** Research supported by DFG grants and the Hausdorff grant EXC59-1.
† Research supported by the Polish NSC grant N201 604 940 and the NSF grant DMS-1102086.
‡ Research supported by the Polish NSC grant N206 565 740.

R Ravi and I.L. Gørtz (Eds.): SWAT 2014, LNCS 8503, pp. 380–391, 2014.
© Springer International Publishing Switzerland 2014

(3, 3)-hypergraphs or (3, 3)-graphs for short). This class of hypergraphs includes also so-called 3D hypergraphs, that is, (3,3)-graphs that are 3-partite. In [11], based on a generalization of the canonical path method of Jerrum and Sinclair [10], we established a fully polynomial time randomized approximation scheme (FPRAS) for counting matchings in the classes of k-uniform hypergraphs without structures called 3-combs. However, the status of the problem in arbitrary (3, 3)-graphs was left wide open among with other general problems for 3-, 4- and 5-uniform hypergraphs (for $k \geq 6$ it is known to be hard, see Sec. 2). In particular, the existence of an FPRAS for counting matchings in (3, 3)-graphs was unknown.

In this paper we design the first fully polynomial time approximation scheme (FPTAS) for arbitrary (3, 3)-graphs. The method of solution depends on the general correlation decay technique and some new structural analysis of underlying intersections graphs based on an extension of the classical claw-freeness notion. The proof method used in the analysis of our algorithm could be also of independent interest.

The paper is organized as follows. Section 2 contains some basic notions and preparatory discussions. In Sec. 3 we formulate our main results and provide the proofs. Finally, Sec. 4 is devoted to the summary and an outlook for future research.

2   Preliminaries

A *hypergraph* $H = (V, E)$ is a finite set of vertices V together with a family E of distinct, nonempty subsets of vertices called edges. In this paper we consider k-*uniform hypergraphs* (called further k-*graphs*) in which, for a fixed $k \geq 2$, each edge is of size k. A *matching* in a hypergraph is a set (possibly empty) of disjoint edges.

Counting matchings is a #P-complete problem already for graphs ($k = 2$) as proved by Valiant [16]. In view of this hardness barrier, researchers turned to approximate counting, which initially has been accomplished via probabilistic techniques.

Given a function C and a random variable Y (defined on some probability space), and given two real numbers $\epsilon, \delta > 0$, we say that Y is an (ϵ, δ)-*approximation* of C if the probability $\mathbb{P}(|Y(x) - C(x)| \geq \epsilon C(x)) \leq \delta$. A *fully polynomial randomized approximation scheme* (FPRAS) for a function f on $\{0, 1\}^*$ is a randomized algorithm which, for every triple (ϵ, δ, x), with $\epsilon > 0$, $\delta > 0$, and $x \in \{0, 1\}^*$, returns an (ϵ, δ)-approximation Y of $f(x)$ and runs in time polynomial in $1/\epsilon$, $\log(1/\delta)$, and $|x|$.

In this paper we investigate the problem of counting the number of matchings in hypergraphs and try to determine the status of this problem for k-graphs with bounded degrees.

Let $deg_H(v)$ be the degree of vertex v in a hypergraph H, that is, the number of edges of H containing v. We denote by $\Delta(H)$ the maximum of $deg_H(v)$ over all v in H. We call a k-graph H a (k, r)-*graph* if $\Delta(H) \leq r$. Let $\#M(k, r)$ be the problem of counting the number of matchings in (k, r)-graphs.

Our inspiration comes from new results (both positive and negative) that emerged for approximate counting of the number of independent sets in graphs with bounded degree and shed some light on the problem $\#M(k,r)$.

Let $\#IS(d)$ [$\#IS(\leq d)$] be the problem of counting the number of all independent sets in d-regular graphs [graphs of maximum degree bounded by d, that is, $(2,d)$-graphs]. Luby and Vigoda [13] established an FPRAS for $\#IS(\leq 4)$. This was complemented later by the approximation hardness results for the higher degree instances by Dyer, Frieze and Jerrum [6]. The subsequent progress has coincided with the revival of a deterministic technique – the spatial correlation decay method – based on early papers of Dobrushin [5] and Kelly [12]. It resulted in constructing deterministic approximation schemes for counting independent sets in several classes of graphs with degree (and other) restrictions, as well as for counting matchings in graphs of bounded degree.

Definition 1. *A fully polynomial time approximation scheme (**FPTAS**) for a function f on $\{0,1\}^*$ is a deterministic algorithm which for every pair (ϵ, x) with $\epsilon > 0$, and $x \in \{0,1\}^*$, returns a number $y(x)$ such that*

$$|y(x) - f(x)| \leq \epsilon f(x),$$

and runs in time polynomial in $1/\epsilon$, and $|x|$.

In 2007 Weitz [17] found an FPTAS for $\#IS(\leq 5)$, while, more recently, Sly [14] and Sly and Sun [15] complemented Weitz's result by proving the approximation hardness for $\#IS(6)$, that is, proving that unless NP=RP, there exists no FPRAS (and thus, no FPTAS) for $\#IS(6)$. By applying two reductions: from $\#IS(6)$ to $\#M(6,2)$ (taking the dual hypergraph of a 6-regular graph), and from $\#M(k,2)$ to $\#IS(k)$ (taking the intersection graph of a $(k,2)$-graph) for $k = 3, 4, 5$, we conclude that

(i) (unless NP=RP) there exists no FPRAS for $\#M(6,2)$;
(ii) there is an FPTAS for $\#M(k,2)$ with $k \in \{3,4,5\}$.

Note that the first reduction results, in fact, in a *linear* $(6,2)$-graph, so the class of hypergraphs in question is even narrower. (A hypergraph is called *linear* when no two edges share more than one vertex.) On the other hand, by the same kind of reduction it follows from a result of Greenhill [8] that *exact* counting of matchings is $\#P$-complete already in the class of linear $(3,2)$-graphs.

Facts (i) and (ii) above imply that the only interesting cases for the positive results are those for (k,r)-graphs with $k = 3, 4, 5$ and $r \geq 3$, and thus, the smallest one among them is that of $(3,3)$-graphs. Our main result establishes an FPTAS for counting the number of matchings in this class of hypergraphs.

3 Main Result and the Proof

The following theorem is the main result of this paper.

Theorem 2. *The algorithm CountMatchings given in Section 3.2 provides an FPTAS for $\#M(3,3)$ and runs in time $O\left(n^2(n/\epsilon)^{\log_{50/49} 144}\right)$.*

The intersection graph of a hypergraph H is the graph $G = L(H)$ with vertex set $V(G) = E(H)$ and edge set $E(G)$ consisting of all intersecting pairs of edges of H. When H is a graph, the intersection graph $L(H)$ is called *the line graph* of H. Graphs which are line graphs of some graphs are characterized by 9 forbidden induced subgraphs [3], one of which is the *claw*, an induced copy of $K_{1,3}$. There is no similar characterization for intersection graphs of k-graphs. Still, it is easy to observe that for any k-graph H, its intersection graph $L(H)$ does not contain an induced copy of $K_{1,k+1}$. We shall call such graphs $(k+1)$-*claw-free*.

Our proof of Thm. 2 begins with an obvious observation that counting the number of matchings in a hypergraph H is equivalent to counting the number of independent sets in the intersection graph $G = L(H)$. More precisely, let $Z_M(H)$ be the number of matchings in a hypergraph H and, for a graph G, let $Z_I(G)$ be the number of independent sets in G. (Note that both quantities count the empty set.) Then $Z_M(H) = Z_I(L(H))$.

To approximately count the number of independent sets in a graph $G = L(H)$ for a $(3,3)$-graph H, we apply some of the ideas from [2] (the preliminary version of this paper appeared in [1]) and [7]. In [2] two new instances of FPTAS were constructed, both based on the spatial correlation decay method. First, for $\#M(2,r)$ with any given r. Then, still in [2], the authors refined their approach to yield an FPTAS for counting independent sets in claw-free graphs of bounded clique number which contain so called *simplicial cliques*. The last restriction has been removed by an ingenious observation in [7].

Papers [2,7] inspired us to seek adequate methods for $(3,3)$-graphs. Indeed, for every $(3,3)$-graph H its intersection graph $G = L(H)$ is 4-claw-free and has $\Delta(G) \leq 6$. This turned out to be the right approach, as we deduced our Thm. 2 from a technical lemma (Lem. 3 below) which constructs an FPTAS for the number of independent sets in $K_{1,4}$-free graphs G with $\Delta(G) \leq 6$ and an additional property stemming from their being intersection graphs of $(3,3)$-graphs.

3.1   Proof of Theorem 2 – Sketch and Preliminaries

We deduce Thm. 2 from a technical lemma. The assumptions of this lemma reflect some properties of the intersection graphs of $(3,3)$-graphs.

Lemma 3. *There exists an FPTAS for the problem of counting independent sets in every 4-claw-free graph with maximum degree at most 6 and such that the neighborhood of every vertex of degree $d \geq 5$ induces a subgraph with at most $6 - d$ isolated vertices.*

Proof (of Thm. 2). Given a $(3,3)$-graph H, consider its intersection graph G. Then G is 4-claw-free, has maximum degree at most 6 and every vertex neighborhood of size $d \geq 5$ must span in G a matching of size $\lfloor d/2 \rfloor$. This means that Lem. 3 applies to G and there is an FPTAS for counting independent sets of G which is the same as counting matchings in H. □

It remains to prove Lem. 3. We begin with underlining some properties of 4-claw-free graphs which are relevant for our method. First, we introduce the notion of a *simplicial 2-clique* which is a generalization of a simplicial clique introduced in [4] and utilized in [2]. Throughout we assume notation $A \setminus B$ for set differences and, for $A \subset V(G)$, we write $G - A$ for the graph operation of deleting from G all vertices belonging to A. In other words, $G - A = G[V(G) \setminus A]$. Also, for any graph G, we use $\delta(G)$ to denote its minimum vertex degree and $\alpha(G)$ for the size of the largest independent set in G.

Definition 4. *A set $K \subseteq V(G)$ is a* 2-clique *if $\alpha(G[K]) \leq 2$. A 2-clique is simplicial if for every $v \in K$, $N_G(v) \setminus K$ is a 2-clique in $G - K$.*

For us a crucial property of simplicial 2-cliques is that if G is a connected graph containing a nonempty simplicial 2-clique K then it is easy to find another simplicial 2-clique in the induced subgraph $G - K$, and consequently, the whole vertex set of G can be partitioned into blocks which are simplicial 2-cliques in suitable nested sequence of induced subgraphs of G (see Claim 8).

However, in the proof of Lem. 3 we shall use a special class of 2-cliques.

Definition 5. *A 2-clique K in a graph G is called a* block *if $|K| \leq 4$ and $\delta(G[K]) \geq 1$ whenever $|K| = 4$. A block K is simplicial if for every $v \in K$ the set $N_G(v) \setminus K$ is a block in $G - K$.*

Next, we state a trivial but useful observation which follows straight from the above definition. (We consider the empty set as a block too.)

Fact 6. *If K is a (simplicial) block in G then for every $V' \subseteq V(G)$ the set $K \cap V'$ is a (simplicial) block in the induced subgraph $G[V']$ of G.*

Let a graph G satisfy the assumptions of Lem. 3. The next claim provides a vital, "self-reproducing" property of blocks in G.

Claim 7. *If K is a simplicial block in G, then for every $v \in K$ the set $N_G(v) \setminus K$ is a simplicial block in $G - K$.*

Proof. Set $K_v := N_G(v) \setminus K$ for convenience. By definition of K, K_v is a block. It remains to show that K_v is simplicial. Let $u \in K_v$ and let $K_u = N_G(u) \setminus (K \cup N_G(v))$. Suppose there is an independent set I in $G[K_u]$ of size $|I| = 3$. Then u, v and the vertices of I would form an induced $K_{1,4}$ in G with u in the center. As this is a contradiction, we conclude that K_u is a 2-clique.

To show that K_u is indeed a block, note first that, by the assumptions that $\Delta(G) \leq 6$, we have $|K_u| \leq 5$. However, if $|K_u| = 5$ then v would be an isolated vertex in $G[N_G(u)]$ – a contradiction with the assumption on the structure of the neighborhoods in G. For the same reason, if $|K_u| = 4$ then regardless of the degree of u in G (which might be 5 or 6) there can be no isolated vertex in $G[K_u]$. $\qquad\square$

Our next claim asserts that once there is a nonempty block in G, one can find a suitable partition of $V(G)$ into sets which are blocks in a nested sequence of induced subgraphs of G defined by deleting these sets one after another.

Claim 8. *Let K be a nonempty simplicial block in G. If, in addition, G is connected then there exists a partition $V(G) = K_1 \cup \cdots \cup K_m$ such that $K_1 = K$ and for every $i = 2, \ldots, m$, K_i is a nonempty, simplicial block in $G_i := G - \bigcup_{j=1}^{i-1} K_j$.*

Proof. Suppose we have already constructed disjoint sets $K_1 \cup \cdots \cup K_s$, for some $s \geq 1$, such that $K_1 = K$, for every $i = 2, \ldots, s$, K_i is a nonempty, simplicial block in $G_i := G - \bigcup_{j=1}^{i-1} K_j$, and that $R_s := V(G) \setminus \bigcup_{i=1}^{s} K_s \neq \emptyset$. Since G is connected, there is an edge between a vertex in R_s and a vertex $v \in K_i$ for some $1 \leq i \leq s$. Since K_i is a simplicial block in G_i, by Fact 6, it is also simplicial in its subgraph $G_i[V']$, where $V' = K_i \cup R_s$, that is the subgraph of G_i obtained by deleting all vertices of $K_{i+1} \cup \cdots \cup K_{s-1}$. Now apply Claim 7 to $G_i[V']$, K_i, and v, to conclude that $N_G(v) \cap R_s$ is a simplicial block in $G_{s+1} := G - \bigcup_{i=1}^{s} K_i$. □

Let $K_1, K_2, \ldots, K_m$ be as in Claim 8. Then,

$$Z_I(G) = \frac{Z_I(G_1)}{Z_I(G_2)} \cdot \frac{Z_I(G_2)}{Z_I(G_3)} \cdot \ldots \cdot \frac{Z_I(G_i)}{Z_I(G_{i+1})} \cdot \ldots \cdot \frac{Z_I(G_m)}{Z_I(G_{m+1})}, \tag{1}$$

where $G_{m+1} = \emptyset$ and $Z_I(G_{m+1}) = 1$. Observe that for each i, $G_{i+1} = G_i - K_i$ and the reciprocal of each quotient in (1) is precisely the probability

$$\mathbb{P}_{G_i}(K_i \cap \mathbf{I} = \emptyset) = \frac{Z_I(G_i - K_i)}{Z_I(G_i)}, \tag{2}$$

where $\mathbf{I}$ is an independent set of G_i chosen uniformly at random. In view of this, the main step in building an FPTAS for $Z_I(G)$ will be to approximate the probability $\mathbb{P}_G(K_i \cap \mathbf{I} = \emptyset)$ within $1 \pm \frac{\varepsilon}{n}$ (see Sec. 3.2 and Algorithm 2 therein).

But what if G is disconnected or does not contain a simplicial block to start with? First, if $G = \bigcup_{i=1}^{c} G_i$ consists of c connected components $G_1, \ldots, G_c$, then, clearly

$$Z_I(G) = \prod_{i=1}^{c} Z_I(G_i) \tag{3}$$

and the problem reduces to that for connected graphs.

As for the second obstacle, Fadnavis [7] proposed a very clever observation to cope with it. Let G be a connected graph satisfying the assumptions of Lem. 3 and let $v \in V(G)$ be such that $G - v$ is connected. By considering the fate of vertex v, we obtain the recurrence

$$Z_I(G) = Z_I(G - v) + Z_I(G^v), \tag{4}$$

where $G^v = G - N_G[v]$ and $N_G[v] = N_G(v) \cup \{v\}$. Let $G^v = \bigcup_{i=1}^{c} G_i^v$ be the partition of G^v into its connected components. For each i let $u_i \in N_G(v)$ be such that $N_G(u_i) \cap V(G_i^v) \neq \emptyset$. Owing to the connectedness of $G - v$, a vertex u_i must exist. Set $K_i = N_G(u_i) \cap V(G_i^v)$.

Claim 9. *The set K_i is a simplicial block in G_i^v.*

Proof. The proof is quite similar to that of Claim 7. We first prove that K_i is a block. Suppose there is an independent set I in $G[K_i]$ of size $|I| = 3$. Then u_i, v and the vertices of I would form an induced $K_{1,4}$ in G with u_i in the center. As this is a contradiction, we conclude that K_i is a 2-clique. To prove that K_i is, in fact, a block, notice that there is no edge between v and K_i. Thus, we cannot have $|K_i| = 5$ because then v would be an isolated vertex in $G[N(u_i)]$ – a contradiction with the assumption on G. If, however, $|K_i| = 4$ then v is the (only) isolated vertex in $G[N(u_i)]$ and, consequently, $\delta(G[K_i]) \geq 1$.

It remains to show that the block K_i is simplicial, that is, for every $w \in K_i$, the set $N_{G_i^v}(w) \setminus K_i$ is a block in $G_i^v - K_i$. This, however, can be proved mutatis mutandis as in the proof of Claim 7.   □

For the first term of recurrence (4) we apply (4) recursively. In view of Claim 9, to the second term of recurrence (4) one can apply formula (3) and then each term $Z_I(G_i^v)$ can be approximated based on (1) and (2).

3.2   The Remainder of the Proof of Lemma 3

Hence, it remains to approximate $\mathbb{P}_G(K \cap \mathbf{I} = \emptyset) = \frac{Z_I(G-K)}{Z_I(G)}$ within $1 \pm \frac{\epsilon}{n}$, where K is a simplicial block in G. We set $N_v := N_G(v)$ and formulate the following recurrence relation by considering how an independent set may intersect K:

$$Z_I(G) = Z_I(G - K) + \sum_{v \in K} Z_I(G - (N_v \cup K)) + \frac{1}{2} \sum_{uv \notin G[K]} Z_I(G - (N_u \cup N_v \cup K))$$

or equivalently, after dividing sidewise by $Z_I(G - K)$,

$$\frac{Z_I(G)}{Z_I(G-K)} = 1 + \sum_{v \in K} \frac{Z_I(G - (N_v \cup K))}{Z_I(G-K)} + \frac{1}{2} \sum_{uv \notin G[K]} \frac{Z_I(G - (N_u \cup N_v \cup K))}{Z_I(G-K)}.$$

Here and throughout the inner summation ranges over all *ordered* pairs of *distinct* vertices of K such that $\{u, v\} \notin G[K]$. At this point, in view of symmetry, it seems redundant to consider ordered pairs (and consequently have the factor of $\frac{1}{2}$ in front of the sum), but we break the symmetry right now as we further observe that

$$\frac{Z_I(G - (N_u \cup N_v \cup K))}{Z_I(G-K)} = \frac{Z_I(G - (N_u \cup N_v \cup K))}{Z_I(G - (N_v \cup K))} \cdot \frac{Z_I(G - (N_v \cup K))}{Z_I(G-K)}.$$

By Claim 7, $N_v \setminus K$ is a simplicial block in $G - K$. We need to show that, similarly, $N_u \setminus (N_v \cup K)$ is a simplicial block in $G - (N_v \cup K)$.

Claim 10. *Let K be a simplicial block in G and let $u, v \in K$ be such that $u \neq v$ and $uv \notin G[K]$. Further, let $H := G - (N_G(v) \cup K)$. Then $N_H(u)$ is a simplicial block in H.*

Proof. By Claim 7, the set $N_G(u) \setminus K$ is a simplicial block in $G - K$. Apply Fact 6 to $N_G(u) \setminus K$ and $G - K$ with $V' = V(H)$.   □

Let

$$\Pi_G(K) := \mathbb{P}(K \cap \mathbf{I} = \emptyset) = \frac{Z_I(G - K)}{Z_I(G)},$$

where $\mathbf{I}$ is a random independent set of G. Finally, setting $K_v := N_v \setminus K$ and $K_{uv} := N_u \setminus (N_v \cup K)$, and rewriting $G - (N_v \cup K) = G - K - K_v$, we get the recurrence for the probabilities:

$$\Pi_G^{-1}(K) = 1 + \sum_{v \in K} \Pi_{G-K}(K_v) \left(1 + \frac{1}{2} \sum_{uv \notin G[K]} \Pi_{G-K-K_v}(K_{uv}) \right).$$

This recurrence, in principle, allows one to compute $\Pi_G(K)$ exactly, but only in an exponential number of steps. Instead, we will approximate it by a function $\Phi_G(K,t)$, also defined recursively, which "mimics" $\Pi_G(K)$ but has a built-in time counter t.

Definition 11. *For every graph G, every simplicial block K in G and an integer $t \in \mathbb{Z}_+$, the function $\Phi_G(K,t)$ is defined recursively as follows: $\Phi_G(K,0) = \Phi_G(K,1) = 1$ as well as $\Phi_G(\emptyset,t) = 1$, while for $t \geq 2$ and $K \neq \emptyset$*

$$\Phi_G^{-1}(K,t) = 1 + \sum_{v \in K} \Phi_{G-K}(K_v, t-1) \left(1 + \frac{1}{2} \sum_{uv \notin G[K]} \Phi_{G-K-K_v}(K_{uv}, t-2) \right).$$

Now we are ready to state the algorithm *CountMatchings* for computing $Z_M(H)$ for any connected $(3,3)$-graph H and its subroutine *CountIS* for computing $Z_I(G)$ in a subgraph of $G = L(H)$ containing a simplicial block K.

Algorithm 1. *CountMatchings*(H, t)

1: $G := L(H)$.
2: $Z_M := 1$, $F := G$.
3: **while** $F \neq \emptyset$ **do**
4:     Pick $v \in V(F)$ s.t. $F - v$ is connected.
5:     $F^v := F - N_F[v]$
6:     If $F^v = \emptyset$ then $Z_M = Z_M + 1$ and go to Line 3.
7:     $F^v = \bigcup_{i=1}^c F_i^v$, where F_i^v are connected components of F^v.
8:     **for** $i := 1$ to c **do**
9:         Find K_i as in Claim 9
10:    **end for**
11:    $Z_M := Z_M + \prod_{i=1}^c CountIS(F_i^v, K_i, t)$
12:    $F := F - v$
13: **end while**
14: Return Z_M

Algorithm 2. $CountIS(G, K, t)$

1: Let $V(G) = \bigcup_{i=1}^{m} K_i$ be a partition of $V(G)$ as in Claim 8 with $K_1 = K$.
2: $Z_I := 1, F := G$
3: **for** $i = 1$ to m **do**
4:     $Z_I := \frac{Z_I}{\Phi_F(K_i, t)}$
5:     $F := F - K_i$
6: **end for**
7: Return Z_I

We will show that already for $t = \Theta(\log n)$, when Φ can be easily computed in polynomial time, the two functions become close to each other.

Note that both quantities, $\Pi_G(K)$ and $\Phi_G(K, t)$, fall into the interval $[\frac{1}{9}, 1]$. The lower bound is due to the fact that a block has at most 4 vertices and each of them has degree at most 2 in G^c, so that the total number of terms in the denominator is at most nine, five of them do not exceed 1, while eight of them do not exceed $\frac{1}{2}$. Our goal is to approximate $\Pi_G(K)$ by $\Phi_G(K, t)$, for a suitably chosen t, within the multiplicative factor of $1 \pm \epsilon/n$. In view of the above lower bound, it suffices to show that $|\Pi_G(K) - \Phi_G(K, t)| \leq \frac{\epsilon}{9n}$.

To achieve this goal, we will use the correlation decay technique which boils down to establishing a recursive bound on the above difference (cf. [2]). The success of this method depends on the right choice of a pair of functions g and h, with $g : [0, 1] \to \Re$, such that they are inverses of each other, that is, $g \circ h \equiv 1$. Then we define a function f_K of $|K| + 2e(G^c[K])$ variables, one for each vertex and each (ordered) non-edge of $G[K]$, as follows. Let $\mathbf{z} = (z_1, \ldots, z_{|K|}, z_{uv} : uv \notin G[K])$ be a vector of variables of that function. For ease of notation, we denote the set of all indices of the coordinates of function f_K by J, that is, we set $J := K \cup \{(u, v) : \{u, v\} \notin G[K]\}$. Then

$$f_K(\mathbf{z}) := f(\mathbf{z}) = g\left(\left\{1 + \sum_{v \in K} h(z_v)\left(1 + \frac{1}{2}\sum_{uv \notin G[K]} h(z_{uv})\right)\right\}^{-1}\right). \tag{5}$$

To understand the reason for this set-up, put $x := g(\Pi_G(K))$, $x_v := g(\Pi_{G-K}(K_v))$, $x_{uv} := g(\Pi_{G-K-K_v}(K_{uv}))$, and, correspondingly,

$$y := g(\Phi_G(K, t)) \quad y_v := g(\Phi_{G-K}(K_v, t-1)) \quad y_{uv} := g(\Phi_{G-K-K_v}(K_{uv}, t-2)).$$

Then, $f(\mathbf{x}) = x$ and $f(\mathbf{y}) = y$, and so the difference we are after can be expressed as $|x - y| = |f(\mathbf{x}) - f(\mathbf{y})|$. Thus, we are in position to apply the Mean Value Theorem to f and conclude that there exists $\alpha \in [0, 1]$ such that, setting $\mathbf{z}_\alpha = \alpha\mathbf{x} + (1 - \alpha)\mathbf{y}$,

$$|f(\mathbf{x}) - f(\mathbf{y})| = |\nabla f(\mathbf{z}_\alpha)(\mathbf{x} - \mathbf{y})| \leq |\nabla f(\mathbf{z}_\alpha)| \times \max_{\kappa \in J}|x_\kappa - y_\kappa|.$$

It remains to bound $\max_z |\nabla f(\mathbf{z})|$ from above, uniformly by a constant $\gamma < 1$. Then, after iterating at most t but at least $t/2$ times, we will arrive at a triple

(G', K', t'), where G' is an induced subgraph of G, K' is a block in G', and $t' \in \{0, 1\}$. At this point, setting $\mu_g := |g(1)| + |\max_s g(s))|$, we will obtain the ultimate bound

$$|x - y| \le \gamma^{t/2} \times |g(\Pi_{G'}(K')) - g(1)| \le \gamma^{t/2} \times \mu_g \le \frac{\epsilon}{9n},$$

$$\text{for} \qquad t \ge 2\log((9\mu_g n)/\epsilon)/\log(1/\gamma). \tag{6}$$

In [2], to estimate $|\nabla f(\mathbf{z})|$ for a similar function f, the authors chose $g(s) = \log s$ and $h(s) = e^s$. This choice, however, does not work for us. Instead, we set $g(s) = s^{1/4}$ and $h(s) = s^4$. Then, $\mu_g = 2$ and

$$|\nabla f(\mathbf{z})| \le \sum_{\kappa \in J} \left| \frac{\partial f(\mathbf{z})}{\partial z_\kappa} \right| = \frac{\sum\limits_{v \in K} \left\{ z_v^3 + \frac{1}{2} \sum\limits_{uv \notin G[K]} (z_v^3 z_{uv}^4 + z_v^4 z_{uv}^3) \right\}}{\left\{ 1 + \sum\limits_{v \in K} z_v^4 \left(1 + \frac{1}{2} \sum\limits_{uv \notin G[K]} z_{uv}^4 \right) \right\}^{5/4}}.$$

Observe that f_K depends only on the isomorphism type of $G[K]$, a graph on up to 4 vertices, with no independent set of size 3, and with no isolated vertex when $|K| = 4$. Let us call all these graphs *block graphs*. One block graph is given in Figure 1 below.

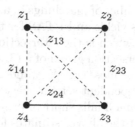

Fig. 1. The essential block graph

In a sense we just need to consider this one block graph. Indeed, the complement of every block graph is contained in the complement of the block graph in Figure 1. Hence, it suffices to maximize $|\nabla f(\mathbf{z})|$ just for this graph. Our computational task is, therefore, to bound from above

$$F(\mathbf{z}) = \|\nabla(\mathbf{z})\|_1 = \frac{1}{4}\left(1 + z_1^4 + z_2^4 + z_3^4 + z_4^4 + \right.$$

$$\left.\frac{1}{2}\left(z_{14}^4\left(z_1^4 + z_4^4\right) + z_{13}^4\left(z_1^4 + z_3^4\right) + z_{23}^4\left(z_2^4 + z_3^4\right) + z_{24}^4\left(z_2^4 + z_4^4\right)\right)\right)^{-5/4} \times$$

$$\left(2z_1^3\left(2 + z_{14}^4 + z_{13}^4\right) + 2z_2^3\left(2 + z_{23}^4 + z_{24}^4\right) + 2z_3^3\left(2 + z_{13}^4 + z_{23}^4\right) + 2z_4^3\left(2 + z_{14}^4 + z_{24}^4\right) + \right.$$

$$\left. 2z_{14}^3\left(z_1^4 + z_4^4\right) + 2z_{13}^3\left(z_1^4 + z_3^4\right) + 2z_{23}^3\left(z_2^4 + z_3^4\right) + 2z_{24}^3\left(z_2^4 + z_4^4\right)\right).$$

One can show (using, e.g., *Mathematica*) that $F(\mathbf{z}) < 0.971$ for $0 \le z_i \le 1$ and $0 \le z_{ij} \le 1$. Thus, we have (6) with $\mu_g = 2$ and, say, $\gamma = 0.98 = \frac{49}{50}$. Summarizing, the running time of computing $\Phi_G(K, t)$ in Step 4 of Algorithm 2 is 12^t since there at most 12 expressions to compute in each step of the recurrence relation (see Def. 11). Also, *CountIS* takes at most $|V(F_i^v)|12^t$ steps and hence, Line 11 of *CountMatchings* takes $n12^t$ steps and is invoked at most n times. Consequently, with $t = 2\lceil \log((18n)/\epsilon)/\log(50/49)\rceil$ we get the running time of our algorithm of order $O\left(n^2(n/\epsilon)^{\log_{50/49} 144}\right)$.

Remark 12. With basically the same proof we can construct an FPTAS for calculating the partition function $Z_M(H, \lambda) = \sum_M \lambda^{|M|}$, where the sum runs over all matchings in H, for any constant $\lambda \in (0, 1.077]$. The λ factor will appear in front of each summation in (5), which one can neutralize by setting $h(s) = \frac{s^4}{\lambda}$ and $g(s) = (\lambda s)^{1/4}$.

4   Summary, Discussion, and Further Research

The main result of this paper (Thm. 2) establishes an FPTAS for the problem $\#M(3,3)$ of counting the number of matchings in a $(3,3)$-graph. A reformulation of Thm. 2 in terms of graphs yields an FPTAS for the problem of counting independent sets in every graph which is the intersection graph of a $(3,3)$-graph. As mentioned earlier, every intersection graph of a $(3,3)$-graph is 4-claw-free. Moreover, its maximum degree is at most six. We wonder if there exists an FPTAS for the problem of counting independent sets in every 4-claw-free graph with maximum degree at most 6. Lemma 3 falls short of proving that. The missing part is due to our inability to repeat the above estimates for 2-cliques of size five.

In an earlier paper [11] three of the authors have found an FPRAS for the number of matchings in k-graphs without 3-combs. As their intersection graphs are claw-free, it follows from the above mentioned result on independent sets in [2,7] that there is also an FPTAS for the number of matchings in (k, r)-graphs without 3-combs, for any fixed r. In view of this conclusion and Thm. 2, we raise the question if for all $k \le 5$ and r there is an FPTAS (or at least FPRAS) for the problem $\#M(k, r)$. The first open instance is that of $(3, 4)$-graphs. For $k = 4, 5$, to avoid recurrences of depth $k - 1 \ge 3$, as an intermediate step, one could first consider the restriction of the class of (k, r)-graphs to those without a 4-comb, that is, to those whose intersection graphs are 4-claw-free. Here, the first open instance is that of $(4,3)$-graphs without 4-combs. In general, it would be also very interesting to elucidate the status of the problem for arbitrary k-graphs for $k = 3, 4$ and 5, or for some generic subclasses of them.

Acknowledgements. We thank Martin Dyer and Mark Jerrum for stimulating discussions on the subject of this paper and the referees for their valuable comments.

References

1. Bayati, M., Gamarnik, D., Katz, D., Nair, C., Tetali, P.: Simple deterministic approximation algorithms for counting matchings. In: STOC 2007—Proceedings of the 39th Annual ACM Symposium on Theory of Computing, pp. 122–127. ACM (2007)
2. Bayati, M., Gamarnik, D., Katz, D., Nair, C., Tetali, P.: Simple deterministic approximation algorithms for counting matchings (2008),
   http://people.math.gatech.edu/~tetali/PUBLIS/BGKNT_final.pdf
3. Beineke, L.W.: Characterizations of derived graphs. J. Combin. Theory 9, 129–135 (1970)
4. Chudnovsky, M., Seymour, P.: The roots of the independence polynomial of a clawfree graph. J. Combin. Theory Ser. B 97(3), 350–357 (2007)
5. Dobrushin, R.: Prescribing a system of random variables by conditional distributions. Theor. Probab. Appl. 15, 458–486 (1970)
6. Dyer, M., Frieze, A., Jerrum, M.: On counting independent sets in sparse graphs. SIAM J. Comput. 31(5), 1527–1541 (2002)
7. Fadnavis, S.: Approximating independence polynomials of claw-free graphs (2012),
   http://www.math.harvard.edu/~sukhada/IndependencePolynomial.pdf
8. Greenhill, C.: The complexity of counting colourings and independent sets in sparse graphs and hypergraphs. Comput. Complexity 9(1), 52–72 (2000)
9. Heilmann, O.: Existence of phase transitions in certain lattice gases with repulsive potential. Lett. Al Nuovo Cimento Series 2 3(3), 95–98 (1972)
10. Jerrum, M., Sinclair, A.: Approximating the permanent. SIAM J. Comput. 18(6), 1149–1178 (1989)
11. Karpiński, M., Ruciński, A., Szymańska, E.: Approximate counting of matchings in sparse uniform hypergraphs. In: 2013 Proceedings of the Workshop on Analytic Algorithmics and Combinatorics (ANALCO), pp. 72–79. SIAM (2013)
12. Kelly, F.P.: Stochastic models of computer communication systems. J. Roy. Statist. Soc. Ser. B 47(3), 379–395, 415–428 (1985)
13. Luby, M., Vigoda, E.: Fast convergence of the Glauber dynamics for sampling independent sets. Random Structures Algorithms 15(3-4), 229–241 (1999)
14. Sly, A.: Computational transition at the uniqueness threshold. In: 2010 IEEE 51st Annual Symposium on Foundations of Computer Science FOCS 2010, pp. 287–296 (2010)
15. Sly, A., Sun, N.: The computational hardness of counting in two-spin models on d-regular graphs. In: FOCS, pp. 361–369 (2012),
   http://arxiv.org/abs/1203.2602
16. Valiant, L.G.: The complexity of enumeration and reliability problems. SIAM J. Comput. 8(3), 410–421 (1979)
17. Weitz, D.: Counting independent sets up to the tree threshold. In: STOC 2006: Proceedings of the 38th Annual ACM Symposium on Theory of Computing, pp. 140–149. ACM (2006)

Author Index